Water Resources Development and Management

Indexed by Scopus

Each book of this multidisciplinary series covers a critical or emerging water issue. Authors and contributors are leading experts of international repute. The readers of the series will be professionals from different disciplines and development sectors from different parts of the world. They will include civil engineers, economists, geographers, geoscientists, sociologists, lawyers, environmental scientists and biologists. The books will be of direct interest to universities, research institutions, private and public sector institutions, international organisations and NGOs. In addition, all the books will be standard reference books for the water and the associated resource sectors.

José María Ponce-Ortega ·
Fabricio Nápoles-Rivera ·
Luis Fernando Lira-Barragán ·
César Ramírez-Márquez

Sustainable Water Systems

Advances in Recycling, Reuse, and Optimization

Springer

José María Ponce-Ortega
Universidad Michoacana de San Nicolás de Hidalgo
Morelia, Michoacán, Mexico

Fabricio Nápoles-Rivera
Universidad Michoacana de San Nicolás de Hidalgo
Morelia, Michoacán, Mexico

Luis Fernando Lira-Barragán
Universidad Michoacana de San Nicolás de Hidalgo
Morelia, Michoacán, Mexico

César Ramírez-Márquez
Universidad Michoacana de San Nicolás de Hidalgo
Morelia, Michoacán, Mexico

ISSN 1614-810X ISSN 2198-316X (electronic)
Water Resources Development and Management
ISBN 978-3-031-72497-8 ISBN 978-3-031-72498-5 (eBook)
https://doi.org/10.1007/978-3-031-72498-5

This Springer imprint is published by the registered company Springer Nature Switzerland AG
The registered company address is: Gewerbestrasse 11, 6330 Cham, Switzerland

Preface

This book presents several optimization strategies for the proper use of water, encompassing mathematical programming approaches for the optimal use and reuse of water at different levels, including domestic, industrial, and macroscopic. The work herein is the culmination of over fifteen years of research conducted at the Universidad Michoacana de San Nicolás de Hidalgo in Mexico, in collaboration with various institutions. We extend our heartfelt gratitude to Prof. Mahmoud M. El-Halwagi from Texas A&M University for his invaluable contributions and for inspiring us to delve into this research area.

In the introduction, we emphasize the importance of sustainable water systems, the evolution of water management practices, and key concepts such as recycling, reuse, and optimization. This foundational chapter also outlines the methodology and approach used throughout the book, providing a robust framework for the subsequent chapters.

The book then explores water recycling and reuse networks, detailing the principles of water recycling, current technologies, design and implementation challenges, and success stories from various case studies. It also examines regulatory and policy frameworks and future trends in water recycling and reuse.

A significant portion of the book is dedicated to the comprehensive optimization of multi-contaminant wastewater process integration networks. This section addresses the complexity of handling multiple contaminants, optimization models and approaches, and integration techniques for efficiency and presents global applications through case studies. It also discusses technical and regulatory challenges and future directions in wastewater management.

The sustainable management of water in macroscopic systems is another crucial focus. Here, we define macroscopic water systems, present strategies for sustainable management, and analyze large-scale applications through case studies. Policy and governance issues, sustainability challenges, and potential solutions are also thoroughly examined.

Urban water system design incorporating future projections for optimal configuration is discussed in detail, covering the dynamics of urban water systems, principles for sustainable design, and the implementation of future-oriented designs. Case studies in urban settings highlight challenges and opportunities in this area.

The book further analyzes the environmental and economic benefits of optimal water reuse in residential complexes. We present a framework for environmental analysis, economic impact assessments, strategies for optimal reuse, and detailed case studies. Policy implications, technological advancements, and recommendations are also provided.

A chapter on Mixed-Integer Nonlinear Programming (MINLP) models outlines the optimal site selection for new industrial facilities, integrating economic and environmental considerations. The methodology, case studies, and challenges in model application are thoroughly explored.

Designing water networks with a focus on the sustainability of adjacent watersheds is covered comprehensively. We discuss watershed sustainability principles, design approaches, case studies, technological and methodological approaches, and policy and regulatory aspects.

The integration of water network characteristics and watershed sustainability into industrial facility site selection is another important topic. Criteria for site selection, the role of water networks in decision-making, watershed considerations, and analytical models and tools are discussed, alongside case studies and challenges in site selection.

Eco-industrial parks and their interaction with adjacent watersheds are explored, highlighting the concept and importance of eco-industrial parks, design and planning strategies, and relevant case studies.

Selecting optimal sites for biorefineries with a focus on environmental sustainability is discussed in detail. Criteria for biorefinery location, environmental integration considerations, optimization models, and case studies are presented.

Optimizing the placement of new industrial facilities dynamically with watershed sustainability considerations involves principles of dynamic optimization, methodologies, models, and case studies, along with challenges and future perspectives.

The book also addresses optimizing the recycling of flowback water in hydraulic fracturing, considering seasonal and environmental limitations. We examine flowback wastewater characteristics, seasonal variations, case studies, and economic and regulatory considerations.

Strategic water management for shale gas production under uncertainty is another focal point. This section covers uncertainty factors, optimization techniques, case studies, environmental and economic impacts, and policy challenges.

Comprehensive planning for optimal management of different types of shale gas wastewater involves strategic management approaches, case studies, and environmental and economic considerations.

Lastly, the book delves into optimizing the water-energy relationship in hydraulic fracturing by utilizing CO_2 and water as fracturing agents. It covers the water-energy nexus, CO_2 utilization, optimization models, case studies, environmental impact, and sustainability.

We are deeply grateful for the contributions of numerous former students from the Chemical Engineering Department of the Universidad Michoacana de San Nicolás de Hidalgo, including Dr. Dulce Celeste López-Díaz, Dr. Tania Itzel Serrano-Arévalo, Dr. Moises Ferreyra-Quiroz, Dr. Ma. Guadalupe Rojas-Torres, M. C. Mariana García-Montoya, Dr. Andrea Bocanegra-Martínez, and many others who have enthusiastically contributed to this research.

We also extend our thanks to the professors who have contributed to this book, especially Prof. Eusiel Rubio-Castro, Prof. Arturo Jiménez-Gutiérrez, Prof. Medardo Serna-González, Prof. Fenqi You, Prof. Gonzalo Guillén-Gosálbez, and Prof. Agustín Jaime Castro-Montoya, among many others. Their support and expertise have been invaluable in the completion of this work.

Morelia, Michoacán, Mexico
August 2024

José María Ponce-Ortega
Fabricio Nápoles-Rivera
Luis Fernando Lira-Barragán
César Ramírez-Márquez

Contents

Chapter 1
Introduction

1.1 Importance of Sustainable Water Systems

The term "sustainable" originates from the Latin verb "sustinere", which means to uphold. The corresponding English verb, "to sustain", has been in use since the late Middle Ages and carries meanings such as: to maintain, to keep going, to support life, and to provide for bodily needs (Butler et al. 2017). These meanings are encapsulated in the concept of "sustainable development", a term broadly used in contemporary discourse. Sustainable development, as articulated in the Brundtland Report of 1987, aims to ensure that development meets the needs of the present without compromising the ability of future generations to meet their own needs (Koudstaal et al. 1992; Kundzewicz 1997). This goal should be achieved while minimizing losses and maximizing gains to economic, social, and environmental systems (Ajami et al. 2008).

Water, as the fundamental element of the life support system of our planet, is indispensable for sustaining all forms of life and virtually every human activity. The annual runoff into the oceans exceeds 40,000 km^3, but the withdrawals, currently reaching approximately 3800 km^3, constitute only about 9% of this total (Kundzewicz 1997). While this might suggest an abundance of water, these global figures can be misleading. The real challenge lies in the availability of water at smaller, more localized scales.

Since the early twentieth century, global water consumption has increased about sevenfold, driven by both population growth and increased per capita water use (Brown et al. 2015). The continuing rise in population, coupled with aspirations for improved living conditions, will undoubtedly escalate the demand for water further. Ensuring an adequate and reliable supply of water of proper quality for the entire global population, while preserving the hydrological, biological, and chemical functions of ecosystems, remains a distant goal. Currently, water shortages already pose significant challenges in various locations, a situation exacerbated by natural variabilities in temperature and precipitation.

J. M. Ponce-Ortega et al., *Sustainable Water Systems*, Water Resources Development and Management, https://doi.org/10.1007/978-3-031-72498-5_1

Water scarcity is projected to be the most dominant water-related issue in the forthcoming century, posing a severe threat to sustainable development. By 2025, it is estimated that the number of countries experiencing water scarcity, defined as availability below 500 m^3 per capita per year, will increase from 12 to 19 (Nápoles-Rivera et al. 2013). Some assessments suggest that up to 35% of the global human population may face water scarcity by 2025 (Nápoles-Rivera et al. 2013). In countries experiencing dynamic population growth, the per capita availability of water has already dramatically decreased, and this trend is expected to worsen, falling below the current threshold of scarcity.

The United Nations Water Conference in Mar del Plata in 1977 emphasized the vital role of water in national development (Koudstaal et al. 1992). The conference produced a Plan of Action focusing on improving conditions for low-income groups and initiated the International Drinking Water Supply and Sanitation Decade (IDWSSD) in the 1980s (Koudstaal et al. 1992). Despite the recognized link between properly managed water resources, economic development, and social well-being, sustainable water resource policies have largely failed to materialize. Consequently, the quantity and quality of surface and groundwater resources continue to deteriorate in many regions.

Historically, water resources management has focused on the development of water resources to meet growing demands and mitigate the impacts of droughts and floods. However, contemporary water resources management requires a more integrated approach. National policies in both developed and developing countries are increasingly based on integrated water resources management (IWRM) (Vairavamoorthy et al. 2001). This approach is supported by international organizations and is reflected in the New Delhi Statement on safe water and sanitation for the 1990s, which emphasizes integrated management of water, liquid, and solid wastes, as well as institutional reforms in the water sector (United Nations 1990).

Integrated water resources management considers the interaction between the quantity, quality, and biological aspects of both groundwater and surface water. It takes into account the water demands and residual generation by different sectors of the economy, considers on-site and off-site environmental issues, and emphasizes the long-term carrying capacity of natural systems (Pahl-Wostl et al. 2008). Public participation, including the roles of women and traditional water use practices, is a crucial component, alongside implementation aspects such as financing, monitoring, and control (Pahl-Wostl et al. 2013).

The importance of sustainable water systems is further underscored by the pressing need for international cooperation in managing shared water resources. With increasing competition and potential conflicts over transboundary water resources, international guidelines and agreements are essential for addressing water quantity and quality issues. The draft Law of the Non-Navigational Use of International Water Courses by the International Law Commission is a step in this direction, although it has not been universally adopted (Rijsberman and Van de Ven 2000).

Environmental degradation poses another significant challenge to sustainable water management. Over the past decades, the recognition of environmental quality as a crucial component of economic development has grown. In developed countries, efforts have focused on reducing wastewater and coordinating environmental management programs. However, in developing countries, the integration of environmental issues into water resources management is still in its early stages.

Economic and financial constraints also complicate sustainable water management. Both free market and planned economies often fail to reflect the true value or scarcity of natural resources like water. Governments typically prioritize short-term production targets over long-term sustainable use. Water resource managers must demonstrate the value of sustainable water resource development to decision-makers operating under tight financial constraints.

Sustainable water systems are vital for ensuring the long-term availability and quality of water resources necessary for life and human activities. The challenge lies in implementing integrated water resources management practices that consider environmental, social, and economic factors. Achieving this requires international cooperation, public participation, and the development of policies that promote sustainable use and development of water resources.

1.2 Evolution of Water Management Practices

The evolution of water management practices reflects humanity's continuous quest to harness and manage one of its most vital resources. From the earliest irrigation systems in ancient civilizations to the advanced technologies of the modern era, the approaches to water management have grown increasingly sophisticated, addressing both the practical needs and the environmental challenges of each era (Cech 2018).

In ancient times, civilizations such as those in the Middle East, Egypt, and the Indus Valley pioneered early water management techniques (Ahmed et al. 2020). These included the construction of canals, reservoirs, and wells to support agriculture and daily life in arid regions. As urban centers developed, the need for efficient water distribution became paramount. The Greeks and Romans made significant strides with the development of aqueducts, which transported water from distant sources to cities, ensuring a reliable supply for public health and sanitation (Angelakis et al. 2023).

The understanding of water's role in public health evolved over time. Ancient writings, including those of Hippocrates, emphasized the importance of clean water in preventing diseases (Angelakis et al. 2020). However, it was not until the development of modern water treatment technologies that significant progress was made in ensuring safe drinking water and reducing waterborne diseases.

In the modern era, water management has continued to advance, driven by technological innovations and a deeper understanding of the complexities involved. One of the most significant shifts has been from focusing solely on increasing water supply to managing water demand. This approach recognizes that water use is influenced by numerous factors, including economic, technological, and social considerations.

Technological innovations have been at the forefront of this evolution. Smart water management systems, which leverage technologies such as the Internet of Things (IoT), artificial intelligence (AI), and remote sensing, have revolutionized traditional approaches (Olatunde et al. 2024). These technologies enable real-time monitoring of water systems, providing valuable data on water quality, flow rates, and consumption patterns. This data can be analyzed to detect leaks, predict demand, and optimize distribution, significantly improving the efficiency and sustainability of water management practices (Rane 2023).

IoT technology has enabled the deployment of sensors throughout water systems. These sensors collect real-time data on various parameters, such as flow rates, pressure, and water quality. This information allows for immediate detection of issues like leaks or contamination, enabling swift corrective actions. Additionally, AI and machine learning are increasingly used to analyze water usage patterns, predict system failures, and optimize treatment processes. AI models can forecast the impacts of weather on water availability and demand, enhancing the adaptability of water systems to climate variability (Alahi et al. 2023).

Remote sensing technology, utilizing satellites and drones, offers comprehensive data on water resources, land use, and environmental conditions. This technology is crucial for monitoring large-scale water basins, tracking changes in water availability, and assessing the health of aquatic ecosystems. The integration of these advanced technologies facilitates a more holistic and adaptive approach to water management.

Policy and regulatory frameworks have also evolved to support these technological advancements. International initiatives emphasize the importance of integrated water resources management, promoting the use of technology and the implementation of policies that ensure equitable access to clean water (Algaba et al. 2024). These efforts are particularly vital in regions facing severe water scarcity and those impacted by climate change.

Climate change poses significant challenges to water management by altering precipitation patterns and increasing the frequency and severity of droughts and floods. Addressing these challenges requires adaptive and resilient strategies that can respond to changing environmental conditions. Advanced technologies play a crucial role in these efforts, enabling more precise monitoring and management of water resources.

The evolution of water management practices highlights humanity's ongoing efforts to secure and manage a vital resource. From ancient irrigation systems to modern smart technologies, these practices have continually adapted to meet the needs of societies. Ensuring the sustainable and equitable management of water resources remains a critical global priority, demanding continuous innovation and cooperation across all sectors.

1.3 Key Concepts: Recycling, Reuse, Optimization

Chemical engineering significantly contributes to minimizing water use and effluent generation through several key methodologies. These methodologies focus on the broader principles of efficient resource management, the integration of processes, and the optimization of systems.

Recycling in the context of water management involves the treatment and reprocessing of wastewater so that it can be reused within the same facility or in different applications (Jhansi et al. 2013). This process reduces the dependency on freshwater sources and minimizes the amount of wastewater that needs to be discharged. For example, semiconductor fabrication plants in water-stressed regions utilize advanced recycling techniques to ensure a steady supply of water, even during drought conditions (Chan and Wu 2022). This not only conserves water but also aligns with regulatory requirements and sustainability goals.

Reuse pertains to the use of treated wastewater for purposes other than its original application (Ofori et al. 2021). This can include using treated industrial water for agricultural irrigation, landscaping, or even within different industrial processes. High-tech manufacturing facilities, such as those in Taiwan, have stringent water reuse criteria to ensure that a significant portion of water is reclaimed and reused, thus reducing the overall water footprint (Den et al. 2018). This practice is essential for maintaining water sustainability and meeting environmental regulations.

Optimization involves the systematic design and operation of water networks to maximize efficiency and minimize costs (Ahmad et al. 2020; Duhbacı et al. 2021; Reis et al. 2023). This includes the use of advanced techniques such as process integration and mathematical programming to identify the best configurations for water treatment and distribution networks. Optimization aims to balance various factors, including water quality, treatment costs, and environmental impact. For instance, optimization strategies might involve the use of concentration potential-based linear programming to enhance in-plant water networks, ensuring that water reuse and recycling are maximized while freshwater consumption is minimized.

The integration of these key concepts—recycling, reuse, and optimization—enables industries to develop sustainable water management practices. By focusing on these methodologies, industries can significantly reduce their water consumption and environmental impact, ensuring a more sustainable future. In the following chapters of this book, we will explore these concepts in greater detail and demonstrate their application in various contexts to promote effective and sustainable water management (Fig. 1.1).

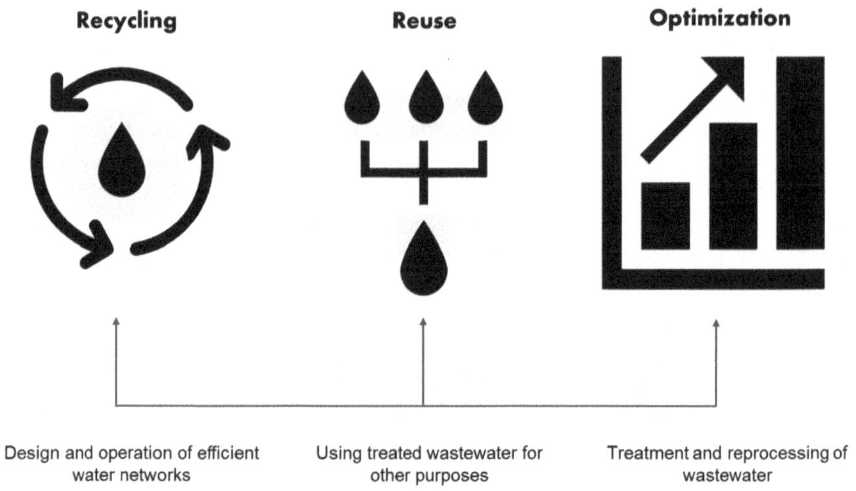

Fig. 1.1 Key concepts in water management

1.4 Methodology and Approach

The methodology for addressing water management challenges in industrial and urban systems involves a structured approach that integrates several key stages. Each stage builds on the previous one to create a comprehensive water management strategy that includes recycling, reuse, and optimization. Below is a general overview of this approach.

Methodology

Stage 1: Identification and Characterization of Water Sources
The first stage involves identifying and characterizing all available water sources, such as deep wells, springs, and dams. These sources are crucial for meeting the water demands of a city or industrial facility. Key activities include:

- Assessing the capacity and quality of each water source.
- Monitoring natural replenishment processes, such as rainfall and tributary inflows.
- Categorizing water sources based on their suitability for different uses (e.g., agricultural, domestic).

Stage 2: Segregation and Initial Treatment
In this stage, water from identified sources is segregated and directed to appropriate treatment processes based on its intended use. This stage includes:

- Diverting water to main systems designated for agricultural or domestic use.

- Implementing preliminary treatment to meet basic quality standards before further processing.

Stage 3: Recycling and Reuse of Water
The third stage focuses on minimizing freshwater intake and reducing waste through recycling and reuse. This involves:

- Treating wastewater generated from various processes to make it suitable for reuse.
- Designing recycling loops that allow treated wastewater to be reused within the same system or in other applications (e.g., irrigation, cooling).

Stage 4: Optimization of Water Networks
Optimization is crucial for efficient water management. This stage involves:

- Developing a comprehensive optimization model representing all possible pathways for water flow within the system.
- Performing component and property balances to ensure water quality meets required standards.
- Selecting the most efficient routes and treatment options using optimization techniques to minimize costs and environmental impact.

Stage 5: Storage and Distribution
Efficient storage and distribution balance water supply and demand. This involves:

- Installing storage tanks and artificial ponds to store excess water during periods of low demand.
- Developing distribution networks to transport water efficiently to various sinks (e.g., agricultural fields, domestic users).
- Optimizing storage solutions for capacity and location to minimize costs and maximize efficiency.

Stage 6: Environmental Compliance
Ensuring compliance with environmental regulations is integral to the methodology. Activities include:

- Treating wastewater to remove hazardous compounds before discharge.
- Monitoring and maintaining discharge water quality within permissible limits.
- Integrating environmental constraints into the optimization model to balance industrial needs with ecological preservation.

Stage 7: Economic Optimization
The final stage involves optimizing the economic aspects of water management to ensure sustainability. This includes:

- Maximizing gross profit from water sales while considering treatment, distribution, and storage costs.
- Developing an objective function that balances revenue generation with cost minimization.
- Ensuring proposed solutions are both environmentally and economically sustainable.

The approach taken in developing this methodology is systematic and iterative, ensuring that all aspects of water management are considered and optimized. Key elements of the approach include:

1. **Comprehensive Assessment**: Begin with a thorough assessment of all water sources, demands, and existing infrastructure.
2. **Integrated Planning**: Use an integrated planning framework that considers all stages of water management, from source identification to economic optimization.
3. **Stakeholder Involvement**: Engage stakeholders, including local authorities, industries, and communities, to ensure that the water management strategy meets all needs and constraints.
4. **Iterative Optimization**: Apply iterative optimization techniques to refine the water network configuration, storage solutions, and treatment processes continuously.
5. **Monitoring and Adaptation**: Implement robust monitoring systems to track water quality and availability, allowing for adaptive management in response to changing conditions.

This chapter outlines a general methodology for water management, highlighting each stage from source identification to economic optimization. The systematic approach ensures that all aspects of water management are considered and optimized.

Subsequent chapters will detail specific methodologies and case studies, demonstrating the application of these principles to achieve sustainable water management solutions tailored to particular scenarios.

Chapter 2 describes a mathematical model to optimize direct recycle-reuse networks and wastewater treatment processes. It uses disjunctive programming reformulated as a MINLP model to minimize the total annual cost, including fresh sources, piping, and waste treatment costs.

Chapter 3 presents a convex mathematical programming model for optimizing recycle/reuse water networks. The model includes configurations like segregation, mixing, recycling, bypass, and stream treatment, avoiding bilinear terms to simplify optimization and minimize annual costs.

Chapter 4 focuses on sustainable water management in macroscopic systems, considering alternate water sources like rainfall and reclaimed water. The multi-period model aims to maximize profit by optimizing the distribution and storage of water resources, ensuring sustainability.

Chapter 5 addresses water scarcity by proposing strategies like water harvesting and reclamation. A multi-period mathematical model considers population growth, economic changes, and precipitation patterns to optimize water distribution and storage in a macroscopic system.

Chapter 6 introduces an optimization-based method for rainwater harvesting systems in residential complexes. The model aims to meet domestic water demands while minimizing costs related to freshwater, catchment areas, storage facilities, and maintenance.

Chapter 7 presents a mixed integer nonlinear programming model for optimally allocating new industrial units, considering economic and environmental factors to ensure sustainable water quality in nearby watersheds.

Chapter 8 introduces a mathematical programming model for synthesizing recycle and reuse water networks, integrating the nearby watershed and water network system to optimize the placement of new industrial facilities.

Chapter 9 describes a mathematical programming approach for selecting industrial facility locations, incorporating water usage and disposal while considering environmental impacts on surrounding watersheds.

Chapter 10 presents a model for water integration in eco-industrial parks, combining material flow analysis with water recycling techniques to minimize environmental impacts and costs associated with fresh water and wastewater treatment.

Chapter 11 focuses on optimizing biorefining systems to manage water resources, considering feedstock choices, facility locations, and conversion technologies to maximize water use efficiency and minimize environmental impacts.

Chapter 12 introduces a dynamic approach for synthesizing water treatment networks, considering sustainability and environmental constraints in integrating new industrial facilities' wastewater discharges with the surrounding watershed.

Chapter 13 presents a model for managing flowback water in shale gas wells, considering treatment, storage, reuse, and disposal options to minimize costs and environmental impacts while addressing seasonal freshwater availability.

Chapter 14 describes a mathematical programming formulation for designing water networks in shale gas hydraulic fracturing operations, accounting for system uncertainties and optimizing costs related to water usage and treatment technologies.

Chapter 15 introduces a strategic planning approach for managing wastewater from hydraulic fracturing, considering treatment, storage, and reuse options to minimize costs and freshwater usage while addressing wastewater properties' variability.

Chapter 16 presents an alternative approach to reducing freshwater usage in shale gas production by using CO_2 as a fracturing fluid. The model optimizes water and CO_2 networks to minimize costs while reducing freshwater dependency.

Chapter 17 outlines an optimization strategy for constructing effective water networks for shale gas production, focusing on reducing freshwater consumption and effluents while balancing economic and environmental goals.

References

Ahmad S, Jia H, Chen Z, Li Q, Xu C (2020) Water-energy nexus and energy efficiency: a systematic analysis of urban water systems. Renew Sustain Energy Rev 134:110381

Ahmed AT, El F, GoharyTzanakakis VA, Angelakis AN (2020) Egyptian and Greek water cultures and hydro-technologies in ancient times. Sustainability 12(22):9760

Ajami NK, Hornberger GM, Sunding DL (2008) Sustainable water resource management under hydrological uncertainty. Water Resour Res 44(11):W11406

Alahi MEE, Sukkuea A, Tina FW, Nag A, Kurdthongmee W, Suwannarat K, Mukhopadhyay SC (2023) Integration of IoT-enabled technologies and artificial intelligence (AI) for smart city scenario: recent advancements and future trends. Sensors 23(11):5206

Algaba MHP, Huyghe W, Van K, LeeuwenKoop S, Eisenreich S (2024) Assessment and actions to support integrated water resources management of Seville (Spain). Environ Dev Sustain 26(3):7347–7375

Angelakis AN, Antoniou GP, Yapijakis C, Tchobanoglous G (2020) History of hygiene focusing on the crucial role of water in the hellenic asclepieia (i.e., ancient hospitals). Water 12(3):754

Angelakis AN, Capodaglio AG, Passchier CW, Valipour M, Krasilnikoff J, Tzanakakis VA, Sürmelihindi G, Baba A, Kumar R, Haut B, Roubelakis MG, Min Z, Dercas N (2023) Sustainability of water, sanitation, and hygiene: from prehistoric times to the present times and the future. Water 15(8):1614

Brown CM, Lund JR, Cai X, Reed PM, Zagona EA, Ostfeld A, Hall J, Characklis GW, Yu W, Brekke L (2015) The future of water resources systems analysis: toward a scientific framework for sustainable water management. Water Resour Res 51(8):6110–6124

Butler D, Ward S, Sweetapple C, Astaraie-Imani M, Diao K, Farmani R, Fu G (2017) Reliable, resilient and sustainable water management: the safe & SuRe approach. Global Chall 1(1):63–77

Cech TV (2018) Principles of water resources: history, development, management, and policy. Wiley, Hoboken

Chan SS, Wu JH (2022) Municipal-to-industrial water reuse via multi-stage and multi-pass reverse osmosis systems: a step from water scarcity towards sustainable development. Water 14(3):362

Den W, Chen CH, Luo YC (2018) Revisiting the water-use efficiency performance for microelectronics manufacturing facilities: using Taiwan's Science Parks as a case study. Water-Energy Nexus 1(2):116–133

Duhbacı TB, Özel S, Bulkan S (2021) Water and energy minimization in industrial processes through mathematical programming: a literature review. J Clean Prod 284:124752

Jhansi SC, Mishra SK (2013) Wastewater treatment and reuse: sustainability options. Consilience 10:1–15

Koudstaal R, Rijsberman FR, Savenije H (1992) Water and sustainable development. In: Natural resources forum, vol 16, no 4. Blackwell Publishing Ltd., Oxford, pp 277–290

Kundzewicz ZW (1997) Water resources for sustainable development. Hydrol Sci J 42(4):467–480

Nápoles-Rivera F, Serna-González M, El-Halwagi MM, Ponce-Ortega JM (2013) Sustainable water management for macroscopic systems. J Clean Prod 47:102–117

Ofori S, Puškáčová A, Růžičková I, Wanner J (2021) Treated wastewater reuse for irrigation: pros and cons. Sci Total Environ 760:144026

Olatunde TM, Adelani FA, Sikhakhane ZQ (2024) A review of smart water management systems from Africa and the United States. Eng Sci Technol J 5(4):1231–1242

Pahl-Wostl C, Tàbara D, Bouwen R, Craps M, Dewulf A, Mostert E, Taillieu T (2008) The importance of social learning and culture for sustainable water management. Ecol Econ 64(3):484–495

Pahl-Wostl C, Arthington A, Bogardi J, Bunn SE, Hoff H, Lebel L, Nikitina E, Palmer M, Poff LRN, Richards K, Schlüter M, Schulze R, St-Hilaire A, Tharme R, Tockner K, Tsegai D (2013) Environmental flows and water governance: managing sustainable water uses. Curr Opin Environ Sustain 5(3–4):341–351

Rane N (2023) Integrating leading-edge artificial intelligence (AI), internet of things (IOT), and big data technologies for smart and sustainable architecture, engineering and construction (AEC) industry: challenges and future directions. In: Engineering and construction (AEC) industry: challenges and future directions, 24 Sept 2023
Reis AL, Lopes MA, Andrade-Campos A, Antunes CH (2023) A review of operational control strategies in water supply systems for energy and cost efficiency. Renew Sustain Energy Rev 175:113140
Rijsberman MA, Van de FH, Ven (2000) Different approaches to assessment of design and management of sustainable urban water systems. Environ Impact Assess Rev 20(3):333–345
United Nations (1990) New Delhi statement. Global consultation on safe water and sanitation
Vairavamoorthy K, Akinpelu E, Lin Z, Ali M (2001) Design of sustainable water distribution systems in developing countries. In: Bridging the gap: meeting the world's water and environmental resources challenges, pp 1–10.

Chapter 2
Recycle and Reuse Water Networks

2.1 Introduction

Mass integration has been widely studied because of its economic and environmental benefits. Mass integration can be used to decrease the consumption of fresh sources and to minimize the waste materials sent to the environment. Reviews about mass integration can be found in literature (El-Halwagi 1997, 2006). Mass integration mainly consists of minimizing freshwater usage and wastewater discharge through recycle/reuse strategies. Wang and Smith (1994, 1995), Dhole et al. (1996), El-Halwagi and Spriggs (1996), Sorin and Bedard (1999), Kuo and Smith (1998), Polley and Polley (2000), Hallale (2002), El-Halwagi et al. (2003) presented conceptual methods to address the fresh water-wastewater allocation problem. Takama et al. (1980), Doyle and Smith (1997), Alva-Argaez et al. (1998, 1999), Galan and Grossmann (1998), Huang et al. (1999), Benko et al. (2000), Savelsky and Bagajewicz (2000, 2001), Hernandez-Suarez et al. (2004), Gabriel and El-Halwagi (2005), and Karuppiah and Grossmann (2006) reported mathematical programming approaches to address water allocation problems. Furthermore, Gunaratnam et al. (2005) reported an approach to designing water systems considering water use and wastewater treatment processes, and Kiperstok and Sharratt (1997) reported an approach for the simultaneous optimization of waste minimization and end-of-pipe treatments. Furthermore, Bagajewicz (2000) reviewed procedures for designing water networks.

Previous approaches provided contributions to the synthesis of water recycling/reusing networks. Nevertheless, most of them have a couple of limitations:

1. In-plant recycling with constraints limited to process units and without considering the effects of the resulting wastewater streams and the environmental regulations (Fig. 2.1). Wastewater streams require to be treated before being discharged to the environment; this situation may increase the total cost associated with the recycle-reuse process. For example, when the solution that minimizes fresh source consumption yields a high wastewater treatment cost. It may be better to use more fresh sources and lower the wastewater treatment cost. Hence, it is

J. M. Ponce-Ortega et al., *Sustainable Water Systems*, Water Resources Development and Management, https://doi.org/10.1007/978-3-031-72498-5_2

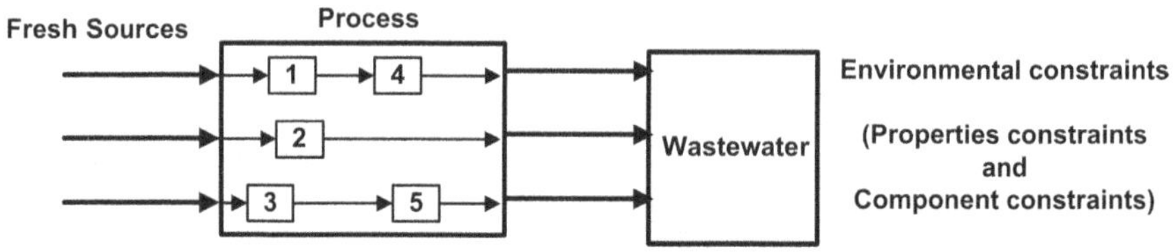

Fig. 2.1 The traditional strategy for mass integration. Printed with permission of "*A property-based optimization of direct recycle networks and wastewater treatment process*", *AIChE Journal*, Copyright (2009 American Institute of Chemical Engineers), John Wiley & Son

important to consider simultaneously the optimization of the process integration together with the wastewater treatment process to consider the trade-offs between the involved factors.

2. Process and recycle constraints have been based on compositions. There are several cases when such constraints are based on properties. Examples include applications when the performance of the process units is affected by the properties of its feed and cases when environmental regulations impose limits for specific properties of waste streams such as toxicity, theoretical oxygen demand (*ThOD*), *pH*, temperature, color, odor, viscosity, and density. These properties are difficult to quantify as a function of composition because of the many components of the process streams. It is also difficult to track properties throughout the process because they are not conserved. To overcome this limitation, Shelley and El-Halwagi (2000) introduced the concept of property-based componentless design by defining surrogate properties (called clusters) that enable the conserved tracking of properties. El-Halwagi et al. (2004) generalized it to the concept of property integration defined as "a functionality-based holistic approach for the allocation and manipulation of streams and processing units, which is based on functionality tracking, adjustment, and assignment throughout the process." For a network with up to three properties, graphical tools may be used to guide the synthesis and analysis tasks (El-Halwagi et al. 2004; Kazantzi and El-Halwagi 2005); and when the number of properties is larger, algebraic tools are used (Qin et al. 2004). Eden et al. (2004), Eljack et al. (2005), and Grooms et al. (2005) incorporated the concept of property integration into the product design strategy. Kazantzi et al. (2007) and Eljack et al. (2007) presented graphical approaches for simultaneous process and molecular design.

This chapter presents a mathematical programming model to optimize direct recycle networks together with the wastewater treatment process to satisfy a set of process and environmental constraints (Ponce-Ortega et al. 2009). An optimization formulation is developed based on disjunctive programming (Raman and Grossmann 1994). The model is formulated to consider environmental-constrained properties such as toxicity, *ThOD*, *pH*, color, odor, and any other property that may cause environmental pollution, in addition to the mass and composition constraints for hazardous compounds in the waste stream. The problem is formulated with proper constraints as an MINLP problem that minimizes the total annual cost, which includes the cost for the fresh sources, piping cost, and the wastewater treatment cost.

2.2 Problem Statement

The addressed problem can be stated as follows. Given a set of processes and fresh sources with known flowrates, compositions, and properties. Similarly, there is a set of sinks (process units) with constraints for the inlet flowrates and allowed compositions and properties. There is a set of constraints given by the environmental regulations for the waste streams discharged to the environment. The problem consists of determining the optimal water network that includes a direct recycling strategy that meets the environmental regulations for the waste streams and minimizes the total annual cost.

The used sets are defined as follows. *NSINKS* represents the process sinks, and *NSOURCES* and *NFRESH* correspond to the process and fresh sources. *NCOMP* represents the involved components, and *NPROP* is the properties constrained for the sinks and environmental regulations.

Figure 2.2 shows a representation of the addressed problem which corresponds to a direct recycle approach, which considers constraints for the components and properties of the sinks and the waste treatment process needed for the waste to be discharged.

2.3 Mathematical Model Formulation

The mathematical model consists of a set of equations for the process constraints and treatment constraints. The formulation is based on a direct recycle configuration, and the waste treatment constraints are based on a disjunctive programming formulation. First, the process constraints to model the direct recycle-mass integration, which includes component and properties constraints given by the process units, are given, followed by the constraints for the waste stream treatment process. The objective function is given at the end of the formulation.

2.3.1 Process Constraints

Splitting of process source streams. The flow rate from every process source i is segregated and sent to sinks j and waste.

$$W_i = \sum_{j \in NSINKS} w_{i,j} + waste_i, \quad i \in NSOURCES \tag{2.1}$$

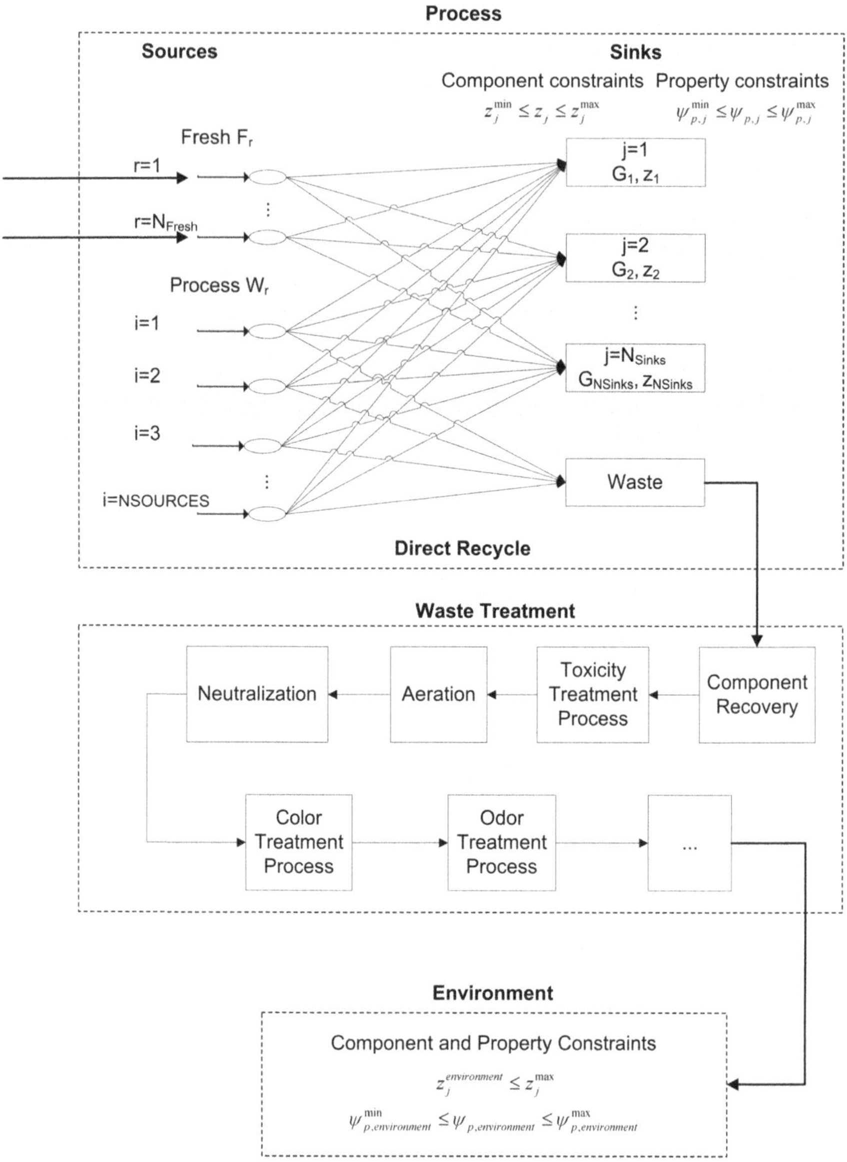

Fig. 2.2 Source-sink representation for mass and property wastewater integration including waste treatment. Printed with permission of "*A property-based optimization of direct recycle networks and wastewater treatment process*", *AIChE Journal*, Copyright (2009 American Institute of Chemical Engineers), John Wiley & Son

W_i is the total flowrate from source i, $w_{i,j}$ is the flowrate segregated from source i and sent to sink j, and $waste_i$ is the flowrate segregated from source i and sent to the waste stream.

Splitting of fresh source streams. Fresh water sources are segregated and directed to sinks.

$$F_r = \sum_{j \in NSINKS} f_{r,j}, \quad r \in NFRESH \tag{2.2}$$

F_r is the total flowrate consumed from fresh source r and $f_{r,j}$ is the flowrate segregated from fresh source r and sent to sink j. Fresh sources cannot be sent to the waste stream.

Total waste. The total waste is the sum of the waste streams from all sources.

$$WASTE = \sum_{i \in NSOURCES} waste_i \tag{2.3}$$

Mass balances at the mixing point before each sink. The following water balance is needed to determine the inlet flowrate at each sink.

$$G_j = \sum_{i \in NSOURCES} w_{i,j} + \sum_{r \in NFRESH} f_{r,j}, \quad j \in NSINKS \tag{2.4}$$

where G_j is the total inlet flowrate at sink j.

Component balances at the mixing point before each sink. The following component balance is needed to determine the composition of a stream inlet to any sink.

$$z_{j,c}^{in} G_j = \sum_{i \in NSOURCES} \left[z_{i,c} \, w_{i,j} \right] + \sum_{r \in NFRESH} \left[z_{r,c} f_{r,j} \right], \quad j \in NSINKS, \quad c \in NCOMP \tag{2.5}$$

$z_{j,c}^{in}$ is the inlet composition for component c at the inlet of sink j, and $z_{i,c}$ and $z_{r,c}$ are the compositions for the process and fresh sources, respectively.

Property balance at the mixing point before each sink. A property balance is required to determine the inlet properties to each sink.

$$\psi_p\left(p_{j,p}^{in}\right) G_j = \sum_{i \in NSOURCES} \left[\psi_p(p_{i,p}) w_{i,j} \right] + \sum_{r \in NFRESH} \left[\psi_p(p_{r,p}) f_{r,j} \right], \quad j \in NSINKS, \quad p \in NPROP \tag{2.6}$$

$\psi_p\left(p_{j,p}^{in}\right)$ is the property operator at the inlet conditions of sink j for property p, and $\psi_p(p_{i,p})$ and $\psi_p(p_{r,p})$ are property operators for process and fresh sources i

and r, respectively. Shelley and El-Halwagi (2000) stated that the property operators follow a mixing rule that is assumed to be proportional to each mass contribution. The property operators can be calculated from first principles or estimated through empirical or semi-empirical methods. For properties like composition, toxicity, chemical oxygen demand, and odor the property operator is $\psi_P(P) = P$, whereas for the pH, $\psi_{pH}(pH) = 10^{pH}$, for the density, $\psi_\rho(\rho) = 1/\rho$, for the viscosity $\psi_\mu(\mu) = \log(\mu)$, and the Reid vapor pressure, $\psi_{RVP}(RVP) = RVP^{1.44}$.

Sinks composition constraints. Every sink is limited to receive the inlet stream within composition constraints,

$$z_{j,c}^{min} \leq z_{j,c}^{in} \leq z_{j,c}^{max}, \quad j \in NSINKS, \quad c \in NCOMP \tag{2.7}$$

Sinks properties constraints. It is necessary to include limits for specific properties that are difficult to quantify with component constraints when there are many components in the stream. Sinks property-based constraints are given as follows,

$$p_{p,j}^{min} \leq p_{p.j}^{in} \leq p_{p.j}^{max}, \quad j \in NSINKS, \quad p \in NPROP \tag{2.8}$$

2.3.2 Wastewater Treatment Constraints

Wastewater needs to meet environmental constraints before discharge into the environment. Constraints are formulated from the environmental regulations for hazardous materials. Whenever possible, they are based on component constraints, but for properties that are difficult to characterize because of the large number of components in some waste streams, they are formulated as property constraints.

The approach presented in this chapter is based on a direct recycle strategy avoiding many complexities that are obtained when the treatment of the streams before being sent to the process sinks, these complexities include many bilinear products for the property balances that cause a lot of computational problems in the optimization process. Because the waste is the only stream that may be treated in the formulation presented in this chapter, the order for the treatment units may be easily fixed based on the industrial knowledge, experience, and characteristics of the streams for the analyzed process; this way, the model becomes less complex, and it can be solved easier.

Ratio waste stream. The contribution of each source stream to the total waste is,

$$Ratio_i = \frac{waste_i}{\sum_{i \in SOURCES} waste_i} \tag{2.9}$$

$Ratio_i$ is the fractional contribution of each process source i for the waste stream.

Density for the waste stream. The density of the waste stream (ρ_{waste}) is calculated through the following mixing rule,

$$\frac{1}{\rho_{waste}} = \sum_{i \in SOURCES} \left[\frac{Ratio_i}{\rho_i} \right] \tag{2.10}$$

2.3.2.1 Recovery Process

It is necessary to include a recovery process in the waste treatment process to satisfy the environmental regulations associated with the discharge of a specific hazardous compound. This recovery process depends on the mass load and concentration of the specific components. Then, it is required to determine the composition and the mass load for the undesirable compounds before being discharged to the environment. For each compound u to be recovered, there exist technologies that may be used with a recovery efficiency EFF_u.

The mass flow inlet to the recovery process is,

$$WINREC = WASTE \tag{2.11}$$

The component balance inlet to the recovery process is,

$$WINRECU_u = \sum_{i \in NSOURCES} \left(waste_i \, z_{i,u}^{inrecov} \right) \tag{2.12}$$

and

$$z_u^{inrecov} \, WINREC = WINRECU_u \tag{2.13}$$

The recovery of component u is required only if its concentration is higher than the maximum value allowed to be discharged into the environment. Otherwise, the recovery process is not required. This is formulated through the following disjunction,

$$\begin{bmatrix} Y_u^{Recov} \\ z_u^{inrecov} \geq z_u^{MaxEnv} \\ W_u^{Recovery} = WINRECU_u \, EFF_u \\ Cost_u^{Recoveryu} = Cost_u^{Recu} \, W_u^{Recovery} \, H_Y \end{bmatrix} \vee \begin{bmatrix} Y_u^{Recov} \\ z_u^{inrecov} \leq z_u^{MaxEnv} \\ W_u^{Recovery} = 0 \\ Cost_u^{Recoveryu} = 0 \end{bmatrix}$$

Y_u^{Recov} is a logical variable that is true when $z_u^{inrecov} \geq z_u^{MaxEnv}$; in this case, it is required to include the recovery process, and the cost associated is a function of the recovery flowrate for component u. Otherwise, when $z_u^{inrecov} \leq z_u^{MaxEnv}$, the logical variable Y_u^{Recov} becomes false, and in this case the recovery process is not needed. The binary variable y_u^{Recov} is used to model the logical variable Y_u^{Recov}; therefore, y_u^{Recov} is one when Y_u^{Recov} is true, otherwise y_u^{Recov} is zero when Y_u^{Recov} is false.

Previous disjunction can be algebraically formulated as a linear set of relationships as follows,

$$z_u^{inrecov} = z_u^{inrecov1} + z_u^{inrecov2} \tag{2.14}$$

$$WINRECU_u = WINRECU_u^1 + WINRECU_u^2 \tag{2.15}$$

$$z_u^{inrecov1} \geq z_u^{Max} \, y_u^{Recov} \tag{2.16}$$

$$W_u^{Recovery} = WINRECU_u^1 \, EFF_u \tag{2.17}$$

$$Cost_u^{Recoveryu} = Cost_u^{Recu} \, W_u^{Recovery} \, H_Y \tag{2.18}$$

$$z_u^{inrecov2} \leq z_u^{Max} \left(1 - y_u^{Recov}\right) \tag{2.19}$$

$$z_u^{inrecov1} \leq M_u^{y^{inrecov}} \, y_u^{Recov} \tag{2.20}$$

$$WINRECOVU_u^1 \leq M_u^{WINRECOV} \, y_u^{Recov} \tag{2.21}$$

$$Cost_u^{Recoveryu} \leq M_u^{Cost^{Recov}} \, y_u^{Recov} \tag{2.22}$$

$$W_u^{Recovery} \leq M_u^{W^{Recovery}} \, y_u^{Recov} \tag{2.23}$$

$$z_u^{inrecov2} \leq M_u^{y^{inrecov}} \left(1 - y_u^{Recov}\right) \tag{2.24}$$

$$WINRECOVU_u^2 \leq M_u^{WINRECOV} \left(1 - y_u^{Recov}\right) \tag{2.25}$$

Mass balance at the recovery process. It is required to calculate the flowrate after the recovery process with a mass balance,

$$W1 = WINREC - \sum_u W_u^{Recovery} \tag{2.26}$$

Component balances. The composition after the recovery process is calculated as follows,

$$z1_u \, W1 = WINRECU_u - W_u^{Recovery} \tag{2.27}$$

2.3.2.2 Toxicity Treatment

Toxicity is a property that defines the effect of hazardous chemicals on biological organisms. Chemicals such as phenol, cyanides, and ionic metals, among others, are considered toxic because they may cause death to organisms exposed to them. To quantify the lethal response of a toxic chemical t, the toxicity is represented by the following equation,

$$Y_t(\%Mortality) = k_{1_t} + k_{2_t}\log(c_t) \tag{2.28}$$

Y_t is the percentage of organisms that would die if exposed to a concentration c_t (in ppm) of chemical t. Usually, the environmental regulations state maximum mass loads and compositions allowed to be discharged to the environment to yield a mortality of 0% for the organisms exposed to the toxic compounds. Therefore, the toxicity of a chemical compound can be characterized by the flowrate and composition of a toxic compound.

The flowrates of the toxic compounds at the inlet of the toxicity treatment process are given by,

$$W_t^{BT} = WINRECU_t - W_t^{Recovery}, \quad t \in TOX \tag{2.29}$$

When there is a toxic compound in the waste stream, an environmental constraint must be satisfied. The environmental regulations are given in terms of the concentration and the mass flowrate of the toxic compounds discharged to the environment. Four scenarios can be considered for the toxicity treatment process of the waste stream, which can be modeled using the following disjunction,

$$\begin{bmatrix} Y_{A_t}^{Tox} \\ \begin{bmatrix} Y_{1_t}^{Tox} \\ z1_t \geq z_{Tox_t}^{Max} \\ W_t^{BT} \leq D_{Tox_t}^{Max} \\ W_t^{Rem} = W_t^{BT} - z_{Tox_t}^{Max}\, W1 \end{bmatrix} \vee \begin{bmatrix} Y_{2_t}^{Tox} \\ z1_t \leq z_{Tox_t}^{Max} \\ W_t^{BT} \geq D_{Tox_t}^{Max} \\ W_t^{Rem} = W_t^{BT} - D_{Tox_t}^{Max} \end{bmatrix} \vee \begin{bmatrix} Y_{3_t}^{Tox} \\ z1_t \geq z_{Tox_t}^{Max} \\ W_t^{BT} \geq D_{Tox_t}^{Max} \\ W_t^{Rema} = W_t^{BT} - z_{Tox_t}^{Max}\, W1 \\ W_t^{Remb} = W_t^{BT} - D_{Tox_t}^{Max} \\ \begin{bmatrix} Y_{3a_t}^{Tox} \\ W_t^{Rema} \geq W_t^{Remb} \\ W_t^{Rem} = W_t^{Rema} \end{bmatrix} \vee \begin{bmatrix} Y_{3b_t}^{Tox} \\ W_t^{Rema} \leq W_t^{Remb} \\ W_t^{Rem} = W_t^{Remb} \end{bmatrix} \end{bmatrix} \\ Cost_t^{Toxicity} = W_t^{Rem}\, Cost_t^{Toxict\,H_Y} \end{bmatrix}$$

$$\vee \begin{bmatrix} Y_{B_t}^{Tox} \\ z1_t \leq z_{Tox_t}^{Max} \\ W_t^{BT} \leq D_{Tox_t}^{Max} \\ W_t^{Rem} = 0 \\ Cost_t^{Toxicity} = 0 \end{bmatrix}, \quad t \in TOX$$

$Y_{A_t}^{Tox}$ and $Y_{B_t}^{Tox}$ are Boolean variables that are used to model the existence or not the existence of the toxicity treatment process. $Y_{1_t}^{Tox}$, $Y_{2_t}^{Tox}$ and $Y_{3_t}^{Tox}$ are Boolean variables used to model three different scenarios when the toxicity treatment process is needed.

The first scenario, $Y_{1_t}^{Tox}$, is for the case when the composition of the toxic compound t in the waste stream is bigger than the maximum allowed given by the toxicity constraint and the mass flowrate for the toxic compound t is lower than the one given by the toxicity constraint, in this case, the mass removed for the toxic compound is dominated by the composition. The second scenario, $Y_{2_t}^{Tox}$, models the situation when the composition satisfies the toxicity constraint, but the mass flowrate does not, in this case, the mass removed for the toxic t is given by the mass flow rate. The scenario $Y_{3_t}^{Tox}$ models the situation when both the composition and mass flowrate of t do not satisfy the toxicity constraints; in this case, there are two inner scenarios to determine the mass of t removed modeled with the logical variables $Y_{3a_t}^{Tox}$ and $Y_{3b_t}^{Tox}$ that are used to model when the composition or the mass flowrate of t is the dominant in the toxicity constraint. The problem is modeled in this form to yield linear terms inside the disjunctions, and the following set of linear equations can be written.

First, we have the following logical relations,

$$y_{A_t}^{Tox} + y_{B_t}^{Tox} = 1, \quad t \in TOX \tag{2.30}$$

$$y_{A_t}^{Tox} = y_{1_t}^{Tox} + y_{2_t}^{Tox} + y_{3_t}^{Tox}, \quad t \in TOX \tag{2.31}$$

$$y_{3_t}^{Tox} = y_{3a_t}^{Tox} + y_{3b_t}^{Tox}, \quad t \in TOX \tag{2.32}$$

The disaggregated variables are,

$$z1_t = z1_t^1 + z1_t^2 + z1_t^3 + z1_t^4, \quad t \in TOX \tag{2.33}$$

$$W_t^{BT} = W_t^{BT1} + W_t^{BT2} + W_t^{BT3} + W_t^{BT4}, \quad t \in TOX \tag{2.34}$$

$$W1 = W1^1 + W1^2 + W1^3 + W1^4, \quad t \in TOX \tag{2.35}$$

$$W_t^{Rem} = W_t^{Rem1} + W_t^{Rem2} + W_t^{Rem3}, \quad t \in TOX \tag{2.36}$$

$$W_t^{Rem3} = W_t^{Rem3a} + W_t^{Rem3b}, \quad t \in TOX \tag{2.37}$$

$$W_t^{Rema3} = W_t^{Rema3a} + W_t^{Rema3b}, \quad t \in TOX \tag{2.38}$$

$$W_t^{Remb3} = W_t^{Remb3a} + W_t^{Remb3b}, \quad t \in TOX \tag{2.39}$$

The algebraic relationships for the first disjunction are,

$$z1_t^1 \geq z_{Tox_t}^{Max} y_{1_t}^{Tox}, \quad t \in TOX \tag{2.40}$$

$$W_t^{BT1} \le D_{Tox_t}^{Max}\, y_{1_t}^{Tox}, \quad t \in TOX \tag{2.41}$$

$$W_t^{Rem1} = W_t^{BT1} - z_{Tox_t}^{Max}\, W1^1, \quad t \in TOX \tag{2.42}$$

For the second disjunction,

$$z1_t^2 \le z_{Tox_t}^{Max}\, y_{2_t}^{Tox}, \quad t \in TOX \tag{2.43}$$

$$W_t^{BT2} \ge D_{Tox_t}^{Max}\, y_{2_t}^{Tox}, \quad t \in TOX \tag{2.44}$$

$$W_t^{Rem2} = W_t^{BT2} - D_{Tox_t}^{Max}\, y_{2_t}^{Tox}, \quad t \in TOX \tag{2.45}$$

For the third disjunction, the following model applies,

$$z1_t^3 \ge z_{Tox_t}^{Max}\, y_{3_t}^{Tox}, \quad t \in TOX \tag{2.46}$$

$$W_t^{BT3} \ge D_{Tox_t}^{Max}\, y_{3_t}^{Tox}, \quad t \in TOX \tag{2.47}$$

$$W_t^{Rema3} = W_t^{BT3} - z_{Tox_t}^{Max}\, W1^3, \quad t \in TOX \tag{2.48}$$

$$W_t^{Remb3} = W_t^{BT3} - D_{Tox_t}^{Max}\, y_{3_t}^{Tox}, \quad t \in TOX \tag{2.49}$$

where for subsection 3a,

$$W_t^{Rema3a} \ge W_t^{Remb3a}, \quad t \in TOX \tag{2.50}$$

$$W_t^{Rem3a} = W_t^{Rema3a}, \quad t \in TOX \tag{2.51}$$

and for subsection 3b,

$$W_t^{Rema3b} \le W_t^{Remb3b}, \quad t \in TOX \tag{2.52}$$

$$W_t^{Rem3b} = W_t^{Remb3b}, \quad t \in TOX \tag{2.53}$$

The total cost for the toxicity treatment is,

$$Cost_t^{Toxicity} = W_t^{Rem}\, Cost_t^{Toxic}\, H_Y, \quad t \in TOX \tag{2.54}$$

For the fourth disjunction, we have,

$$z1_t^4 \le z_{Tox_t}^{Max} y_{B_t}^{Tox}, \quad t \in TOX \tag{2.55}$$

$$W_t^{BT4} \le D_{Tox_t}^{Max} y_{B_t}^{Tox}, \quad t \in TOX \tag{2.56}$$

Upper limits for the disaggregated variables are incorporated:

$$z1_t^1 \le M_t^{z_{Tox}^{Max}} y_{1_t}^{Tox}, \quad t \in TOX \tag{2.57}$$

$$W_t^{BT1} \le M_t^{W_{Tox}^{Max}} y_{1_t}^{Tox}, \quad t \in TOX \tag{2.58}$$

$$W1^1 \le M^{W1^{Max}} y_{1_t}^{Tox}, \quad t \in TOX \tag{2.59}$$

$$W_t^{Rem1} \le M_t^{W^{Rem}} y_{1_t}^{Tox}, \quad t \in TOX \tag{2.60}$$

$$z1_t^2 \le M_t^{z_{Tox}^{Max}} y_{2_t}^{Tox}, \quad t \in TOX \tag{2.61}$$

$$W_t^{BT2} \le M_t^{W_{Tox}^{Max}} y_{2_t}^{Tox}, \quad t \in TOX \tag{2.62}$$

$$W1^2 \le M^{W1^{Max}} y_{2_t}^{Tox}, \quad t \in TOX \tag{2.63}$$

$$W_t^{Rem2} \le M_t^{W^{Rem}} y_{2_t}^{Tox}, \quad t \in TOX \tag{2.64}$$

$$z1_t^3 \le M_t^{z_{Tox}^{Max}} y_{3_t}^{Tox}, \quad t \in TOX \tag{2.65}$$

$$W_t^{BT3} \le M_t^{W_{Tox}^{Max}} y_{3_t}^{Tox}, \quad t \in TOX \tag{2.66}$$

$$W1^3 \le M^{W1^{Max}} y_{3_t}^{Tox}, \quad t \in TOX \tag{2.67}$$

$$W_t^{Rem3} \le M_t^{W^{Rem}} y_{3_t}^{Tox}, \quad t \in TOX \tag{2.68}$$

$$W_t^{Rem3a} \le M_t^{W^{Rem}} y_{3a_t}^{Tox}, \quad t \in TOX \tag{2.69}$$

$$W_t^{Rema3a} \le M_t^{W^{Rem}} y_{3a_t}^{Tox}, \quad t \in TOX \tag{2.70}$$

$$W_t^{Remb3a} \le M_t^{W^{Rem}} y_{3a_t}^{Tox}, \quad t \in TOX \tag{2.71}$$

$$W_t^{Rem3b} \leq M_t^{W^{Rem}} y_{3b_t}^{Tox}, \quad t \in TOX \tag{2.72}$$

$$W_t^{Rema3b} \leq M_t^{W^{Rem}} y_{3b_t}^{Tox}, \quad t \in TOX \tag{2.73}$$

$$W_t^{Remb3b} \leq M_t^{W^{Rem}} y_{3b_t}^{Tox}, \quad t \in TOX \tag{2.74}$$

$$z1_t^4 \leq M_t^{z_{Tox}^{Max}} y_{B_t}^{Tox}, \quad t \in TOX \tag{2.75}$$

$$W_t^{BT4} \leq M_t^{W_{Tox}^{Max}} y_{B_t}^{Tox}, \quad t \in TOX \tag{2.76}$$

$$W1^4 \leq M^{W1^{Max}} y_B^{Tox}, \quad t \in TOX \tag{2.77}$$

Mass balance in the unit for toxicity treatment. It is necessary to formulate a mass balance around the toxicity treatment unit to find the flowrate at the exit conditions,

$$W2 = W1 - \sum_{t \in TOX} W_t^{Rem} \tag{2.78}$$

Component balances. A component balance is used to determine the composition at the exit of the toxicity treatment unit,

$$W2\, z2_t = W1\, z1_t - W_t^{Rem}, \quad t \in TOX \tag{2.79}$$

2.3.2.3 Aeration Process

Theoretical oxygen demand (*ThOD*) is the amount of oxygen required to oxidize an organic compound to its final oxidation products. *ThOD* in wastewater streams has been strictly regulated since the discharge of high values of *ThOD* would significantly reduce the oxygen available and cause adverse effects on the aquatic flora and fauna. Because the *ThOD* in the waste stream depends on the many compounds present, it would be difficult to quantify it with the composition of the waste stream. In this case, the *ThOD* in the waste stream is calculated using a property balance, since the property *ThOD* for each process source discharged to the waste stream is known before the optimization process. Therefore, the theoretical oxygen demand for the waste stream can be calculated from the following equation,

$$ThOD^{Total} = \sum_{i \in SOURCES} (Ratio_i\, ThOD_i) \tag{2.80}$$

$ThOD_i$ is the theoretical oxygen demand for process source i, which is a parameter known before the optimization process.

To get linear disjunctions, there are defined the following variables,

$$W^{ThOD} = \frac{ThOD^{Total}\, W2}{\rho_{waste}} \tag{2.81}$$

and

$$W^{ThOD^{Reg}} = \frac{ThOD^{Reg}\, W2}{\rho_{waste}} \tag{2.82}$$

The aeration process is needed only when the $ThOD$ for a waste stream is higher than the value stated by the environmental regulation. Therefore, the following disjunction applies to the aeration process,

$$\begin{bmatrix} Y^{ThOD} \\ W^{ThOD} \geq W^{ThOD^{Reg}} \\ Cost^{ThOD} = \left(W^{ThOD} - W^{ThOD^{Reg}}\right)\left[\frac{29}{32*0.21*10^3}\right]Cost^{Aeration}\, H_Y \end{bmatrix} \vee \begin{bmatrix} Y^{ThOD} \\ W^{ThOD} \leq W^{ThOD^{Reg}} \\ Cost^{ThOD} = 0 \end{bmatrix}$$

Y^{ThOD} is a logical variable that is true when $W^{ThOD} \geq W^{ThOD^{Reg}}$; in that case, the cost of the aeration process is calculated (the factors in parenthesis are used to get proper units). Otherwise, when the $ThOD$ for the waste stream is lower than the one given by the environmental regulation, the cost of the aeration process is zero. This disjunction can be algebraically modeled as follows. First, the continuous variables are disaggregated,

$$W^{ThOD} = W^{ThOD^1} + W^{ThOD^2} \tag{2.83}$$

$$W^{ThOD^{Reg}} = W^{ThOD^{Reg^1}} + W^{ThOD^{Reg^2}} \tag{2.84}$$

The constraints for the first disjunction are,

$$W^{ThOD^1} \geq W^{ThOD^{Reg^1}} \tag{2.85}$$

$$Cost^{ThOD} = \left(W^{ThOD^1} - W^{ThOD^{Reg^1}}\right)\left[\frac{29}{32*0.21*10^3}\right]Cost^{Aeration}\, H_Y \tag{2.86}$$

and for the second disjunction,

$$W^{ThOD^2} \leq W^{ThOD^{Reg^2}} \tag{2.87}$$

Finally, the limits for the disaggregated variables are implemented,

$$W^{ThOD^1} \leq M^{W^{ThOD}} y^{ThOD} \tag{2.88}$$

$$W^{ThOD^{Reg^1}} \leq M^{W^{ThOD}} y^{ThOD} \tag{2.89}$$

$$Cost^{ThOD} \leq M^{Cost^{ThOD}} y^{ThOD} \tag{2.90}$$

$$W^{ThOD^2} \leq M^{W^{ThOD}} \left(1 - y^{ThOD}\right) \tag{2.91}$$

$$W^{ThOD^{Reg^2}} \leq M^{W^{ThOD}} \left(1 - y^{ThOD}\right) \tag{2.92}$$

The reformulation of the previous disjunction is linear.

2.3.2.4 Neutralization Process

pH is an important property to be regulated in waste discharges. High and low values of *pH* for the waste streams are considered particularly hazardous; therefore, waste streams need to be treated before being discharged into the environment to satisfy *pH* environmental regulations. The *pH* approximation for a waste stream with many compounds is a difficult task; in this case, the property integration formulation by El-Halwagi et al. (2004) are calculated as follows,

$$10^{pH_{mean}} = \sum_{i \in NSOURCES} \left(Ratio_i \, 10^{pH_i}\right) \tag{2.93}$$

pH_i is the individual *pH* for process source *i* that is known before the optimization process.

Then, the *pOH* for the waste stream is,

$$pOH_{mean} = 14 - pH_{mean} \tag{2.94}$$

and,

$$W3 = W2 \tag{2.95}$$

The concentration of ions H_3O^+ and OH^- in the waste stream are determined as follows,

$$(WH3O) = 10^{-pH_{mean}} \tag{2.96}$$

$$(WOH) = 10^{-pOH_{mean}} \tag{2.97}$$

To avoid singularities, the following constraints are used since the highest value for pH and pOH is 14,

$$WH3O \geq 10^{-14} \tag{2.98}$$

$$WOH \geq 10^{-14} \tag{2.99}$$

The total moles of H_3O^+ and OH^- discharged into the waste stream are,

$$Total^{\mathrm{H_3O}} = \frac{W3\, WH3O}{\rho_{waste}} \tag{2.100}$$

$$Total^{\mathrm{OH}} = \frac{W3\, WOH}{\rho_{waste}} \tag{2.101}$$

Three cases may occur for the neutralization process. The pH of the waste stream can be lower than the value given by the environmental regulation; in this case, it is necessary to add an amount of base that must be determined. The second case arises when the pH of the waste stream is higher than the one given by the environmental regulation; in this case, it is necessary to add an amount of acid. Ultimately, if the pH of the waste stream meets the environmental regulation, the neutralization process is not needed. Therefore, to calculate the cost of acid or base added to a waste stream, as needed to satisfy the environmental regulations, the following disjunction can be written,

$$\begin{bmatrix} Y_B^{Neu} \\ pH_{mean} \leq 5.5 \\ pH_{discharged} = 5.5 \\ Base = \frac{Total^{\mathrm{H_3O}} - 10^{-5.5}\, Waste4}{\mathrm{OH}_{conc}} \\ Acid = 0 \\ Cost^{Neu} = Base\; Cost^{Base} H_Y \end{bmatrix} \vee \begin{bmatrix} Y_A^{Neu} \\ pH_{mean} \geq 9 \\ pH_{discharged} = 9 \\ Base = 0 \\ Acid = \frac{Total^{\mathrm{OH}} - 10^{-5}\, Waste4}{\mathrm{H_3O}_{conc}} \\ Cost^{Neu} = Acid\; Cost^{Acid} H_Y \end{bmatrix}$$

$$\vee \begin{bmatrix} Y_C^{Neu} \\ pH_{mean} \geq 5.5 \\ pH_{mean} \leq 9 \\ pH_{discharged} = pH_{mean} \\ Base = 0 \\ Acid = 0 \\ Cost^{Neu} = 0 \end{bmatrix}$$

Y_B^{Neu}, Y_A^{Neu}, and Y_C^{Neu} are logical variables for the base treatment, acid treatment, and no neutralization treatment, respectively. $pH_{discharged}$ is the pH for the waste stream discharged to the environment, and *Acid* and *Base* are the amounts of acid and base added in the neutralization process to yield $pH_{discharged}$. This disjunction can be modeled as follows,

$$y_B^{Neu} + y_A^{Neu} + y_C^{Neu} = 1 \tag{2.102}$$

Then, it is necessary to define the following disaggregated variables,

$$pH_{mean} = pH_{mean}^B + pH_{mean}^A + pH_{mean}^C \tag{2.103}$$

$$pH_{discharged} = pH_{discharged}^B + pH_{discharged}^A + pH_{discharged}^C \tag{2.104}$$

$$Total^{\mathrm{H_3O}} = Total^{\mathrm{H_3O}^B} + Total^{\mathrm{H_3O}^A} + Total^{\mathrm{H_3O}^C} \tag{2.105}$$

$$Total^{\mathrm{OH}} = Total^{\mathrm{OH}^B} + Total^{\mathrm{OH}^A} + Total^{\mathrm{OH}^C} \tag{2.106}$$

$$Waste4 = Waste4^B + Waste4^A + Waste4^C \tag{2.107}$$

$$Cost^{Neu} = Cost^{Neu^B} + Cost^{Neu^A} \tag{2.108}$$

For the first disjunction,

$$pH_{mean}^B \leq 5.5\, y_B^{Neu} \tag{2.109}$$

$$pH_{discharged}^B = 5.5\, y_B^{Neu} \tag{2.110}$$

$$Base = \frac{Total^{\mathrm{H_3O}^B} - 10^{-5.5}\, Waste4^B}{\mathrm{OH}_{conc}} \tag{2.111}$$

$$Cost^{Neu^B} = Base\; Cost^{Base} H_Y \tag{2.112}$$

For the second disjunction,

$$pH_{mean}^A \geq 9\, y_A^{Neu} \tag{2.113}$$

$$pH_{discharged}^A = 5.5\, y_A^{Neu} \tag{2.114}$$

$$Acid = \frac{Total^{OH^A} - 10^{-5}\,Waste4^A}{H_3O_{conc}} \tag{2.115}$$

$$Cost^{Neu^A} = Acid\;Cost^{Acid}\,H_Y \tag{2.116}$$

And, for the third disjunction,

$$pH_{mean}^{C} \leq 9\,y_C^{Neu} \tag{2.117}$$

$$pH_{mean}^{C} \geq 5.5\,y_C^{Neu} \tag{2.118}$$

$$pH_{discharged}^{C} = pH_{mean}^{C} \tag{2.119}$$

Finally, the following upper limits for the disaggregated variables are needed,

$$pH_{mean}^{B} \leq M^{pH}\,y_B^{Nue} \tag{2.120}$$

$$Total^{H_3O^B} \leq M^{Total^{H_3O}}\,y_B^{Nue} \tag{2.121}$$

$$Total^{OH^B} \leq M^{Total^{OH}}\,y_B^{Nue} \tag{2.122}$$

$$Waste4^B \leq M^{Waste4}\,y_B^{Nue} \tag{2.123}$$

$$pH_{discharged}^{B} \leq M^{pH}\,y_B^{Nue} \tag{2.124}$$

$$Base \leq M^{Base}\,y_B^{Neu} \tag{2.125}$$

$$Cost^{Neu^B} \leq M^{Cost^{Neu}}\,y_B^{Nue} \tag{2.126}$$

$$pH_{mean}^{A} \leq M^{pH}\,y_A^{Nue} \tag{2.127}$$

$$Total^{H_3O^A} \leq M^{Total^{H_3O}}\,y_A^{Nue} \tag{2.128}$$

$$Total^{OH^A} \leq M^{Total^{OH}}\,y_A^{Nue} \tag{2.129}$$

$$Waste4^A \leq M^{Waste4}\,y_A^{Nue} \tag{2.130}$$

$$pH_{disch\,arg\,ed}^{A} \leq M^{pH}\,y_A^{Nue} \tag{2.131}$$

$$Acid \leq M^{Acid} y_A^{Neu} \tag{2.132}$$

$$Cost^{Neu^A} \leq M^{Cost^{Neu}} y_A^{Nue} \tag{2.133}$$

$$pH_{mean}^{C} \leq M^{pH} y_C^{Nue} \tag{2.134}$$

$$Total^{\mathrm{H_3O}^C} \leq M^{Total^{\mathrm{H_3O}}} y_C^{Nue} \tag{2.135}$$

$$Total^{\mathrm{OH}^C} \leq M^{Total^{\mathrm{OH}}} y_C^{Nue} \tag{2.136}$$

$$Waste4^{C} \leq M^{Waste4} y_C^{Nue} \tag{2.137}$$

$$pH_{discharged}^{C} \leq M^{pH} y_C^{Nue} \tag{2.138}$$

The total flowrate after the neutralization process is,

$$W4 = W3 + Base\ \rho_{Base} + Acid\ \rho_{Acid} \tag{2.139}$$

$$Waste4 = \frac{W3}{\rho_{waste}} + Base + Acid \tag{2.140}$$

2.3.2.5 Color

Industrial waste discharges (e.g. forestry products, textile, food, tannery) frequently produce effluents that present a strong level of coloration. The level of color in industrial waste is regulated by legislation. The color may be quantified in terms of the absorbance of the stream. The absorbance of the individual stream i can be determined before the optimization process, then it is possible to calculate the index for the color (or absorbance) through a linear combination of the streams discharged,

$$Color_{mean} = \sum_{i \in NSOURCES} (Ratio_i\ Color_i) \tag{2.141}$$

To get a linear disjunction, the following variables are defined,

$$W4^{Color} = W4\ Color_{mean} \tag{2.142}$$

$$W4^{ColorReg} = W4\ Color_{Reg} \tag{2.143}$$

The color treatment process is applied when the color index in the waste steam is higher than the value stated by the environmental regulation. Then, the following disjunction is formulated,

$$\begin{bmatrix} Y^{Color} \\ W4^{Color} \geq W4^{ColorReg} \\ Cost^{Color} = \left(W4^{Color} - W4^{ColorReg}\right)Costu^{Color}H_Y \end{bmatrix} \vee \begin{bmatrix} Y^{Color} \\ W4^{Color} \leq W4^{ColorReg} \\ Cost^{Color} = 0 \end{bmatrix}$$

Y^{Color} is a logical variable that is true when the index of color for the waste stream is higher than the one given by the environmental regulation. This disjunction can be modeled as follows,

$$W4^{Color} = W4^{Color^1} + W4^{Color^2} \tag{2.144}$$

$$W4^{ColorReg} = W4^{ColorReg^1} + W4^{ColorReg^2} \tag{2.145}$$

$$W4^{Color^1} \geq W4^{ColorReg^1} \tag{2.146}$$

$$Cost^{Color} = \left(W4^{Color^1} - W4^{ColorReg^1}\right)Costu^{Color}H_Y \tag{2.147}$$

$$W4^{Color^2} \leq W4^{ColorReg^2} \tag{2.148}$$

$$W4^{Color^1} \leq M^{W4^{Color}} y^{Color} \tag{2.149}$$

$$W4^{ColorReg^1} \leq M^{W4^{Color}} y^{Color} \tag{2.150}$$

$$W4^{Color^2} \leq M^{W4^{Color}} \left(1 - y^{Color}\right) \tag{2.151}$$

$$W4^{ColorReg^2} \leq M^{W4^{Color}} \left(1 - y^{Color}\right) \tag{2.152}$$

2.3.2.6 Odor

Odor nuisance from different industrial sources has become a serious environmental concern. One reason for the increasing number of odor complaints is that industrial plants have been located nearer to urbanized and residential areas. The odor compounds often do not induce serious human health problems; however, they cause nose, eye, and throat irritation, headaches, nausea, diarrhea, vomiting, cough, sleep disturbances, and loss of appetite (Bruno et al. 2007). In most cases, odor substances are sulfides, nitrogen compounds, organic acids, aldehydes, and ketones, but measurement of odor emissions is difficult because of the many compounds present in the waste streams that cause odor. The European Norm EN 13725 (2002) and the ASTM Standard of Practice E679-91 (1991) measure the total odor impact in terms of odor units (ou/m^3 or ou/ft^3) and do not quantify the specific chemicals responsible for the odor. The instrument used to quantify the odor is called olfactometer. Therefore, the odor of the individual waste stream is known before the optimization process, and it is possible to calculate the total odor of the waste stream using the property integration technique as follows,

$$Odor_{mean} = \sum_{i \in NSOURCES} (Ratio_i \, Odor_i) \tag{2.153}$$

$$W5^{Odor} = \frac{W4 \, Odor_{mean}}{\rho_{waste}} \tag{2.154}$$

$$W5^{OdorReg} = \frac{W4 \, Odor_{Reg}}{\rho_{waste}} \tag{2.155}$$

The odor treatment process is applied to the waste stream only when the odor index is higher than the environmental regulations. Otherwise, when the waste stream satisfies the environmental regulation for the odor index, the associated cost for the odor process is set as zero. Therefore, the following disjunction is formulated,

$$\begin{bmatrix} Y^{Odor} \\ W5^{Odor} \geq W5^{OdorReg} \\ Cost^{Odor} = \left(W5^{Odor} - W5^{OdorReg}\right) Costu^{Odor} H_Y \end{bmatrix} \vee \begin{bmatrix} Y^{Odor} \\ W5^{Odor} \leq W5^{OdorReg} \\ Cost^{Odor} = 0 \end{bmatrix}$$

The disjunction is modeled as follows,

$$W5^{Odor} = W5^{Odor^1} + W5^{Odor^2} \tag{2.156}$$

$$W5^{OdorReg} = W5^{OdorReg^1} + W5^{OdorReg^2} \tag{2.157}$$

$$W5^{Odor^1} \geq W5^{OdorReg^1} \tag{2.158}$$

$$Cost^{Odor} = \left(W5^{Odor^1} - W5^{OdorReg^1}\right) Costu^{Odor} H_Y \tag{2.159}$$

$$W5^{Odor^2} \leq W5^{OdorReg^2} \tag{2.160}$$

$$W5^{Odor^1} \leq M^{W5^{Odor}} y^{Odor} \tag{2.161}$$

$$W5^{OdorReg^1} \leq M^{W5^{Odor}} y^{Odor} \tag{2.162}$$

$$W5^{Odor^2} \leq M^{W5^{Odor}} \left(1 - y^{Odor}\right) \tag{2.163}$$

$$W5^{OdorReg^2} \leq M^{W5^{Odor}} \left(1 - y^{Odor}\right) \tag{2.164}$$

2.3.2.7 Any Property

Lastly, it is possible to state that any other specific property that may cause pollution to the environment can be modeled in the same way. Thus, for any property, the property integration produces,

$$f_{Property}(Property_{mean}) = \sum_{i \in NSOURCES} f_{Property}(Ratio_i\, Property_i) \tag{2.165}$$

$$Wn^{Property} = Wn\, Property_{mean} \tag{2.166}$$

$$Wn^{PropertyReg} = Wn\, Property_{Reg} \tag{2.167}$$

The following disjunction can be formulated,

$$\begin{bmatrix} Y^{Property} \\ Wn^{Property} \geq Wn^{PropertyReg} \\ Cost^{Property} = \left(Wn^{Property} - Wn^{PropertyReg}\right) Costu^{Property} H_Y \end{bmatrix} \vee \begin{bmatrix} Y^{Property} \\ Wn^{Property} \leq Wn^{PropertyReg} \\ Cost^{Property} = 0 \end{bmatrix}$$

which can be algebraically modeled as,

$$Wn^{Property} = Wn^{Property^1} + Wn^{Property^2} \tag{2.168}$$

$$Wn^{PropertyReg} = Wn^{PropertyReg^1} + Wn^{PropertyReg^2} \tag{2.169}$$

$$Wn^{Property^1} \geq Wn^{PropertyReg^1} \tag{2.170}$$

$$Cost^{Property} = \left(Wn^{Property^1} - Wn^{PropertyReg^1}\right) Costu^{Property} H_Y \tag{2.171}$$

$$Wn^{Property^2} \leq Wn^{PropertyReg^2} \tag{2.172}$$

$$Wn^{Property^1} \leq M^{Wn^{Property}} y^{Property} \tag{2.173}$$

$$Wn^{PropertyReg^1} \leq M^{Wn^{Property}} y^{Property} \tag{2.174}$$

$$Wn^{Property^2} \leq M^{Wn^{Property}} \left(1 - y^{Property}\right) \tag{2.175}$$

$$Wn^{PropertyReg^2} \leq M^{Wn^{Property}} \left(1 - y^{Property}\right) \tag{2.176}$$

2.3.3 Objective Function

The objective function consists of the minimization of the total annual cost associated with both the water integration and the wastewater treatment processes. The costs of the water network include the cost associated to the fresh sources and the piping costs to build the mass integration network. Piping costs are important to be accounted for, and they are a function of the layout (which is known) and the flowrate of the streams. To meet the environmental regulations, it may be necessary to treat the wastewater streams. The costs associated with the wastewater treatment process include the cost for the recovery of specific compounds in the waste streams, the cost to remove the toxic compounds, the cost of aeration to satisfy the theoretical oxygen demand regulation, the cost for the neutralization process, the cost to decrease color, odor, and any other property that may cause contamination. The objective function can be formulated as,

$$\begin{aligned} TAC = & \sum_{r \in FRESH} Cost_r^{Fresh} F_r H_Y \\ & + \sum_{\substack{i \in SOURCES \\ j \in SINKS}} pip_{i,j} w_{i,j} + \sum_{\substack{r \in FRESH \\ j \in SINKS}} pip_{r,j} f_{r,j} \\ & + \sum_{u \in RECOV} Cost_u^{Recoveryu} + \sum_{t \in TOX} Cost_t^{Toxicity} \\ & + Cost^{ThoD} + Cost^{Neu} + Cost^{Color} + Cost^{Odor} + \cdots + Cost^{Property} \end{aligned} \tag{2.177}$$

Even though the model presented is a no convex mixed integer non linear programming problem with an adequate initialization for the optimization variables techniques like the outer approximation (1990) may yield a good solution. To yield the global optimal solution solvers like BARON (1996) can be used.

2.4 Case Study

This section presents the application of the proposed approach to a case study. The stream data for the example problem were taken from Hortua (2007). The process consists of the production of phenol from cumene, yielding acetone as a subproduct. This process was selected because phenol is particularly considered a hazardous and toxic material.

For the water integration process, two fresh sources are available with a phenol impurity of 0 and 0.012 mass fraction (kg-phenol/kg-total), and with unit costs of $0.00132277/kg and $0.00088185/kg respectively. In addition, there are three process sources; the stream data for these sources are given in Table 2.1. The data for the sinks are given in Table 2.2.

The environmental regulations applied to this specific process are as follows. The maximum concentration allowed to be discharged to the environment for acetone and phenol is 0.005 kg-phenol/kg-total. The toxicity must be zero, and the maximum concentration and discharge load of phenol, the toxic material, are 0.0000011 kg-phenol/kg-total and 0.00245393 kg/h, respectively (Soboslay 1987). The theoretical oxygen demand of the waste stream must be lower than 75 mg O_2/l, and the *pH* must be between 5.5 and 9.0 (Nemerow 1978).

The unit costs for the recovery processes are 0.14330047 $/kg and $0.07275255 $/kg for phenol and acetone, respectively. The unit cost for aeration is 0.01322774 $/kg air diffused, and the unit costs for the H_2SO_4 and NaOH used for the neutralization

Table 2.1 Sources data for the example

Source, i	W_i [kg/h]	$z_{i,phenol}$	$z_{i,acetone}$	$ThOD$ [gO_2/l]	pH	ρ [kg/l]
1	3666.46	0.016	0.000	0.187	5.4	1.00017
2	1769.18	0.024	0.010	48.850	5.1	1.00017
3	1487.77	0.22	0.028	92.100	4.8	1.00017

Table 2.2 Sinks data for the example

Sink, j	G_j [kg/h]	z_j^{max}
1	2721.55	0.015
2	1129.44	0.1
3	1995.80	0.015

process are \$46/l and \$31/l, respectively. Table 2.3 presents the unit piping cost for the water integration process.

Figure 2.3 presents the results for the case when only process constraints are considered, without including the waste treatment process, which is referred to as Solution A, where the waste stream does not satisfy the environmental constraints; therefore, it will be necessary to include a waste treatment process to the solution obtained here.

Figure 2.4 presents the solution of the example problem including all the environmental constraints for this example (Solution B), where the total waste is reduced in the solution that incorporates the waste treatment process by 11.1% concerning the solution that considered only the process constraints. In addition, in the optimal solution shown in Fig. 2.4, the total consumption of fresh source 1 (with an impurity concentration of phenol of 0) is increased by 76.1% concerning Solution A of Fig. 2.3; this yields a decrease in the concentration of phenol and acetone in the waste stream by 22.2% and 30%, respectively. The theoretical oxygen demand in the waste stream for the optimal solution that considers the waste treatment process is reduced by 27.4% concerning the simplified solution of Fig. 2.3. The *pH* in the optimal solution B is increased by 0.93% concerning the simplified Solution A.

Table 2.4 presents a summary of the costs associated with both solutions. To have a point of reference, the costs of the waste treatment processes that would be needed for Solution A were calculated after the optimization procedure (Notice that the sequential solution is a very common industrial practice). Notice in Table 2.4 that the sum of the costs associated with the process (fresh sources and piping costs) in solution B is 8% higher than those required for solution A. In addition, the total costs associated with the waste treatment process for the optimal solution B is 41.3% lower than those required by the simplified solution A. Therefore, the solution that simultaneously optimizes the process and the waste treatment process yields a total annual cost with savings of 25.4% concerning the solution that only optimizes the process. This result confirms the advantage of simultaneously considering water and property integration and trading off direct recycling, usage of fresh resources, and

Table 2.3 Piping costs for the example[a]

Sink, j	Sources				
	Process, i			Fresh, r	
	1	2	3	1	2
1	11.0231	4.4092	6.6138	9.9208	5.5115
2	7.7161	2.2046	11.0231	6.6138	2.2046
3	4.4092	8.8184	4.4092	7.7161	3.3069

[a]Units in [(\$ h)/(kg year)]

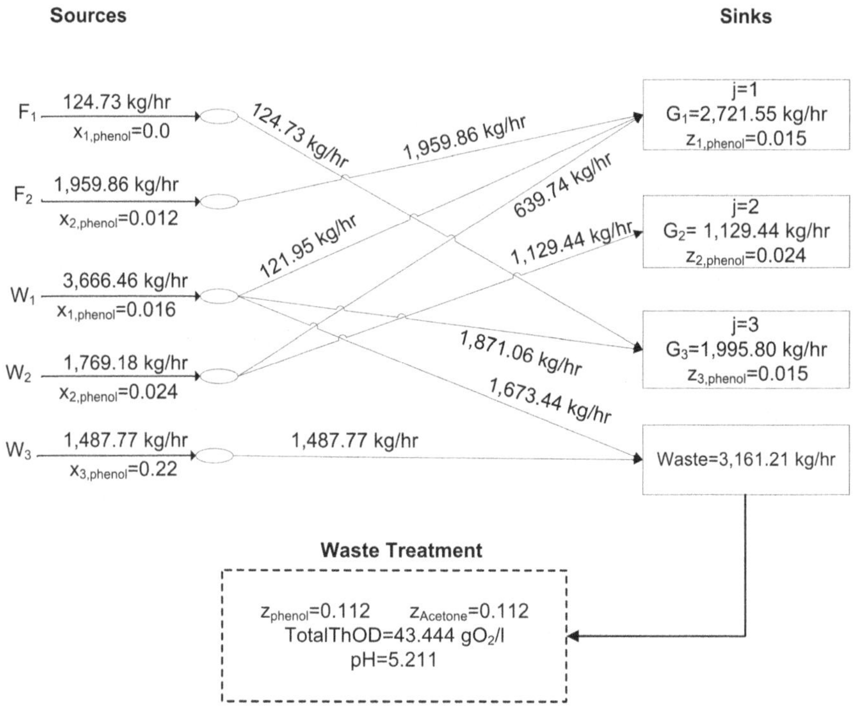

Fig. 2.3 Water integration without environmental constraints (solution A). Printed with permission of "*A property-based optimization of direct recycle networks and wastewater treatment process*", *AIChE Journal*, Copyright (2009 American Institute of Chemical Engineers), John Wiley & Son

treatment of waste streams while satisfying process constraints and environmental regulations.

This example has shown that if one does not consider the waste treatment cost to satisfy the environmental regulations as part of the mass integration problem, the model can lead to sub-optimum solutions. It is shown how a simultaneous model can be formulated, which provides a better basis for the optimization of these types of systems.

2.5 Concluding Remarks

An optimization approach has been presented to (1) integrate the design of in-plant recycle/reuse networks with the end-of-pipe waste treatment facilities, (2) consider composition- and property-based constraints and process models. An optimization formulation has been developed to implement the devised framework. The model is based on disjunctive programming and considers the technologies that are commonly used to treat wastewater streams so that they meet environmental regulations. The

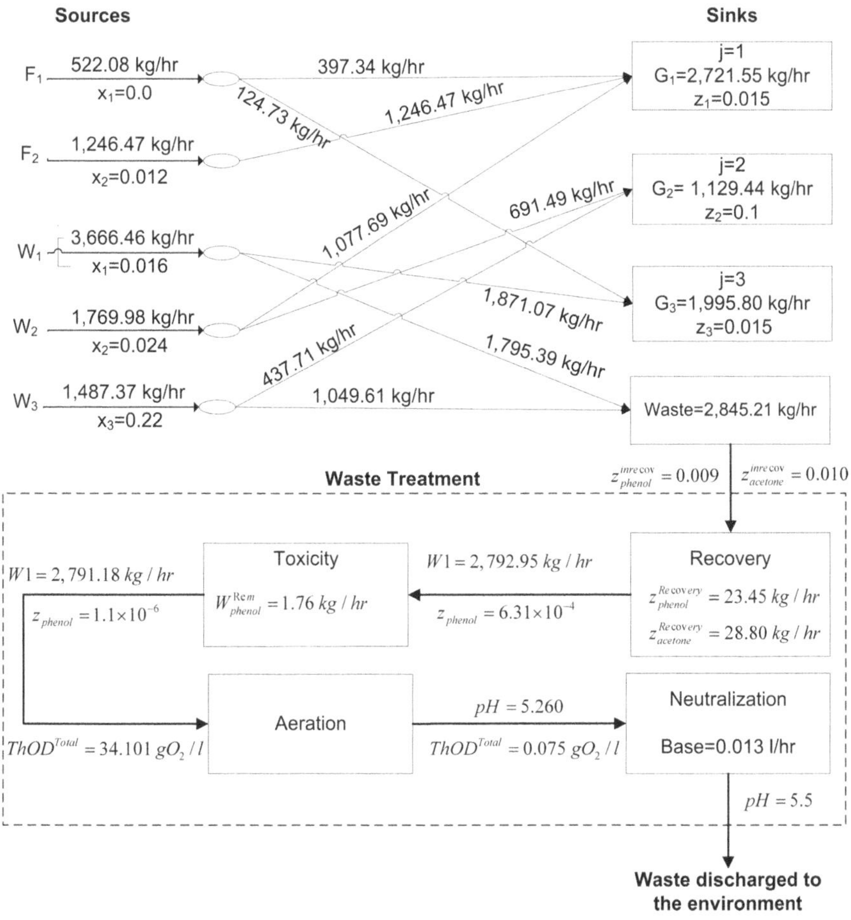

Fig. 2.4 Water integration including waste treatment process (solution B). Printed with permission of "*A property-based optimization of direct recycle networks and wastewater treatment process*", *AIChE Journal*, Copyright (2009 American Institute of Chemical Engineers), John Wiley & Son

model is formulated with linear relationships within each one of the disjunctions, which aids its numerical solution. In addition to standard mass balance and composition constraints, the model incorporates property-based models, property mixing rules, and environmental constraints based on properties such as toxicity, theoretical oxygen demand, *pH*, color, and odor, and it is generalized to any other property that may cause pollution. The methodology has also shown how property integration can be used for cases in which a given property is difficult to estimate when there are many compounds in a waste stream. The formulation results in an MINLP model, which has been used to minimize the total annual cost of the system. The solution of the proposed model shows that the simultaneous consideration of mass and property

Table 2.4 Comparison of results

Concept	With process constraints (Solution A)	With process and environmental constraints (Solution B)
Waste [kg/h]	3161.21	2845.21
Fresh sources cost [$/year]	15,146	14,318
Piping cost [$/year]	26,669	31,128
Recovery cost [$/year]	*61,859*	43,661
Toxicity cost [$/year]	*7226*	5098
Aeration cost [$/year]	*161,083*	43,363
Neutralization cost [$/year]	*4567*	3237
Total process costs [$/year]	41,816	45,446
Total waste treatment costs [$/year]	*134,734*	95,360
TAC [$/year]	*176,550*	140,807

integration and the tradeoff between in-plant recycle/reuse with wastewater treatment sections can yield important economic savings. The model did not show any numerical problems, and the CPU times required for the solutions of the example problem were relatively small.

Nomenclature

$Acid$	Acid added to a waste stream to meet environmental regulation for pH
$Base$	Base added to a waste stream to meet environmental regulation for pH
c_t	Concentration of toxic compound t
$Color_i$	Index for the color of source i
$Color_{Reg}$	Maximum index for color allowed by environmental regulation
$Color_{mean}$	Index for the color of waste stream
$Cost^{Aeration}$	Unit cost for the aeration process
$Cost^{Acid}$	Unit cost for the acid
$Cost^{Base}$	Unit cost for the base
$Cost^{Color}$	Total cost for the color treatment
$Cost_r^{Fresh}$	Unitary cost for the fresh source r
$Cost^{Neu}$	Total cost for the neutralization process

$CCost^{Neu^B}$, $Cost^{Neu^C}$	Disaggregated variables for $Cost^{Neu}$
$Cost^{Odor}$	Total cost for the odor treatment process
$Cost^{Property}$	Total cost for the property treatment process
$Cost_u^{Recoveryu}$	Total cost for recovery component u
$Cost_u^{Recu}$	Unit cost for recovery component u
$Cost^{ThOD}$	Total cost for the theoretical oxygen demand treatment process
$Cost_t^{Toxicity}$	Total cost for the toxicity treatment process for toxic t
$Cost_t^{Toxic}$	Unit cost for the toxicity treatment process for toxic t
$Costu^{Color}$	Unit cost for the color treatment process
$Costu^{Odor}$	Unit cost for the odor treatment process
$Costu^{Property}$	Unit cost for the property treatment process
$D_{Tox_t}^{Max}$	Maximum mass flowrate allowed to be discharged for toxic compound t
EFF_u	Efficiency for the recovery process for component u
F_r	Total flowrate consumed from fresh source r
$f_{r,j}$	Segregated flowrate from fresh source r to sink j
G_j	Total flowrate inlet to sink j
H_Y	Operating hours per year
k_{1_t}	Constant for the toxicity relationship of t
k_{2_t}	Constant for the toxicity relationship of t
M^{var}	Upper limit for variable var
$NFRESH$	Set for the fresh sources
$NPROP$	Set for the number of properties considered
$NSINKS$	Set for the number of sinks
$NSOURCES$	Set for the number of sources
$Odor_i$	Index for the odor for process source i
$Odor_{mean}$	Index for the odor of waste stream
$Odor_{Reg}$	Maximum index for the odor allowed by environmental regulation
pH	Potential hydrogen
$pH_{discharged}$	pH Discharged to the environment for the waste stream
$pH_{discharged}^B$, $pH_{discharged}^A$, $pH_{discharged}^C$	Disaggregated variables for $pH_{discharged}$
pH_{mean}	pH Mean for the waste stream
pH_{mean}^B, pH_{mean}^A, pH_{mean}^C	Disaggregated variables for pH_{mean}
pip	Unit piping costs from any source to any sink
pOH	Potential hydroxyl
pOH_{mean}	Mean value of pOH for the waste stream

$Property_i$	Property for process source i
$Property_{mean}$	Mean property for a waste stream
$Ratio_i$	Fractional contribution for each process source i to the waste stream
TAC	Total yearly cost
$ThOD^{Total}$	Total theoretical oxygen demand for a waste stream
$ThOD_i$	Theoretical oxygen demand for stream i
$Total^{H_3O}$	Total moles of H_3O^+ in a waste stream
$Total^{H_3O^B}$, $Total^{H_3O^A}$, $Total^{H_3O^C}$	Disaggregated variables for $Total^{H_3O}$
$Total^{OH}$	Total moles of OH^- in a waste stream
$Total^{OH^B}$, $Total^{OH^A}$, $Total^{OH^C}$	Disaggregated variables for $Total^{OH}$
TOX	Set for toxic compounds
W_i	Total mass flowrate from process source i
$w_{i,j}$	Segregated mass flowrate from process source i to sink j
$waste_i$	Segregated mass flowrate from process source i to the waste stream
$WASTE$	Total flowrate for a waste stream
$Waste4$	Volumetric flowrate for the stream after neutralization process
$Waste4^B$, $Waste4^A$, $Waste4^C$	Disaggregated variables for $Waste4$
$WINREC$	Inlet mass flowrate to the recovery process
$WINREC_u$	Inlet mass flowrate for component u to the recovery process
$WINRECU_u^1$, $WINRECU_u^2$	Disaggregated variables for $WINREC_u$
W_t^{BT}	Mass flowrate for the toxic compound t before the toxicity treatment
W_t^{BT1}, W_t^{BT2}, W_t^{BT3}, W_t^{BT4}	Disaggregated variables for W_t^{BT}
$W_u^{Recovery}$	Mass flowrate recovered for component u
W_t^{Rem}	Mass flowrate removed of toxic t
W_t^{Rem1}, W_t^{Rem2}, W_t^{Rem3}, W_t^{Rem4}	Disaggregated variables for W_t^{Rem}
W_t^{Rem3a}, W_t^{Rem3b}	Disaggregated variables for W_t^{Rem3}
W_t^{Rema3a}, W_t^{Rema3b}	Disaggregated variables for W_t^{Rema3}
W_t^{Remb3a}, W_t^{Remb3b}	Disaggregated variables for W_t^{Remb3}
W^{ThOD}	Product of theoretical oxygen demand times flowrate
W^{ThOD^1}, W^{ThOD^2}	Disaggregated variables for W^{ThOD}
$W^{ThOD^{Reg}}$	Product of theoretical oxygen demand regulation times flowrate
$W^{ThOD^{Reg^1}}$, $W^{ThOD^{Reg^2}}$	Disaggregated variables for $W^{ThOD^{Reg}}$
$W1$	Mass flowrate for the waste stream after the recovery process
$W1^1$, $W1^2$, $W1^3$, $W1^4$	Disaggregated variables for $W1$

$W2$	Mass flowrate for the waste stream after toxicity treatment process
$W3$	Mass flowrate inlet to the aeration process
$W4$	Mass flowrate for the waste stream after the neutralization process
$W4^{Color}$	Mass flowrate by the color index for the waste stream
$W4^{Color^1}$, $W4^{Color^2}$	Disaggregated variables for $W4^{Color}$
$W4^{ColorReg}$	Mass flow rate times the color index given by the environmental regulation
$W4^{ColorReg^1}$, $W4^{ColorReg^2}$	Disaggregated variables for $W4^{ColorReg}$
$W5^{Odor}$	Flowrate times odor index for the waste stream
$W5^{Odor^1}$, $W5^{Odor^2}$	Disaggregated variables for $W5^{Odor}$
$W5^{OdorReg}$	Flowrate times odor index given by the environmental regulation
$W5^{OdorReg^1}$, $W5^{OdorReg^2}$	Disaggregated variable for $W5^{OdorReg}$
$WH3O$	Concentration of ions H_3O^+ in the waste stream
$Wn^{Property}$	Flowrate times property for the waste stream
$Wn^{Property^1}$, $Wn^{Property^2}$	Disaggregated variables for $Wn^{Property}$
$Wn^{PropertyReg}$	Flowrate times property given by the environmental regulation
$Wn^{PropertyReg^1}$, $Wn^{PropertyReg^2}$	Disaggregated variables for $Wn^{PropertyReg}$
WOH	Concentration of OH^- ions in a waste stream
Y_u^{Recov}	Logic variable for the recovery process of component u
Y_t	Percentage of mortality for toxic compound t
Y^{Color}	Logic variable for the color treatment process
Y^{Odor}	Logic variable for the odor treatment process
Y_A^{Neu}	Logic variable for the neutralization process with acid addition
Y_B^{Neu}	Logic variable for the neutralization process with base addition
Y_C^{Neu}	Logic variable for the neutralization process with no acid or base addition
$Y^{Property}$	Logic variable for property
Y^{ThOD}	Logic variable for the theoretical oxygen demand
$Y_{A_t}^{Tox}$	Logic variable for toxic treatment process for toxic t
$Y_{B_t}^{Tox}$	Logic variable to denote that toxic treatment process is not needed for toxic t

$Y_{1_t}^{Tox}$	Logic variable for case 1 for toxic treatment process for t
$Y_{2_t}^{Tox}$	Logic variable for case 2 for toxic treatment process for t
$Y_{3_t}^{Tox}$	Logic variable for case 3 for toxic treatment process for t
$Y_{3a_t}^{Tox}$	Logic variable for case 3a for toxic treatment process for t
$Y_{3b_t}^{Tox}$	Logic variable for case 3b for toxic treatment process for t
$z_{i,c}$	Given composition for process source i for component c
$z_{j,c}^{in}$	Inlet composition for component c inlet to sink j
$z_{r,c}$	Given composition for fresh source r for component c
$z_{j,c}^{max}$	Maximum concentration allowed in sink j for component c
$z_{j,c}^{min}$	Minimum concentration allowed in sink j for component c
$z_{i,u}^{inrecov}$	Inlet composition to the recovery process for component u from source i
$z_{u}^{inrecov1}$, $z_{i,u}^{inrecov2}$	Disaggregated variables for $z_{i,u}^{inrecov}$
z_{u}^{MaxEnv}	Maximum discharge allowed by environmental regulation for component u
$z_{Tox_t}^{Max}$	Maximum composition allowed to be discharged for toxic compound t
$z1_u$	Component composition after the recovery process
$z1_t^1, z1_t^2, z1_t^3, z1_t^4$	Disaggregated variables for $z1_t$
$z2_t$	Composition of waste stream after toxicity treatment process for toxic compound t

Greek Symbols

ρ_{Acid}	Density of the acid used in the neutralization process
ρ_{Base}	Density of the base used in the neutralization process
ρ_{waste}	Density of the waste stream
$\psi_p\left(p_{j,p}^{in}\right)$	Property operator for property p at the inlet condition to sink j

$\psi_p(p_{i,p})$	Property operator for property p for process source i
$\psi_p(p_{r,p})$	Property operator for property p for fresh source r
$\psi_p^{max}(p_{p,j})$	Maximum relation for property p in sink j
$\psi_p^{min}(p_{p,j})$	Minimum relation for property p in sink j

Indices

i	Process sources
in	Inlet conditions
j	Sinks
max	Maximum value
min	Minimum value
p	Properties
r	Fresh sources
t	Toxic compound
u	Hazardous component to be recovery
waste	Waste stream

References

Alva-Argaez A, Kokossis AC, Smith R (1998) Wastewater minimization of industrial systems using an integrated approach. Comput Chem Eng 22(Suppl):S741–S744

Alva-Argaez A, Vallianatos A, Kokossis A (1999) A multi-contaminant transshipment model for mass exchange networks and wastewater minimization problems. Comput Chem Eng 23(10):1439–1453

ASTM E679-91 (1991) Standard practice for determination of odor and thresholds by a forced-choice ascending concentration series method of limits. American Society for Testing and Materials, Philadelphia, PA

Bagajewicz M (2000) A review of recent design procedure for water networks in refineries and process plants. Comput Chem Eng 24(9–10):2093–2113

Benko N, Rev E, Fonyo Z (2000) The use of nonlinear programming to optimal water allocation. Chem Eng Commun 178:67–101

Bruno P, Caselli M, de Gennaro G, Solito M, Tutino M (2007) Monitoring of odor compounds produced by solid waste treatment plants with diffusive samplers. Waste Manage 27(4):539–544

Dhole VR, Ramchandani N, Tainsh RA, Wasilewski M (1996) Make your process water pay for itself. Chem Eng 103(1):100–103

Doyle SJ, Smith R (1997) Targeting water reuse with multiple contaminants. Process Saf Environ Prot 75(B3):181–189

Eden MR, Jorgensen SB, Gani R, El-Halwagi MM (2004) A novel framework for simultaneous separation process and product design. Chem Eng Process 43(5):595–608

El-Halwagi MM (1997) Pollution prevention through process integration: systematic design tools. Academic Press, San Diego

El-Halwagi MM (2006) Process integration. Academic Press, New York

El-Halwagi M.M, Spriggs HD (1996) An integrated approach to cost and energy efficient pollution prevention. In: Proceedings of the fifth world congress of chemical engineering, San Diego, vol 1, pp 675–680

El-Halwagi MM, Gabriel F, Harell D (2003) Rigorous graphical targeting for resource conservation via material recycle/reuse networks. Ind Eng Chem Res 42(19):4319–4328

El-Halwagi MM, Glasgow IM, Qin XY, Eden MR (2004) Property integration: componentless design techniques and visualization tools. Am Inst Chem Eng J 50(8):1854–1869

Eljack FT, Abdelhady AF, Eden MR, Gabriel FB, Qin XY, El-Halwagi MM (2005) Targeting optimum resource allocation using reverse problem formulations and property clustering techniques. Comput Chem Eng 29(11–12):2304–2317

Eljack FT, Eden MR, Kazantzi V, Qin XY, El-Halwagi MM (2007) Simultaneous process and molecular design a property based approach. Am Inst Chem Eng J 53(5):1232–1239

EN 13725 (2002) Air quality—determination of odour concentration by dynamic olfactometry. European Committee for Standardization, Brussels, Belgium

Gabriel FB, El-Halwagi MM (2005) Simultaneous synthesis of waste interception and material reuse networks: problem reformulation for global optimization. Environ Prog 24(2):171–180

Galan B, Grossmann IE (1998) Optimal design of distributed wastewater treatment networks. Ind Eng Chem Res 37(10):4036–4048

Grooms D, Kazantzi V, El-Halwagi M (2005) Optimal synthesis and scheduling of hybrid dynamic/steady property integration networks. Comput Chem Eng 29(11–12):2318–2325

Gunaratnam M, Alva-Argaez A, Kokossis A, Kim JK, Smith R (2005) Automated design of total water systems. Ind Eng Chem Res 44(3):588–599

Hallale N (2002) A new graphical targeting method for water minimization. Adv Environ Res 6(3):377–390

Hernandez-Suarez R, Castellanos-Fernandez J, Zamora JM (2004) Superstructure decomposition and parametric optimization approach for the synthesis of distributed wastewater treatment networks. Ind Eng Chem Res 43(9):2175–2191

Hortua AC (2007) Chemical process optimization and pollution prevention via mass and property integration. M.S. thesis, Texas A&M University

Huang CH, Chang CT, Ling HC, Chang CC (1999) A mathematical programming model for water usage and treatment network design. Ind Eng Chem Res 38(7):2666–2679

Karuppiah R, Grossmann IE (2006) Global optimization for the synthesis of integrated water systems in chemical processes. Comput Chem Eng 30(4):650–673

Kazantzi V, El-Halwagi MM (2005) Targeting material reuse via property integration. Chem Eng Prog 101(8):28–37

Kazantzi V, Qin XY, El-Halwagi M, Eljack F, Eden M (2007) Simultaneous process and molecular design through property clustering techniques: a visualization tool. Ind Eng Chem Res 46(10):3400–3409

Kiperstok A, Sharratt PN (1997) Optimization of pollution control operations in industrial sites considering decay capabilities of the receptors. Comput Chem Eng 21(Suppl):S977–S982

Kuo WCJ, Smith R (1998) Designing for the interactions between water-use and effluent treatment. Chem Eng Res Des 76(A3):287–301

Nemerow NL (1978) Industrial water pollution. Addison-Wesley Publishing, Menlo Park

Polley GT, Polley HL (2000) Design better water networks. Chem Eng Prog 96(2):47–52

Ponce-Ortega JM, Hortua AC, El-Halwagi MM, Jiménez-Gutiérrez A (2009) A property-based optimization of direct recycle networks and wastewater treatment processes. AIChE J 55(9):2329–2344

Qin X, Gabriel F, Harell D, El-Halwagi MM (2004) Algebraic techniques for property integration via componentless design. Ind Eng Chem Res 43(14):3792–3798

Raman R, Grossmann IE (1994) Modeling and computational techniques for logic based integer programming. Comput Chem Eng 18(7):563–578

Savelsky MJ, Bagajewicz MJ (2000) On the optimality conditions of water utilization systems in process plants with single contaminants. Chem Eng Sci 55(21):5035–5048
Savelsky MJ, Bagajewicz MJ (2001) Algorithmic procedure to design water utilization systems featuring a single contaminant in process plants. Chem Eng Sci 56(5):1897–1911
Shelley MD, El-Halwagi MM (2000) Componentless design of recovery and allocation systems: a functionality-based clustering approach. Comput Chem Eng 24(9–10):2081–2091
Soboslay EG (1987) Toxicity of tiocyanate, phenol and their mixtures to freshwater aquatic organisms: a modeling approach. Ph. Thesis, University of Pittsburgh, Pittsburgh, PA
Sorin M, Bedard S (1999) The global pinch point in water reuse networks. Process Saf Environ Prot 77(B5):305–308
Takama N, Kuriyama T, Shiroko K, Umeda T (1980) Optimal water allocation in a petroleum refinery. Comput Chem Eng 4(4):251–258
Wang YP, Smith R (1994) Wastewater minimization. Chem Eng Sci 49(7):981–1006
Wang YP, Smith R (1995) Wastewater minimization with flowrate constraints. Chem Eng Res Des 73(A8):889–904

Chapter 3
Comprehensive Optimization for Multi-contaminant Wastewater Process Integration Networks

3.1 Introduction

The industrial consumption of natural resources has affected their availability, which has translated into significant increases in the prices of raw materials. The accumulated environmental damage has promoted stricter environmental regulations for the waste streams discharged by industries. This has promoted the development of methodologies for the rational use of resources and the minimization of wastewater discharged to the environment (El-Halwagi 2006). One approach is the implementation of mass integration strategies. Mass integration is a holistic approach used to optimize the allocation, transformation, and separation of species and streams. As part of the mass integration strategies, the recycle/reuse networks have received special attention. Such networks consider the distribution of waste streams (sources) to process equipment (sinks) and/or treatment units before disposal. Figure 3.1 shows a schematic representation of this type of network.

Several methodologies have been reported for resource conservation using recycle/reuse networks (see El-Halwagi 2006; Dunn and El-Halwagi 2003). Foo (2009) presents an exhaustive revision of methodologies that use the pinch-point approach for the synthesis of water networks. Jeżowski (2010) presented a review of the main developments in the last years in the field of water network design using heuristic and optimization approaches. Takama et al. (1980) reported a superstructure and a mathematical programming model for the optimal water allocation problem. Quesada and Grossmann (1995) presented a global optimization model for designing water networks consisting of splitters, mixers, and linear process units that involve multicomponent streams when bilinear terms appear; a linear reformulation was used to get a valid lower bound for the global optimum solution, followed using a spatial branch and bound search. Galan and Grossmann (1998) reported a mathematical programming model for the synthesis of distributed wastewater networks; a search procedure was implemented based on successive solutions of a relaxed linear model and the original nonconvex nonlinear problem. Lee and Grossmann

J. M. Ponce-Ortega et al., *Sustainable Water Systems*, Water Resources Development and Management, https://doi.org/10.1007/978-3-031-72498-5_3

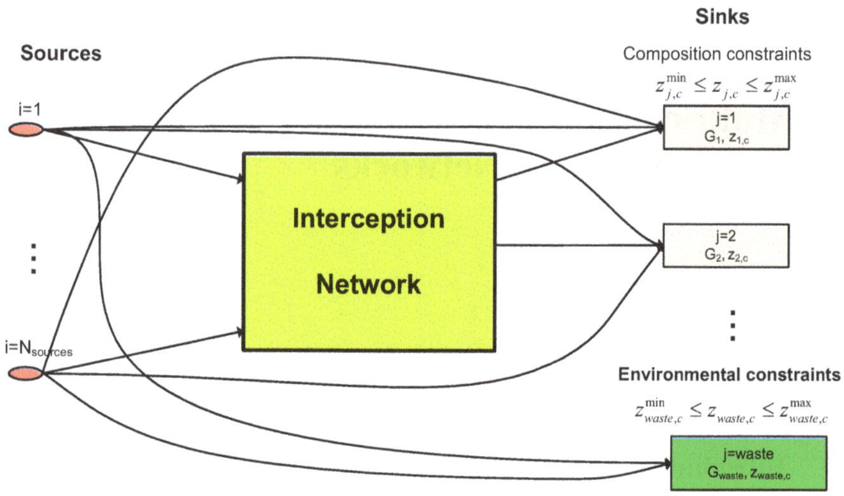

Fig. 3.1 Schematic representation of recycle/reuse networks. Printed with permission of "*An optimization approach for the synthesis of recycle and reuse water integration networks*", *Clean Technologies and Environmental Policy*, Copyright (2012), Springer

(2003) implemented a global optimization algorithm based on generalized disjunctive programming for the synthesis of wastewater networks. Karuppiah and Grossmann (2006) addressed the problem of optimal synthesis of an integrated water system, where water-using processes and water treatment operations are combined into a single network; they implemented a global optimization technique using piecewise under- and over-estimators to approximate the nonconvex terms in the original model to obtain a convex relaxation, whose solution gives a lower bound on the global optimum. Savelski and Bagajewicz (2000, 2001, 2003) presented linear models for designing water-using systems to avoid nonconvex formulations. Teles et al. (2008) reported a method for the optimal design of water-using networks considering an initialization procedure based on a set of solutions of LP problems that are then used as initial values for a general NLP problem. Chakroborty (2009) implemented topological constraints in the synthesis of direct recycle networks based on an MILP formulation. Hul et al. (2007), and Tan et al. (2008) used metaheuristic techniques for the synthesis of water networks to yield near-global optimal solutions. Tan et al. (2009) presented a superstructure for the optimization of water networks with partition interceptors (i.e., interceptors with multiple outlet streams), which was formulated as a nonlinear model limited to single contaminant problems. Gabriel and El-Halwagi (2005) presented a simplified formation that eliminates the mixing of different streams to avoid bilinear terms and to yield a convex problem formulation in synthesizing water networks. Although this approach avoids the bilinear terms of the mass balances, it oversimplifies the problem and does not consider configurations of potential interest. In addition, this formulation may produce a network with a high number of treatment units and complicated piping arrangements. Ponce-Ortega et al. (2010) improved the model by Gabriel and El-Halwagi (2005) to consider constraints based on different properties instead of dealing only with compositions. Ponce-Ortega et al. (2009) and Nápoles-Rivera et al. (2010) addressed the problem of global optimization of recycle and reuse water networks based on properties using NLP and MINLP formulations. Other papers have been reported to handle the nonconvex terms within the models, specifically the bilinear terms that appear as products of flowrates times compositions, through the use of deterministic optimization techniques (see Hasan and Karimi 2010; Misener and Floudas 2010; Almutairi and Elhedhli 2009; Gounaris et al. 2009; Pham et al. 2009; Wicaksono and Karimi 2008; Liberti and Pantelides 2006; Meyer and Floudas 2006; Faria and Bagajewicz 2010).

This chapter presents a mathematical programming model for synthesizing recycle and reuse water networks (Ponce-Ortega et al. 2012). A general superstructure without simplifications is used as a basis, and the model is formulated as a convex mixed integer linear programming model (MILP). Process unit constraints related to their capacities and maximum/minimum allowed compositions as well as environmental constraints (maximum/minimum allowed compositions) are included. The model is based on using component flowrates as optimization variables instead of total flowrates and compositions; this formulation avoids the bilinear nonconvex terms that appear in mixers of other formulations (i.e., products of flowrates and compositions, both continuous variables). A proper formulation for the nonlinear cost of treatment units is used to yield a convex problem.

3.2 Problem Statement

The problem addressed in this chapter consists of the optimal synthesis of a recycle and reuse network for resource conservation given a set of process sources *NSOURCES*, which can be treated in a set of treatment units *NUNITS* to satisfy the constraints related to the concentrations of a set of components *NCOMP* for a set of process sinks *NSINKS*. A set of fresh sources *NFRESH* is available to satisfy the composition constraints in the process sinks. The model also must satisfy the environmental constraints for the waste streams discharged to the environment.

The model formulation is based on the general superstructure shown in Fig. 3.2. The implemented superstructure considers the segregation of process sources to be sent directly to process sinks, to available treatment units to remove specific components, and/or to waste. Once the streams are processed in the units, they are segregated and directed to process sinks, to the waste, or recycled to other units for further treatment. Fresh water sources are segregated and sent only to the process sinks.

For the presented model, i, j and c represent indexes for the process sources, sinks, and components, respectively, u is an index used to denote the treatment units, and r is for the fresh sources. The main variables used in the model are presented

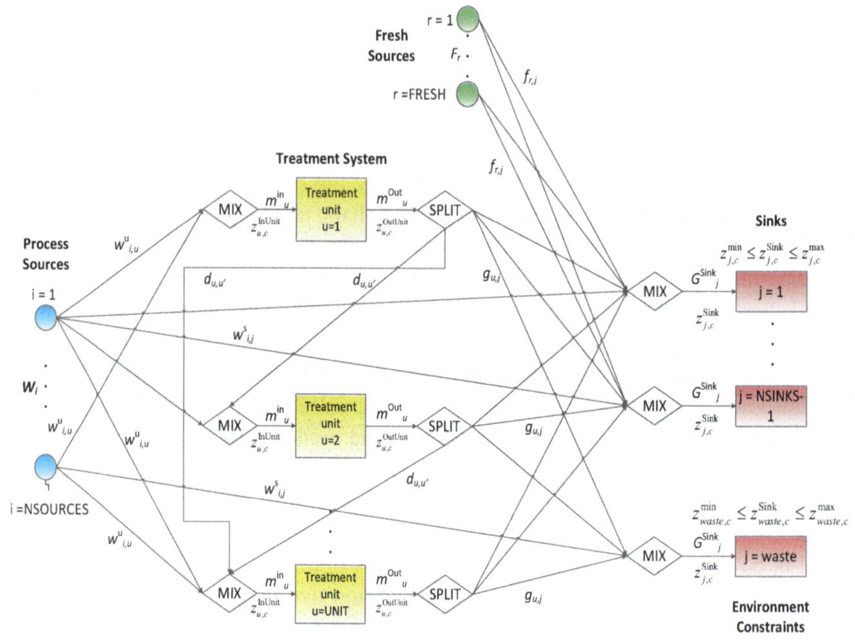

W_i – Process source i (kg/hr)

$w^{u}_{i,u}$ – Flowrate from source i to treatment unit u (kg/hr)

$w^{s}_{i,u}$ – Flowrate from source i to sink j (kg/hr)

m^{In}_{u} – Flowrate at the entrance of treatment unit u (kg/hr)

m^{Out}_{u} – Flowrate at the exit of treatment unit u (kg/hr)

$d_{u,u'}$ – Flowrate from unit u to unit u' (kg/hr)

$g_{u,j}$ – Flowrate from unit u to sink j (kg/hr)

$f_{r,j}$ – Flowrate from fresh source r to sink j (kg/hr)

G^{Sink}_{j} – Flowrate in sink j (kg/hr)

$z^{InUnit}_{u,c}$ – Composition at the entrance of unit u for component c

$z^{OutUnit}_{u,c}$ – Composition at the exit of unit u for component c

$z^{Sink}_{u,c}$ – Composition at the entrance of sink j for component c

$z^{min}_{u,c}$ – Minimum allowed composition in sink j for component c

$z^{max}_{u,c}$ – Maximum allowed composition in sink j for component c

Fig. 3.2 Superstructure implemented for the recycle and reuse network. Printed with permission of "*An optimization approach for the synthesis of recycle and reuse water integration networks*", *Clean Technologies and Environmental Policy*, Copyright (2012), Springer

in the superstructure of Fig. 3.2, and the nomenclature section presents a detailed description of all symbols.

Treatment units are characterized before the optimization process. A set of treatment options are available as part of the design problem. Given the knowledge of the design details of such options, one can then simulate them to obtain the output characteristics and set these values as part of the network synthesis model. This pre-synthesis stage provides the data for the performance of each treatment unit which is modeled as follows:

$$z_{u,c}^{OutUnit} = f(z_{u,c}^{InUnit}, \text{Design and Operating Parameters}), \quad u \in NUNITS, \quad c \in NCOMP$$

This relationship provides the composition of all components at the exit of each treatment unit, with the convenience of formulating the system as a linear model. The optimization model should then determine the selection of treatment units from the set of available units.

3.3 Model Formulation

The main idea of the model of this chapter is to consider component balances for all components (including the solvent) instead of total balances (the sum of all component balances yields the total balance for splitters, mixers, and treatment units). All the components not constrained in the process sinks and by environmental regulations are added to the composition of the solvent; as a result, balances for all components subject to constraints plus a pseudo-solvent are included. Also, the model considers as optimization variables the flowrates for the specific species (including the pseudo-solvent) instead of compositions. Therefore, the bilinear terms in the mixers due to the products of flowrates and compositions are avoided.

The model formulation consists of the following constraints.

3.3.1 Component Balances for Splitting Fresh Sources

Fresh sources can be segregated and sent to the process sinks but not to the waste stream.

$$F_r z_{r,c}^{\text{Fresh}} = \sum_{\substack{j \in NSIKS \\ j \neq waste}} f_{r,j,c}, \quad r \in NFRESH, \quad c \in NCOMP \tag{3.1}$$

F_r and $z_{r,c}^{\text{Fresh}}$ are parameters known before the optimization process, and $f_{r,j,c}$ is an optimization variable.

3.3.2 Component Balances for Splitting Process Sources

Process sources can be segregated and directed to treatment units or sent directly to sinks.

$$W_i z_{i,c}^{\text{InSource}} = \sum_{u \in NUNITS} w_{i,u,c}^u + \sum_{j \in NSINKS} w_{i,j,c}^s, \quad i \in NSOURCES, \quad c \in NCOMP \tag{3.2}$$

W_i and $z_{i,c}^{\text{InSource}}$ are known parameters, while the individual flowrates $w_{i,u,c}^u$ and $w_{i,j,c}^s$ are optimization variables.

3.3.3 Component Balances in the Mixers Before the Treatment Units

The inlet mass flowrate for each component to each treatment unit is equal to the flowrate from the process sources plus the flowrate from other treatment units for each component.

$$m_{u,c}^{\text{In}} = \sum_{i \in NSOURCES} w_{i,u,c}^u + \sum_{\substack{u' \in NUNITS \\ u' \neq u}} d_{u',u,c}, \quad u \in NUNITS, \quad c \in NCOMP \tag{3.3}$$

Recycling of streams to the same treatment unit is not allowed.

3.3.4 Component Balances in the Splitters at the Exit of Treatment Units

The flowrate at the exit of each treatment unit for each component can be segregated and sent to the sinks and/or recycled to another treatment unit.

$$m_{u,c}^{\text{Out}} = \sum_{j \in NSINKS} g_{u,j,c} + \sum_{\substack{u' \in NUNITS \\ u' \neq u}} d_{u,u',c}, \quad u \in NUNITS, \quad c \in NCOMP \tag{3.4}$$

3.3.5 Component Balances for the Mixers Before Any Sink

The flowrate inlet to any process sink for each component is equal to the sum of the flowrates from the fresh sources, the flowrate from the treatment units, and the bypassing flowrate from the process sources.

$$G_{j,c}^{\text{Sink}} = \sum_{r \in NFRESH} f_{r,j,c} + \sum_{u \in NUNITS} g_{u,j,c} + \sum_{i \in NSOURCES} w_{i,u,c}^s, \quad j \in NSINK, \quad c \in NCOMP \tag{3.5}$$

The following constraint is included since the fresh sources cannot be sent to the waste stream:

$$f_{r,j,c} = 0, \quad r \in NFRESH, \quad j = waste, \quad c \in NCOMP \tag{3.6}$$

3.3.6 Constraints for the Flowrate Inlet to Sinks

The flowrate inlet to any process sink must be between given lower and upper limits ($G_j^{\min}$ and $G_j^{\max}$). In this optimization model, the component flowrates are the optimization variable. The total flowrate inlet to the sinks is equal to the sum of all component flowrates (including the pseudo-solvent), the flowrate sink constraints are stated as follows.

$$G_{\mathrm{j}}^{\min} \leq \sum_{c \in NCOMP} G_{j,c}^{\mathrm{Sink}} \leq G_{\mathrm{j}}^{\max}, \quad j \in NSINKS, \quad j \neq waste \tag{3.7}$$

Limits on the flowrate discharged to the environment (through the waste sink) are usually not enforced and therefore are not considered here.

3.3.7 Constraints for the Composition in Sinks

The streams inlet to sinks must satisfy the requirements for the composition of the stream components; these limits are given in terms of lower and upper limits for the composition of specific components ($\mathrm{z}_{j,c}^{\min}$ and $\mathrm{z}_{j,c}^{\max}$). Since in the formulation the flowrate of each component, and not the concentration, is considered, the composition sink constraints are stated as follows.

$$z_{j,c}^{\min} \sum_{c \in NCOMP} G_{j,c}^{\mathrm{Sink}} \leq G_{j,c}^{\mathrm{Sink}} \leq z_{j,c}^{\max} \sum_{c \in NCOMP} G_{j,c}^{\mathrm{Sink}}, \quad j \in NSINKS, \quad c \in NCOMP \tag{3.8}$$

$z_{j,c}^{\min}$ and $z_{j,c}^{\max}$ are known before the optimization process; thus, this relationship is linear. These constraints are given for the process sinks and for the waste stream discharged to the environment (which is restricted by environmental regulations).

3.3.8 Balances in Treatment Units

The following component balances are required to determine the pollutants removed for each treatment unit:

$$m_{u,c}^{\text{remov}} = m_{u,c}^{\text{In}} - m_{u,c}^{\text{Out}}, \quad u \in NUNITS, \quad c \in NCOMP \tag{3.9}$$

and the total flowrate removed in each treatment unit is given as follows,

$$m_{u}^{\text{unit}} = \sum_{c \in NCOMP} m_{u,c}^{\text{remov}}, \quad u \in NUNITS \tag{3.10}$$

The above equations correspond to single-pass type interceptors (Tan et al. 2009; Ng et al. 2009); to include partitioning interceptors, the equations must be replaced by the ones reported in Appendix.

3.3.9 Composition Constraints for Splitters

Since component balances are considered in the optimization model, the composition of the streams exiting any splitter can be different from each other because the stream segregation implies that the flowrate for a specific component can be sent to any sink, which would then act as a splitter. To avoid this, the following composition constraints for each splitter of the superstructure must be included.

3.3.9.1 Fresh Sources Splitters

$$f_{r,j,c} = z_{r,c}^{\text{Fresh}} \sum_{c \in NCOMP} f_{r,j,c}, \quad r \in NFRESH, \quad j \in NSINKS, \quad c \in NCOMP \tag{3.11}$$

3.3.9.2 Process Source Splitters

$$w_{i,u,c}^{\text{u}} = z_{i,c}^{\text{InSource}} \sum_{c \in NCOMP} w_{i,u,c}^{\text{u}}, \quad i \in NSOURCES, \quad u \in NUNITS, \quad c \in NCOMP \tag{3.12}$$

$$w_{i,j,c}^{\text{s}} = z_{i,c}^{\text{InSource}} \sum_{c \in NCOMP} w_{i,j,c}^{\text{s}}, \quad i \in NSOURCES, \quad j \in NSINKS, \quad c \in NCOMP \tag{3.13}$$

3.3.9.3 Exit Treatment Unit Splitters

$$g_{u,j,c} = z_{u,c}^{\mathrm{OutUnit}} \sum_{c \in NCOMP} g_{u,j,c}, \quad u \in NUNITS, \ j \in NSINKS, \ c \in NCOMP \tag{3.14}$$

$$d_{u,u',c} = z_{u,c}^{\mathrm{OutUnit}} \sum_{c \in NCOMP} d_{u,u',c}, \quad u \in NUNITS, \ u' \neq u, \ c \in NCOMP \tag{3.15}$$

$z_{r,c}^{\mathrm{InFresh}}$, $z_{i,c}^{\mathrm{InSource}}$, and $z_{u,c}^{\mathrm{OutUnit}}$ are parameters known before the optimization process (in Eq. 3.15, for instance, the exit composition for a given treatment unit is fixed from composition data taken from the pre-synthesis task), which yields these relationships as linear. These equations are formulated for any split stream to ensure that the same proportion (or composition) of the chemical components is obtained in all the branches of the stream being split.

3.3.10 Cost of Treatment Units

There is a set of treatment units available to remove specific components. Each treatment unit can remove the components up to some specified composition levels, $z_{u,c}^{\mathrm{OutUnit}}$, and the removed components for each unit are calculated using Eqs. (3.9) and (3.10). The cost for each treatment unit depends on the flowrate removed and it follows an exponential rule as follows:

$$Cost_u^{\mathrm{Unit}} = \mathrm{FCost}_\mathrm{u}^{\mathrm{Unit}} + \mathrm{VCost}_\mathrm{u}^{\mathrm{Unit}} \left(m_u^{\mathrm{unit}}\right)^n, \quad u \in NUNITS \tag{3.16}$$

Typically, $0.5 \leq n \leq 0.9$, this yields a nonlinear term with the typical behavior shown in Fig. 3.3. The unit costs for Eq. (3.16) are obtained from simulations or provided by the manufacturer of the specific technology. To avoid nonlinear terms in the model, a piecewise linear formulation for the cost function can be used (i.e., the nonlinear function is approximated with a series of linear functions as shown in Fig. 3.3). Then, the following disjunction is implemented:

$$\begin{bmatrix} Y_u^1 \\ m_u^{\mathrm{unit}} \geq \mathrm{m_u^{unit^{min1}}} \\ m_u^{\mathrm{unit}} \leq \mathrm{m_u^{unit^{max1}}} \\ Cost_u^{\mathrm{Unit}} = \mathrm{FCost_u^{Unit^1}} + \mathrm{VCost_u^{Unit^1}}\, m_u^{\mathrm{unit}} \end{bmatrix} \vee \begin{bmatrix} Y_u^2 \\ m_u^{\mathrm{unit}} \geq \mathrm{m_u^{unit^{min2}}} \\ m_u^{\mathrm{unit}} \leq \mathrm{m_u^{unit^{max2}}} \\ Cost_u^{\mathrm{Unit}} = \mathrm{FCost_u^{Unit^2}} + \mathrm{VCost_u^{Unit^2}}\, m_u^{\mathrm{unit}} \end{bmatrix}$$

$$\vee \cdots \vee \begin{bmatrix} Y_u^N \\ m_u^{\mathrm{unit}} \geq \mathrm{m_u^{unit^{min\,N}}} \\ m_u^{\mathrm{unit}} \leq \mathrm{m_u^{unit^{max\,N}}} \\ Cost_u^{\mathrm{Unit}} = \mathrm{FCost_u^{Unit^{\mathit{N}}}} + \mathrm{VCost_u^{Unit^{\mathit{N}}}}\, m_u^{\mathrm{unit}} \end{bmatrix}, \quad u \in NUNITS$$

$\mathrm{FCost}_u^{\mathrm{Unit}^N}$ and $\mathrm{VCost}_u^{\mathrm{Unit}^N}$ can be estimated using linear regressions. The disjunction is reformulated as a set of linear relationships using convex-hull reformulations (Raman and Grossmann 1994; Ponce-Ortega et al. 2008) as follows:

$$\sum_{n \in N} y_u^n = 1, \quad u \in NUNITS \tag{3.17}$$

$$m_u^{\mathrm{unit}} = \sum_{n \in N} m_u^{\mathrm{unit}^n}, \quad u \in NUNITS \tag{3.18}$$

$$Cost_u^{\mathrm{Unit}} = \sum_{n \in N} Cost_u^{\mathrm{Unit}^n}, \quad u \in NUNITS \tag{3.19}$$

$$m_u^{\mathrm{unit}^n} \geq \mathrm{m}_u^{\mathrm{unit}^{\mathrm{min}^n}} y_u^n, \quad u \in NUNITS, \quad n \in N \tag{3.20}$$

$$m_u^{\mathrm{unit}^n} \leq \mathrm{m}_u^{\mathrm{unit}^{\mathrm{max}^n}} y_u^n, \quad u \in NUNITS, \quad n \in N \tag{3.21}$$

$$Cost_u^{\mathrm{Unit}^n} = \mathrm{FCost}_u^{\mathrm{Unit}^n} y_u^n + \mathrm{VCost}_u^{\mathrm{Unit}^n} m_u^{\mathrm{unit}^n}, \quad u \in NUNITS, \quad n \in N \tag{3.22}$$

$$Cost_u^{\mathrm{unit}^n} \leq \mathrm{Cost}_u^{\mathrm{max}} y_u^n, \quad u \in NUNITS, \quad n \in N \tag{3.23}$$

These relationships can be explained as follows. The first segment (n equals one) corresponds to the case when no pollutants are removed in the treatment units, while the other segments reflect the need for treatment. Only one segment is selected; then, the other segments do not apply, and their associated variables are set to zero with relationships (3.21) and (3.23).

3.3.11 Objective Function

The objective function consists of the minimization of the total annual cost, which involves the costs for fresh sources, treatment units, and the piping required for the water network.

$$\begin{aligned} Min\, TAC = {} & \mathrm{H_Y} \sum_{r \in NFRESH} \left(\mathrm{Cost}_r^{\mathrm{Fresh}} F_r \right) \\ & + \sum_{u \in NUNITS} \left(Cost_u^{\mathrm{Unit}} \right) \\ & + \mathrm{H_Y} \sum_{r \in NFRESH} \sum_{j \in NSINKS} \left(\mathrm{pip}_{r,j}^{\mathrm{Fresh}} \sum_{c \in NCOMP} f_{r,j,c} \right) \end{aligned}$$

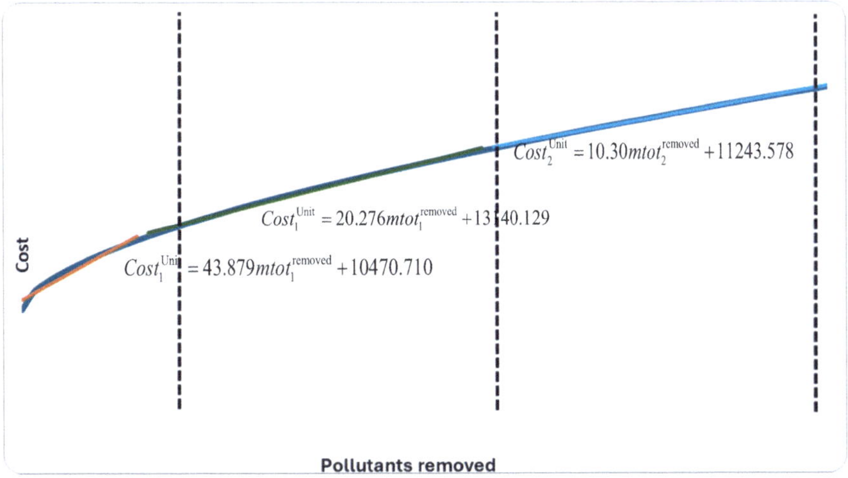

Fig. 3.3 Usual treatment unit costs behavior. Printed with permission of "*An optimization approach for the synthesis of recycle and reuse water integration networks*", *Clean Technologies and Environmental Policy*, Copyright (2012), Springer

$$
\begin{aligned}
&+ \mathrm{H_Y} \sum_{i \in NSOURCE} \sum_{u \in NUNITS} \left(\mathrm{pip}_{i,u}^{\mathrm{In}} \sum_{c \in NCOMP} w_{i,u,c}^{\mathrm{u}} \right) \\
&+ \mathrm{H_Y} \sum_{i \in NSOURCE} \sum_{j \in NSINKS} \left(\mathrm{pip}_{i,j}^{\mathrm{By}} \sum_{c \in NCOMP} w_{i,j,c}^{\mathrm{s}} \right) \\
&+ \mathrm{H_Y} \sum_{u' \in NUNITS} \sum_{u \in NUNITS} \left(\mathrm{pip}_{u',u}^{\mathrm{Units}} \sum_{c \in NCOMP} d_{u',u,c} \right) \\
&+ \mathrm{H_Y} \sum_{u \in NUNITS} \sum_{j \in NSINKS} \left(\mathrm{pip}_{u,j}^{\mathrm{Exit}} \sum_{c \in NCOMP} g_{u,j,c} \right)
\end{aligned}
\tag{3.24}
$$

$\mathrm{pip}_{r,j}^{\mathrm{Fresh}}$ corresponds to the unit piping cost for the fresh sources, $\mathrm{pip}_{i,u}^{\mathrm{In}}$ is the unit piping cost from the process sources to the treatment units, $\mathrm{pip}_{i,j}^{\mathrm{By}}$ is the unit piping cost for the segments from the process sources to the process sinks, $\mathrm{pip}_{u',u}^{\mathrm{Units}}$ is the unit piping cost for the piping between treatment units, and $\mathrm{pip}_{u,j}^{\mathrm{Exit}}$ is the unit piping cost for the segments from treatment units to sinks. All data for the unit piping costs are parameters known before the optimization process.

3.3.12 Model Remarks

- The model is a convex MILP problem; thus, its global optimal solution can be guaranteed.
- The model is based on component balances, and all components (including the pseudo-solvents) are considered in the balances.
- A pre-synthesis task must be implemented to obtain the feasible sets of outlet compositions from the treatment units. The use of models such as Kremser equations, simulations, or results from experimental works provides such values.
- Equations (3.11) to (3.15) ensure the consistency of outlet compositions when any given stream is split.
- The number of continuous variables, binary variables, and constraints depends on *NCOMP*, *NSINKS*, *NSOURCE*, *NFRESH*, and *N* as follows:

$$
\begin{aligned}
\#Continuous\ Variables = {} & NUNITS(4 + 2N + 3NCOMP + NSOURCE \times NCOMP \\
& + NUNIT \times NCOMP + NSINKS \times NCOMP) \\
& + NFRESH \times NSINKS \times NCOMP \\
& + NSOURCE \times NSINKS \times NCOMP + NSINKS \\
& \times NCOMP + NFRESH + 1 \\
\#Binary\ Variables = {} & NUNITS * N \\
\#Constraints = {} & NUNITS(4 + 4 \times NCOMP + NSOURCE \times NCOMP \\
& + NSINKS \times NCOMP + NUNITS \times NCOMP + 4 \times N) \\
& + NCOMP(NSOURCE + 3 \times NSINKS + NSOURCE \\
& \times NSINKS + NFRESH \times NSINKS \\
& + NFRESH) + 2 \times NSINKS + 1
\end{aligned}
$$

3.4 Case Studies

Four case studies are used to show the application of the presented model. The first three examples are theoretical whereas the fourth one is a practical case that was taken from Al Zarooni and Elshorbagy (2006). For all example problems, the solver CPLEX implemented in the General Algebraic Modeling System (GAMS) was used (Brooke et al. 2006). The operating time for all examples was assumed as 8000 h/year, and all problems were solved in a Core i7 CPU at 2.67 GHz with 9.00 GB of RAM.

Table 3.1 Sources and sinks characteristics for Example 3.1

Process sources			Fresh sources			Sinks				
Sources i	F (kg/h)	z_{phenol}	Source r	z_{phenol}	Cost ($/kg)	Sinks j	$G_j^{\min}$ (kg/h)	$G_j^{\max}$ (kg/h)	$z_{j,phenol}^{\min}$	$z_{j,phenol}^{\max}$
1	2900	0.033	1	0.000	0.0009	1	3000	3000	0	0.013
2	2450	0.022				2	1900	1900	0	0.011
						Waste	0	–	0	0.005

Printed with permission of "*An optimization approach for the synthesis of recycle and reuse water integration networks*", *Clean Technologies and Environmental Policy*, Copyright (2012), Springer

3.4.1 Example 1

This example corresponds to a simple problem used as motivation, and it consists of two process sources, two process sinks, and one fresh source with the data given in Table 3.1. The unit piping costs are presented in Table 3.2. The cost data for the treatment units are segregated into four intervals, one for the case in which no pollutant is removed with the associated cost of zero, and three other intervals that are modeled with suitable linear relationships. The cost data for treatment units 1 and 2 are displayed in Figs. 3.3 and 3.4; the composition at the exit of each unit and their cost functions are presented in Table 3.2.

This problem consists of 90 continuous variables, 8 binary variables, and 133 constraints and it was solved in 0.0001 s of CPU time. The optimal network is presented in Fig. 3.5, and Table 3.3 shows a summary of the costs for such a solution. The installation of the treatment unit type 2 is required, and there are no needed fresh sources. The existence of bypassing streams to satisfy the sink constraints at the lowest cost.

3.4.2 Example 2

This example consists of four process sources and two fresh sources with the stream data presented in Table 3.4. The constraints that must be satisfied in the sinks are presented in Table 3.5. The compositions of three different pollutants are restricted in this case. The unit piping costs associated with the problem are given in Table 3.6. The outlet concentrations and the equations for the linear approximations in the four intervals considered to model the cost function of the treatment units are shown in Table 3.7. The formulation proposed in this chapter yields a problem with 243 continuous variables, 8 binary variables and 329 constraints. The CPU time required to solve this problem was 0.03 s.

Figure 3.6 presents the optimal network, where the four process sources must be treated in unit 1. Source 4 needs to be partially treated as one fraction can be sent directly to sinks 1 and 2. Although the most efficient treatment unit is used, there is

Table 3.2 Data for the unitary piping costs for Example 3.1

$pip_{r,j}^{Fresh}$			$pip_{i,u}^{In}$			$pip_{i,j}^{By}$			$pip_{u',u}^{Units}$			$pip_{u,j}^{Exit}$		
Sources r	Sink j	$pip_{r,j}^{Fresh}$ \$/kg × 10^{-4}	Source i	Unit u	$pip_{i,u}^{In}$ \$/kg × 10^{-4}	Source i	Sink j	$pip_{i,j}^{By}$ \$/kg × 10^{-4}	Unit u	Unit u'	$pip_{u',u}^{Units}$ \$/kg × 10^{-4}	Unit u	Sink j	Cost \$/kg × 10^{-4}
1	1	1.7	1	1	1	1	1	1	1	2	0.2	1	1	1.5
	2	1.9	1	2	1.3	1	2	1.3	2	1	0.7	1	2	1.2
			2	1	0.9	1	Waste	0.4				1	Waste	0
			2	2	1.2	2	1	0.9				2	1	1.2
						2	2	1.2				2	2	0.9
						2	Waste	1.5				2	Waste	0

Printed with permission of "*An optimization approach for the synthesis of recycle and reuse water integration networks*", *Clean Technologies and Environmental Policy*, Copyright (2012), Springer

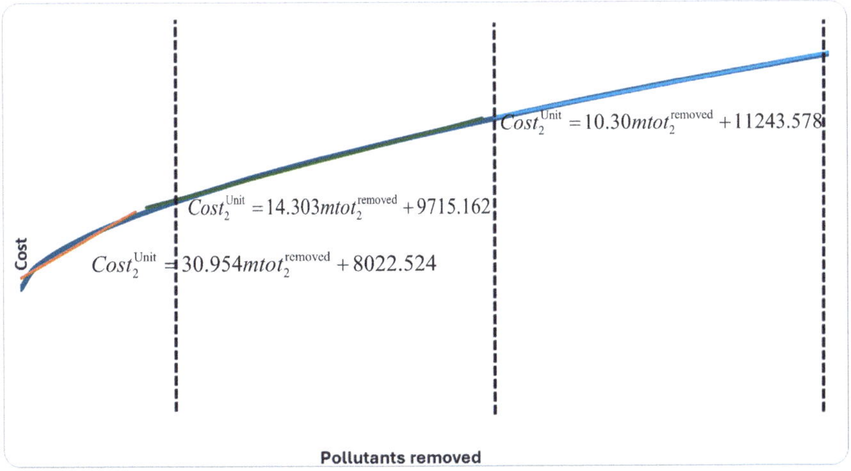

Fig. 3.4 Cost data for the treatment unit 2 for Example 3.1. Printed with permission of "*An optimization approach for the synthesis of recycle and reuse water integration networks*", *Clean Technologies and Environmental Policy*, Copyright (2012), Springer

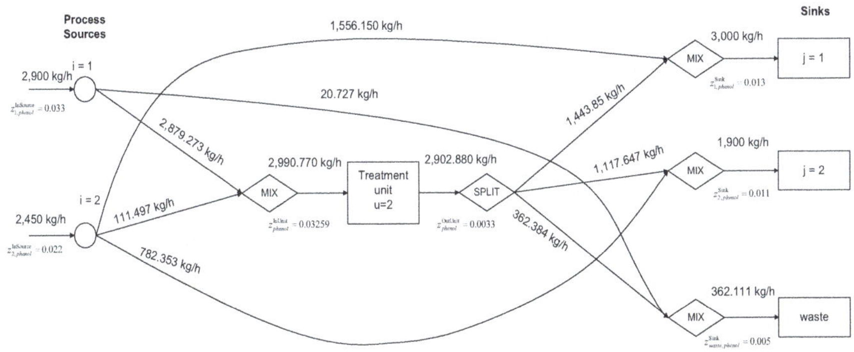

Fig. 3.5 Optimal network for Example 3.1. Printed with permission of "*An optimization approach for the synthesis of recycle and reuse water integration networks*", *Clean Technologies and Environmental Policy*, Copyright (2012), Springer

Table 3.3 Results for Example 3.1

Concept	Optimal solution
Fresh source cost (\$/year)	0
Treatment unit cost (\$/year)	10,743
Piping costs (\$/year)	7170
TAC (\$/year)	17,913

Printed with permission of "*An optimization approach for the synthesis of recycle and reuse water integration networks*", *Clean Technologies and Environmental Policy*, Copyright (2012), Springer

Table 3.4 Process and fresh sources characteristics for Example 3.2

Process sources					Fresh sources				
Sources i	F (kg/h)	Z_a	Z_b	Z_c	Source r	Z_a	Z_b	Z_c	Cost ($/kg)
1	688	0.0015	0.0087	0.0145	1	0	0	0	0.0012
2	413	0.0097	0.0121	0.0194	2	0.001	0.003	0.004	0.0003
3	342.5	0.0117	0.0161	0.0263					
4	1820	0.0016	0.0027	0.0066					

Printed with permission of "*An optimization approach for the synthesis of recycle and reuse water integration networks*", *Clean Technologies and Environmental Policy*, Copyright (2012), Springer

Table 3.5 Flowrate and composition constraints for the sinks for Example 3.2

Sinks j	G_j^{min} (kg/h)	G_j^{max} (kg/h)	$z_{j,a}^{min}$	$z_{j,a}^{max}$	$z_{j,b}^{min}$	$z_{j,b}^{max}$	$z_{j,c}^{min}$	$z_{j,c}^{max}$
1	808	808	0	0.0019	0	0.0037	0	0.0043
2	726.7	726.7	0	0.0014	0	0.0037	0	0.0041
3	1238.5	1238.5	0	0.0004	0	0.0020	0	0.0032
Waste	0	–	0	0.0005	0	0.0010	0	0.002

Printed with permission of "*An optimization approach for the synthesis of recycle and reuse water integration networks*", *Clean Technologies and Environmental Policy*, Copyright (2012), Springer

still a need for 264.71 kg/h of fresh source 1 because the use of the unit cannot yield the pollutant concentrations within the limits established for sink 3. The total waste is 680.28 kg/h, and it meets the environmental regulations for the three pollutants. Table 3.8 shows the economic results of this problem. The operating cost of treating units contributes to 70% of the total annual cost of the network.

3.4.3 Example 3

In this case, eight process sources are available, with the data presented in Table 3.9. Four sinks are involved, and the constraints for maximum and minimum capacities and allowed compositions of two key pollutants in the sinks and in the waste are presented in Table 3.10. The concentrations are in ppm, and the conversion to mass fraction is straightforward. For instance, consider component a in stream 1, the volumetric flow of this stream is 6309 l/h, and using a density like water (1 kg/l, a valid assumption in wastewater streams), one can calculate the mass of component a in stream 1 as follows:

$$Mass\,a = (15\text{ ppm})\left(6309\ \frac{\text{l}}{\text{h}}\right) = \left(15\ \frac{\text{mg}}{\text{l}}\right)\left(6309\ \frac{\text{l}}{\text{h}}\right)$$

Table 3.6 Data for the unitary piping costs for Example 3.2

$pip_{r,j}^{Fresh}$			$pip_{i,u}^{In}$			$pip_{i,j}^{By}$			$pip_{u',u}^{Units}$			$pip_{u,j}^{Exit}$		
Sources r	Sink j	$pip_{r,j}^{Fresh}$ \$/kg × 10^{-4}	Source i	Unit u	$pip_{i,u}^{In}$ \$/kg × 10^{-4}	Source i	Sink j	$pip_{i,j}^{By}$ \$/kg × 10^{-4}	Unit u	Unit u'	$pip_{u',u}^{Units}$ \$/kg × 10^{-4}	Unit u	Sink j	Cost \$/kg × 10^{-4}
1	1	1.1	1	1	0.9	1	1	1.1	1	2	0.2	1	1	0.9
1	2	1.2	1	2	1.0	1	2	1.2	2	1	0.2	1	2	1.0
1	3	1.3	2	1	1.0	1	3	1.2				1	3	1.0
2	1	1.2	2	2	0.9	1	Waste	1.3				1	Waste	1.1
2	2	1.3	3	1	1.1	2	1	1.1				2	1	1.0
2	3	1.4	3	2	1.0	2	2	1.0				2	2	0.9
			4	1	1.2	2	3	1.1				2	3	1.0
			4	2	1.1	2	Waste	1.2				2	Waste	1.1
						3	1	1.2						
						3	2	1.1						
						3	3	1.1						
						3	Waste	1.2						
						4	1	1.3						
						4	2	1.0						
						4	3	1.0						
						4	Waste	1.0						

Printed with permission of "*An optimization approach for the synthesis of recycle and reuse water integration networks*", *Clean Technologies and Environmental Policy*, Copyright (2012), Springer

Table 3.7 Minimum outlet concentrations and cost functions for treatment units for Example 3.2

Unit	z_a^{OutUnit}	z_b^{OutUnit}	z_c^{OutUnit}
1	0.0004995	0.000962	0.001258
2	0.00108	0.00208	0.00272
Cost functions unit 1 ($/year)			
Segment 1	0		
Segment 2	$Cost_1^{\text{Unit}} = 158.651 mtot_1^{\text{removed}} + 10{,}498.616$		
Segment 3	$Cost_1^{\text{Unit}} = 83.538 mtot_1^{\text{removed}} + 11{,}845.212$		
Segment 4	$Cost_1^{\text{Unit}} = 65.59 mtot_1^{\text{removed}} + 12{,}701.396$		
Cost functions unit 2 ($/year)			
Segment 1	0		
Segment 2	$Cost_2^{\text{Unit}} = 97.695 mtot_2^{\text{removed}} + 8307.052$		
Segment 3	$Cost_2^{\text{Unit}} = 51.441 mtot_2^{\text{removed}} + 9136.262$		
Segment 4	$Cost_2^{\text{Unit}} = 40.3901 mtot_2^{\text{removed}} + 9663.491$		

Printed with permission of "*An optimization approach for the synthesis of recycle and reuse water integration networks*", *Clean Technologies and Environmental Policy*, Copyright (2012), Springer

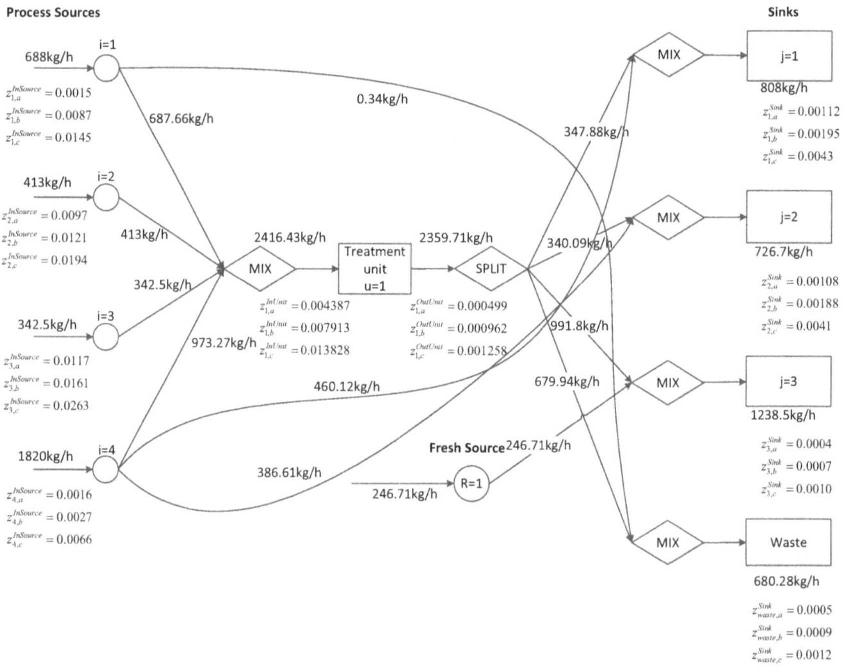

Fig. 3.6 Optimal network for Example 3.2. Printed with permission of "*An optimization approach for the synthesis of recycle and reuse water integration networks*", *Clean Technologies and Environmental Policy*, Copyright (2012), Springer

Table 3.8 Results for Example 3.2

Concept	Optimal solution
Fresh source cost ($/year)	2368
Treatment unit cost ($/year)	16,422
Piping costs ($/year)	4790
TAC ($/year)	23,580

Printed with permission of "*An optimization approach for the synthesis of recycle and reuse water integration networks*", *Clean Technologies and Environmental Policy*, Copyright (2012), Springer

$$= 94{,}635\ \frac{\text{mg}}{\text{h}} = 9.4635 \times 10^{-2}\ \frac{\text{kg}}{\text{h}}$$

The mass fraction is then obtained from:

$$Mass\ Fraction\ a = \frac{9.4635 \times 10^{-2}\ \frac{\text{kg}}{\text{h}}}{6309\ \frac{\text{kg}}{\text{h}}} = 1.5 \times 10^{-5}$$

Table 3.11 provides the unit piping cost, and four units are considered for the treatment of stream sources, whereas the cost function for these units is approximated using five intervals of load removal, with the first one corresponding to the case of no

Table 3.9 Process and fresh sources characteristics for Example 3.3

Process sources

Sources i	F (kg/h)	Z_a (ppm)	Z_a Mass fraction	Z_b (ppm)	Z_b Mass fraction
1	6309	15	1.5×10^{-5}	15	1.5×10^{-5}
2	3154	30	3.0×10^{-5}	1200	1.2×10^{-3}
3	157,725	12	1.2×10^{-5}	12	1.2×10^{-5}
4	3154	30	3.0×10^{-5}	60	6.0×10^{-5}
5	39,431	120	1.2×10^{-4}	72	7.2×10^{-5}
6	11,040	8.5	8.5×10^{-6}	8.5	8.5×10^{-6}
7	50,472	2	2.0×10^{-6}	30	3.0×10^{-5}
8	283,905	30	3.0×10^{-5}	33	3.3×10^{-5}

Fresh sources

Source r	Z_a (ppm)	Z_a Mass Fraction	Z_b (ppm)	Z_b Mass fraction	Cost ($/kg)
1	0	0	0	0	0.0009

Printed with permission of "*An optimization approach for the synthesis of recycle and reuse water integration networks*", *Clean Technologies and Environmental Policy*, Copyright (2012), Springer

Table 3.10 Flowrate and composition constraints for the sinks for Example 3.3

Sinks j	G_j^{min} (kg/h)	G_j^{max} (kg/h)	$z_{j,a}^{min}$ (ppm)	$z_{j,a}^{max}$ (ppm)	$z_{j,a}^{max}$ Mass fraction	$z_{j,b}^{min}$ (ppm)	$z_{j,b}^{max}$ (ppm)	$z_{j,b}^{max}$ Mass fraction
1	60,000	60,000	0	5	5.0×10^{-6}	0	20	2.0×10^{-5}
2	160,000	160,000	0	1	1.0×10^{-6}	0	20	2.0×10^{-5}
3	80,000	80,000	0	3	3.0×10^{-6}	0	30	3.0×10^{-5}
4	70,000	70,000	0	3	3.0×10^{-6}	0	30	3.0×10^{-5}
Waste	0	–	0	0.1	1.0×10^{-7}	0	10	1.0×10^{-5}

Printed with permission of "*An optimization approach for the synthesis of recycle and reuse water integration networks*", *Clean Technologies and Environmental Policy*, Copyright (2012), Springer

removal. Table 3.12 presents the minimum outlet composition and the cost functions for the treatment units in each interval.

The model for this problem consists of 20 binary variables, 448 continuous variables and 566 constraints. The optimal configuration provided by the model is presented in Fig. 3.7, where all the mass balances are satisfied. The constraints in the sinks for the maximum and minimum allowed compositions are satisfied for both components. However, the most stringent constraints are the maximum allowed values for all sinks. Two treatment units, 1 and 4, are used, with unit 1 being the cheapest. Once unit 1 reaches its minimum feasible concentration, part of the exit flowrate is sent to unit 4, which can achieve lower concentration levels for both components; the rest of the flowrate from unit 1 is segregated and sent to the sinks. The output from unit 4 is sent to the waste and two process sinks. Most of the flowrate sent to the waste comes from unit 4 because this unit is the only one that can reach the constraint for the waste composition of component a. Table 3.13 presents the costs associated with the network. There is no need for fresh sources, and the cost of the piping required to implement the network represents 47% of the total annual cost. The CPU time required to solve this problem was 0.062 s.

3.4.4 *Example 4*

This example was taken from Al Zarooni and Elshorbagy (2006). Nine wastewater streams from the Al Ruwais refinery are considered (Fig. 3.8); the characteristics of the wastewater and fresh sources are shown in Table 3.14. Three sinks are used, with the maximum allowed compositions and flows presented in Table 3.15, it is considered that no minimum composition is required for the considered pollutants components (PAH, phenols, and sulfur components) neither in the sinks nor in the waste discharge. Other relevant parameters such as piping costs, minimum outlet concentrations, and unit costs are presented in Tables 3.16 and 3.17. This problem

Table 3.11 Data for the unit piping costs for Example 3.3

$pip^{In}_{i,u}$ (\$/kg × 10^{-4})					
i	u				
	1	2	3	4	
1	0.9	1	1.1	1.1	
2	0.9	1	1.1	1.1	
3	0.9	1	1.1	1.1	
4	1	0.9	0.9	1	
5	1	0.9	0.9	1	
6	1.1	1.1	1	0.9	
7	1.1	1.1	1	0.9	
8	1.1	1.1	1	0.9	
$pip^{By}_{i,j}$ (\$/kg × 10^{-4})					
i	j				
	1	2	3	4	5
1	1	1.1	1.1	1.1	1
2	1	1.1	1.1	1.1	1
3	1	1.1	1.1	1.1	1
4	1.1	1	1	1.1	1
5	1.1	1	1	1.1	1
6	1.2	1.1	1	1	1
7	1.2	1.1	1	1	1
8	1.2	1.1	1	1	1
$pip^{Units}_{u',u}$ (\$/kg × 10^{-4})					
u'	u				
	1	2	3	4	
1	0	0.2	0.3	0.5	
2	0.2	0	0.4	0.5	
3	0.3	0.4	0	0.3	
4	0.5	0.5	0.3	0	
$pip^{Exit}_{u,j}$ (\$/kg × 10^{-4})					
i	j				
	1	2	3	4	5
1	0.9	1	1	1.1	1.1
2	1	0.9	1	1.1	1.1
3	1.1	1	0.9	1	1.1
4	1.1	1.1	1	1	1

(continued)

Table 3.11 (continued)

$pip^{Fresh}_{r,j}$ ($/kg × 10^{-4})					
r	*j*				
	1	2	3	4	
1	1.1	1.2	1.3	1.4	

Printed with permission of "*An optimization approach for the synthesis of recycle and reuse water integration networks*", *Clean Technologies and Environmental Policy*, Copyright (2012), Springer

consists of 20 binary variables, 571 continuous variables, and 709 constraints, and the CPU time required for the solution was 0.373 s.

Figure 3.9 presents the optimal configuration for this case and Table 3.18 presents the main results, where the *TAC* is $2,221,400/year, with the treatment system contributing the highest share. 2900 kg/h of fresh source 1 are required to satisfy the constraints given in the sinks. Unit 4 is needed to meet the environmental constraints since fresh sources are not to be mixed with waste sources. The waste before integration is 135 ton/year, and after integration, it is reduced to 15.36 ton/year, which represents a decrease of nearly 90%. Furthermore, the generated waste satisfies the environmental regulations because of the model formulation.

3.5 Concluding Remarks

Mass integration network is a practical tactic for transformation processes that can operate sustainably and have a major positive impact on the environment. A global optimum synthesis formulation for recycle/reuse water networks has been given using MILP. Based on component balances for every component (including a pseudo-solvent), the model uses the individual flowrates as optimization variables. When the compositions and total flowrates are utilized as optimization variables, this formulation conveniently avoids the bilinear terms that show up in the mixer balances, yielding a simple (global) optimal solution.

The fundamental superstructure that underpins the model formulation considers the configuration networks of interest, including mixing, segregation, direct recycling, and process stream treatment to meet sink restrictions. Four example issues were solved using the suggested paradigm, demonstrating its applicability with the least amount of computational work. This gives a strong reason to consider this kind of formulation, as it is a major improvement over other global optimization techniques that have been previously reported and may require a significant amount of CPU time, with which it might be challenging to find a global optimal solution for large problems.

Table 3.12 Minimum outlet concentrations and cost functions for treatment units for Example 3.3

Unit	z_a^{OutUnit} (ppm)	z_b^{OutUnit} (ppm)
1	3.5	10
2	0.15	10
3	0.0975	9.9
4	0.09	9
Cost functions unit 1 ($/year)		
Segment 1	0	
Segment 2	$Cost_1^{\text{Unit}} = 7578.582 mtot_1^{\text{removed}} + 200{,}807.139$	
Segment 3	$Cost_1^{\text{Unit}} = 3247.872 mtot_1^{\text{removed}} + 209{,}473.889$	
Segment 4	$Cost_1^{\text{Unit}} = 2114.395 mtot_1^{\text{removed}} + 218{,}681.318$	
Segment 5	$Cost_1^{\text{Unit}} = 1660.138 mtot_1^{\text{removed}} + 227{,}349.503$	
Cost functions unit 2 ($/year)		
Segment 1	0	
Segment 2	$Cost_2^{\text{Unit}} = 15{,}157.165 mtot_2^{\text{removed}} + 301{,}614.278$	
Segment 3	$Cost_2^{\text{Unit}} = 6495.745 mtot_2^{\text{removed}} + 318{,}947.778$	
Segment 4	$Cost_2^{\text{Unit}} = 4228.790 mtot_2^{\text{removed}} + 337{,}362.635$	
Segment 5	$Cost_2^{\text{Unit}} = 3320.276 mtot_2^{\text{removed}} + 354{,}699.006$	
Cost functions unit 3 ($/year)		
Segment 1	0	
Segment 2	$Cost_3^{\text{Unit}} = 18{,}946.457 mtot_3^{\text{removed}} + 402{,}010.784$	
Segment 3	$Cost_3^{\text{Unit}} = 8119.682 mtot_3^{\text{removed}} + 423{,}684.722$	
Segment 4	$Cost_3^{\text{Unit}} = 5285.988 mtot_3^{\text{removed}} + 446{,}703.294$	

(continued)

Table 3.12 (continued)

Unit	z_a^{OutUnit} (ppm)	z_b^{OutUnit} (ppm)
Segment 5	$Cost_3^{\text{Unit}} = 4150.345 mtot_3^{\text{removed}} + 468{,}373.758$	
Cost functions unit 4 ($/year)		
Segment 1	0	
Segment 2	$Cost_4^{\text{Unit}} = 37{,}892.914 mtot_4^{\text{removed}} + 704{,}035.695$	
Segment 3	$Cost_4^{\text{Unit}} = 16{,}239.364 mtot_4^{\text{removed}} + 747{,}369.445$	
Segment 4	$Cost_4^{\text{Unit}} = 10{,}571.9763 mtot_4^{\text{removed}} + 793{,}406.588$	
Segment 5	$Cost_4^{\text{Unit}} = 8300.6903 mtot_4^{\text{removed}} + 836{,}747.516$	

Nomenclatures

Cost	Unitary cost ($/kg)
d	Component flowrate between treatment units (kg/h)
f	Component flowrate for fresh sources (kg/h)
F	Total flowrate for fresh source (kg/h)
FCost	Fixed cost for treatment unit ($/year)
g	Component flowrate from treatment unit directed to sinks (kg/h)
G	Flowrate for process sink (kg/h)
H_Y	Operating time of the plant, h per year (h/year)
m	Component flowrate (kg/h)
N	Total number
NCOMP	Set for the components, $\{c \mid c = 1, \ldots, N_{\text{Components}}\}$
NFRESH	Set for the fresh sources, $\{r \mid r = 1, \ldots, N_{\text{Fresh}}\}$
NSINKS	Set for the sinks, $\{u \mid u = 1, \ldots, N_{\text{Units}}\}$
NSOURCES	Set for the process sources, $\{i \mid i = 1, \ldots, N_{\text{Sources}}\}$
NUNITS	Set for the treatment units, $\{u \mid u = 1, \ldots, N_{\text{Units}}\}$
pip	Unitary piping cost ($/kg)
TAC	Total annual cost ($/year)
VCost	Variable cost for treatment unit ($/year h/kg)
W	Total flowrate for the process sources (kg/h)

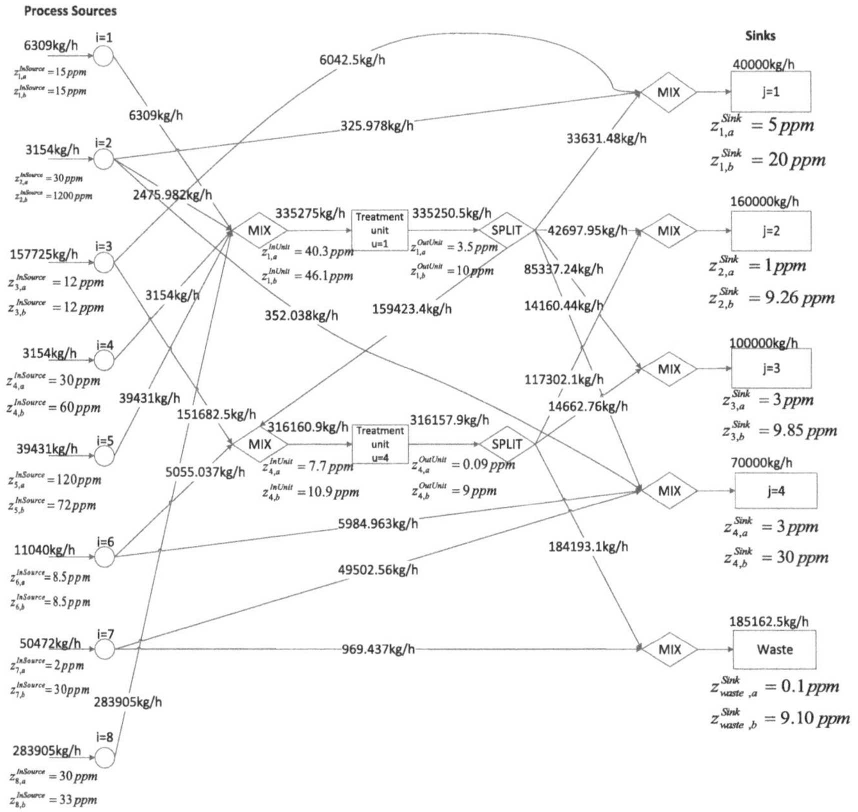

Fig. 3.7 Optimal network for Example 3.3. Printed with permission of "*An optimization approach for the synthesis of recycle and reuse water integration networks*", *Clean Technologies and Environmental Policy*, Copyright (2012), Springer

Table 3.13 Results for Example 3.3

Concept	Optimal solution
Fresh source cost ($/year)	0
Treatment unit cost ($/year)	1,064,100
Piping costs ($/year)	943,100
TAC ($/year)	2,007,200

Printed with permission of "*An optimization approach for the synthesis of recycle and reuse water integration networks*", *Clean Technologies and Environmental Policy*, Copyright (2012), Springer

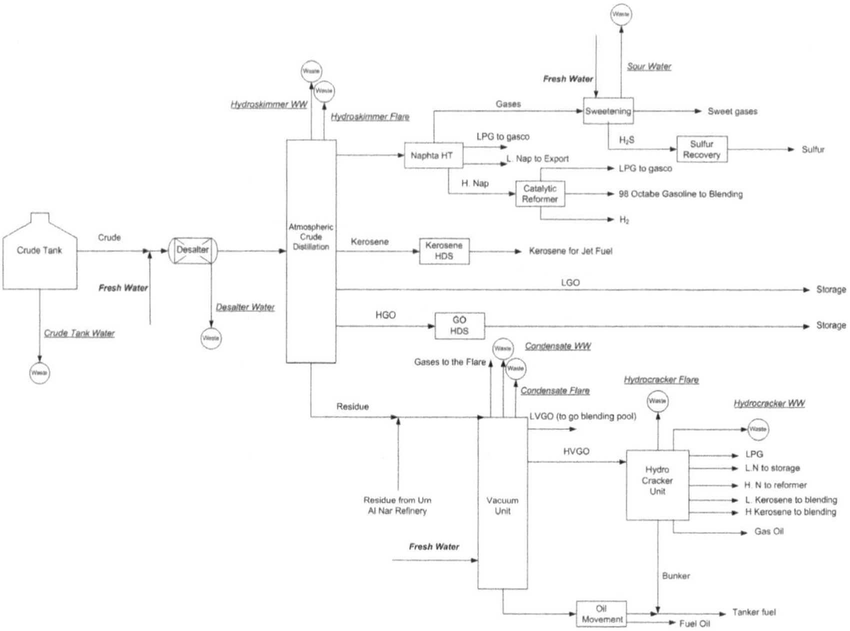

Fig. 3.8 Al Ruwais refinery major process components before integration. Printed with permission of "*An optimization approach for the synthesis of recycle and reuse water integration networks*", *Clean Technologies and Environmental Policy*, Copyright (2012), Springer

w^{s}	Component flowrate for process source directed to process sink (kg/h)
w^{u}	Component flowrate for process source directed to the treatment unit (kg/h)
Y_u^n	Boolean variable for the cost unit u for segment n
y_u^n	Binary variable for the cost unit u for segment n
z	Composition

Greek Symbols

α	Exponent for the cost of the treatment units

Indices

By	Bypass
c	Component

Table 3.14 Process and fresh sources characteristics for Example 3.4

Process sources

Sources i	F (kg/h)	z_{PAH} (ppm)	z_{PAH} Mass fraction	$z_{Phenols}$ (ppm)	$z_{Phenols}$ Mass fraction	$z_{Sulfur\ Comp}$ (ppm)	$z_{Sulfur\ Comp}$ Mass fraction
Crude tank	3000	0.31	3.1×10^{-7}	0.2	2×10^{-7}	559	5.59×10^{-4}
Desalter	22,500	4	4.0×10^{-6}	1.5	1.5×10^{-6}	137	1.37×10^{-4}
Hydroskimmer	22,500	32	3.2×10^{-5}	5	5.0×10^{-6}	111	1.11×10^{-4}
Condensate	22,500	5	5.0×10^{-6}	7	7.0×10^{-6}	136	1.36×10^{-4}
Condensate flare	6000	70	7.0×10^{-5}	5	5.0×10^{-6}	75	7.5×10^{-5}
Sour wáter	30,000	3	3.0×10^{-6}	1.21	1.21×10^{-6}	410	4.1×10^{-4}
Hydrocracker	22,500	35	3.5×10^{-5}	3	3.0×10^{-6}	100	1.0×10^{-4}
Hydrocracker flare	3000	1.7	1.7×10^{-6}	1	1.0×10^{-6}	20	2.0×10^{-5}
Hydroskimmer flare	3000	7	7.0×10^{-6}	4	4.0×10^{-6}	30	3.0×10^{-5}

Fresh sources

Source r	z_{PAH} (ppm)	z_{PAH} Mass fraction	$z_{Phenols}$ (ppm)	$z_{Phenols}$ Mass fraction	$z_{Sulfur\ Comp}$ (ppm)	$z_{Sulfur\ Comp}$ Mass fraction	Cost ($/kg)
1	0	0	0	0			0.0009
2	1	1.0×10^{-6}	0.1	1.0×10^{-7}	50	5.0×10^{-5}	0.0005

Table 3.15 Flowrate and composition constraints for the sinks for Example 3.4

Sinks j	G_j^{min} (kg/h)	G_j^{max} (kg/h)	$z_{j,PHA}^{max}$ (ppm)	$z_{j,PHA}^{max}$ Mass fraction	$z_{j,Phenols}^{max}$ (ppm)	$z_{j,Phenols}^{max}$ Mass fraction	$z_{j,Sulfur\ Comp}^{max}$	$z_{j,Sulfur\ Comp}^{max}$
Desalter	27,500	27,500	3.5	3.5×10^{-6}	1	1.0×10^{-6}	30	3.0×10^{-5}
Sweetening	45,000	45,000	7.5	7.5×10^{-6}	1	1.0×10^{-6}	52	5.2×10^{-5}
Quenching	50,000	50,000	2.5	2.5×10^{-6}	0.5	5.0×10^{-7}	35	3.5×10^{-5}
Waste	0	–	1	1.0×10^{-6}	0.1	1.0×10^{-7}	100	1.0×10^{-4}

Components	Components
Exit	Exit from treatment units
Fresh	Fresh sources
i	Process source
In	Inlet
j	Sink

Table 3.16 Data for the unit piping costs for Example 3.4

$pip^{In}_{i,u}$ ($/kg × 10^{-4}), *i*	*u*			
	1	2	3	4
Crude tank	0.9	1	1.1	1.1
Desalter	0.9	1	1.1	1.1
Hydroskimmer	0.9	1	1.1	1.1
Condensate	1	0.9	0.9	1
Condensate flare	1	0.9	0.9	1
Sour water	1.1	1.1	1	0.9
Hydrocracker	1.1	1.1	1	0.9
Hydrocracker flare	1.1	1.1	1	0.9
Hydroskimmer flare	1.2	1.1	1	0.9
$pip^{By}_{i,j}$ ($/kg × 10^{-4}), *i*	*j*			
	Desalter	Sweetening	Quenching	Waste
Crude tank	1	1.1	1.1	1
Desalter	1	1.1	1.1	1
Hydroskimmer	1	1.1	1.1	1
Condensate	1.1	1	1	1
Condensate flare	1.1	1	1	1
Sour water	1.2	1.1	1	1
Hydrocracker	1.2	1.1	1	1
Hydrocracker flare	1.2	1.1	1	1
Hydroskimmer flare	1.2	1.1	1	1
$pip^{Units}_{u',u}$ ($/kg × 10^{-4}), *u′*	*u*			
	1	2	3	4
1	0	0.2	0.3	0.5
2	0.2	0	0.4	0.5
3	0.3	0.4	0	0.3
4	0.5	0.5	0.3	0
$pip^{Exit}_{u,j}$ ($/kg × 10^{-4}), *u*	*j*			
	Desalter	Sweetening	Quenching	Waste
1	0.9	1	1	1.1
2	1	0.9	1	1.1
3	1.1	1	0.9	1
4	1.1	1.1	1	1
$pip^{Fresh}_{r,j}$ ($/kg × 10^{-4}), *r*	*j*			
	1	2	3	4

(continued)

Table 3.16 (continued)

1	1.1	1.2	1.3	1.4
2	1.1	1.2	1.3	1.4

L	Lean stream
max	Maximum
min	Minimum
n	Segment for the cost function
Out	Out
r	Fresh source
R	Rich stream
remov	Mass removed for the pollutant
Sink	Sink
Source	Source
u	Treatment unit
Unit	Units
waste	Waste discharged to the environment

Appendix: Partition Interceptors

To take into account partition interceptors (Tan et al. 2009), when more than one stream leaves the treatment unit (Fig. 3.10), the following equations have to be used.

Component balances in the interception devices:

$$m_{u,c}^{\mathrm{In}} = m_{u,c}^{\mathrm{Out\,L}} + m_{u,c}^{\mathrm{Out\,R}}, \quad u \in NUNITS, \quad c \in NCOMP \tag{3.25}$$

Component balances in the mixers before the interception devices:

$$m_{u,c}^{\mathrm{In}} = \sum_{i \in NSOURCES} w_{i,u,c}^{\mathrm{u}} + \sum_{\substack{u' \in NUNITS \\ u' \neq u}} d_{u',u,c}^{\mathrm{L}} + \sum_{\substack{u' \in NUNITS \\ u' \neq u}} d_{u',u,c}^{\mathrm{R}}, \quad u \in NUNITS, \quad c \in NCOMP \tag{3.26}$$

This means that Eq. (3.3) must be replaced by Eq. (3.26) for partition interceptors.

For the component balance in the splitters after the partition interceptions, replace Eq. (3.4) with the following set of relationships:

Table 3.17 Minimum outlet concentrations and cost functions for treatment units for Example 3.4

Unit	$z_{PAH}^{OutUnit}$ (ppm)	$z_{Phenols}^{OutUnit}$ (ppm)	$z_{Sulfur\ Comp}^{OutUnit}$ (ppm)
1	3	1.5	50
2	2.5	0.5	25
3	1.5	0.1	25
4	0.75	0.075	100
Cost functions unit 1 ($/year)			
Segment 1	0		
Segment 2	$Cost_1^{Unit} = 41{,}682.205 mtot_1^{removed} + 404{,}439.265$		
Segment 3	$Cost_1^{Unit} = 17{,}863.3 mtot_1^{removed} + 452{,}106.389$		
Segment 4	$Cost_1^{Unit} = 11{,}629.173 mtot_1^{removed} + 502{,}747.247$		
Segment 5	$Cost_1^{Unit} = 9130.759 mtot_1^{removed} + 550{,}422.267$		
Cost functions unit 2 ($/year)			
Segment 1	0		
Segment 2	$Cost_2^{Unit} = 55{,}323.654 mtot_2^{removed} + 555{,}892.115$		
Segment 3	$Cost_2^{Unit} = 23{,}709.471 mtot_2^{removed} + 619{,}159.389$		
Segment 4	$Cost_2^{Unit} = 15{,}435.085 mtot_2^{removed} + 686{,}373.619$		
Segment 5	$Cost_2^{Unit} = 12{,}119.007 mtot_2^{removed} + 749{,}651.373$		
Cost functions unit 3($/year)			
Segment 1	0		
Segment 2	$Cost_3^{Unit} = 69{,}722.962 mtot_3^{removed} + 707{,}425.679$		
Segment 3	$Cost_3^{Unit} = 29{,}880.43 mtot_3^{removed} + 787{,}159.775$		
Segment 4	$Cost_3^{Unit} = 19{,}452.436 mtot_3^{removed} + 871{,}868.123$		
Segment 5	$Cost_3^{Unit} = 15{,}273.27 mtot_3^{removed} + 951{,}615.429$		

(continued)

Table 3.17 (continued)

Unit	$z_{PAH}^{OutUnit}$ (ppm)	$z_{Phenols}^{OutUnit}$ (ppm)	$z_{Sulfur\ Comp}^{OutUnit}$ (ppm)
Cost functions unit 4 ($/year)			
Segment 1	0		
Segment 2	$Cost_4^{Unit} = 68{,}965.103 mtot_4^{removed} + 977{,}344.965$		
Segment 3	$Cost_4^{Unit} = 29{,}555.643 mtot_4^{removed} + 1{,}056{,}212.38$		
Segment 4	$Cost_4^{Unit} = 19{,}240.996 mtot_4^{removed} + 1{,}139{,}999.991$		
Segment 5	$Cost_4^{Unit} = 15{,}107.256 mtot_4^{removed} + 1{,}218{,}880.478$		

Printed with permission of "*An optimization approach for the synthesis of recycle and reuse water integration networks*", *Clean Technologies and Environmental Policy*, Copyright (2012), Springer

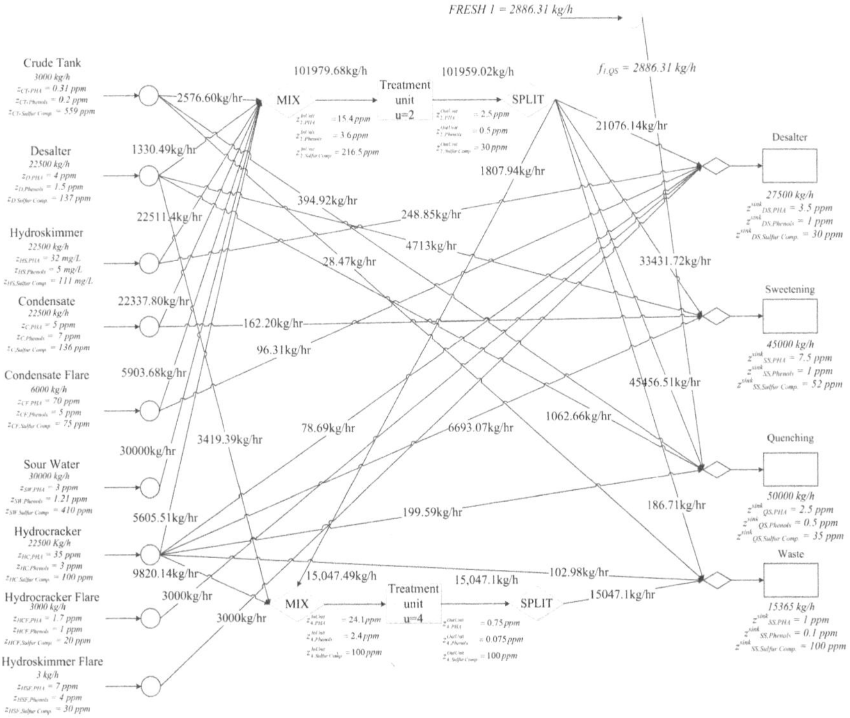

Fig. 3.9 Optimal network for Example 3.4. Printed with permission of "*An optimization approach for the synthesis of recycle and reuse water integration networks*", *Clean Technologies and Environmental Policy*, Copyright (2012), Springer

Table 3.18 Results for Example 3.4

Concept	Optimal solution
Fresh source cost ($/year)	20,800
Treatment unit cost ($/year)	2,000,000
Piping costs ($/year)	200,600
TAC ($/year)	2,221,400

Printed with permission of "*An optimization approach for the synthesis of recycle and reuse water integration networks*", *Clean Technologies and Environmental Policy*, Copyright (2012), Springer

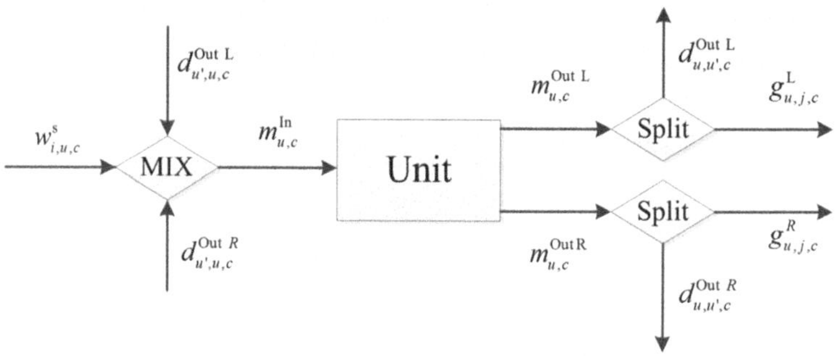

Fig. 3.10 Schematic representation of a partitioning regenerator. Printed with permission of "*An optimization approach for the synthesis of recycle and reuse water integration networks*", *Clean Technologies and Environmental Policy*, Copyright (2012), Springer

$$m_{u,c}^{\text{OutL}} = \sum_{j \in NSINKS} g_{u,j,c}^{\text{L}} + \sum_{\substack{u' \in NUNITS \\ u' \neq u}} d_{u,u',c}^{\text{L}}, \quad u \in NUNITS, \quad c \in NCOMP \tag{3.27}$$

$$m_{u,c}^{\text{OutR}} = \sum_{j \in NSINKS} g_{u,j,c}^{\text{R}} + \sum_{\substack{u' \in NUNITS \\ u' \neq u}} d_{u,u',c}^{\text{R}}, \quad u \in NUNITS, \quad c \in NCOMP \tag{3.28}$$

The fraction of the feed that enters the treatment unit and exists as lean stream is referred to as the liquid recovery (Tan et al., 2009) and it is given by $m_{u,c}^{\text{OutL}} = m_{u,c}^{\text{in}}\alpha$, where α is the recovery factor. An assumption in this formulation to calculate the cost of the unit is that the load removed is the difference between the loads of the inlet stream and the lean stream.

$$m_{u,c}^{\text{remov}} = m_{u,c}^{\text{In}} - m_{u,c}^{\text{OutL}}, \quad u \in NUNITS, \quad c \in NCOMP \tag{3.29}$$

$$m_{u}^{\text{unit}} = \sum_{c \in NCOMP} m_{u,c}^{\text{remov}}, \quad u \in NUNITS \tag{3.30}$$

This means that Eq. (3.9) has to be replaced by (3.29) for partition units and Eq. (3.10) is the same that Eq. (3.30)

To ensure that the concentrations of the split streams are the same, the following constraints are needed:

$$g^{\mathrm{L}}_{u,j,c} = z^{\mathrm{OutUnitL}}_{u,c} \sum_{c \in NCOMP} g^{\mathrm{L}}_{u,j,c'}, \quad u \in NUNITS, \ j \in NSINKS, \ c \in NCOMP \tag{3.31}$$

$$g^{\mathrm{R}}_{u,j,c} = z^{\mathrm{OutUnitR}}_{u,c} \sum_{c \in NCOMP} g^{\mathrm{R}}_{u,j,c'}, \quad u \in NUNITS, \ j \in NSINKS, \ c \in NCOMP \tag{3.32}$$

$$d^{\mathrm{L}}_{u,u',c} = z^{\mathrm{OutUnitL}}_{u,c} \sum_{c \in NCOMP} d^{\mathrm{L}}_{u,u',c'}, \quad u \in NUNITS, u' \neq u, \ c \in NCOMP \tag{3.33}$$

$$d^{\mathrm{R}}_{u,u',c} = z^{\mathrm{OutUnitR}}_{u,c} \sum_{c \in NCOMP} d^{\mathrm{R}}_{u,u',c'}, \quad u \in NUNITS, u' \neq u, \ c \in NCOMP \tag{3.34}$$

Thus, Eqs. (3.14) and (3.15) are replaced by (3.31)–(3.34) for partition units, and the outlet concentrations of the streams are given by the efficiency values of the units; under this set of equations, the model formulation remains linear.

Lastly, Eq. (3.5) is replaced by:

$$G^{\mathrm{Sink}}_{j,c} = \sum_{r \in NFRESH} f_{r,j,c} + \sum_{u \in NUNITS} g^{\mathrm{L}}_{u,j,c} + \sum_{u \in NUNITS} g^{\mathrm{R}}_{u,j,c} + \sum_{i \in NSOURCES} w^{\mathrm{S}}_{i,u,c'}, \quad j \in NSINK, \ c \in NCOMP \tag{3.35}$$

References

Al Zarooni M, Elshorbagy W (2006) Characterization and assessment of Al Ruwais refinery wastewater. J Hazard Mater 136:398–405

Almutairi H, Elhedhli S (2009) A new Lagrangean approach to the pooling problem. J Global Optim 45(2):237–257

Brooke A, Kendrick D, Meeruas A, Raman R (2006) GAMS-language guide. GAMS Development Corporation, Washington, D.C.

Chakraborty A (2009) A globally convergent mathematical model for synthesizing topologically constrained water recycle networks. Comput Chem Eng 33(7):1279–1288

Dunn RF, El-Halwagi MM (2003) Process integration technology review: background and applications in the chemical process industry. Chem Technol Biotech 78(9):1011–1021

El-Halwagi MM (2006) Process integration. Elsevier, San Diego

Faria DC, Bagajewicz MJ (2010) Novel bound contractor procedure for global optimization of bilinear MINLP problems with applications to water management problems. Comput Chem Eng 35(3):446–455

Foo DCY (2009) State-of-the-art review of pinch analysis techniques for water network synthesis. Ind Eng Chem Res 48(11):5125–5159

Gabriel FB, El-Halwagi MM (2005) Simultaneous synthesis of waste interception and material reuse networks: problem reformulation for global optimization. Environ Prog 24(2):171–180

Galan B, Grossmann IE (1998) Optimal design of distributed wastewater treatment networks. Ind Eng Chem Res 37(10):4036–4048

Gounaris CE, Misener R, Floudas CA (2009) Computational compression of piecewise-linear relaxations of pooling problems. Ind Eng Chem Res 48(12):5742–5766

Hasan MF, Karimi IA (2010) Piecewise linear relaxation of bilinear programs using bivariate partitioning. AIChE J 56(7):1880–1893

Hul S, Tan RR, Auresenia J, Fuchino T, Foo CYD (2007) Water network synthesis using mutation-enhanced particle swarm optimization. Process Saf Environ Prot 85(6):507–514

Jeżowski J (2010) Review of water network design methods with literature annotations. Ind Eng Chem Res 49(10):4475–4516

Karuppiah R, Grossmann IE (2006) Global optimization for the synthesis of integrated water systems in chemical processes. Comput Chem Eng 30(4):650–673

Lee S, Grossmann IE (2003) Global optimization of nonlinear generalized disjunctive programming with bilinear equality constraints: applications to process networks. Comput Chem Eng 27(11):1557–1575

Liberti L, Pantelides CC (2006) An exact reformulation algorithm for large nonconvex NLPs bilinear terms. J Global Optim 36(2):161–189

Meyer CA, Floudas CA (2006) Global optimization of a combinatorially complex generalized pooling problem. AIChE J 52(3):1027–1037

Misener R, Floudas CA (2010) Global optimization of large-scale generalized problems: quadratic constrained MINLP models. Ind Eng Chem Res 49(11):5424–5438

Nápoles-Rivera F, Ponce-Ortega JM, El-Halwagi MM, Jiménez-Gutiérrez A (2010) Global optimization of mass and property integration networks with in-plant property interceptors. Chem Eng Sci 65(15):4363–4377

Ng DKS, Foo DCY, Tan RR, Pau CH, Tan YL (2009) Automatic targeting for conventional and bilateral property-based resource conservation network. Chem Eng J 149(1–3):87–101

Pham V, Laird C, El-Halwagi MM (2009) Convex hull discretization approach to the global optimization of pooling problems. Ind Eng Chem Res 48(4):1973–1979

Ponce-Ortega JM, Jiménez-Gutiérrez A, Grossmann IE (2008) Optimal synthesis of heat exchanger networks involving isothermal streams. Comput Chem Eng 32(8):1918–1942

Ponce-Ortega JM, Hortua AC, El-Halwagi MM, Jiménez-Gutiérrez A (2009) A property-based optimization of direct recycle networks and wastewater treatment processes. AIChE J 55(9):2329–2344

Ponce-Ortega JM, El-Halwagi MM, Jiménez-Gutiérrez A (2010) Global optimization of property-based recycle and reuse networks including environmental constraints. Comput Chem Eng 34(3):318–330

Ponce-Ortega JM, Nápoles-Rivera F, El-Halwagi MM, Jiménez-Gutiérrez A (2012) An optimization approach for the synthesis of recycle and reuse water integration networks. Clean Technol Environ Policy 14:133–151

Quesada I, Grossmann IE (1995) A global optimization algorithm for linear fractional and bilinear programs. J Global Optim 6(1):39–76

Raman R, Grossmann IE (1994) Modeling and computational techniques for logic-based integer programming. Comput Chem Eng 18(7):563–578

Savelski MJ, Bagajewicz MJ (2000) On the optimality conditions of water utilization systems in process plants with single contaminants. Chem Eng Sci 55(21):5035–5048

Savelski MJ, Bagajewicz MJ (2001) Algorithmic procedure to design water utilization systems featuring a single contaminant in process plants. Chem Eng Sci 56(5):1897–1911

Savelski MJ, Bagajewicz MJ (2003) On the necessary conditions of optimality of water utilization systems in process plants with multiple contaminants. Chem Eng Sci 58(23–24):5349–5362

Takama N, Kuriyama T, Shiroko K, Umeda U (1980) Optimal water allocation in a petroleum refinery. Comput Chem Eng 4(4):251–258

Tan RR, Col-Long KJ, Foo DCY, Hul S, Ng DKS (2008) A methodology for the design of efficient resource conservation networks using adaptive swarm intelligence. J Clean Prod 16(7):822–832

Tan RR, Ng DKS, Foo DCY, Aviso KB (2009) A superstructure model for the synthesis of single-contaminant water networks with partitioning regenerators. Process Saf Environ Prot 7(3):197–205

Teles J, Castro PM, Novals AQ (2008) LP-based solution strategies for the optimal design of industrial water networks with multiple contaminants. Chem Eng Sci 63(2):376–394

Wicaksono DS, Karimi IA (2008) Piecewise MILP under- and over-estimators for global optimization of bilinear programs. AIChE J 54(4):991–1008

Chapter 4
Sustainable Management of Water in Macroscopic Systems

4.1 Introduction

One of the biggest issues facing most of the world's nations is water scarcity. Studies predict that by 2025, two-thirds of the world's population may be experiencing water stress, and 1.8 billion people will reside in nations or areas with complete water scarcity (UN-Water 2007). As such, measures must be taken to mitigate this circumstance. Various optimization techniques have been put forth to address the management of macroscopic and industrial water systems. The literature has several recent reviews (Foo 2009; Foo et al. 2011; El-Halwagi 2012). A sampling of contributions in the field is highlighted in the discussion that follows. To take into consideration the physical, chemical, and biological activity occurring within watersheds, material flow analysis (MFA) has been utilized in conjunction with systems integration models (El-Baz 2005a, b; Lovelady et al. 2009; Lira-Barragán et al. 2011a, b; 2012; Burgara-Montero et al. 2012). A model for water integration in industrial processes utilizing a water pinch-based approach was provided by Dvarioniene and Stasiskiene (2007). A water pinch-based approach was used by Dakwala et al. (2009) to address a multi-contaminant case in the starch sector. They discovered that employing process integration might cut the amount of freshwater used in the procedure by up to 64%. Afterward, a mathematical programming model for the global optimization of mass and property integration networks was presented by Nápoles-Rivera et al. (2010). This strategy uses decentralized treatment units to reduce the amount of freshwater used. Several strategies for water integration in eco-industrial parks were presented by Spriggs et al. (2004), Chew et al. (2008), Lovelady and El-Halwagi (2009), and Rubio-Castro et al. (2010). In these parks, wastewater can be recycled into various industries to reduce both the total amount of waste produced and the amount of freshwater used. To maximize the retrofit of water conservation networks, Sotelo-Pichardo et al. (2011) advocated modifying the current processes to meet new requirements and/or environmental limitations while reducing waste production and freshwater consumption.

J. M. Ponce-Ortega et al., *Sustainable Water Systems*, Water Resources Development and Management, https://doi.org/10.1007/978-3-031-72498-5_4

Various strategies for conserving water have been put forth, considering the usage of other water sources such desalinated water, rainfall collection, and wastewater recycling. Mugistein et al. (2003) investigated the feasibility of establishing power/ desalination facilities in Israel to meet the growing energy (about 7.2% per year) and water requirements in this regard. This project's primary goals were to decrease the amount of fossil fuels used to generate electricity and to raise the amount of potable water available each year by 400–600 Mm^3. They demonstrated that the investment in the desalination process is only warranted if the plant runs continuously for extended periods of time, and that dual-purpose facilities are less efficient than energy production plants alone from the standpoint of energy generation. Hadadin et al. (2010) examined several options for water resources, including desalinated water, for use in Jordan. They discovered that the primary impediment to this strategy is the high cost of the desalination process in addition to the distribution costs (piping and pumping), which become significant because the water needs to be pumped at distances of up to 350 km and with static charges ranging from 0 to 1000 m. Jordan has experienced previous droughts, and the scarcity of natural resources makes desalination an appealing alternative even with its high cost. It should be noted that the primary drawback of desalination methods at the moment is the high cost of both the process and distribution, which can occasionally become unaffordable.

Rainwater harvesting is another alternative source of water. Appan (1999) conducted a feasibility study for the installation of a rainwater collection system for non-potable uses at Nanyang Technological University in Singapore. He discovered that the system could save S$18,500 in water consumption costs. This study shows that rainwater is suitable for non-potable usage, but it may also be utilized as potable water with the right treatment. A case study of a rainwater recovery system constructed in a large-roofed supermarket, where the water is used for household cold water requirements (toilet flushing), was presented by Chilton et al. (1999). According to their calculations, this initiative will pay for itself in 12 years at a collecting efficiency of 57.4%. According to Abdulla and Al-Shareef (2009), rainwater harvesting (RWH) can meet up to 5.6% of Jordan's total water demand. In addition, they discovered that the water's quality typically meets the World Health Organization's (WHO) standards for drinking water quality. Gikas and Angelakis (2009) examined the situation of the Aegean Islands and Crete, highlighting the significance of employing unconventional water sources like desalination and RWH, particularly in areas with physical constraints like the islands. To acquire environmentally friendly procedures, Partzsch (2009) provided the case of Germany, where rainwater is collected and distributed utilizing centralized facilities. This work emphasized the significance of adopting decentralized methods. Furthermore, the application of RWH to meet various water demands was examined in the works of Li et al. (2010), Farreny et al. (2011), Domènech and Saurí (2011), Domènech et al. (2012), and Morales-Pinzón et al. (2012).

Few works deal with integrated macroscopic systems, for which conventional and alternative sources of water can be employed to meet demands, even though various authors have looked at water saving through process integration and the use of other water sources. To manage water resources in this situation, Liu et al. (2011) created a mixed integer linear programming (MILP) model. They presented two case studies for the islands of Syros and Paros-Antiparos, where the total capital and operating costs of the water production and distribution were obtained. Desalinated water and reclaimed water are used to satisfy the demands. A system integration technique was used by Atilhan et al. (2012) to determine the best way to distribute water in Qatar. Notably, Atilhan et al. (2012) considered the required storage capacity but not its location or cost, whereas Liu et al. (2011) did not consider the option of storing water for future use or make a differentiation between the possible water uses. Because of the restrictions on catchment areas in the islands and the low annual precipitation of 100 mm in Qatar, neither of these works examined the use of rainwater as a substitute supply. Thus, in situations where the amount of rainwater is considerable, these earlier methods cannot be used. This chapter presents a mathematical programming approach for the optimal management (distribution and storage) of non-conventional (rainwater and treated wastewater) and conventional (superficial and groundwater) water sources to get beyond these restrictions. The goal is to determine the best timetable for distributing water to various consumers in a macroscopic system while taking sustainability and economic factors into account (Nápoles-Rivera et al. 2013).

4.2 Problem Statement

Several water sources, including deep wells, springs, and dams, are involved in the addressed problem. Rain and tributaries are two ways that these sources might be replenished. In normal circumstances, a city's demands are met by these sources. The water sources are first separated and transferred to a main for use in agriculture or home settings. Depending on the purpose, the water is treated there. Water from the main for agriculture is routed to agricultural sinks, while water from the main for domestic use is sent to a domestic sink (i.e., domestic user) once it has been treated. A centralized treatment facility can handle the leftover water from household sinks. The remaining treated water is released into the environment, with a portion of it being retrieved for use in agriculture. It is possible to create artificial ponds and a series of storage tanks in this work. These containers can store water for a long time and use it for residential or agricultural uses. The challenge is to determine the best timetable for managing the water resources to maximize revenue from water sales while minimizing capital and operating expenses related to distribution and storage. This involves satisfying the needs of a macroscopic system and preserving sustainable levels of natural resources. It is important to note that determining the ideal storage capacity and the placement of the storage tanks and artificial pounds comprise a portion of the challenge.

4.3 Mathematical Model

This section describes the presented mathematical model for the optimal water storage and distribution in a macroscopic system. The implemented superstructure is shown in Fig. 4.1. Primary, the indexes used are presented: i is the natural sources, m represents the number of tributaries recharging the natural sources and t corresponds to the period in which the balance is performed (in this work, the chosen period is one month). l and n are the possible locations of storage tanks and artificial ponds, respectively. Lastly, j is the domestic sinks, and h is the agricultural sinks. So, the implemented model is described as follows.

4.3.1 *Balances for the Natural Sources of Water*

$$G_{i,t} - G_{i,t-1} = \sum_{m \in M} r_{m,i,t} + p^{g}_{i,t} - g^{d}_{i,t} - g^{a}_{i,t} - v^{g}_{i,t}, \quad i \in I, \quad t \in T \tag{4.1}$$

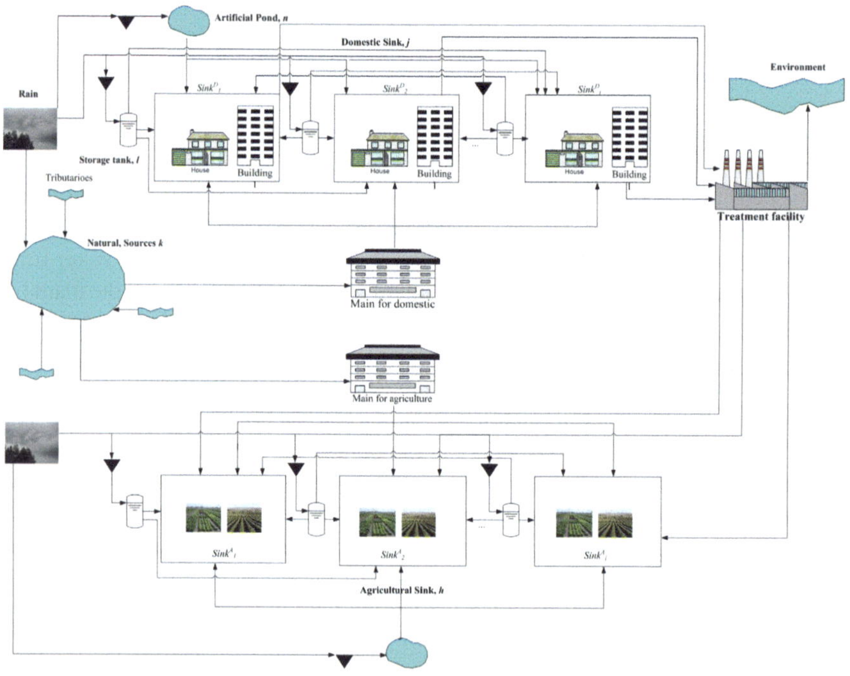

Fig. 4.1 Superstructure for water distribution in a macroscopic system. Printed with permission of "*Sustainable water management for macroscopic systems*", *Journal of Cleaner Production*, Copyright (2013), Elsevier

The accessible water in a location ($G_{i,t}$) at a given time t minus the accumulated at the previous period ($G_{i,t-1}$) is equal to the sum of all the tributaries to source i ($r_{m,i,t}$) plus the contribution from direct precipitation and runoff water (grouped in the parameter $p^g_{i,t}$) minus the water distribution to domestic ($g^d_{i,t}$) and agricultural ($g^a_{i,t}$) uses and water lost ($v^g_{i,t}$). The precipitation information can be collected from weather stations in each location at any period t, while the runoff water can be determined using indirect methods such as the one reported by the National Council of Water of Mexico (CONAGUA) in the norm NOM-011-CNA-2000, which states that if no data is available to estimate the runoff water ($ROWV_{k,t}$), it can be determined as a function of the precipitation, runoff coefficient and area of collection as follows:

$$ROWV_{i,t} = P_t \cdot A_i^{ROW} \cdot Ce, \quad i \in I, \quad t \in T \tag{4.2}$$

the runoff coefficient Ce (a dimensionless number) depends on the annual precipitation and the type and use of terrain as follows:

$$Ce = \frac{K\left(P^{Total} - 250\right)}{2000}, \quad K \leq 0.15 \tag{4.3}$$

$$Ce = \frac{K\left(P^{Total} - 250\right)}{2000} + \frac{K - 0.15}{1.5}, \quad K \geq 0.15 \tag{4.4}$$

the parameter K is calculated as suggested in the above-mentioned norm. This parameter is a weighted average of the type and use of the terrain and P^{Total} is the total annual precipitation in mm H_2O.

The water collected from direct precipitation ($DPWV_{k,t}$) in the natural bodies is determined as follows:

$$DPWV_{i,t} = P_t \cdot A_i^{DPW}, \quad i \in I, \quad t \in T \tag{4.5}$$

A_i^{DPW} is the existing area for the natural sources. For modeling purposes, deep wells cannot be recharged from direct precipitation which means that their area for collection is equal to zero. Hence, the recharge of natural bodies is the sum of the runoff water and the direct recharge:

$$p^g_{i,t} = ROWV_{i,t} + DPWV_{i,t}, \quad i \in I, \quad t \in T \tag{4.6}$$

There might be several sources such as dams, springs, and wells, which can be sent either to the main for domestic distribution ($g^d_{i,t}$) or the main for agricultural use ($g^a_{i,t}$). In the main for domestic use, the water is treated so it can be used as tap water. The water for agricultural use might require a less demanding treatment or no treatment; nonetheless, it must be collected in the main for its distribution. The last term in the source balance equation involves the losses due to natural processes such as filtration and evaporation ($v^g_{i,t}$). This parameter is taken as 20% of the total collected water:

$$v_{i,t}^{g} = 0.2\left(\sum_{m\in M} r_{m,i,t} + p_{i,t}^{g}\right), \quad i \in I, \quad t \in T \tag{4.7}$$

4.3.2 Balance for Precipitation in Artificial Water Storages

The precipitation (P) is a season and location-dependent. In this case, the collected water depends directly on the area of collection, the runoff coefficient, and the precipitation as follows:

$$P = f(Area, Ce, Precipitation)$$

The roofs of the houses and buildings can be used as collectors; therefore, the total cost for water collectors consists of the rain gutters installation (although many buildings already have rain gutters) and the piping and pumping costs to send the water either to a storage tank or to an artificial pond. In non urbanized zones the total area for collection is considerably less than in urban areas; however, the water can be redirected to the storage sites increasing, at least artificially, the area for collection. Now, the collected water is split in water for storage tanks and water for artificial ponds. Storage tanks are installed in urbanized areas where the space for installation is limited, while the artificial ponds must be installed in locations with larger available areas; in this case, the runoff coefficient for roofs used as collectors ranges from 0.7 to 0.95 depending on the type of roof and material of construction (Farreny et al. (2011) compiled the values previously reported). In this work, a value of 0.8 is selected because this is a common value for any roof in general:

$$P_{l,t}^{s} = P_t \cdot A_l^{s} \cdot Ce^{s}, \quad l \in L, t \in T \tag{4.8}$$

$$P_{n,t}^{a} = P_t \cdot A_n^{a} \cdot Ce^{a}, \quad n \in N, \quad t \in T \tag{4.9}$$

$P_{l,t}^{s}$ represents the collected water in storage tank l at time t. $P_{n,t}^{a}$ corresponds to the collected water in artificial pond n at time t. A_l^{s} and A_n^{a} are the available areas of collection for each type of storage. After the total available precipitation is determined for each type of storage, the balance for the collected water becomes:

$$P_{l,t}^{s} = s_{l,t}^{in} + v_{l,t}^{s}, \quad l \in L, \quad t \in T \tag{4.10}$$

$$P_{n,t}^{a} = a_{n,t}^{in} + v_{n,t}^{a}, \quad n \in N, \quad t \in T \tag{4.11}$$

The total precipitation at each period (recovered by artificial means) should be sent to storage tanks ($s_{l,t}^{in}$) or to artificial ponds ($a_{n,t}^{in}$). If no storage capacity is available, so the remaining water can be discarded. The last term of these equations considers this possibility ($v_{l,t}^{s}$ and $v_{n,t}^{a}$). The difference between a storage tank and an artificial pond for this formulation will be explained later in the paper.

4.3.3 Balance in Storage Tanks

Several tanks can be installed if they are required. The total capacity of these tanks is limited; nevertheless, they can be installed near the consumption site, thus the piping and pumping costs are lower than the cost required for an artificial pond, and its total cost is lower. The balance in a storage tank at any given period depends on the inventory of that tank (i.e., the water from the previous period, $S_{l,t-1}$), the received water from precipitation ($s_{l,t}^{in}$), and the used water either in domestic sinks ($s_{l,j,t}^{out,d}$) or agricultural sinks ($s_{l,h,t}^{out,a}$) as follows:

$$S_{l,t} - S_{l,t-1} = s_{l,t}^{in} - \sum_{j \in J} s_{l,j,t}^{out,d} - \sum_{h \in H} s_{l,h,t}^{out,a}, \quad l \in L, \quad t \in T \tag{4.12}$$

4.3.4 Balance in Artificial Ponds

For artificial ponds, the balance is like the one required for the storage tanks. The stored water ($A_{n,t}$) depends on the previous inventory ($A_{n,t-1}$) and the inlets (from rain $a_{n,t}^{in}$) and outlets (domestic use $a_{n,j,t}^{out,d}$ or agricultural use $a_{n,h,t}^{out,a}$) of the site. The artificial ponds have a greater storage capacity; nevertheless, because of space limitations in the cities, they cannot be installed near the sinks. This limitation increases their overall cost. Consequently, the balance for an artificial pond is stated as follows:

$$A_{n,t} - A_{n,t-1} = a_{n,t}^{in} - \sum_{j \in J} a_{n,j,t}^{out,d} - \sum_{h \in H} a_{n,h,t}^{out,a}, \quad n \in N, \quad t \in T \tag{4.13}$$

4.3.5 Balance in Main for Domestic Use

The main sources must be treated in a centralized facility, which distributes the water into the domestic sinks as follows:

$$\sum_{i} g_{i,t}^{d} = \sum_{j} f_{j,t}, \quad t \in T \tag{4.14}$$

$g_{i,t}^{d}$ is the flow sent from natural source i at time t to the main for domestic use, and $f_{j,t}$ is the flow sent from the main for domestic use to domestic sink j. In usual conditions, this water is enough to satisfy the demands of the city, but as the demand increases it becomes necessary to find alternative sources of water resources.

4.3.6 Balance in Main for Agricultural Use

Likewise, the main for agriculture receives the water obtained from the sources ($g_{i,t}^{a}$) and distributes it to the final users ($r_{h,t}$) as follows:

$$\sum_{i \in I} g_{i,t}^{a} = \sum_{h \in H} r_{h,t}, \quad t \in T \tag{4.15}$$

$r_{h,t}$ is the flow sent from the agricultural main to agricultural sink h at period t.

4.3.7 Balance in Domestic Sinks

The domestic sinks can receive water from the main for domestic use, the storage tanks, and the artificial ponds. All the water received in each domestic sink should satisfy the demand for it during the period t, $D_{j,t}^{ds}$. The demand for each sink is a function of the population and the season of the year (the demand increases during the hottest months of the year and decreases during the coldest months). The balance for any domestic sink is stated as follows:

$$f_{j,t} + \sum_{l \in L} s_{l,j,t}^{out,d} + \sum_{n \in N} a_{n,j,t}^{out,d} = D_{j,t}^{ds}, \quad j \in J, \quad t \in T \tag{4.16}$$

The exit flow of the domestic sinks is divided into two terms; the first one ($cw_{j,t}^{d}$) is the consumption of water and the losses (i.e., water used for gardening, losses by leaks, water stored, evaporated, among others), and the second term ($int_{j,t}^{in}$) is the water that exits the domestic sinks as wastewater. This water should be sent to a waste treatment facility, where can be reconditioned for use in agriculture or directed to its final disposal:

$$D_{j,t}^{ds} = cw_{j,t}^{d} + int_{j,t}^{in}, \quad j \in J, \quad t \in T \tag{4.17}$$

The losses ($cw_{j,t}^{d}$) for simulation are assumed as 50% of the water received in each domestic sink.

4.3.8 Treatment Plant Balance

The treatment plant receives the wastewater from domestic sinks that can be split and reused for agricultural (int_t^{out}) purposes or final disposal. The second term (cw_t^{tp}) of the balance involves the flow of water not recovered during period t (i.e., water sent to final disposal):

$$\sum_j int_{j,t}^{in} = int_t^{out} + cw_t^{tp}, \quad t \in T \tag{4.18}$$

The treated water can be directed to any agricultural sink:

$$int_t^{out} = \sum_h int_{h,t}^{out,ag}, \quad t \in T \tag{4.19}$$

$int_{h,t}^{out,ag}$ represents the flow of water sent from the treatment facility to any agricultural sink during period t.

4.3.9 Balance in Agricultural Sinks

The most important sources for this use are the main for agriculture ($r_{h,t}$), the storage tanks ($s_{l,h,t}^{out,a}$), the artificial ponds ($a_{n,h,t}^{out,a}$), and the water recovered from the treatment plant ($int_{h,t}^{out,ag}$). All the flow received in the agricultural sinks is spent in the production process and must meet the demands ($D_{h,t}^{as}$):

$$r_{h,t} + \sum_{l \in L} s_{l,h,t}^{out,a} + \sum_{n \in N} a_{n,h,t}^{out,a} + int_{h,t}^{out,ag} = D_{h,t}^{as}, \quad h \in H, \quad t \in T \tag{4.20}$$

4.3.10 Disjunctions for Installing or Not Storage Tanks or an Artificial Pond

To choose whether to install or not a storage tank or an artificial pond in a given site, a set of disjunctions must be implemented, which are given in terms of a new set of variables (S_l^{max} and A_n^{max}) that express the maximum capacity of the storage devices. For these variables, the following relationships are needed:

$$S_l^{max} \geq S_{l,t}, \quad l \in L, \quad t \in T \tag{4.21}$$

$$A_n^{max} \geq A_{n,t}, \quad n \in N, \quad t \in T \tag{4.22}$$

$$S_l^{max} \geq s_{l,t}^{in}, \quad l \in L, \quad t \in T \tag{4.23}$$

$$A_n^{max} \geq a_{n,t}^{in}, \quad n \in N, \quad t \in T \tag{4.24}$$

The first two relationships indicate that the stored water at any period must be less than the maximum capacity of the storage device. The last two equations state that the devices cannot receive more water than the maximum capacity at any period.

The existence or not existence of storage tanks is modeled with the following disjunction: if storage is required at any period, then the binary variable z_l^s is equal to one, and the storage tank is installed in location l. The capacity of storage tanks must lie within maximum ($\delta_l^{s,\max}$) and minimum bounds ($\delta_l^{s,\min}$). Thus, the corresponding cost is considered ($Cost_l^s$). If the binary variable is equal to zero, then no tank is installed, and the cost is also zero.

$$\begin{bmatrix} z_l^s \\ S_l^{\max} \geq \delta_l^{s,\min} \\ S_l^{\max} \leq \delta_l^{s,\max} \\ Cost_l^s = A + B\left(S_l^{\max}\right)^{\alpha} \end{bmatrix} \vee \begin{bmatrix} z_l^s \\ S_l^{\max} = 0 \\ Cost_l^s = 0 \end{bmatrix}, \quad l \in L \tag{4.25}$$

A and B are constants used to account for the fixed and variable costs of the tank, and α is a positive exponent that is less than 1.0 to consider the economies of scale.

For artificial ponds, a disjunction is required to model the existence or not existence of these facilities as follows:

$$\begin{bmatrix} z_n^a \\ A_n^{max} \geq \delta_n^{a,min} \\ A_n^{max} \leq \delta_n^{a,max} \\ Cost_n^a = C + D\left(A_n^{max}\right)^{\alpha} \end{bmatrix} \vee \begin{bmatrix} z_n^a \\ A_n^{max} = 0 \\ Cost_n^a = 0 \end{bmatrix}, \quad n \in N \tag{4.26}$$

$\delta_n^{a,min}$ and $\delta_n^{a,max}$ are the maximum and minimum capacity of artificial ponds, C and D are constants for the fixed and variable costs of artificial ponds, and $Cost_n^a$ is the annualized cost of the artificial pond (included only if the binary variable z_n^a associated with location n is equal to one). The nonlinearities in previous disjunctions can be eliminated by linearizing the cost functions to obtain the following linear relationships:

$$Cost_l^s = A' + B' S_l^{max}, \quad l \in L \tag{4.27}$$

$$Cost_n^a = C' + D' A_n^{max}, \quad n \in N \tag{4.28}$$

A', B', C' and D' are constants of the linearized cost functions for the storage tanks and artificial ponds. The disjunctions (in terms of the linearized cost functions) can be algebraically reformulated using the Big-M reformulation (Veccietti et al. 2003) as follows:

$$S_l^{max} \geq \delta_l^{s,min} \cdot z_l^s, \quad l \in L \tag{4.29}$$

$$S_l^{max} \leq \delta_l^{s,max} \cdot z_l^s, \quad l \in L \tag{4.30}$$

$$Cost_l^s = A' \cdot z_l^s + B' \cdot S_l^{max}, \quad l \in L \tag{4.31}$$

$$A_n^{max} \geq \delta_n^{a,min} \cdot z_l^a, \quad n \in N \tag{4.32}$$

$$A_n^{max} \leq \delta_n^{a,max} \cdot z_n^a, \quad n \in N \tag{4.33}$$

$$Cost_n^a = C' \cdot z_n^a + D' \cdot A_n^{max}, \quad n \in N \tag{4.34}$$

The proposed locations for the artificial pounds are planned considering the smaller damage possible to the land, accounting for the existing use for the land and avoiding the damage to the environment.

4.3.11 Objective Function

The objective function considers the maximization of the gross profit for the sales of the water used in the city. The water is managed and distributed by the local government; thus, the revenue is used as part of the governmental budget for infrastructure development and maintenance. The implemented objective function considers the sales of water (*revenue*) for domestic and agricultural uses minus the cost associated with the treatment (*treatment cost*) and distribution of the water (*pp cost*), and the cost associated with the installation and operation of storage tanks and/or artificial ponds (*storage cost*). The objective function is stated as follows:

$$TAR = revenue - treatment\ cost - storage\ cost - pp\ cost \tag{4.35}$$

The sale price of the water is independent of the source; nonetheless, the treatment cost changes depending on the source and the use. The revenue can be obtained as follows:

$$revenue = \left(\sum_i \sum_t g_{i,t}^d + \sum_l \sum_j \sum_t s_{l,j,t}^{out,d} + \sum_n \sum_j \sum_t a_{n,j,t}^{out,d} \right) DSC$$
$$+ \left(\sum_i \sum_t g_{i,t}^a + \sum_l \sum_h \sum_t s_{l,h,t}^{out,a} + \sum_n \sum_h \sum_t a_{n,h,t}^{out,a} + \sum_t int_t^{out} \right) ASC \quad (4.36)$$

DSC represents the water sale price for domestic use and *ASC* corresponds to the water sale price for agricultural purposes.

Different sources have different qualities, and the different uses require diverse treatments. Therefore, depending on the source and use, the water has a different total treatment cost, which is given as follows:

$$treatment\ cost = \left(\sum_i \sum_t g_{i,t}^d CTND + \sum_i \sum_t g_{i,t}^a CTNA \right)$$
$$+ \left(\sum_l \sum_j \sum_t s_{l,j,t}^{out,d} + \sum_n \sum_j \sum_t a_{n,j,t}^{out,d} \right) CTAD$$
$$+ \left(\sum_l \sum_h \sum_t s_{l,h,t}^{out,a} + \sum_n \sum_h \sum_t a_{n,h,t}^{out,a} \right) CTAA$$
$$+ \sum_t int_t^{out} CTPA + \sum_t cw_t^{tp} CTPE \quad (4.37)$$

CTND represents the cost of natural sources for domestic use, *CTNA* the cost of natural sources for agricultural use, *CTAD* and *CTAA* correspond to the treatment costs of rainwater for domestic and agricultural uses, respectively, *CTPA* represents the treatment cost for regeneration of wastewater for agricultural use and *CTPE* the wastewater treatment cost for final disposal into the environment.

The storage cost is obtained from the sum of the annualized costs of artificial storage devices ($Cost_l^s$ and $Cost_n^a$ for ponds and tanks, respectively):

$$storage\ cost = \left(\sum_l Cost_l^s + \sum_n Cost_n^a \right) \quad (4.38)$$

The piping and pumping cost (*pp cost*) for the transportation of water from one location to another is given in terms of the following parameters. $PCSTD_{l,j}$ represents the piping and pumping unit cost from storage tank l to domestic sink j, $PCASD_{n,j}$ the piping and pumping unit cost from artificial pond n to domestic sink j, $PCSTA_{l,h}$ corresponds to the piping and pumping unit cost from storage tank l to agricultural sink h, $PCASA_{n,h}$ the piping and pumping unit cost from artificial pond n to agricultural sink h, $PCND_i$ and $PCNA_i$ are the piping and pumping unit costs from natural source i to domestic and agricultural mains, respectively, and lastly $PCTW_h$ represents the piping and pumping unit cost from treatment plant to agricultural sinks h.

So, the total piping and pumping cost is determined as follows:

$$pp\,cost = \sum_l \sum_j \sum_t s_{l,j,t}^{out,d} PCSTD_{l,j} + \sum_n \sum_j \sum_t a_{n,j,t}^{out,d} PCASD_{n,j}$$
$$+ \sum_l \sum_h \sum_t s_{l,h,t}^{out,a} PCSTA_{l,h} + \sum_n \sum_h \sum_t a_{n,h,t}^{out,a} PCASA_{n,h}$$
$$+ \sum_i \sum_t g_{i,t}^{d} PCND_i + \sum_i \sum_t g_{i,t}^{a} PCNA_i + \sum_h \sum_t int_{h,t}^{out,ag} PCTW_h \tag{4.39}$$

The piping and pump costs consider the capital and operating costs, which are functions of the flowrate, which include the costs associated with the new pipelines and the pumps; while the operating costs include mainly the cost for the electricity required to run the associated pump.

Lastly, the proposed model formulation is a mixed integer linear programming problem, where the global optimal solution is guaranteed.

4.4 Case Study

The selected case study is the water distribution in the city of Morelia, México. Morelia is the capital city of the state of Michoacán. One of the main concerns of the current municipal government that administrates the water supply is that the level of the city aquifer has an annual deficit of water of 2.46 m (Avila-Oliveira and Garduño-Monroy 2007). This is mainly caused by the irrational exploitation of the aquifer and the lack of alternative sources of water. Morelia has a population of 729,279 inhabitants (INEGI 2010a). According to data from CONAGUA (National Council of Water-Mexico), the water consumption in Morelia for public uses is 90,168,374.77 m^3/year (CONAGUA 2011), which represents a daily consumption per capita of 338.75 lcd. This is an average consumption, but it is acknowledged that the water consumption is dependent on the socioeconomic stratum (Morales-Pinzón et al. 2012). For simulation, the average daily consumption per capita is used. The agricultural activity in Morelia is one of the lowest in the state; nonetheless, 21,348,208.28 m^3/year are required for this activity in the city. This gives a total annual consumption for public and agricultural uses of 111,516,583.05 m^3/year, which is met by 105 deep wells (43.93%), springs (33.41%) and the Cointzio dam (22.66%) (Avila-Olivera and Garduño-Monroy 2007). The problem consists in determining the optimal schedule for the distribution of water to meet the domestic and agricultural demands in the city during a year, maintaining a sustainable level of water in the natural sources, for which the following conditions must be satisfied:

- The level of the springs and dam should not be less than 35% of their maximum capacity at any period, and at the end of the planning period, it must not be less than 45% of their maximum capacity.
- The level of the deep wells should not be less than 80% of their maximum capacity at any period.
- The amount of water used from each source should not exceed 20% of its content at any period.

For simplicity, the city is divided into five sectors: Center (CE), North East (NE), North West (NW), South East (SE), and South West (SW) (Fig. 4.2) considering data from INEGI (2010b).

Figure 4.3 (INEGI 2010b) presents the water bodies that feed the city. The four most important water bodies are the springs La Mintzita, La Higuera, El Salto, and San Miguel. For modeling purposes, the main springs are grouped and treated as one source. The 105 deep wells are clustered in 10 groups according to their location, and the Cointzio dam is treated as a separate source. The dam and the springs have 50% of their capacity as initial inventory and the deep wells have their maximum current capacity.

The water demands in each of the five zones according to their population and season of the year are presented in Table 4.1, where the water consumption is higher during the hottest months. In the city, 17,668 ha are used for agricultural purposes,

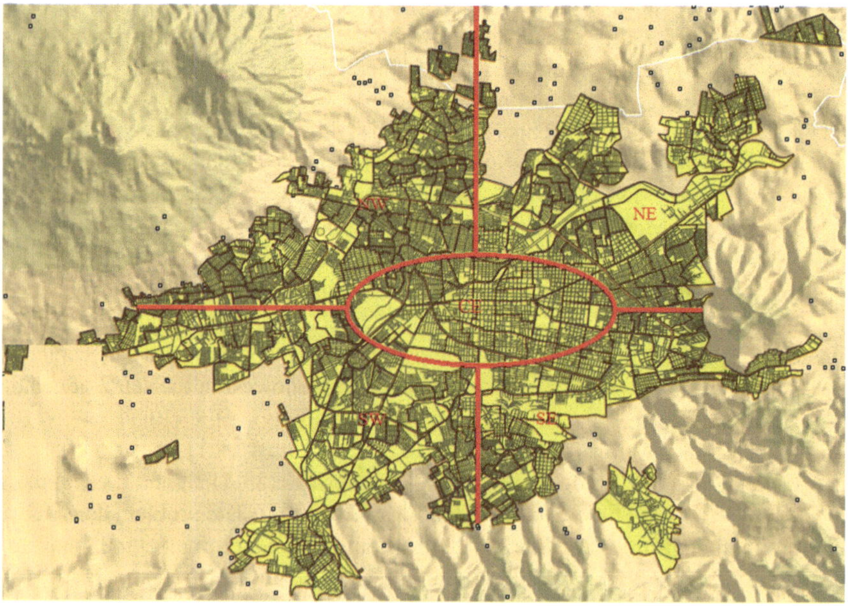

Fig. 4.2 Considered division of Morelia city. Printed with permission of "*Sustainable water management for macroscopic systems*", *Journal of Cleaner Production*, Copyright (2013), Elsevier

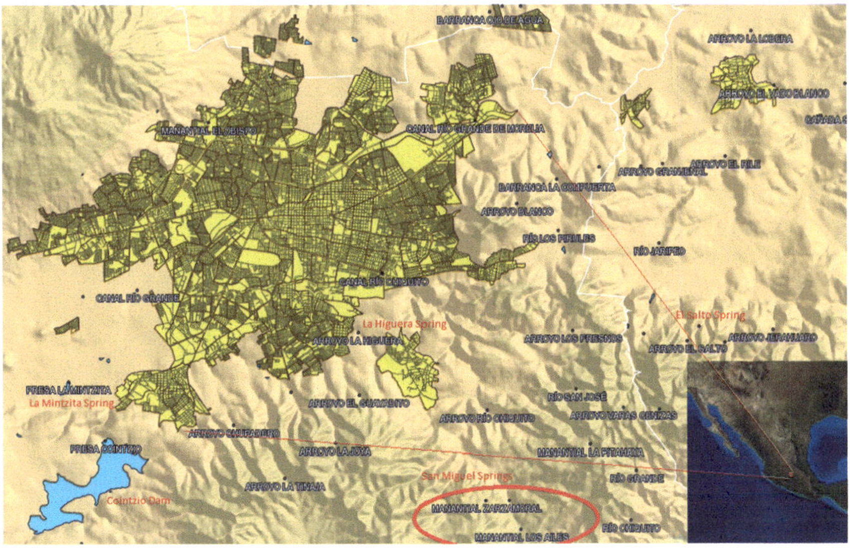

Fig. 4.3 Water sources for the city of Morelia. Printed with permission of "*Sustainable water management for macroscopic systems*", *Journal of Cleaner Production*, Copyright (2013), Elsevier

from which 1251 ha are irrigation lands, and the rest are seasonal (INEGI 2010a); the water consumption for agricultural purposes is presented in Table 4.2.

There are different sale rates for the water depending on the socioeconomic zone of the city and level of consumption; nonetheless, because the socioeconomic zones are mixed, the most common rate will be used ($13.86 MXN per cubic meter). This is equivalent approximately to $1 USD per cubic meter for domestic purposes. For agricultural uses the cost is $0.5 USD per cubic meter. These costs are independent of the source of water. Whereas the treatment costs depend not only on the final use but also on the source of water. In this case, the costs are $0.5 USD for domestic use, $0.3 USD for agricultural purposes and $0.1 for wastewater. The cost for harvested rainwater treatment is $0.4 USD for domestic use, and $0.2 USD for agricultural use. All the costs were taken from the 2012 Official Bulletin of the State of Michoacan, except for the treatment cost of rainwater. This is because currently this source of water is not considered in the OOAPAS legislation. The OOAPAS is the local branch of the National Council of Water, which is the national administrator of the water resources in México. Hence, the rainwater treatment costs were assumed to be $0.1 USD less than the cost required for treatment of natural sources; this cost is mainly associated to the maintenance of the collection and the treatment of water for its final use.

20 tanks can be installed in the city: tanks 1–6 can be installed in section CE, tanks 7–10 in NW, 11–14 in NE, tanks 15–17 in SW, and storage tanks 18–20 can be installed in section SE. Also, 6 artificial ponds can be constructed; due to space limitations these ponds can only be built outside the city: ponds 1–2 in NE, ponds

Table 4.1 Seasonal water demand in domestic sinks (m^3/month)

	January	February	March	April	May	June	July	August	September	October	November	December
CE	2,245,950	2,093,000	2,388,550	2,346,000	2,566,800	2,484,000	2,531,150	2,495,500	2,380,500	2,352,900	2,242,500	2,245,950
NW	1,367,100	1,274,000	1,453,900	1,428,000	1,562,400	1,512,000	1,540,700	1,519,000	1,449,000	1,432,200	1,365,000	1,367,100
NE	976,500	910,000	1,038,500	1,020,000	1,116,000	1,080,000	1,100,500	1,085,000	1,035,000	1,023,000	975,000	976,500
SW	1,269,450	1,183,000	1,350,050	1,326,000	1,450,800	1,404,000	1,430,650	1,410,500	1,345,500	1,329,900	1,267,500	1,269,450
SE	1,269,450	1,183,000	1,350,050	1,326,000	1,450,800	1,404,000	1,430,650	1,410,500	1,345,500	1,329,900	1,267,500	1,269,450

Table 4.2 Seasonal water demand in agricultural sinks (m^3/month)

	January	February	March	April	May	June	July	August	September	October	November	December
NW	917,568	1,060,938	946,242	900,363.6	843,015.6	286,740	189,248.4	172,044	344,088	659,502	1,032,264	1,066,673
NE	458,784	530,469	473,121	450,181.8	421,507.8	143,370	94,624.2	86,022	172,044	329,751	516,132	533,336.4
SW	688,176	795,703.5	709,681.5	675,272.7	632,261.7	215,055	141,936.3	129,033	258,066	494,626.5	774,198	800,004.6
SE	229,392	265,234.5	236,560.5	225,090.9	210,753.9	71,685	47,312.1	43,011	86,022	164,875.5	258,066	266,668.2

3–4 in NW, pond 5 in SE, and pond 6 in SW. It is assumed that there is no difference between the considered tanks and ponds, so all the storage devices will be constructed if storage capacity is required. If not all the storage devices are needed, the selection of the device to be constructed is arbitrary (i.e., there is no difference in constructing storage tanks CE 1 or CE 2). This might lead to degeneracy in the model; to avoid it, different types of devices can be considered (for example, different storage capacities and costs for each tank and artificial pond available for each section of the city) or making a distinction between the possible locations within a specific section of the city.

In this case, only the transportation cost of the water is considered because the annualized capital cost of pipelines and pumps is overweighed by the operation cost (this is considering that the current piping network can be used, and the only new piping required is from storage devices to existing piping network); this is because of the high local cost of energy. However, the formulation allows to consider these costs if they are significant. The transportation unit costs to send water from a tank to any section or from a pond to any section are given in Tables 4.3, 4.4, 4.5, and 4.6.

Table 4.3 Piping and pumping costs from storage tanks to domestic sinks ($\$ \times 10^2/m^3$)

ST/DS	CE	NW	NE	SW	SE
1	1	2	2	2	2
2	1	2	2	2	2
3	1	2	2	2	2
4	1	2	2	2	2
5	1	2	2	2	2
6	1	2	2	2	2
7	2	1	2	2	4
8	2	1	2	2	4
9	2	1	2	2	4
10	2	1	2	2	4
11	2	2	1	4	2
12	2	2	1	4	2
13	2	2	1	4	2
14	2	2	1	4	2
15	2	2	4	1	2
16	2	2	4	1	2
17	2	2	4	1	2
18	2	4	2	2	1
19	2	4	2	2	1
20	2	4	2	2	1

Table 4.4 Piping and pumping costs from storage tanks to agricultural sinks ($\$ \times 10^2/m^3$)

ST/AS	NW	NE	SW	SE
1	2.5	2.5	2.5	2.5
2	2.5	2.5	2.5	2.5
3	2.5	2.5	2.5	2.5
4	2.5	2.5	2.5	2.5
5	2.5	2.5	2.5	2.5
6	2.5	2.5	2.5	2.5
7	1	2.5	2.5	5
8	1	2.5	2.5	5
9	1	2.5	2.5	5
10	1	2.5	2.5	5
11	2.5	1	5	2.5
12	2.5	1	5	2.5
13	2.5	1	5	2.5
14	2.5	1	5	2.5
15	2.5	5	1	2.5
16	2.5	5	1	2.5
17	2.5	5	1	2.5
18	5	2.5	2.5	1
19	5	2.5	2.5	1
20	5	2.5	2.5	1

Table 4.5 Piping and pumping costs from artificial ponds to domestic sinks ($\$ \times 10^2/m^3$)

AP/DS	CE	NW	NE	SW	SE
1	2.3	2.3	1	4.6	2.3
2	2.3	2.3	1	4.6	2.3
3	2.3	1	2.3	2.3	4.6
4	2.3	1	2.3	2.3	4.6
5	2.3	4.6	2.3	2.3	1
6	2.3	2.3	4.6	1	2.3

Table 4.6 Piping and pumping cost from artificial pond to agricultural sink ($\$ \times 10^2/m^3$)

AP/AS	NW	NE	SW	SE
1	2.5	1	5	2.5
2	2.5	1	5	2.5
3	1	2.5	2.5	5
4	1	2.5	2.5	5
5	5	2.5	2.5	1
6	2.5	5	1	2.5

Printed with permission of "*Sustainable water management for macroscopic systems*", *Journal of Cleaner Production*, Copyright (2013), Elsevier

Table 4.7 Average precipitation for the city of Morelia (1951–2010)

	Precipitation in Morelia (1951–2010) mm H_2O
January	15.6
February	7.7
March	8.6
April	10.1
May	41.7
June	150.7
July	167.5
August	170.4
September	129.9
October	52.8
November	10
December	3.9
Annual	768.9

Printed with permission of "*Sustainable water management for macroscopic systems*", *Journal of Cleaner Production*, Copyright (2013), Elsevier

The average monthly precipitation in Morelia for the period 1951–2010 according to the National Meteorological System (SMN 2010) is presented in Table 4.7. To calculate the runoff coefficient for the city of Morelia, K is taken as 0.263, which represents the distribution of land presented by CONAGUA (2009). For this value of K and the average annual precipitation, the runoff coefficient is given as follows:

$$Ce = \frac{0.263(768.9 - 250)}{2000} + \frac{0.263 - 0.15}{1.5} = 0.1435$$

The maximum capacity and cost functions of the storage tanks and ponds are presented in Table 4.8. The cost function involves fixed and variable costs for units and the exponent 0.8 considers the economies of scale. To avoid the nonlinear terms,

Table 4.8 Capacity and cost functions of storage tanks and artificial ponds

	Maximum capacity (m^3)	Cost function
Storage tank	50,000	$20{,}000 + 17.5\left(S_l^{\max}\right)^{0.8}$
Artificial pond	600,000	$50{,}000 + 14.0\left(A_n^{\max}\right)^{0.8}$

Printed with permission of "*Sustainable water management for macroscopic systems*", *Journal of Cleaner Production*, Copyright (2013), Elsevier

the cost functions are linearized as follows:

$$Cost_l^s = 13{,}080 + 1.8135\left(S_l^{\max}\right), \quad l \in L$$
$$Cost_n^a = 111{,}968 + 0.8895\left(A_n^{\max}\right), \quad n \in N$$

In both cases, the adjustment is satisfactory with a correlation factor r^2 of 0.998 and 0.997, respectively.

4.5 Results

The presented mathematical model was coded in the software GAMS (Brooke et al. 2012), which involves 2047 constraints, and 26 and 4685 binary and continuous variables, respectively, and it is a mixed integer linear programming problem (MILP) that was solved with the solver CPLEX, using an absolute and relative gap of 1e−5 on a computer with an AMD Athlon 7550 Dual Core Processor 2.5 GHz with 3 Gb of RAM. The solution for each scenario was obtained in an average of 0.109 CPU s, and the global optimal solution is guaranteed because the model is linear.

Four scenarios were analyzed. Scenario A is the base case (i.e., no rainwater or reclaimed water can be used, only natural sources) and corresponds to the current operation. Scenario B allows the use of rainwater but no reclaimed water can be used. In Scenario C no rainwater can be used but reclaimed water can be used. Ultimately, in Scenario D, both rainwater and reclaimed water can be used. The results for these four scenarios are described as follows.

4.5.1 Scenario A

The solution for this scenario provides a total revenue of $24,617,870/year with a total waste generation of 45,026,400 m^3. The total demand is met using exclusively natural sources. The final balance for the natural sources indicates that deep wells provide 58% of the annual demand, 23.8% is met with water from springs, and the rest (18.3%) from the Cointzio dam. This distribution satisfies the sustainability constraints imposed on the level of the water resources.

4.5.2 *Scenario B*

The optimal solution for this case has a total revenue of $24,692,198/year; here there is not much difference in the total revenue concerning Scenario A. Nonetheless, 8.6% of the water used during the year comes from rainwater, which increases the natural sources inventory. To accomplish this task, six storage tanks and four artificial ponds are needed. The location of each storage tank and artificial pond is presented in Fig. 4.4. The capacities for these devices are (ST CE-3) 28,896 m^3, (ST CE-4) 32,336 m^3, (ST NW-8) 42,656 m^3, (ST NE-13) 33,712 m^3, (ST NE-14) 33,272 m^3 and (ST SE-18) 46,096 m^3 for the storage tanks and (AP NE-1) 240,770 m^3, (AP NW-4) 266,110 m^3, (AP SE-5) 244,990 m^3 and (AP SW-6) 295,680 m^3 for the artificial ponds, respectively. Three storage tanks are installed in the northern part of the city, while only one is installed in the southern part. Nonetheless, the biggest artificial pond is in the SW section. The agricultural activity is mainly concentrated in the western part because the natural water resources (the Cointzio dam and La Mintzita spring) are in SW; despite this, the biggest ponds are in NW and SW. Though most of the water in storage tanks and artificial ponds is used for domestic demands, during January artificial ponds 4 and 6 provide 165,984 m^3 for agricultural demands. Figure 4.4 presents a detailed distribution schedule for August. All the available storage devices are recharged at full capacity during this month (maximum precipitation); in the case of artificial ponds 5 and 6, the water is stored and used during the next months. The water in artificial ponds 1 and 4 is used exclusively for domestic purposes, while the water collected in storage tanks is used both for domestic and agricultural uses. However, in this month most of the water for agricultural use is given by natural sources. This planning schedule helps maintain the natural water resources in desirable conditions during the year.

4.5.3 *Scenario C*

The total profit for the solution of this scenario is $26,301,608/year, which represents an increment of 6.5% in the annual revenue concerning Scenario A. It also represents a reduction of 14.2% in the use of fresh sources, which makes it an attractive alternative. Additionally, the total waste is reduced from 45,026,400 to 29,295,840 m^3/year, which is an additional incentive to reuse water in the macroscopic system. Figure 4.5 presents the water distribution in July, which is the month with higher water demand in domestic sinks. Although this month has the highest domestic demand because the highest historic temperatures are registered during this month, it also has one of the highest rates of precipitation, which reduces the requirements

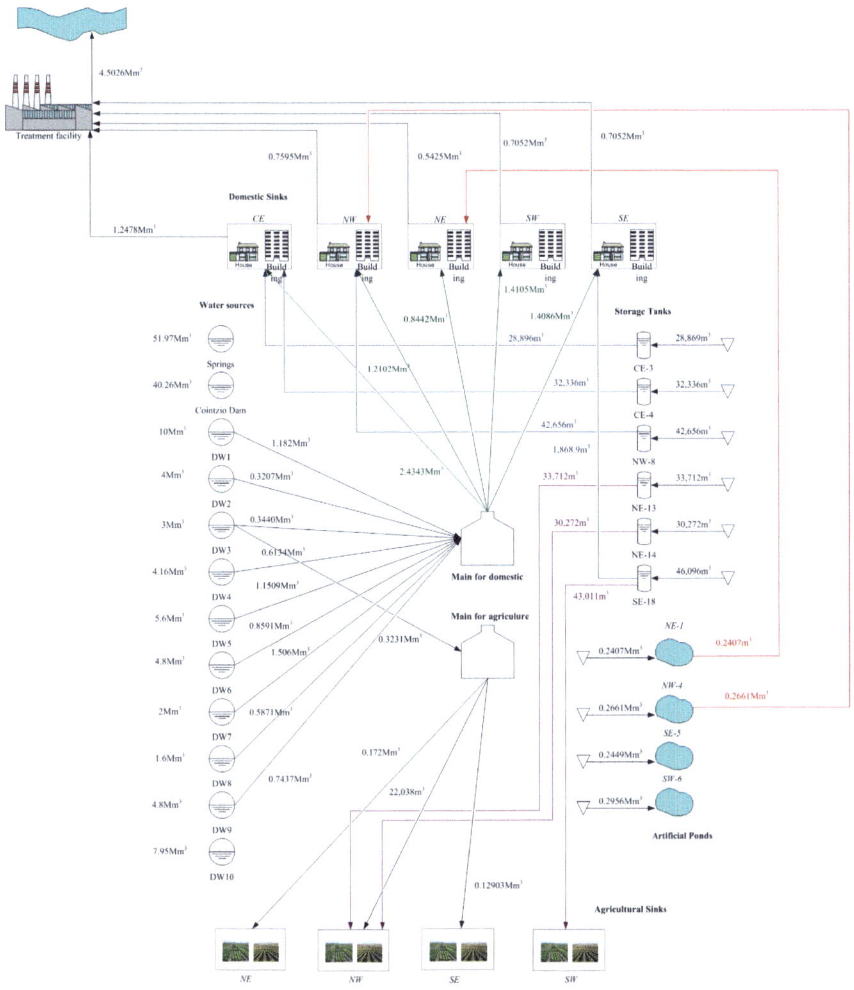

Fig. 4.4 Water distribution in August for Scenario B. Printed with permission of "*Sustainable water management for macroscopic systems*", *Journal of Cleaner Production*, Copyright (2013), Elsevier

for irrigation of agricultural sinks. The total demand for irrigation is 0.4731 Mm^3, from which 0.1892 Mm^3 are provided from treated wastewater; this means that 40% of the demand for irrigation can be obtained from reclaimed water.

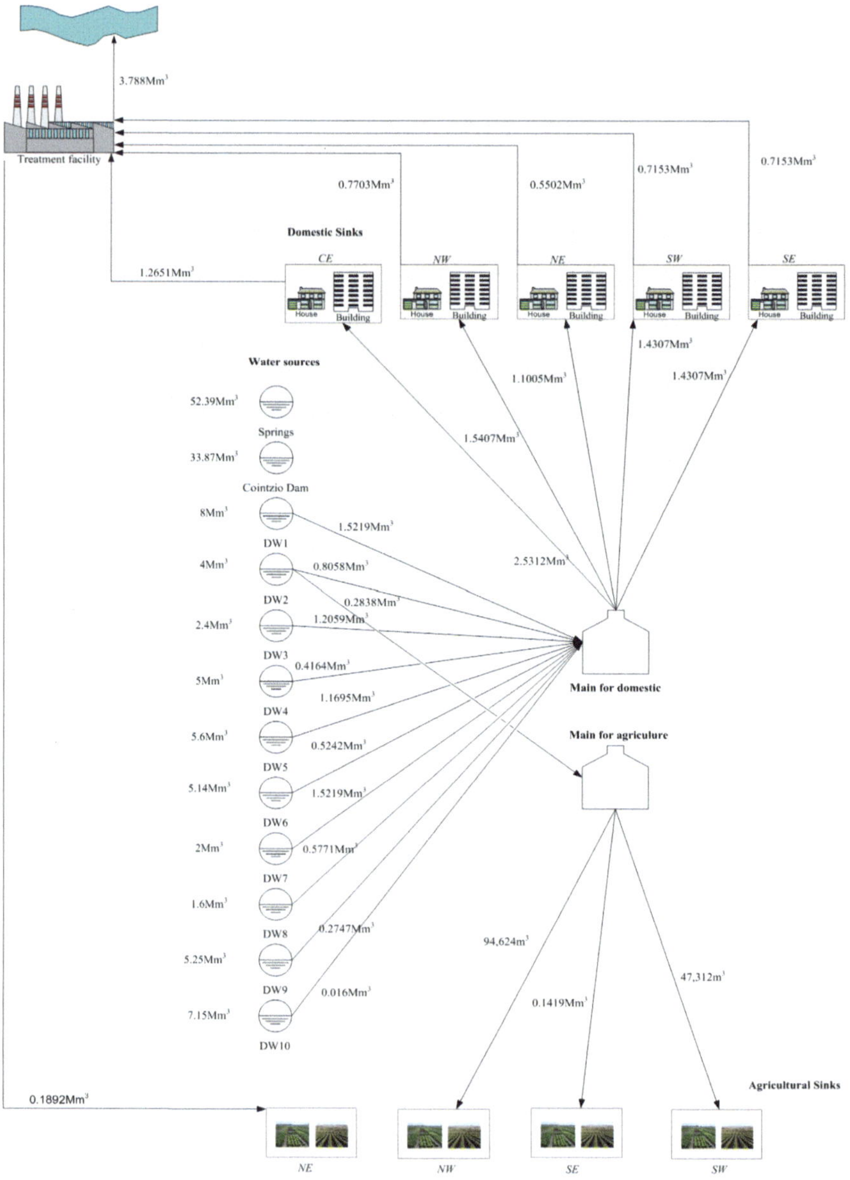

Fig. 4.5 Water distribution in July for Scenario C. Printed with permission of "*Sustainable water management for macroscopic systems*", *Journal of Cleaner Production*, Copyright (2013), Elsevier

4.5.4 Scenario D

The profit for the solution of this case is $26,375,936/year; the reduction in the use of natural sources is 17.1%. Harvested rainwater provides 8.6% of the annual demand, however treated water accounts for 8.5% of the total demand. The springs provide 18.4%, the Cointzio dam 11.2%, and the deep wells provide 53.3%. The location and capacity of the storage tanks and artificial ponds are the same as in Scenario B. Figures 4.6 and 4.7 present examples of the distribution of water for two months in this scenario, whereas Fig. 4.6 presents the distribution in the month with maximum precipitation (August), while Fig. 4.7 shows the distribution in the month with the lowest precipitation (December). In August, the rainwater used was 1,069,889 m^3 for domestic and agricultural uses, which represents 12.8% of the water required during that month. In December only 1.78% is provided by rainwater and 16% is obtained from reclaimed water from the treatment plant. During July, artificial pond NE-1 collects its maximum capacity of water (240,700 m^3) and it uses 191,450 m^3, which represents an inventory of 49,321 m^3 that was used in August. During August, the water taken from this pond was 49,321 m^3 and again it receives an amount of water that is equal to its maximum capacity. Thus, this pond has a reservoir of water, which was used in September and October together with the water received in these two months.

As can be seen from the four scenarios, the systematic management, storage, and distribution of water in macroscopic systems yields sustainable solutions, maintaining acceptable levels in the natural water bodies even in cases in which no alternative water sources can be used (Scenario A). Nevertheless, the use of alternative water sources further improves the sustainability of the overall process reducing, at the same time, the waste generation, and the natural bodies depletion. Soon, these reasons should overweight the economic aspect in the decision-making process and become the basis for governmental policies aiming for the conservation of natural resources. Lastly, the maximum volume of the artificial ponds is 500,000 m^3, which can be constructed in small terrains; thus, the environmental impact for the change in use of land might not be important.

4.6 Concluding Remarks

A model has been created to represent the best way to manage water in macroscopic systems. The model can be used to discover the best water-management techniques globally since it is a mixed integer linear issue. To facilitate the creation of scheduling solutions, a multi-period approach is utilized to represent the variations in water sources and needs throughout the year. For macroscopic systems, the presented methodology aids in creating the best possible scenarios for inventory control and water distribution. The case study's findings demonstrate that using alternate water sources is encouraged by sustainability and economic factors. The answers found

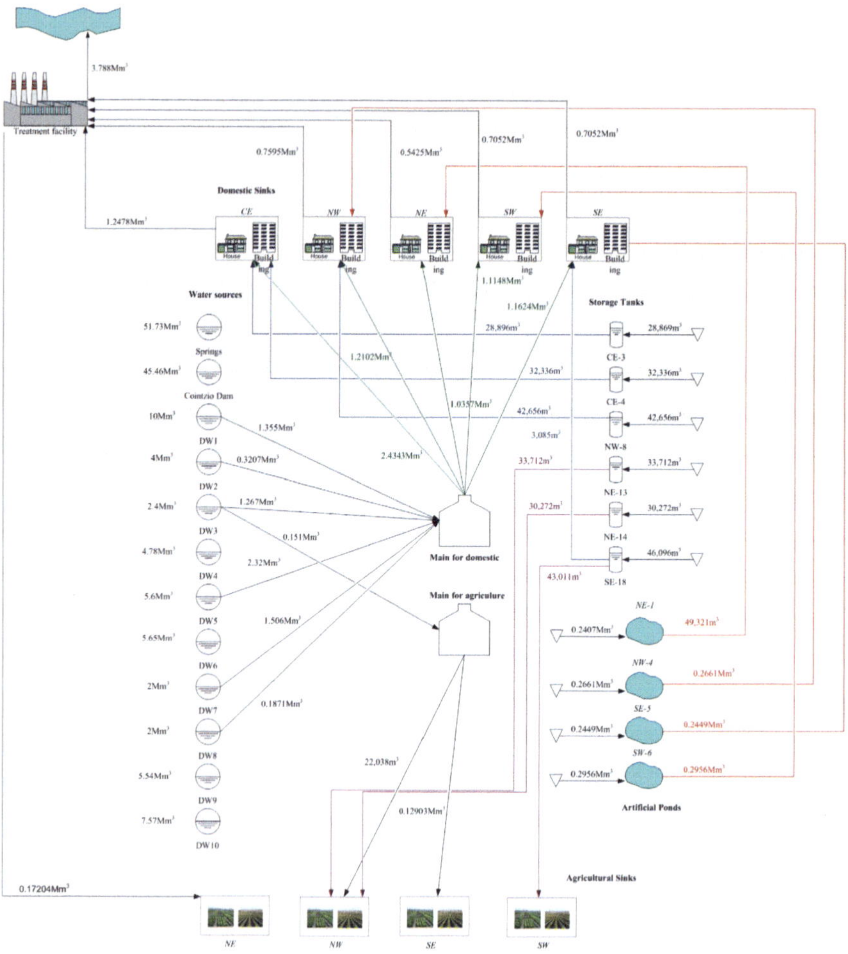

Fig. 4.6 Water distribution in August for Scenario D. Printed with permission of "*Sustainable water management for macroscopic systems*", *Journal of Cleaner Production*, Copyright (2013), Elsevier

for the Morelia City case study show that using alternate water sources reduces overuse of natural resources while producing revenues that are as least as high as or higher than those gained without using alternative sources. The water aquifers' sustainability is further increased by the remedies found.

The broad concept applies to all cities, even those with little precipitation. Rainwater collection would not be feasible in this situation; hence no storage structures will be built. Reclaimed water would also be the option in this case in place of natural resources. Furthermore, the primary benefit of utilizing this model is that the distribution of natural resources would be ideal (as in scenarios A, B, and C) even in the absence of one or more alternate water sources. The presented method can

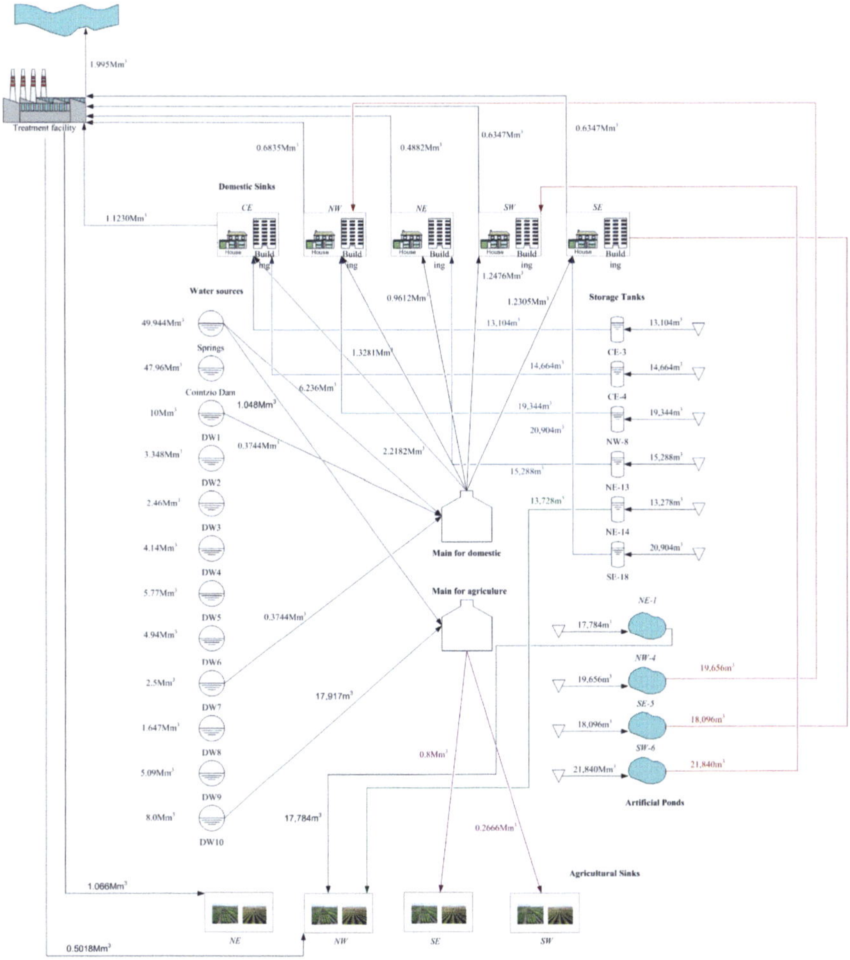

Fig. 4.7 Water distribution in December for Scenario D. Printed with permission of "*Sustainable water management for macroscopic systems*", *Journal of Cleaner Production*, Copyright (2013), Elsevier

also be used in urban areas where domestic demand is less than agricultural demand. To generate long-term sustainable solutions, it is vital to incorporate population increase and, consequently, water demand growth in a multi-annual scenario, even though the model can be applied to any city that has the relevant data. To enhance the outcomes, future research should incorporate into the model other features such as the system's dynamics in a shorter amount of time, various recycling/reusing scenarios for wastewater, the usage of desalinated seawater, and effective water-use methods in each home.

Nomenclatures

Parameters

A_n^{max}	Maximum capacity of artificial pond *n*
A_n^a	Available area of collection for artificial ponds in location *n*
A_l^s	Available area of collection for storage tanks in location *l*
A_i^{DPW}	Area of collection for source *k* for direct precipitation
A_i^{ROW}	Area of collection for source *k* for runoff water
ASC	Water sale cost for agricultural use
Ce^s	Runoff coefficient for storage tanks catchment areas
Ce^a	Runoff coefficient for artificial ponds catchment areas
$CTAA$	Treatment cost of rainwater for agricultural use
$CTAD$	Treatment cost of rainwater for domestic use
$CTNA$	Treatment cost of natural sources for agricultural use
$CTND$	Treatment cost of natural sources for domestic use
$CTPA$	Treatment cost of regenerating wastewater for agricultural use
$CTPE$	Treatment cost of wastewater before final disposal
$D_{j,t}^{ds}$	Water demand in domestic sink *j* at time *t*
$D_{h,t}^{as}$	Water demand in agricultural sink *h* at time *t*
DSC	Water sale cost for domestic use
$DPWV_{i,t}$	Water collected from direct precipitation in source *k* at time *t*
K	Pondered average of the type and use of terrain
P_t	Precipitation in time *t*
P^{total}	Annual precipitation
$p_{i,t}^g$	Water collected from direct precipitation and runoff water in source *k* at time *t*
$PCASA_{n,h}$	Piping and pumping cost from artificial pond *n* to agricultural sink *h*
$PCASD_{n,j}$	Piping and pumping cost from artificial pond *n* to domestic sink *j*
$PCNA_i$	Piping and pumping cost from natural source *k* to agricultural main
$PCND_i$	Piping and pumping cost from natural source *k* to domestic main
$PCSTA_{l,h}$	Piping and pumping cost from storage tank *l* to agricultural sink *h*
$PCSTD_{l,j}$	Piping and pumping cost from storage tank *l* to domestic sink *j*
$PCTW_h$	Piping and pumping cost from the treatment plant to the agricultural sink *h*
$r_{m,i,t}$	Water recharged from tributary *m* to source *k* at time *t*
S_l^{max}	Maximum capacity of storage tank *l*
$ROWV_{i,t}$	Runoff water collected in source *k* at time *t*

Binary Variables

z_l^s	Binary variable used to select the existence of a storage tank in location l
z_n^a	Binary variable used to select the existence of an artificial pond in location n

Variables

$A_{n,t}$	Water in an artificial pond n at time t
$a_{n,t}^{in}$	Inlet flowrate to artificial pond n at time t
$a_{n,j,t}^{out,d}$	Water sent from artificial pond n to the domestic sink j at time t
$a_{n,h,t}^{out,a}$	Water sent from the artificial pond n to the agricultural sink h at time t
Ce	Runoff coefficient
$Cost_n^a$	Cost of artificial pond n
$Cost_l^s$	Cost of storage tank l
$cw_{j,t}^d$	Water used or lost in domestic sink j at time t
cw_t^{tp}	Water sent to final disposal at time t from treatment plant
$f_{j,t}$	Water sent from domestic main to domestic sink j at time t
$g_{i,t}^a$	Water sent from source k to agricultural main at time t
$g_{i,t}^d$	Water sent from source k to domestic main at time t
$G_{i,t}$	Water in source k at time t
$int_{j,t}^{in}$	Water sent from domestic sink j to treatment facility at time t
int_t^{out}	Water recovered for agricultural use in treatment plant at time t
$int_{h,t}^{out,ag}$	Water sent from treatment plant to agricultural sink h at time t
$P_{n,t}^a$	Available precipitation for artificial ponds in location n at time t
$P_{l,t}^s$	Available precipitation for storage tanks in location l at time t
$r_{h,t}$	Water sent from agricultural main to agricultural sink h at time t
$S_{l,t}$	Water in storage tank l at time t
$s_{l,t}^{in}$	Inlet flowrate to storage tank l at time t
$s_{l,j,t}^{out,d}$	Water sent from storage tank l to domestic sink j at time t
$s_{l,h,t}^{out,a}$	Water sent from storage tank l to agricultural sink h at time t
TAR	Total annual revenue
$v_{n,t}^a$	Not collected water in artificial pond n at time t
$v_{l,t}^s$	Not collected water in storage tank l at time t
$v_{i,t}^g$	Water lost due to natural processes in source k at time t

Greek Symbols

α	Exponent to take into account the economies of scale
$\delta_l^{s,\min}$	Minimum capacity of storage tank l
$\delta_l^{s,\max}$	Maximum capacity of storage tank l
$\delta_n^{a,min}$	Minimum capacity of artificial pond n
$\delta_n^{a,max}$	Maximum capacity of artificial pond n

Sets

H	Set for agricultural sinks $(h\|h = 1, \ldots, H)$
I	Set for natural sources $(i\|i = 1, \ldots, I)$
J	Set for domestic sinks $(j\|j = 1, \ldots, J)$
L	Set for location of storage tanks $(l\|l = 1, \ldots, L)$
M	Set for tributaries $(m\|m = 1, \ldots, M)$
N	Set for location of artificial ponds $(n\|n = 1, \ldots, N)$
T	Set for time period $(t\|t = 1, \ldots, T)$

References

Abdulla FA, Al-Shareef AW (2009) Roof rainwater harvesting systems for household water supply in Jordan. Desalination 243(1–3):195–207

Appan A (1999) A dual-mode system for harnessing roofwater for non-potable uses. Urban Water 1(4):317–321

Atilhan S, Mahfouz A, Batchelor B, Linke P, Abdel-Wahab A, Nápoles-Rivera F, Jiménez-Gutiérrez A, El-Halwagi MM (2012) A systems-integration approach to the optimization of macroscopic water desalination and distribution networks: a general framework applied to Qatar's water resources. Clean Technol Environ Policy 14(2):161–171

Avila-Olivera JA, Garduño-Monroy VH (2007) Analysis for the decrement of the available water in the aquifers of Morelia. Ciencia Nicolaita 46:105–122

Brooke A, Kendrick D, Meeraus A, Raman R (2012) GAMS a user's guide. GAMS Development Corporation, Washington, DC

Burgara-Montero O, Ponce- Ortega JM, Serna-González M, El-Halwagi MM (2012) Optimal design of distributed treatment systems for the effluents discharged to the rivers. Clean Technol Environ Policy 14(5):925–942

Chew IML, Tan R, Ng DKS, Foo DCY, Majozi T, Gouws J (2008) Synthesis of direct and indirect interplant water network. Ind Eng Chem Res 47(23):9485–9496

Chilton JC, Maidment GG, Marriott D, Francis A, Tobias G (1999) Case study of a rainwater recovery system in a commercial building with a large roof. Urban Water 1(4):345–354

CONAGUA (2009) National Council of Water. Determination of the availability of water in the Morelia-Querendaro aquifer (1602) state of Michoacán

CONAGUA (2011) National Council of Water. Statistic of the water in México, chap 3 appendix. Available at http://www.conagua.gob.mx/CONAGUA07/Contenido/Documentos/SINA/Capitulo3Usosdelagua.zip. Accessed Aug 2012

Dakwala M, Mohanty B, Bhargava R (2009) A process integration approach to industrial water conservation: a case study for an Indian starch industry. J Clean Prod 17(18):1654–1662

Domènech L, Saurí D (2011) A comparative appraisal of the use of rainwater harvesting in single and multi-family buildings of the Metropolitan Area of Barcelona (Spain): social experience, drinking water savings and economic costs. J Clean Prod 19(6–7):598–608

Domènech L, Heijnen H, Saurí D (2012) Rainwater harvesting for human consumption and livelihood improvement in rural Nepal: benefits and risks. Water Environ J 26(4):465–472. https://doi.org/10.1111/j.1747-6593.2011.00305.x

Dvarioniene J, Stasiskiene Z (2007) Integrated water resource management model for process industry in Lithuania. J Clean Prod 15(10):950–957

El-Baz AA, Ewida KT, Shouman MA, El-Halwagi MM (2005a) Material flow analysis and integration of watersheds and drainage systems: I. Simulation and application to ammonium management in Bahr El-Baqar drainage system. Clean Technol Environ Policy 7:51–61

El-Baz AA, Ewida KT, Shouman MA, El-Halwagi MM (2005b) Material flow analysis and integration of watersheds and drainage systems: II. Integration and solution strategy with application to ammonium management in Bahr El-Baqar drainage system. Clean Technol Environ Policy 7:78–86

El-Halwagi MM (2012) Sustainable design through process integration: fundamentals and applications to industrial pollution prevention, resource conservation, and profitability enhancement. Butterworth-Heinemann/Elsevier, Oxford/Amsterdam

Farreny R, Morales-Pinzón T, Guisasola A, Tayà C, Rieradevall J, Gabarrell X (2011) Roof selection for rainwater harvesting: quantity and quality assessments in Spain. Water Res 45(10):3245–3254

Foo DCY (2009) State-of-the-art review of pinch analysis techniques for water network synthesis. Ind Eng Chem Res 48:5125–5159

Foo DCY, El-Halwagi MM, Tan RR (eds) (2011) Recent advances in sustainable process design and optimization. Series on advances in process systems engineering. World Scientific Publishing Company

Gikas P, Angelakis AN (2009) Water resources management in Crete and in the Aegean Islands, with emphasis on the utilization of non-conventional water sources. Desalination 248(1–3):1049–1064

Hadadin N, Qaqish M, Akawwi E, Bdour A (2010) Water shortage in Jordan—sustainable solutions. Desalination 250(1):197–202

INEGI (2010a) National Institute of Statistics and Geography. National census of population and housing. Basic questionnaire. Available at: http://www.inegi.org.mx/sistemas/mexicocifras/default.aspx?e=16. Accessed Aug 2012

INEGI (2010b) National Institute of Statistics and Geography. Flow simulator of water in hydrographic watersheds. Available at http://antares.inegi.org.mx/analisis/red_hidro/SIATL/. Accessed Aug 2012

Li Z, Boyle F, Reynolds A (2010) Rainwater harvesting and greywater treatment systems for domestic application in Ireland. Desalination 260(1–3):1–8

Lira- Barragán F, Ponce-Ortega JM, Serna-González M, El-Halwagi MM (2011a) An MINLP model for the optimal location of a new industrial plant with simultaneous consideration of economic and environmental criteria. Ind Eng Chem Res 50(2):953–964

Lira-Barragán LF, Ponce-Ortega JM, Serna-González M, El-Halwagi MM (2011b) Synthesis of water networks considering the sustainability of the surrounding watershed. Comput Chem Eng 35(12):2837–2852

Lira-Barragán LF, Ponce-Ortega JM, Nápoles-Rivera F, Serna-González M, El-Halwagi MM (2012) Incorporating property-based water networks and surrounding watersheds in site selection of industrial facilities. Ind Eng Chem Res 52(1):91–107. https://doi.org/10.1021/ie3003792

Liu S, Konstantopoulou F, Gikas P, Papageorgiou LG (2011) A mixed integer optimisation approach for integrated water resources management. Comput Chem Eng 35(5):858–875

Lovelady EM, El-Halwagi MM (2009) Design and integration of eco-industrial parks. Environ Prog Sustain Energy 28(2):265–272

Lovelady EM, El-Baz AA, El-Monayeri D, EL-Halwagi MM (2009) Reverse problem formulation for integrating process discharges with watersheds and drainage systems: managing phosphorus in lake Manzala. J Ind Ecol 13(6):914–927

Morales-Pinzón T, Rieradevall J, Gasol CM, Gabarrell X (2012) Potential of rainwater resources based on urban and social aspects in Colombia. Water Environ J 26(4):550–559. https://doi.org/10.1111/j.1747-6593.2012.00316.x

Muginstein A, Cohen Y, Levin L, Frant S (2003) Production of desalinated water and electricity in a dual-purpose plant operating in a dispatchable electricity system—techno-economical analysis. Desalination 156(1–3):361–366

Nápoles-Rivera F, Ponce-Ortega JM, El-Halwagi MM, Jiménez-Gutiérrez A (2010) Global optimization of mass and property integration networks with in-plant property interceptors. Chem Eng Sci 65(15):4363–4377

Nápoles-Rivera F, Serna-González M, El-Halwagi MM, Ponce-Ortega JM (2013) Sustainable water management for macroscopic systems. J Clean Prod 47:102–117

NOM-011-CNA-2000 (2000) Official Mexican norm. Available at http://www.conagua.gob.mx/CONAGUA07/Noticias/NOM-011-CNA-2000.pdf

Official Bulletin of the State of Michoacan (2012) Water costing decree 2012. Available at http://www.ooapas.gob.mx/ooapas_web/docs/decreto_tarifario_2012.pdf

Partzsch L (2009) Smart regulation for water innovation—the case of decentralized rainwater technology. J Clean Prod 17(11):985–991

Rubio-Castro E, Ponce-Ortega JM, Nápoles-Rivera F, El-Halwagi MM, Serna-González M, Jiménez-Gutiérrez A (2010) Water integration of eco-industrial parks using a global optimization approach. Ind Eng Chem Res 49(20):9945–9960

SMN (2010) National meteorological system. Climatologic averages 1951–2010. Available at http://www.cm.colpos.mx/meteoro/progde/norm/n5110/. Accessed Aug 2012

Sotelo-Pichardo C, Ponce-Ortega JM, El-Halwagi MM, Frausto-Hernández S (2011) Optimal retrofit of water conservation networks. J Clean Prod 19(14):1560–1581

Spriggs HD, Lowe EA, Watz J, Lovelady EM, El-Halwagi MM (2004) Design and development of eco-industrial parks, paper #109a. AIChE Spring meeting, New Orleans, Apr 2004

UN-Water (2007) Coping with water scarcity: challenge of the twenty-first century. Available at http://www.fao.org/nr/water/docs/escarcity.pdf. Accessed Aug 2012

Vecchietti A, Lee S, Grossmann IE (2003) Modeling of discrete/continuous optimization problems: characterization and formulation of disjunctions and their relaxations. Comput Chem Eng 27(3):433–448

Chapter 5
Sustainable Urban Water Systems Design Incorporating Future Projections for Optimal Configuration

5.1 Introduction

Recent years have seen an increase in studies on resource augmentation and water conservation due to worries about the overexploitation of water bodies. Human factors including population growth, industrialization, rising living standards, irrational resource consumption, and climate change have contributed to this phenomenon. But also, because of natural factors like geographical restrictions and the unequal distribution of water throughout the planet (for instance, in the case of islands and arid regions). Recent research efforts have been primarily focused on resource conservation and water augmentation through targeted practices such as water regeneration, reuse, and recycling. Several authors have successfully applied these tactics in an industrial setting. Foo (2009) provided a thorough overview of pinch-based methods for creating water networks in continuous processes in this regard. Several authors have proposed mathematical programming approaches to overcome the limitations of pinch-based techniques, which, despite their usefulness, have at least one of the following drawbacks: they are generally limited to the single contaminant case, only work with small or medium-sized problems, do not account for stream regeneration, and/or cannot include multiple fresh utilities. In this way, Gouws et al. (2010) provided an overview of batch process-based water minimization strategies. Additionally, Jezowski (2010) offered an additional study that rewarded water networks using mathematical and graphical methods. The synthesis of water networks has lately been expanded to include industrial complexes (also known as eco-industrial parks). Chen et al. (2010) tackled the interplant water optimization problem with central and decentralized water mains, and they considered two distinct objective functions: lowering water usage and lowering the network's overall annual cost. Next, in an eco-industrial park, Rubio-Castro et al. (2011) also considered the synthesis of multi-contaminant water networks. Tudor and Lavric (2011) presented a method to reduce both operating costs and water use at the same time. The problem

J. M. Ponce-Ortega et al., *Sustainable Water Systems*, Water Resources Development and Management, https://doi.org/10.1007/978-3-031-72498-5_5

of water networks and their impact on surrounding watersheds have been successfully extended by several authors (see, for instance, Burgara-Montero et al. 2013; Lira-Barragan et al. 2010, 2011, 2012; Martínez-Gomez et al. 2013) using mathematical programming approaches. These models are also multi-objective optimization problems that have considered various optimization criteria, such as economic, environmental, and safety factors. Optimal reconfiguration of current water networks and the synthesis of different water networks for multipurpose plants were the topics of certain publications (Sotelo-Pichardo et al. 2011; Rubio Castro et al. 2012). Fu et al. (2012) incorporated global optimality when more stringent models were taken into consideration. Some studies (Nápoles-Rivera et al. 2010; Rubio-Castro et al. 2011; Khor et al. 2012; Teles et al. 2012) focus on methods for synthesizing global optimal solutions. Additionally, Castro and Teles (2013) presented a comparison of the global optimization approaches used in the design of water networks.

In addition to the above-listed methods, using alternative water sources has also been suggested by some authors to save resources; these sources include seawater, collected rainwater, and reclaimed water. Pereira et al. (2002) talked about a few of the problems with managing crop irrigation when there is a water shortage. They investigated the effects of this practice on the environment and human health as well as the usage of low-quality water (processed wastewater and saline water) to meet needs. They emphasized how crucial it is to use cutting-edge irrigation technologies to minimize water usage. Agrafioti and Diamadopoulos (2012), Li et al. (2011), Jhorar et al. (2009), and other writers have also tackled the issue of creating substitutes to meet irrigation needs for farming. However, in spatially constrained areas like the Islands of Syros and Paros-Antiparos, Liu et al. (2011) created a MINLP model for the best placement of desalination plants and treatment facilities needed to meet water demands. They displayed the necessary financial outlay for putting these plans into practice. Howari and Ghrefat (2011) conducted an environmental assessment of Jordan's air, water, and soil. They noted the negative effects of human activity on these three crucial areas, and in the case of water, they emphasized the necessity of using alternative water sources to combat water scarcity, such as treated wastewater, water harvesting, importing water across boundaries, and desalinating brackish and seawater. Next, a system integration strategy was put up by Atilhan et al. (2012) for the best management of desalinated water in Qatar. This technique decreased water distribution costs while concurrently meeting demand. Moreover, Nápoles-Rivera et al. (2012, 2013a, b) suggested using seawater in industrial facilities as a cooling fluid. They discovered the best maintenance schedules and biocide dosages to attain environmentally, economically, and technically sustainable operations. Seawater use is still regarded as a costly choice, even though it is becoming more of an option due to technology advancements, the use of renewable energy sources, and acute water stress in some countries (Compain, 2012; Shatat et al. 2013). Rainwater collection has so been investigated as a potential substitute. Eroksuz and Rahman (2010) suggested using rainwater tanks in multi-unit buildings in this situation; they discovered that, even in dry years, a suitable choice of tank size may result in cost-effective and appealing solutions. Rahman et al. (2012) also discussed the usage of rainwater tanks in detached homes and discovered that to make this practice economically viable, government incentives are needed. To avoid the depletion of natural resources, Nápoles-Rivera et al. (2013a, b) proposed a mathematical programming model for the optimal allocation of water in a macroscopic system that considers rainwater harvesting and water treatment reuse. They discovered that there are efficient storage and distribution schedules for meeting the needs of the city's residents and farmers, even though there is little overall revenue gain from the sale of water. Nevertheless, waste minimization can effectively prevent pollution and conserve resources. The significance of using alternate water sources as a chance for resource conservation in an industrial setting was emphasized by Barrington et al. (2013).

As a result, a mathematical programming model for the sustainable design of macroscopic water systems is put forth in this chapter. It involves the best possible scheduling and planning of a city's water storage and distribution system while considering both natural and alternative water sources, such as reclaimed and harvested rainwater. A mathematical model is created to consider how all the users in a macroscopic system—domestic, agricultural, and industrial users—interact with one another. It also considers the rise in demands over the project's duration and the effect of climate change on precipitation patterns, which has decreased the availability of rainwater. Lastly, the model also considers sustainability restrictions and desires (Rojas-Torres et al. 2014).

5.2 Problem Statement

The presented model is based on the source-sink representation shown in Fig. 5.1. The main components considered in the system are natural water sources (damns, springs, and deep wells), users (industrial, agricultural, and domestic), and alternative water sources (harvested rainwater and reclaimed water). The natural sources can be recharged by direct precipitation, runoff water, and natural tributaries. These sources can be used to satisfy the demands of the different users in the city, firstly they must be treated in central facilities (mains), which distributes the water to the final consumers. Then, it is considered that the industrial and domestic wastewater can be treated in centralized treatment facilities; part of the treated water can be recovered for agricultural purposes, while the rest is disposed of as waste. For the natural sources and recovered wastewater, a set of storage devices can be installed at different times to collect rainwater. These devices can store water during rainy periods and use it to satisfy the different demands as needed. The problem consists of finding the optimal schedule for the management of the resources that satisfies the demands in a macroscopic system maintaining sustainable levels of natural resources while maximizing the revenue from the sales of water minus the capital and operational costs of distribution and storage. The obtained solutions will also determine the optimal location and installation time of storage tanks and artificial ponds. Because the model is a multi-annual scheduling problem, it will be assumed the population growth rate and its relation to water demand, the value of money through time, and the change in the precipitation patterns due to climate change behave in a deterministic way and can be predicted based on historical data.

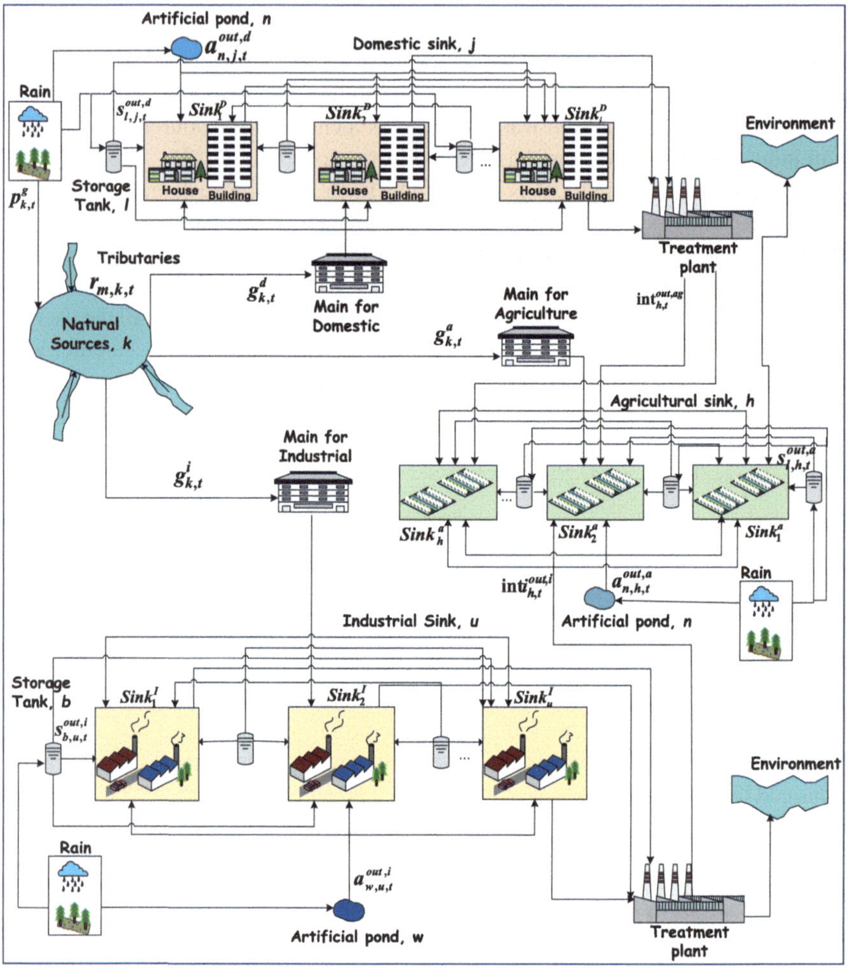

Fig. 5.1 Implemented superstructure for water distribution at the macroscopic level. Printed with permission of "*Optimal design of water systems for cities involving future projections*", *Computers and Chemical Engineering*, Copyright (2014), Elsevier

5.3 Mathematical Model

For a better understanding, the scripts used in the model are defined: k is natural sources, m is the number of tributaries that recharge the natural sources, t is the period. l and n correspond to the potential location for storage tanks and artificial ponds (which are independent of the final user, i.e., domestic, agricultural, or industrial). Lastly, j corresponds to domestic sinks, h agricultural sinks, and u industrial sinks. The model consists of a set of mass balances in the mixing and splitting points of the network and a set of equations needed to select the location and time of installation of storage devices in the case that these are required. Component balances are not considered, instead, it is assumed that the water from natural sources is not mixed before or after treatment, therefore the water leaving each of the mains meets the quality constraints. Therefore, the bilinear terms resulting from component balances are avoided, and the only sources of nonlinearity are the cost functions.

5.3.1 *Mass Balance for Natural Water Sources*

A mass balance for the natural sources is implemented. This balance implies that the accumulation depends on the initial inventory of natural source k ($r_{m,k,t}$) and the inlets and outlets during that period. In Eq. (5.1), $r_{m,k,t}$ are the inlets from all the tributaries to stream k, whereas $p^g_{k,t}$ are the inlets from direct precipitation and runoff water collection. The outlets are the water sent to the different mains: domestic ($g^d_{k,t}$), agricultural ($g^a_{k,t}$), and industrial ($g^i_{k,t}$). Almost all these facilities can be in the same place, although for modeling these are considered as different facilities where diverse levels of treatment are given to the natural sources depending on the final use of the processed water.

$$G_{k,t} - G_{k,t-1} = \sum_{m \in M} r_{m,k,t} + p^g_{k,t} - g^d_{k,t} - g^a_{k,t} - g^i_{k,t} - v^g_{k,t} - Drop^g_{k,t}, \quad \forall k \in K, \quad \forall t \in T \tag{5.1}$$

$p^g_{k,t}$ is determined as the sum of the runoff water collected plus the direct precipitation.

$$p^g_{k,t} = ROW_{k,t} + DPWV_{k,t}, \quad \forall k \in K, \quad \forall t \in T \tag{5.2}$$

Runoff water ($ROW_{k,t}$) is calculated as a function of the runoff coefficient, the total annual precipitation, and the collection area:

$$ROW_{k,t} = P_t A^{ROW}_k Ce, \quad \forall k \in K, \quad \forall t \in T \tag{5.3}$$

Ce is determined with indirect methods as a function of the annual precipitation and the parameter K, which considers the type and use of land (the Mexican Norm states the approach to determine the average annual accessibility of natural water bodies, NOM-001-CNA-2000).

$$Ce = \frac{K\left(P^{Total} - 250\right)}{2000}, \quad K \leq 0.15 \tag{5.4}$$

$$Ce = \frac{K\left(P^{Total} - 250\right)}{2000} + \frac{K - 0.15}{1.5}, \quad K > 0.15 \tag{5.5}$$

The direct precipitation is function of the area of collection, and the monthly precipitation.

$$DPWV_{k,t} = P_t A_k^{DPW}, \quad \forall k \in K, \quad \forall t \in T \tag{5.6}$$

$v_{k,t}^{g}$ and $Drop_{k,t}^{g}$ represent the water losses (due to evaporation, filtration, and losses in the distribution process, here considered as 20% of the collected water) and the excess water (i.e. amount of water that exceeds the maximum capacity of source k), which means an over-flooding:

$$v_{k,t}^{g} = 0.2\left(\sum_{m \in M} r_{m,k,t} + p_{k,t}^{g}\right), \quad \forall k \in K, \quad \forall t \in T \tag{5.7}$$

5.3.2 *Mass Balance for Precipitation in Artificial Water Reservoirs*

The capacity for rainwater harvesting for artificial devices depends on the precipitation in a specific location and season of the year, the available collection area, and the runoff coefficient, this functionality can be expressed as follows:

$$P = f(Area, Ce, Precipitation) \tag{5.8}$$

Precipitation data can be obtained from meteorological stations, numerous authors have reported data on runoff coefficients for roofs, this parameter relates the runoff to the total water received and ranges from 0.7 to 0.95 depending on the roof type and building material (Farreny et al. 2011), and the collection area is taken as the available roof area nearby to the storage device. Two different types of storage devices are considered, storage tanks and artificial ponds, these devices can feed any user. For each device, an equation to determine the amount of water available for recollection is required, which is stated as follows:

$$P^s_{l,t} = P_t A^s_l Ce^s, \quad \forall l \in L, \quad \forall t \in T \tag{5.9}$$

$P^s_{l,t}$ represents the total available water for storage tank l at period t, A^s_l is the collection area for storage tank l. Similarly for any artificial pond n, replacing $P^s_{l,t}$ by $P^a_{n,t}$ (which is the total available water for artificial pond n at period t) and A^s_l by A^a_n (which is the collection area available for the artificial pond n at period t).

Then, the effective collected rainwater is evaluated as follows:

$$P^s_{l,t} = s^{in}_{l,t} + v^s_{l,t}, \quad \forall l \in L, \quad \forall t \in T \tag{5.10}$$

The total available rainfall for each period (t) can be sent to storage tanks ($s^{in}_{l,t}$). If there is not storage capacity, so the excess water can be discarded (the last term in equation 5.10). For the case of artificial ponds, a similar equation applies by replacing $P^s_{l,t}$ by $P^a_{n,t}$, $s^{in}_{l,t}$ by $a^{in}_{n,t}$, and $v^s_{l,t}$ by $v^a_{n,t}$.

5.3.3 Mass Balance in Storage Tanks and Artificial Ponds

A mass balance is needed for the storage tanks, where the accumulation over the period t ($S_{l,t}$) is equal to the stored water in the previous time ($S_{l,t-1}$) plus the inlet and minus the summation of the outlets. The inlet to the storage tanks is harvested rainwater ($s^{in}_{l,t}$), whereas the outlets are the water sent to domestic ($s^{out,d}_{l,j,t}$), agricultural ($s^{out,a}_{l,h,t}$), and industrial users ($s^{out,i}_{l,u,t}$). The existence and capacity of these tanks are calculated in the optimization process:

$$S_{l,t} - S_{l,t-1} = s^{in}_{l,t} - \sum_{j \in J} s^{out,d}_{l,j,t} - \sum_{h \in H} s^{out,a}_{l,h,t} - \sum_{u \in U} s^{out,i}_{l,u,t}, \quad \forall l \in L, \quad \forall t \in T \tag{5.11}$$

The mass balances for the artificial ponds n are like the ones of Eq. (5.11) for the storage tanks, just it is required to replace l by n, $S_{l,t}$ by $A_{n,t}$ (which is the water stored over the period t), $S_{l,t-1}$ by $A_{n,t-1}$ (which is the water stored in the previous time period), $s^{in}_{l,t}$ by $a^{in}_{n,t}$ (which is the harvested rainwater), $s^{out,d}_{l,j,t}$ by $a^{out,d}_{n,j,t}$ (which is the water consumed by domestic users), $s^{out,a}_{l,h,t}$ by $a^{out,a}_{n,h,t}$ (which is the water consumed by agricultural users) and $s^{out,i}_{l,u,t}$ by $a^{out,i}_{n,u,t}$ (which is the water consumed by industrial users).

5.3.4 Mass Balance in Mains

Natural water sources are sent to different mains to receive an appropriate treatment depending on their final use. For example, in the main for domestic users, the sum of the natural water sources ($g_{k,t}^{d}$) sent to this main are equal to the sum of the water sent from the domestic main to the domestic sinks ($f_{j,t}$):

$$\sum_{k \in K} g_{k,t}^{d} = \sum_{j \in J} f_{j,t}, \quad t \in T \tag{5.12}$$

Similar relationships are used for agricultural mains (replacing $g_{k,t}^{d}$ by $g_{k,t}^{a}$ and $f_{j,t}$ by $r_{h,t}$) and for industrial mains (replacing $g_{k,t}^{d}$ by $g_{k,t}^{i}$ and $f_{j,t}$ by $q_{u,t}$). Details for these symbols can be found in the nomenclature section.

5.3.5 *Balance for Sinks and Treatment Plants*

For agricultural users, the demands ($D_{h,t}^{as}$) can be met with streams from the agricultural main ($r_{h,t}$), storage tanks ($s_{l,h,t}^{out,a}$), artificial ponds ($a_{n,h,t}^{out,a}$), reclaimed water from domestic treatment plant ($\text{int}_{h,t}^{out,ag}$) and reclaimed water from industrial treatment plant ($\text{int}_{h,t}^{out,i}$) as follows:

$$D_{h,t}^{as} \leq r_{h,t} + \sum_{l \in L} s_{l,h,t}^{out,a} + \sum_{n \in N} a_{n,h,t}^{out,a} + \text{int}_{h,t}^{out,ag} + \text{inti}_{h,t}^{out,i}, \quad h \in H, \quad t \in T \tag{5.13}$$

The total flow received in the agricultural sinks is consumed in the production process.

Domestic demands ($D_{j,t}^{ds}$) can be satisfied by using water from the domestic main, water from the storage tanks or water from the artificial ponds. The demand in each sink is a function of the population and season of year (demand increases during the hottest months of year and reduces during the coolest months):

$$D_{j,t}^{ds} \leq f_{j,t} + \sum_{l \in L} s_{l,j,t}^{out,d} + \sum_{n \in N} a_{n,j,t}^{out,d}, \quad j \in j, \quad t \in T \tag{5.14}$$

Similar relationships apply for industrial demands, just there is needed to replace j by u, $D_{j,t}^{ds}$ by $D_{u,t}^{di}$, $f_{j,t}$ by $q_{u,t}$, $s_{l,j,t}^{out,d}$ by $s_{l,u,t}^{out,i}$ and $a_{n,j,t}^{out,d}$ by $a_{n,u,t}^{out,i}$.

The outlet flow of domestic sinks is segregated in two terms, the first one ($cw_{j,t}^{d}$) involves the water consumed and/or lost in the domestic sinks, and the second one ($\text{int}_{j,t}^{in}$) is the wastewater generated in domestics sinks, which can be sent to a treatment plant to regenerate it and use in agricultural sinks (int_{t}^{out}) or sent to final disposal (cw_{t}^{tp}):

$$D_{j,t}^{ds} = cw_{j,t}^{d} + \text{int}_{j,t}^{in}, \quad j \in j, \quad t \in T \tag{5.15}$$

$$\sum_j \text{int}_{j,t}^{in} = \text{int}_t^{out} + cw_t^{tp}, \quad t \in T \tag{5.16}$$

The treated water can be sent to any agricultural:

$$\text{int}_t^{out} = \sum_h \text{int}_{h,t}^{out,ag}, \quad t \in T \tag{5.17}$$

For treating and distributing wastewater generated in industrial sinks, Eqs. (5.15)–(5.1.17) applied by replacing j by u, $D_{j,t}^{ds}$ by $D_{u,t}^{di}$, $cw_{j,t}^{d}$ by $cw_{u,t}^{di}$, $\text{int}_{j,t}^{in}$ by $\text{inti}_{u,t}^{in}$, int_t^{out} by inti_t^{out}, cw_t^{tp} by cwi_t^{tpi}, and $\text{int}_{h,t}^{out,ag}$ by $\text{inti}_{h,t}^{out,i}$.

5.3.6 *Equations for Installing Storage Tanks or Artificial Ponds*

The following relationship limits to the maximum capacity of the different storage devices.

$$S_l^{\max} \geq S_{l,t}, \quad l \in L, \quad t \in T \tag{5.18}$$

The previous relationship indicates that the accumulation in a storage device and the water entering such device during a single period must not exceed its maximum capacity. Analogous limits are needed for artificial ponds (using $A_n^{\max}$).

The existence or not existence of storage tanks is simulated by the following equations: if a tank is required at any period, then the binary variable $z_{l,t}^{s}$ is one and the storage tank is installed in location l at period t, other case when the binary variable is zero and the tank is not required, the associate installation cost for this tank must be zero. The same logic is used for the artificial ponds ($z_{n,t}^{a}$). The capacity of the storage devices must be between the upper and lower capacity of the devises. $\delta_l^{s,\max}$ and $\delta_l^{s,\min}$ are the maximum and minimum capacities; for artificial ponds the limits are $\delta_n^{a,\max}$ and $\delta_n^{a,\min}$. The cost functions are function of the location of the storage tanks and artificial ponds, where A and B are constants used to calculate the fixed and variable costs of storage tanks, and C and D are the constants related to the fixed cost and variable cost of artificial ponds. $Cost_{l,t}^{s}$ is the cost of storage tanks. $Cost_{n,t}^{a}$ is the cost for artificial ponds. α is an exponent that is lower than 1, which considers the economies of scale. Furthermore, K_F is used to annualize the investment, and it involves the inflation effect on investment costs and the value of money through the time. *VP* is the net present value, which corresponds to the value of investment that considers the value of money over time of each storage facility. This is simulated as follows.

For storage tanks:

$$\sum_{t} z_{l,t}^{s} \leq 1, \quad \forall l \in L \tag{5.19}$$

$$S_{l,t'} \leq \delta_l^{s,\max} \sum_{t=1}^{t=t'} z_{l,t}^{s}, \quad \forall l \in L, \ \forall t' \in T', \ \forall t' \leq t \tag{5.20}$$

$$s_{l,t'}^{in} \leq \delta_l^{s,\max} \sum_{t=1}^{t=t'} z_{l,t}^{s}, \quad \forall l \in L, \ \forall t' \in T', \ \forall t' \leq t \tag{5.21}$$

$$K_{F_l}^{s} = \sum_{t} K_{F_{l,t}} \cdot VP_{l,t} \cdot z_{l,t}^{s}, \quad \forall l \in L \tag{5.22}$$

$$S_l^{cap} \geq S_{l,t}, \quad l \in L, \ t \in T \tag{5.23}$$

$$Cost_{l,t}^{s} = K_{F_l}^{s}\left[A + B\left((S_l^{cap})\right)^{\alpha}\right], \quad \forall l \in L \tag{5.24}$$

For artificial ponds replacing l by n, $z_{l,t}^{s}$ by $z_{n,t}^{a}$, $S_{l,t'}$ by $A_{n,t'}$, $\delta_l^{s,\max}$ by $\delta_n^{a,\max}$, $s_{l,t'}^{in}$ by $a_{n,t'}^{in}$, $K_{F_l}^{s}$ by $K_{F_n}^{a}$, $K_{F_{l,t}}$ by $K_{F_{n,t}}$, $VP_{l,t}$ by $VP_{n,t}$, $Cost_{l,t}^{s}$ by $Cost_{n,t}^{a}$, A by C, B by D.

The equations for the capacity of the devices are state that if a storage device is installed at period t, it can be used in any other period after this time without adding new costs. This is implemented by using the two sets t and t'.

Additionally, the cost function stated in Eq. (5.24) is the source of nonlinearities in the optimization model. To linearize this equation, a piecewise approach can be used for eliminating the exponent α. Additionally, a bilinear term between $K_{F_l}^{s}$ and S_l^{cap} can be manipulated using a spatial search.

Previous equations allow to account for the value of money through the time. Thus, $K_{F_{l,t}}$ and $VP_{l,t}$ are parameters that annualize the inversion and to consider the present value, respectively; these factors are determined for each year before the optimization process considering the interest rate. So, the binary variable $z_{l,t}^{s}$ indicates the time when a storage device is needed, and this allows to calculate the right value of $K_{F_l}^{s}$ to accurately determine $Cost_{l,t}^{s}$ through Eq. (5.24).

5.3.7 Objective Function

The objective function consists in maximizing the net profit from the sale of the water managed in the city. In several countries, the water is administered and distributed by the local government; the profits are used as part of the budget for the development and maintenance of infrastructure. The considered objective function involves water sales for domestic, agricultural, and industrial purposes (contributions), minus the cost related to treatment (*treatment cost*) and distribution of water (*pp cost*) and the cost for the installation and operation of artificial tanks (storage cost), which is calculated as follows:

$$TAR = contribution - treatment\ cost - storage\ cost - ppcost \tag{5.25}$$

The sale price of water does not depend on its source and just on the final use, thus the revenue can be determined as follows:

$$\begin{aligned} contribution = & \left(\sum_k \sum_t g_{k,t}^d + \sum_l \sum_j \sum_t s_{l,j,t}^{out,d} + \sum_n \sum_j \sum_t a_{n,j,t}^{out,d} \right) DSC \\ & + \left(\sum_k \sum_t g_{k,t}^a + \sum_l \sum_h \sum_t s_{l,h,t}^{out,a} + \sum_t \mathrm{int}_t^{out} \right) ASC \\ & + \left(\sum_k \sum_t g_{k,t}^i + \sum_l \sum_u \sum_t s_{l,u,t}^{out,i} + \sum_n \sum_u \sum_t a_{n,u,t}^{out,i} \right. \\ & \left. + \sum_h \sum_t \mathrm{inti}_{h,t}^{out,i} \right) ISC \end{aligned} \tag{5.26}$$

DSC is the price of water for domestic use, *ASC* the price of water for agricultural purposes, and *ISC* the price of water for industrial users.

Different sources have different qualities, and each user needs dissimilar specification. Therefore, depending on the source and use, the treatment cost is determined as follows:

$$\begin{aligned} treatment\ cost = & \left(\sum_k \sum_t g_{k,t}^d CTND + \sum_k \sum_t g_{k,t}^a CTNA + \sum_k \sum_t g_{k,t}^i CTNI \right) \\ & + \left(\sum_l \sum_j \sum_t s_{l,j,t}^{out,d} + \sum_n \sum_j \sum_t a_{n,j,t}^{out,d} \right) CTAD \\ & + \left(\sum_l \sum_h \sum_t s_{l,h,t}^{out,a} + \sum_n \sum_j \sum_t a_{n,j,t}^{out,a} \right) CTAA \\ & + \left(\sum_b \sum_u \sum_t si_{b,u,t}^{out,i} + \sum_w \sum_u \sum_t a_{w,u,t}^{out,i} \right) CTAI \\ & + \left(\sum_l \sum_u \sum_t s_{l,u,t}^{out,i} + \sum_n \sum_u \sum_t a_{n,u,t}^{out,i} \right) ISC \\ & + \left(\sum_t \mathrm{int}_t^{out} + \sum_h \sum_t \mathrm{int}_{h,t}^{out,i} \right) CTPA \\ & + \left(\sum_t Cw_t^{tp} + \sum_t Cw_t^{tpi} \right) CTPE \end{aligned} \tag{5.27}$$

CTND represents the treatment cost for natural streams for domestic use, *CTNA* is the treatment cost for natural streams for agricultural use, *CTNI* is the treatment cost of natural streams for industrial use, *CTAD* and *CTAA* are the treatment costs of rainwater for domestic and agricultural purposes respectively, furthermore *CTAI* represents the cost of treating rainwater for industrial use, *CTP* the treatment cost for the regeneration of wastewater for agricultural use and *CTPE* the cost of wastewater treatment for disposal in environment.

The storage cost is calculated from the sum of the annualized cost of artificial reservoirs ($Cost^{s}_{l,t}$ and $Cost^{a}_{n,t}$ for tanks and ponds, respectively):

$$storage\ cost = \left(\sum_{l} \sum_{t} Cost^{s}_{l,t} + \sum_{n} \sum_{t} Cost^{a}_{n,t} \right) \tag{5.28}$$

The piping and pumping cost (*pp cost*) to transport water from one site to another one is calculated in terms of the following parameters: *PCSTD* represents the unit transportation cost from storage tank l to domestic sink j, *PCASD* corresponds to the unit pumping cost from artificial pond n to domestic sink j, *PCSTA* represents the unit cost of pipeline and pumping from storage tank l to agricultural sink h, *PCASA* corresponds to the unit cost of piping and pumping from an artificial pond n to agricultural sink h, *PCSTI* represents the unit cost of piping and pumping from storage tank l to industrial sink u, *PCASI* corresponds to the unit cost of piping and pumping from artificial pond n to industrial sink u, *PCND*, *PCNA* and *PCNI* are the unit costs of piping and pumping from natural sources k to domestic, agricultural and industrial mains, respectively, and finally *PCTW* (domestic) and *PCTI* (industrial) are the unit costs of piping and pumping from treatment plants to agricultural sinks h. Therefore, the total cost of piping and pumping is calculated as follows:

$$\begin{aligned} pp\ Costo = & \sum_{l} \sum_{j} \sum_{t} s^{out,d}_{l,j,t} PCSTD + \sum_{n} \sum_{j} \sum_{t} a^{out,d}_{n,j,t} PCASD \\ & + \sum_{l} \sum_{h} \sum_{t} s^{out,a}_{l,h,t} PCSTA + \sum_{n} \sum_{h} \sum_{t} a^{out,a}_{n,h,t} \\ & + \sum_{l} \sum_{u} \sum_{t} s^{out,i}_{l,u,t} PCSTI + \sum_{n} \sum_{u} \sum_{t} a^{out,i}_{n,u,t} PCASI \\ & + \sum_{k} \sum_{t} g^{d}_{k,t} PCND + \sum_{k} \sum_{t} g^{a}_{k,t} PCNA \\ & + \sum_{k} \sum_{t} g^{i}_{k,t} PCNI + \sum_{h} \sum_{t} \text{int}^{out,ag}_{h,t} PCTW \\ & + \sum_{h} \sum_{t} \text{inti}^{out,i}_{h,t} PCTI \end{aligned} \tag{5.29}$$

5.4 Case Study

The addressed problem consists in finding the optimal distribution of water that satisfies the domestic, agricultural, and industrial demands in the period from 2013 to 2023 for the city of Morelia in México, maintaining a sustainable level of water of natural sources. Morelia is the capital of the state of Michoacán; this is the city with the highest unit costs of water in Mexico, it has a serious overutilization problem of its natural resources that has led to an annual deficit of 2.46 m of the city aquifer according to Avila-Oliveira and Garduño-Monroy (2007). The city has a population of 729,279 habitants (INEGI 2013a), and based on data from CONAGUA (National Council of Water-Mexico), the water consumption in Morelia for domestic use during 2009 was 90,168,374.77 m^3/year (CONAGUA 2011), which signified an average consumption of 338.75 liters per capita per day. During the same year, the agricultural activity in Morelia required 21,348,208.28 m^3/year, while industrial activities utilized 26,005,188.32 m^3/year. This yields an annual consumption of 137,521,771.4 m^3/year, which traditionally is satisfied extracting water from deep wells (43.93%), springs (33.41%) and the Cointzio dam (22.66%) (Avila-Oliveira and Garduño-Monroy 2007). According to CONAPO (2008) (National Council of Population) the average urban population growth rate for the years 2010–2025 would be 0.27% per year, and along with the population growth the demands will also be increased. Based on data of the National Meteorological Service, the annual precipitation has diminished over the last years and this tendency is expected to hold (CONAGUA-SMN 2011). This information was taken as basis to make the projection regarding the water demand and precipitation availability during the period 2013–2023 as follows:

Firstly, the considered sinks (users) are described, at this point for simplicity, the city is segregated in five areas: Center (CE), North-East (NE), North-West (NW), South-East (SE) and South-West (SW) (Fig. 5.2); furthermore, additional areas can be considered if required. Each area involves a sink for domestic, agricultural, or industrial user, the demand in a sink is equal to zero if an activity is not present in a particular zone. For domestic users each area involves a given number of BGSA (Basic Geo Statistic Area) defined as a group of 293 blocks. NW has 77 BGSA with 146,327 inhabitants, NE 72 BGSA with 128,867 inhabitants, SE 64 BGSA and 158,061 inhabitants, SW 52 BGSA with 113,784 inhabitants and CE 28 BGSA with 182,240 inhabitants. Area CE corresponds to the downtown and has fewer BGSAs than the rest of the areas, nonetheless it involves the highest number of inhabitants. The average consumption is taken as basis to make the prediction of the demand in each are, moreover, the season of the year can be considered, during the coldest months the demand is reduced up 8% and during the hottest months it increases up to 17%. Table 5.1 presents the percentage of reduction or increase concerning the mean value. The information for BGSA was taken from INEGI (2013b) to determine the population in each section of the city and the annual increase in the population of 0.27% was considered to forecast the increase in the demands, the population growth is uniform in all the areas of the city. Similarly, the industrial sinks and agricultural sinks were defined, in the case of industrial users the industrial activity is concentrated in the North East side of the city, having the only industrial corridor. Though, several factories are in the central part of the city and a pulp mill in the southwest, the pulp mill factory is the most important consumer of water for industrial purposes having a concession for 400 l/s that is an annual consumption of 12,614,400 m^3 (48% of the industrial use), the rest of the water for industrial purposes is separated in the rest of the industries. Lastly, the agricultural activity is manly located in the western part of the city, but there is no agricultural activity in the CE section that is completely in the urban area. There are 40 tanks that can be installed, the first 20 tanks are used for domestic service and agricultural activity, while the rest are offered for industrial service. 6 artificial ponds can be installed for domestic and agricultural users, and 6 artificial ponds for industrial users, for a total of 12 artificial ponds. The material of construction for the tanks was concrete. The material of construction for pipes is PVC. The monthly precipitation in Morelia in the period of 1951–2010 (SMN 2010) is expected to decrease 6% by 2030. The precipitation data of the analyzed period (2013–2023) is presented in Table 5.2.

To determine the runoff coefficient in Morelia, K is 0.263, which represents the ground type and distribution in the city (CONAGUA 2009). For the first year, this runoff coefficient is:

$$Ce = \frac{0.263(768.9 - 250)}{2000} + \frac{0.263 - 0.15}{1.5} = 0.1435$$

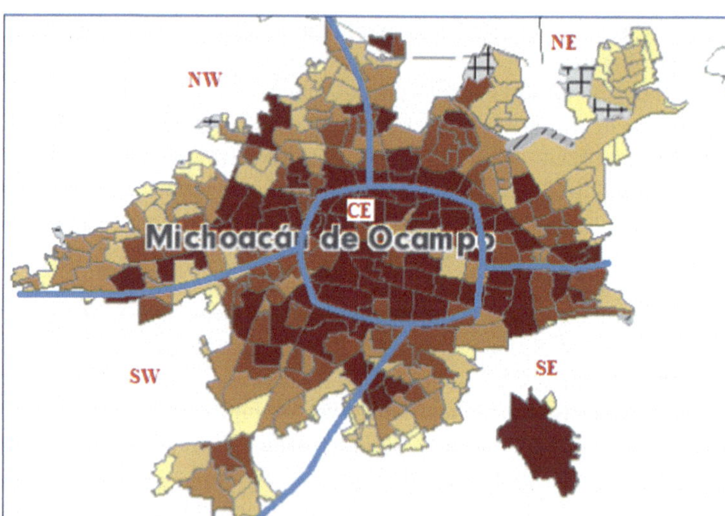

Fig. 5.2 Morelia City geographic division. Printed with permission of "*Optimal design of water systems for cities involving future projections*", *Computers and Chemical Engineering*, Copyright (2014), Elsevier

Table 5.1 Deviation from the mean value of demand for domestic users

Month	%
January	− 8.5
February	4.0
March	4.0
April	17.9
May	14.1
June	6.9
July	0
August	0
September	− 2.2
October	− 2.2
November	− 2.3
December	− 7.9

For the rest of the years, the procedure is the same and only the annual precipitation changes.

The maximum capacity for the storage tanks is 50,000 m^3 and for the artificial ponds is 600,000 m^3, independently of the user. The coefficients of the cost functions are in Table 5.3. While the exponent in the cost function is considered as 1. Piping and pumping costs are calculated in base of preliminary designs considering the distances between the sources and sinks (users), these costs range from $\$1 \times 10^{-3}/m^3$ for neighboring zones (1–3 km) to $8 \times 10^{-3}/m^3$ for distances higher than 10 km. The material of construction depends on location and capacity, and for pipes this is PVC and for storage tanks is concrete.

5.5 Results

The presented model was coded in the software GAMS (Brooke et al. 2013), which consists of 53,473 constrains, 6864 discrete variables, and 89,257 continuous variables and corresponds to a Mixed Integer Non-Linear Programming (MINLP) problem that was solved with the solver DICOPT on a computer with an Intel Core i7 Processor 2.20 GHz with 8 GB of RAM, where the solution of each scenario was obtained in an average of 216,000 s. Three scenarios are presented to find the optimal distribution of water in the city for a given time horizon of 132 month.

Table 5.2 Expected precipitation per month in Morelia, Mexico (mm H_2O)

Month/year	2013*	2014	2015	2016	2017	2018	2019	2020	2021	2022	2023
Jan	15.60	15.51	15.41	15.32	15.23	15.14	15.05	14.96	14.87	14.78	14.69
Feb	7.70	7.65	7.61	7.56	7.52	7.47	7.43	7.38	7.34	7.29	7.25
Mar	8.60	8.55	8.50	8.45	8.40	8.35	8.30	8.25	8.20	8.145	8.10
Apr	10.10	10.04	9.98	9.919	9.86	9.80	9.74	9.68	9.63	9.57	9.51
May	41.70	41.45	41.20	40.95	40.71	40.46	40.22	39.98	39.74	39.50	39.26
June	150.70	149.80	148.90	148.00	147.12	146.23	145.36	144.48	143.62	142.76	141.90
July	167.50	166.50	165.50	164.50	163.52	162.54	161.56	160.59	159.63	158.67	157.72
Aug	170.40	169.38	168.36	167.35	166.35	165.35	164.36	163.37	162.39	161.42	160.45
Sept	129.90	129.12	128.35	127.58	126.81	126.05	125.29	124.54	123.79	123.05	122.31
Oct	52.80	52.48	52.17	51.86	51.54	51.24	50.93	50.62	50.32	50.02	49.72
Nov	10.00	9.94	9.88	9.821	9.76	9.70	9.65	9.59	9.53	9.47	9.42
Dec	3.90	3.88	3.85	3.83	3.81	3.78	3.76	3.74	3.72	3.69	3.67
Total	769	764	760	755	751	746	742	737	733	728	724

*Average precipitation 1951–2010 (SMN 2010)

Table 5.3 Coefficients of cost functions

Thank/Coefficient	A ($)	B ($)	C ($/m^3)	D ($/m^3)
Tanks 1–20	13,080	1.81135		
Tanks 20–40	28,080	4.9134		
Ponds 1–6			111,968	0.8895
Ponds 7–12			151,968	4.9895

5.5.1 *Scenario 1*

In this case, neither storage nor water reclamation is allowed; this is the present operation, except that the planning and scheduling is completed systematically to meet the demands and maintain sustainable levels in the natural water sources. The optimal configuration for Scenario 1 is shown in Fig. 5.3, where the total water withdrawn from each of the sources to meet the demands of all users during the last year of the time horizon. During the last year, the total demand for domestic, industrial, and agricultural users (211,130,387 m^3) is met mostly with water taken from the Cointzio damn and the springs (sources M1 and M2), which contribute with 62.4% of the demands. For the year of 2023 the demand increases by 53% concerning the base demand (from year 2009). The proposed distribution allows to satisfy this demand through recharging natural sources and reducing their overexploitation. However, it is still necessary to find alternatives to further reduce the impact in the natural reservoirs. Furthermore, the total waste generated by domestic users during the last year is expected to be 74.11 × 10^6 m^3 and for industrial users 20.77 × 10^6 m^3, to sum a total of 94.4 × 10^6 m^3, which is an incentive to find options regarding the use of wastewater and reducing freshwater consumption and reducing pollution of the water bodies through direct discharges. The total profit is \$1.1411 × 10^9 at the end of the project.

5.5.2 *Scenario 2*

In this scenario, just rainwater harvesting is considered, no water reclamation is allowed; subsequently, wastewater is discharged directly into the environment after an appropriate treatment. For this scenario, 9 artificial ponds were installed in the first year, one in the third year, and two in the fifth year. From the 12 artificial ponds, 6 are used for agricultural and domestic purposes and the rest are used for industrial service. This way, nine storage tanks were installed in the first year, one in the second year, and one in the sixth year, in this case 9 storage tanks are used for industrial service and the rest for agricultural and domestic activities. The profit is \$1.3723 × 10^9 at the end of the project, which represents an increment of 20.26%

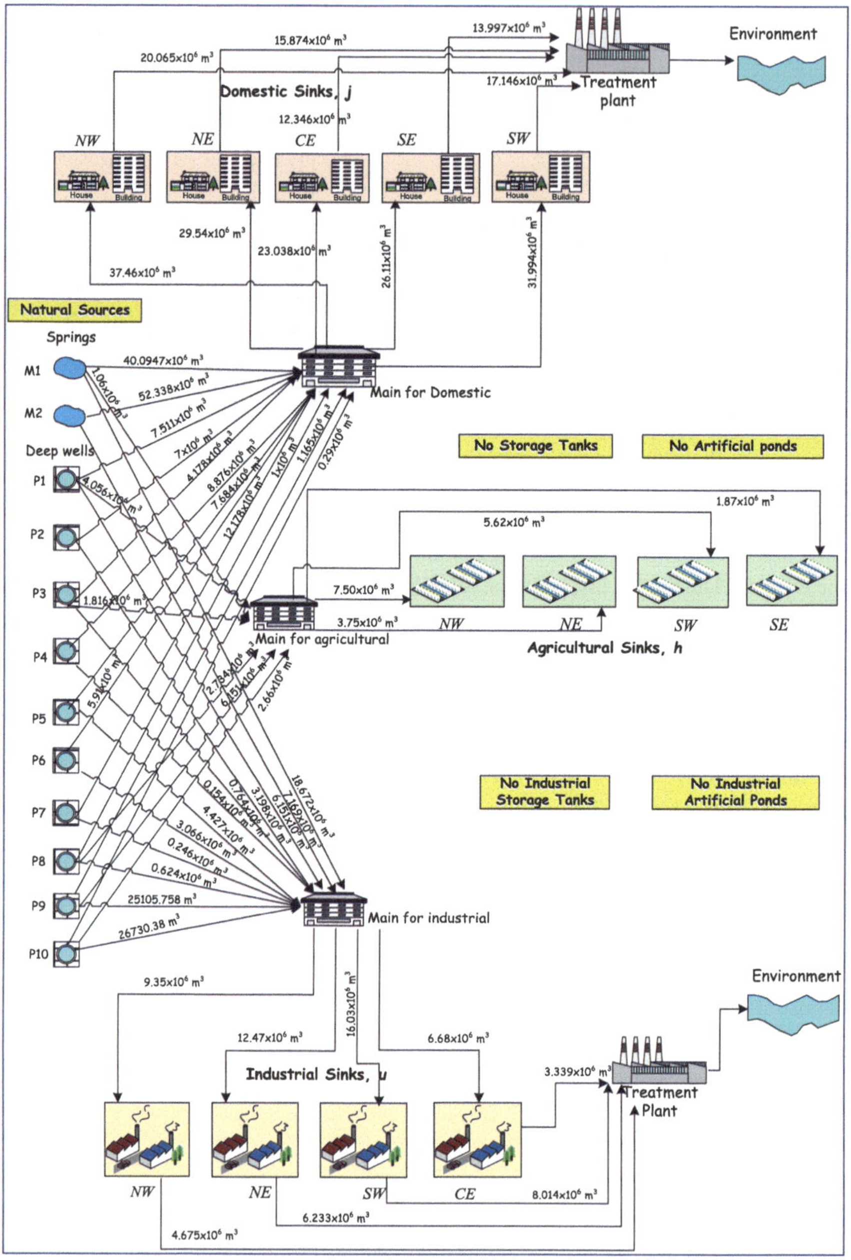

Fig. 5.3 Results of water distribution in the last year in Morelia city for scenario 1. Printed with permission of "*Optimal design of water systems for cities involving future projections*", *Computers and Chemical Engineering*, Copyright (2014), Elsevier

in the profit (with respect to scenario 1). Here, for the sixth year all the required infrastructure must be installed and the total water taken from harvested rainwater is 49.8×10^6 m^3 (i.e. 28.1% of the global demand in the year). Meanwhile during the seventh year the collected water is 48.1×10^6 m^3. This propensity (reduction in the collected water) holds for the rest of the years, due to a reduction in the availability of rainwater, to compensate the effect of the change in the precipitation patterns and take full advantage of the installed infrastructure, additional collection area would be required. Another significant point is that in the last year no water is withdrawn from natural sources P8, P9 and P10, and agricultural demands are fulfilled exclusively with harvested rainwater (Fig. 5.4); therefore, these groups of deep wells can be naturally recharged. The harvested rainwater delivers up to 14.6% of the water demand in the city during the rainy season (June–September), nevertheless the stored water is used during dry periods, this way during the months January–May and October–December the share of contribution of harvested rainwater can reach up to 16.7%. For this scenario, the total water provided by rainwater is 22.7%.

5.5.3 Scenario 3

This scenario involves simultaneously rainwater harvesting and water reclamation, and Fig. 5.5 presents the results, whereas in scenario 2, the storage devices were installed in different years; nevertheless, in this case fewer storages devices are needed. In the first year, three artificial ponds and one storage tank are installed for domestic and agricultural sinks, and for industrial sinks it is necessary to install five artificial ponds in the sixth year and one artificial pond in seventh year, furthermore five storage tanks are required for industrial users, four are installed during the first year and one in the sixth year. In this scenario, reclaimed water delivers all the required water for agricultural purposes in the last year, and the total water obtained from alternative water sources is 3.88×10^6 m^3, which represents 18.4% of the total demand of the last year, 8% more compared with scenario 2. Reclaimed water participates with 12.5% of the total demand during the time of project, and in the harvested rainwater gives 12.7% of the total demand. This means that alternative water sources have the potential to satisfy up to 25.2% of the water demands in the city. This helps to reduce the pressure over the natural water bodies and increase the profit up to $\$1.403 \times 10^9$ at the end of the project (i.e. 22.9% more than the base case). For this scenario 3, the share of contribution of rainwater is reduced from 22.7% to only 12.7%, nevertheless there is no substantial effect of the precipitation diminishment in the recollection capacity (as in scenario 2), this because the available area for collection and the installed storage capacity are less than the available rainwater in any period. An additional advantage of this scenario is the reduction of the waste generated; in the base case (scenario 1), the total waste (domestic and industrial) was 94.891×10^6 m^3 in the last year, and in this case is 73.821×10^6 m^3. This is a reduction of 21.07×10^6 m^3, which assists to reduce the overall impact in the surrounding watersheds for the direct discharge of industrial and domestic

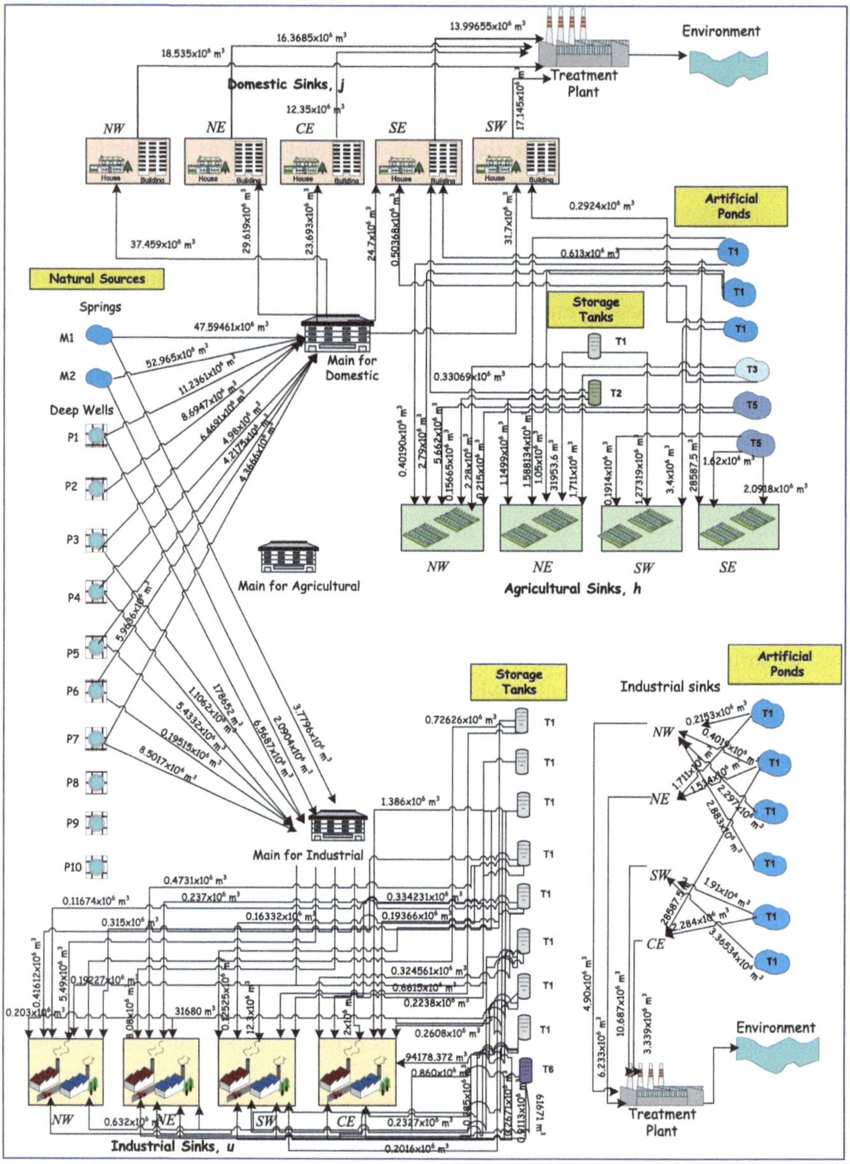

Fig. 5.4 Results of water distribution in the last year in Morelia city for Scenario 2. Printed with permission of "*Optimal design of water systems for cities involving future projections*", *Computers and Chemical Engineering*, Copyright (2014), Elsevier

residual waters. Moreover, the implementation of water reclamation can effectively supplement the use of harvested rainwater providing enough water for agricultural activities in the city even during drought seasons.

Table 5.4 presents a summary of the main results obtained for the different scenarios presented. Lastly, this model might have several local optimal solutions due to the non-convex nature of the bilinear terms in the cost functions. Nonetheless, in

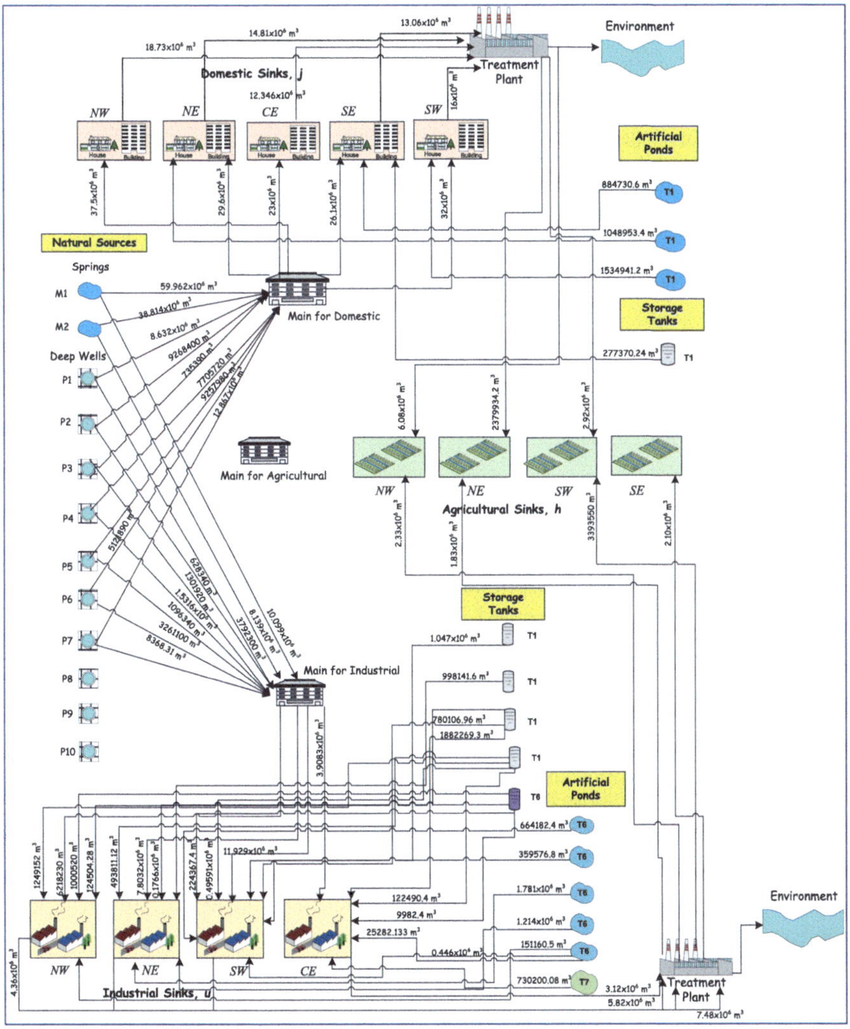

Fig. 5.5 Results of water distribution in the last year in Morelia city for scenario 3. Printed with permission of "*Optimal design of water systems for cities involving future projections*", *Computers and Chemical Engineering*, Copyright (2014), Elsevier

Table 5.4 Summary of results

Item/Case study	1	2	3
Natural sources to domestic main	1.26×10^9	1.19×10^9	1.16×10^9
Natural sources to agricultural main	2.32×10^8	1.88×10^8	5.28×10^5
Natural sources to industrial main	3.55×10^8	2.71×10^8	2.10×10^8
Domestic demand (m^3)	1.26×10^9	1.26×10^9	1.26×10^9
Agricultural demand (m^3)	2.32×10^8	2.32×10^8	2.32×10^8
Industrial demand (m^3)	3.55×10^8	3.55×10^8	3.55×10^8
Total waste (m^3)	1.04×10^9	1.04×10^9	1.02×10^9
Number of storage tanks	0	11	6
Number of artificial ponds	0	12	9
Total alternative water sources (m^3)	0	1.99×10^8	4.70×10^8
Profit ($/year)	1.14×10^9	1.37×10^9	1.40×10^9

the numerical examples several initial points were used, and the solutions converged to the same solutions.

5.6 Concluding Remarks

A general methodology for the optimal planning of water storage and distribution in a macroscopic system was presented. The proposed approach overcomes the limitations of previously reported methodologies, in which important aspects such as population growth, time value of the money and change in the precipitation patterns due to climate change were not considered. The results show the importance of implementing sustainable water management policies, which might help to reduce the pressure over natural water bodies. It can be seen that in opposition to short time projects, which might not be economically attractive due to the high investment cost of implementation, in mid and long time term projects this tendency might be reverted, resulting in economically attractive solutions. An example for the city of Morelia was presented. Three case studies were analyzed considering different policies. Even in the case when no alternative water sources are allowed, a systematic procedure for the water management might help maintain sustainable levels in the natural water sources. For the cases when alternative water sources are allowed, most of the storage device were installed during the first years of the project (requiring a strong initial investment), although additional storage was required to handle the increasing demands. Rainwater harvesting by itself has the potential to provide more than 20% of the water demands, however the change in the precipitation patterns cannot be accurately predicted and long drought periods could make it an unreliable source. In this work, a deterministic prediction regarding the precipitation patterns was taken, for this scenario the change in the precipitation pattern directly affects the amount of collected water, this means that for the last years of the project, an additional collection area is required to completely use the storage capacity. On the other hand, when both harvested rainwater and reclaimed water are used simultaneously, they have potential to reduce the consumption of natural water sources up to a 25.2%. This fact along with the increase in the profit, are incentives to promote policies aiming for the implementation of alternative water sources. For the case of the city of Morelia, one of the main problems consists in the excessively large use of water per capita per day. Thus, local governments should promote the use of alternative water sources through governmental subsides, launch social programs aiming to reduce the bad practices in households which lead to the misuse of the water and finally to invest in new infrastructure in the recollection, treatment, storage, and distribution line.

Nomenclature

Parameters

A_n^a	Collection area in location n for artificial ponds a at period t
$A_n^{\max}$	Maximum capacity of artificial ponds A in location n
A_l^s	Collection area in location l for storage tank s
A_k^{ROW}	Collection area for natural sources k
ASC	Unit water cost for agricultural use
Ce^s	Runoff coefficient for storage tanks s
Ce^a	Runoff coefficient for artificial ponds a
$CTND$	Unit treatment cost for natural sources with domestic use
$CTNA$	Unit treatment cost for natural sources with agricultural use
$CTNI$	Unit treatment cost for natural sources with industrial use
$CTAD$	Unit treatment cost of rainwater for domestic use
$CTAA$	Unit treatment cost of rainwater for agricultural use
$CTAI$	Unit treatment cost of rainwater for industrial use
CTP	Unit treatment cost of wastewater regeneration for agricultural use
$CTPE$	Unit treatment cost of wastewater regeneration for final disposal
$D_{h,t}^{as}$	Agricultural users h demands at time period t
$D_{u,t}^{di}$	Industrial users u demands at time period t
$D_{j,t}^{ds}$	Domestic users j demands at time period t
$DPWV_{k,t}$	Water collected from direct precipitation in natural sources k at time period t
DSC	Unit water sale price for domestic use
ISC	Unit water sale price for industrial use
K	Parameter for the type of land and used land

K_F	Factor to take into account annualized investment
P_t	Precipitation over the period t
P^{total}	Annual precipitation
$PCSTD$	Unit cost of transport from storage tank l to domestic sink j
$PCASD$	Unit cost of pumping from artificial pond n to domestic sink j
$PCSTA$	Unit cost of pipeline and pumping from storage tank in location l to agricultural sink h
$PCASA$	Unit cost of piping and pumping from an artificial pond in location n to agricultural sink h
$PCSTI$	Unit cost of piping and pumping from storage tank in location l to industrial sink u
$PCASI$	Unit cost of piping and pumping from artificial pond in location n to industrial sink u
$PCND$	Unit cost of piping and pumping from natural sources k to domestic main
$PCNA$	Unit cost of piping and pumping from natural sources k to agricultural main
$PCNI$	Unit cost of piping and pumping from natural sources k to industrial main
$PCTW$	Unit cost of piping and pumping from treatment plant to agricultural sinks h
$PCTI$	Unit cost of piping and pumping from industrial treatment plant to agricultural sinks h
$p^g_{k,t}$	Water collected from direct precipitation and runoff water in sources k over time period t
$r_{m,k,t}$	Segregated flowrate from the tributaries m to natural sources k over period t
$ROWV_{k,t}$	Runoff water collection in natural sources k over period t
S^{max}_l	Maximum capacity of storage tanks S in location l
VP	Factor to consider the value of investment

Binary Variables

$z^a_{n,t}$	Binary variable to select the installation of artificial ponds a in location n over the time period t
$z^s_{l,t}$	Binary variable for installing the storage tanks s in location l over time period t

Variables

$A_{n,t}$	Existing water in artificial ponds A in location n over time period t

$A_{n,t-1}$	Existing water in artificial ponds A in location n over the previous time period $t-1$
$a^{in}_{n,t}$	Water obtained from rainfall sent to artificial ponds a in location n over time period t
$a^{out,d}_{n,j,t}$	Segregated flowrate from artificial ponds a in location n sent to domestic users j over time period t
$a^{out,a}_{n,h,t}$	Segregated flowrate from artificial ponds a in location n sent to agricultural users h over time period t
$a^{out,i}_{n,u,t}$	Segregated flowrate from artificial ponds ai in location n sent to industrial users u over time period t
Ce	Runoff coefficient
$Cost^{a}_{n,t}$	Cost of artificial pond a in location n
$Cost^{s}_{l,t}$	Cost of storage tank s in location l
$cw^{d}_{j,t}$	Water consumed and lost in domestic sinks j over time period t
$cw^{di}_{u,t}$	Water consumed and lost in industrial sink u over time period t
cw^{tp}_{t}	Water reclaimed and sent to final disposal at time period t
cwi^{tpi}_{t}	Water reclaimed in industrial treatment plant and sent to final disposal over period t
$Drop^{g}_{k,t}$	Water that exceeds the maximum capacity of natural sources k over period t
$f_{j,t}$	Segregated flowrate sent from the domestic main to the domestic users j over time period t
$G_{k,t}$	Existing water in natural sources k at period t
$G_{k,t-1}$	Existing water in natural sources k in previous period $t-1$
$g^{a}_{k,t}$	Segregated flowrate from the natural sources k to main agricultural a over period t
$g^{d}_{k,t}$	Segregated flowrate from the natural sources k to main domestic d over period t
$g^{i}_{k,t}$	Segregated flowrate from the natural sources k to main industrial i over period t
$\text{int}^{in}_{j,t}$	Wastewater produced in domestic sinks j over period t
$\text{inti}^{in}_{u,t}$	Wastewater sent to industrial treatment plant over period t
int^{out}_{t}	Wastewater sent to treatment plant over period t
$\text{int}^{out,ag}_{h,t}$	Water reclaimed and sent to agricultural sinks h over period t
$\text{inti}^{out,i}_{h,t}$	Water reclaimed in industrial treatment plant and sent to agricultural sinks h over period t
$K^{a}_{F_n}$	Factor to take into account annualized investment index of ponds
$K^{s}_{F_l}$	Factor to take into account annualized investment index of tanks
$P^{a}_{n,t}$	Available precipitation in location n for artificial ponds a over period t
$P^{s}_{l,t}$	Available precipitation in location l for storage tanks s over period t
$q_{u,t}$	Segregated flowrate sent from the industrial main to the industrial users u over period t
$r_{h,t}$	Segregated flowrate sent from the agricultural main to the agricultural users h over period t

$S_{l,t}$	Existing water in storage tanks S in location l over period t
$S_{l,t-1}$	Existing water in storage tanks S in location l over the previous period $t-1$
S_l^{cap}	Capacity for storage tank
$s_{l,t}^{in}$	Water obtained from rainfall sent to storage tanks s in location l over period t
$s_{l,h,t}^{out,a}$	Segregated flowrate from storage tanks s in location l sent to agricultural users h over period t
$s_{l,j,t}^{out,d}$	Segregated flowrate from storage tanks s in location l sent to domestic users j over period t
$si_{l,u,t}^{out,i}$	Segregated flowrate from storage tanks s in location l sent to industrial users u over period t
$v_{n,t}^{a}$	Water losses in artificial ponds a over period t
$v_{w,t}^{ai}$	Water lost in industrial artificial ponds ai over period t
$v_{k,t}^{g}$	Water lost in natural sources k over period t
$v_{l,t}^{s}$	Water lost in storage tanks s over period t
$v_{b,t}^{si}$	Water lost in industrial storage tanks si over time period t

Greek Symbols

α	Exponent to take into account the economies of scale
$\delta_n^{a,\max}$	Maximum flow sent to artificial ponds a in location n over period t
$\delta_n^{a,\min}$	Minimum flow sent to artificial ponds a in location n over period t
$\delta_l^{s,\max}$	Maximum flow sent to storage tanks s in location l over period t
$\delta_l^{s,\min}$	Minimum flow sent to storage tank s in location l over period t

Sets

H	Set for agricultural sinks $(h \mid h = 1, \ldots, H)$
J	Set for domestic sinks $(j \mid j = 1, \ldots, J)$
K	Set for natural sources $(k \mid k = 1, \ldots, K)$
L	Set for location of storage tanks $(l \mid l = 1, \ldots, L)$
M	Set for tributaries $(m \mid m = 1, \ldots, M)$
N	Set for location of artificial ponds $(n \mid n = 1, \ldots, N)$
T	Set for period $(t \mid t = 1, \ldots, T)$
U	Set for industrial sinks $(u \mid u = 1, \ldots, U)$

References

Agrafioti E, Diamadopoulos E (2012) A strategic plan for reuse of treated municipal wastewater for crop irrigation on the Island of Crete. Agric Water Manag 105:57–64

Atilhan S, Mahfouz A, Batchelor B, Linke P, Abdel-Wahab A, Nápoles-Rivera F, Jiménez-Gutiérrez A, El-Halwagi M (2012) A systems-integration approach to the optimization of macroscopic water desalination and distribution networks: a general framework applied to Qatar's water resources. Clean Technol Environ Policy 14:161–171

Avila-Olivera JA, Garduño-Monroy VH (2007) Analysis for the decrement of the available water in the aquifers of Morelia. Ciencia Nicolaita 46:105–122

Barrington DJ, Prior A, Ho G (2013) The role of water auditing in achieving water conservation in the process industry. J Clean Prod 52:356–361

Brooke A, Kendrick D, Meeraus A, Raman R (2013) GAMS. A user's guide, Washington, DC

Burgara-Montero O, Ponce-Ortega JM, Serna-González M, El-Halwagi MM (2013) Incorporation of the seasonal variations in the optimal treatment of industrial effluents discharged to watersheds. Ind Eng Chem Res 52:5145–5160

Castro PM, Teles JP (2013) Comparison of global optimization algorithms for the design of water-using networks. Comput Chem Eng 52:249–261

Chen CL, Hung SW, Lee JY (2010) Design of inter-plant water network with central and decentralized water mains. Comput Chem Eng 34:1522–1531

Compain P (2012) Solar energy for water desalination. Procedia Eng 46:220–227

CONAGUA (2009) National Water Commission. Determination of the availability of water in the Morelia-Querendaro Aquifer (1602). National Council of Water, State of Michoacán

CONAGUA (2011) National Water Commission. Water statistics in Mexico. Available at http://www.conagua.gob.mx/CONAGUA07/Publicaciones/Publicaciones/SGP-1-11-EAM2011.PDF. Accessed June 2013

CONAGUA-SMN (2011) National Water Commission—National Meteorological System, weather report in Mexico. http://smn.cna.gob.mx/climatologia/analisis/reporte/Anual2011.pdf. Accessed June 2013

CONAPO (2008) National Council of Population. Projections of the economically active population of Mexico and the states, 2005–2050. http://www.portal.conapo.gob.mx/00cifras/PEA/pea.pdf. Accessed June 2013

Eroksuz E, Rahman A (2010) Rainwater tanks in multi-unit buildings: a case study for three Australian cities. Resour Conserv Recycl 54:1449–1452

Farreny R, Morales-Pinzón T, Guisasola A, Tayà C, Rieradevall J, Gabarrell X (2011) Roof selection for rainwater harvesting: quantity and quality assessments in Spain. Water Res 45:3245–3254

Foo DCY (2009) State-of-the-art review of pinch analysis techniques for water network synthesis. Ind Eng Chem Res 48:5125–5159

Fu J, Cai T, Xu Q (2012) Coupling multiple water-reuse network designs for agile manufacturing. Comput Chem Eng 45:62–71

Gouws JF, Majozi T, Foo DCY, Chen CL, Lee JY (2010) Water minimization techniques for batch processes. Ind Eng Chem Res 49:8877–8893

Howari FM, Ghrefat HA (2011) Jordan: environmental status of water, soil and air. In: Jerome ON (ed) Encyclopedia of environmental health. Elsevier, Burlington, pp 323–334

INEGI (2013a) National Institute of Statistics and Geography. National census of population and housing. Available at http://www.inegi.org.mx/sistemas/mexicocifras/default.aspx?e=16. Accessed July 2013

INEGI (2013b) National Institute of Statistics and Geography. Available at http://gaia.inegi.org.mx/scince2/viewer.html. Accessed July 2013

Jezowski J (2010) Review of water network design methods with literature annotations. Ind Eng Chem Res 49:4475–4516

Jhorar RK, Smit AAMFR, Roest CWJ (2009) Assessment of alternative water management options for irrigated agriculture. Agric Water Manag 96:975–981

Khor CS, Chachuat B, Shah N (2012) A superstructure optimization approach for water network synthesis with membrane separation-based regenerators. Comput Chem Eng 42:48–63
Li YP, Huang GH, Nie SL, Chen X (2011) A robust modeling approach for regional water management under multiple uncertainties. Agric Water Manag 98:1577–1588
Lira-Barragán LF, Ponce-Ortega JM, Serna-González M, El-Halwagi MM (2010) An MINLP model for the optimal location of a new industrial plant with simultaneous consideration of economic and environmental criteria. Ind Eng Chem Res 50:953–964
Lira-Barragán LF, Ponce-Ortega JM, Serna-González M, El-Halwagi MM (2011) Synthesis of water networks considering the sustainability of the surrounding watershed. Comput Chem Eng 35:2837–2852
Lira-Barragán LF, Ponce-Ortega JM, Nápoles-Rivera F, Serna-González M, El-Halwagi MM (2012) Incorporating property-based water networks and surrounding watersheds in site selection of industrial facilities. Ind Eng Chem Res 52:91–107
Liu S, Konstantopoulou F, Gikas P, Papageorgiou LG (2011) A mixed integer optimisation approach for integrated water resources management. Comput Chem Eng 35:858–875
Martinez-Gomez J, Burgara-Montero O, Ponce-Ortega JM, Nápoles-Rivera F, Serna-González M, El-Halwagi MM (2013) On the environmental, economic and safety optimization of distributed treatment systems for industrial effluents discharged to watersheds. J Loss Prev Process Ind 26(5):908–923
Nápoles-Rivera F, Ponce-Ortega JM, El-Halwagi MM, Jiménez-Gutiérrez A (2010) Global optimization of mass and property integration networks with in-plant property interceptors. Chem Eng Sci 65:4363–4377
Nápoles-Rivera F, Bin Mahfouz A, Jiménez-Gutiérrez A, El-Halwagi MM, Ponce-Ortega JM (2012) MINLP model for optimal biocide dosing and maintenance scheduling of seawater cooled plants. In: Ian David Lockhart B, Michael F (eds) Computer aided chemical engineering, vol 30. Elsevier, Amsterdam, pp 872–876
Nápoles-Rivera F, Serna-González M, Bin-Mahfouz A, Jiménez-Gutiérrez A, El-Halwagi MM, Ponce-Ortega JM (2013) Simultaneous optimization of energy management, biocide dosing and maintenance scheduling of thermally integrated facilities. Energy Convers Manage 68:177–192
Nápoles-Rivera F, Serna-González M, El-Halwagi MM, Ponce-Ortega JM (2013) Sustainable water management for macroscopic systems. J Clean Prod 47:102–117
Pereira LS, Oweis T, Zairi A (2002) Irrigation management under water scarcity. Agric Water Manag 57:175–206
Rahman A, Keane J, Imteaz MA (2012) Rainwater harvesting in greater Sydney: water savings, reliability and economic benefits. Resour Conserv Recycl 61:16–21
Rojas-Torres MG, Nápoles-Rivera F, Ponce-Ortega JM, Serna-González M, El-Halwagi M (2014) Comput Chem Eng 69:1–15
Rubio-Castro E, Ponce-Ortega JM, Serna-González M, Jiménez-Gutiérrez A, El-Halwagi MM (2011) A global optimal formulation for the water integration in eco-industrial parks considering multiple pollutants. Comput Chem Eng 35:1558–1574
Rubio-Castro E, Ponce-Ortega JM, Serna-González M, El-Halwagi MM (2012) Optimal reconfiguration of multi-plant water networks into an eco-industrial park. Comput Chem Eng 44:58–83
Shatat M, Worall M, Riffat S (2013) Opportunities for solar water desalination worldwide: review. Sustain Cities Soc 9:67–80
SMN (2010) National Meteorological System. Climatologic averages 1951–2010. Available at http://www.cm.colpos.mx/meteoro/progde/norm/n5110/. Accessed July 2013
Sotelo-Pichardo C, Ponce-Ortega JM, El-Halwagi MM, Frausto-Hernández S (2011) Optimal retrofit of water conservation networks. J Clean Prod 19:1560–1581
Teles JP, Castro PM, Matos HA (2012) Global optimization of water networks design using multiparametric disaggregation. Comput Chem Eng 40:132–147
Tudor R, Lavric V (2011) Dual-objective optimization of integrated water/wastewater networks. Comput Chem Eng 35:2853–2866

Chapter 6
Rainwater Collecting Systems for a Residential Development

6.1 Introduction

One of the most precious resources on Earth is water, which is essential to life. As a result, water demands have grown significantly in recent years due to population growth and changes in precipitation patterns brought on by climate change. Thus, water scarcity exists in several densely inhabited areas of the world. To meet the water demands, this has led to the development of water harvesting strategies (Abdulla and Al-Shareef 2009; Appan 2000; Chilton et al. 2000). To meet specific demands and reduce the amount of freshwater consumed, many solutions for water reuse, recycling, and regeneration have been presented in the industry (Deng and Feng 2011; Fu et al. 2012; Ng et al. 2010; Yu et al. 2013). Specifically, methods based on mathematical programming have proven to be effective tools for designing industrial water networks in a way that minimizes freshwater consumption and wastewater discharge into the environment (Burgara-Montero et al. 2012; Deng et al. 2013; Khor et al. 2012; Nápoles-Rivera et al. 2010; 2012; Ponce-Ortega et al. 2011; 2012; Rojas-Torres et al. 2012; Rubio-Castro et al. 2010; 2011; 2012; 2013; Sotelo-Pichardo et al. 2011; 2013; Vázquez-Castillo et al. 2013). Conversely, meeting domestic water demands is a significant issue in areas where freshwater is scarce, as household water usage accounts for around 10% of global freshwater demand. Based on heuristic approaches, which often fail to yield an optimal solution for the problem at hand, several strategies have been implemented in this context to meet domestic water demands through rainwater harvesting, as illustrated by Fig. 6.1 (Domenech and Saurí 2011; Domenech et al. 2012; Gikas and Angelakis 2009; Hadadin et al. 2010; Li et al. 2010; Morales-Pinzón 2012). Several methodical strategies have recently been put out to meet household water demands. These strategies are based on the best practices for distributing and storing the available water. A model was presented by Khastagir and Jayasuriya (2010) for the ideal tank size for rainwater used for household purposes. A decision assistance tool for sizing rainwater tanks connected to large roofs was presented by Imteaz et al. (2011). A mixed integer linear programming

J. M. Ponce-Ortega et al., *Sustainable Water Systems*, Water Resources Development and Management, https://doi.org/10.1007/978-3-031-72498-5_6

model was introduced by Lu et al. (2013) to use desalinated water to meet household demands. To account for desalinated water when optimizing water distribution in Qatar, Atilhan et al. (2012) introduced a system-integration strategy. A method for sizing storage tanks for home rainwater collecting was given by Campisano and Modica (2012). Imteaz et al. (2012a, b) analyzed for Australia and Nigeria the use of rainwater tanks. To managing water in cities, Nápoles-Rivera et al. (2013) reported a methodical optimization tool. Rozos and Markropoulos (2011) and Rozo et al. (2010) presented various plans for rainwater-based water recycling. A multi-criteria optimization strategy for Shanghai's sustainable use of rainfall was presented by Lu et al. (2013). An integrated nonlinear stochastic optimization model was presented by Huang et al. (2013) for the assessment and planning of urban water resources. Unfortunately, there is no formal mathematical programming model that ensures the best possible solution for the optimal design of a rainwater harvesting system to meet domestic demands in residential development through a systematic optimization technique. To meet the household water demands of residential development, this chapter presents an approach to rainwater collection system design that is based on formal optimization. To meet water demands at the lowest possible cost and with the least amount of freshwater consumption, this system considers the best way to collect rainwater, condition the collecting area, store it (using a multi-tank system), and install the necessary pipes and pumps. The model takes into consideration the seasonal variation in water demands; typically, during the hottest months of the year, there is an increase in demand, while during the coldest months, there is a decrease. The model also considers the seasonal variations in precipitation, which have an impact on the water availability in the area under consideration all year round. The total cost includes the cost of using fresh water, the capital costs of collecting systems, storing materials, pipes, and pumps, as well as the system's pumping expenses.

6.2 Problem Statement

The addressed problem can be stated as follows. Given a residential development with a known location, geographic distribution, number of houses, and specific water demands during the year. The problem consists of determining the optimal rainwater harvesting system to satisfy domestic demands at the minimum cost and with the minimum freshwater consumption. The solution must determine the collecting water required as well as the storage, pipes, and pumps and the optimal operation of the system throughout the year. The economic objective function involves the capital cost for the collecting system, storage, pipes, and pumps, as well as the fresh water and pumping costs.

To solve this multi-objective optimization model the superstructure presented in Fig. 6.2 is considered, where there is an available area in the buildings that can be conditioned for rainwater harvesting, and the model contemplates the possibility of conditioning all or part of this area to catch rainwater; likewise the implemented model contemplates the possibility to install several water storages (i.e. tanks, dams,

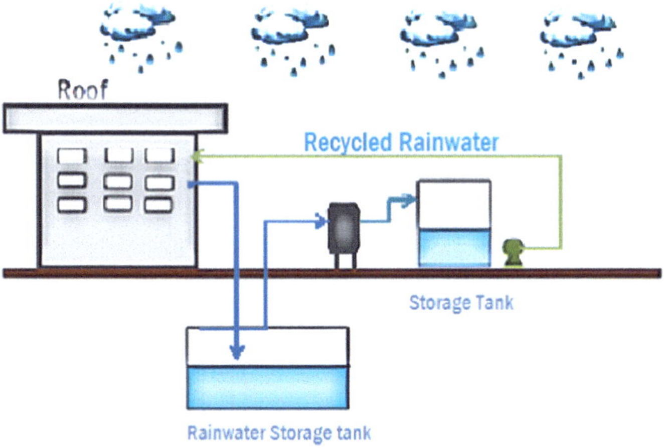

Fig. 6.1 Rainwater harvesting system. Printed with permission of "*Simultaneous design of water reusing and rainwater harvesting systems in a residential complex*", *Computers and Chemical Engineering*, Copyright (2015), Elsevier

etc.) in specific places where it is possible, the maximum size for these storages is controlled by operational, technical and geographic constraints, and the model calculates the number and size for these storage devices, the number and diameter for the pipes, including pipes for the collecting area to the storages and from the storages to the elevated reservoir The model also calculates the optimal size for the elevated reservoir, and the costs linked to the freshwater consumption, and the operational cost for the collecting system, tanks, storages, and pumps. Lastly, the model is a multi-objective optimization approach considering simultaneously the minimization of the total cost and the freshwater consumption.

6.3 Implemented Optimization Model

This section presents the implemented optimization model for rainwater harvesting, storing, and distribution into a residential development based on the superstructure shown in Fig. 6.2, where i is the number of catchment areas that are delineated on the roof of the buildings, m represents the periods considered in the model (months), the processes being modeled are in general better described in shorter time scales, nonetheless considering shorter time scales should increase exponentially the size of the model, thus to keep the model in a manageable size the monthly step is used. j represents the number of storage units that can be installed in the residential development and whose existence, size, and operation are optimized, while l is the water users in the residential development, which have specific demands for each period. So, the implemented mathematical model is presented as follows.

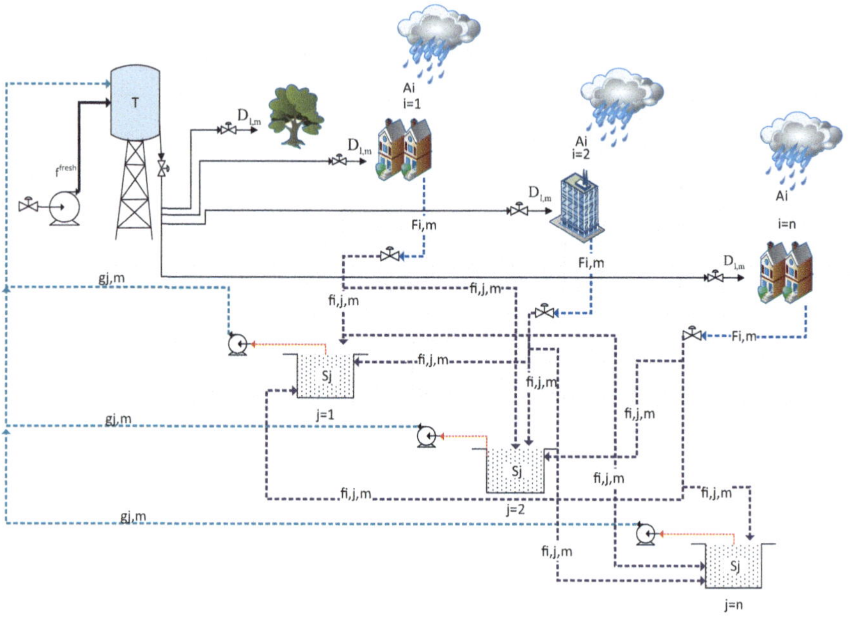

Fig. 6.2 Superstructure for rainwater harvesting, storing, and distribution into a residential development. Printed with permission of "*Simultaneous design of water reusing and rainwater harvesting systems in a residential complex*", *Computers and Chemical Engineering*, Copyright (2015), Elsevier

6.3.1 Balance in the Catchment Areas

Water is collected and sent to the storage; the amount of water collected depends on the collection area and the monthly rainfall measured in meters per month. Thus, the rainwater obtained at the facility i during the period m ($F_{i,m}$) is equal to the conditioned surface (A_i) times the average precipitation during this period ($\mathrm{P_m}$):

$$F_{i,m} = \mathrm{P_m} A_i, \quad \forall i \in I, \quad \forall m \in \mathrm{M} \tag{6.1}$$

the average precipitation is obtained from meteorological reports.

6.3.2 Balances in the Splitters After the Catchment Areas

There are required water balances for the splitters after the catchment areas, where the harvested water ($F_{i,m}$) in each catchment area i is equal to the sum of the flow rate sent to each storage device j ($f_{i,j,m}$) during each month m:

$$F_{i,m} = \sum_{j} f_{i,j,m}, \quad \forall i \in I, \quad \forall m \in \mathrm{M} \tag{6.2}$$

6.3.3 Balances in the Storage Units

There are needed water balances for the potential storage units to calculate their existence, size, and operation. Thus, the accumulated water in the storage j at the end of period m ($S_{j,m}$) is equal to the water available at the end of the previous period ($S_{j,m-1}$) plus the sum of the rainwater sent from the catchment areas i ($f_{i,j,m}$) minus the distributed water to the elevated reservoir ($g_{j,m}$) during the period m:

$$S_{j,m} = S_{j,m-1} + \sum_{i} f_{i,j,m} - g_{j,m}, \quad \forall j \in J, \quad \forall m \in \mathrm{M} \tag{6.3}$$

Equation (6.3a) simulates the balance of water in the storage tank j to guarantee that the water available in the storage tanks at the beginning of the first period is equal to the water available at the end of the last period of the year:

$$S_{j,m=0} = S_{j,m=12}, \quad \forall j \in J \tag{6.3a}$$

6.3.4 Balance in the Elevated Reservoir

The water balance for the elevated reservoir states that the water available at the end of the period m (T_m) is equal to the water available at the end of the previous period (T_{m-1}) plus the sum of water from the storage devices ($g_{j,m}$) and the freshwater consumed (taken from the public network) during each month (f_m^{fresh}) minus the water used by the final consumers in the residential development ($\mathrm{D}_{\mathrm{l,m}}$) during each period:

$$T_m = T_{m-1} + \sum_{j} g_{j,m} + f_m^{fresh} - \sum_{l} \mathrm{D}_{\mathrm{l,m}}, \quad \forall m \in \mathrm{M} \tag{6.4}$$

The water demand for the users ($\mathrm{D}_{\mathrm{l,m}}$) is given and this is calculated as a function of the number of inhabitants in each house and the season of the year.

Equation (6.4a) is the water balance in the elevated reservoir needed to ensure that the water available at the end of the first period is equal to the one of the last period of the year:

$$T_{m=0} = T_{m=12} \tag{6.4a}$$

6.3.5 Existence of the Units

Logical relationships are needed to determine if a potential unit is used in the optimal solution. If a given unit is needed in the optimal solution, then, the associated capital cost must be determined considering the size and existence of the unit. Then, there are needed binary variables to indicate the existence of such units (i.e. the binary variable is one when the unit exists, otherwise is zero).

When a collecting area (A_i) in each building i is conditioned to catch rainwater, afterward the binary variable (y_i^{capt}) is activated:

$$A_i \leq \mathrm{A}_{\mathrm{i}}^{\max} y_i^{capt}, \quad \forall i \in I \tag{6.5}$$

$\mathrm{A}_{\mathrm{i}}^{\max}$ is the maximum area available in building i and this is known before the optimization; next, when A_i is greater than zero the binary variable y_i^{capt} is one.

To calculate the existence of a storage device j, the binary variable y_i^{stg} is used. Then, to activate this binary variable the following relationship is used:

$$S_j^{\max} \leq \mathrm{STG}_{\mathrm{j}}^{\max} y_j^{stg}, \quad \forall j \in J \tag{6.6}$$

$\mathrm{STG}_{\mathrm{j}}^{\max}$ is the maximum capacity for the storage device in place j.

Additionally, to activate the existence for the pipe segments and to determine the fixed charge associated with the capital cost for the pipes, the next logical relationships are used:

$$f_{i,j}^{cap} \leq \mathrm{F}^{\max} z_{i,j}, \quad \forall i \in I, \quad \forall j \in J \tag{6.7}$$

$$g_j^{cap} \leq \mathrm{G}^{\max} x_j, \quad \forall j \in J \tag{6.8}$$

$z_{i,j}$ and x_j are binary variables related to existence of the pipe segments for the links between the harvesting area i-storage j and storage j-elevated reservoir, respectively. Now, $\mathrm{F}^{\max}$ and $\mathrm{G}^{\max}$ are the maximum flow rates allowed in the corresponding pipes.

6.3.6 Maximum Capacity for all Time Periods

The maximum capacity for the storage reservoirs ($S_j^{\max}$) is given by the maximum amount of water that is collected in each period ($S_{j,m}$) and this is determined as follows:

$$S_{j,m} \leq S_j^{\max}, \quad \forall j \in J, \quad m \in \mathrm{M} \tag{6.9}$$

Furthermore, the outflow from the splitters ($f_{i,j,m}$) must be lower than the maximum storage capacity ($S_j^{\max}$):

$$\sum_i f_{i,j,m} \leq S_j^{\max}, \quad \forall j \in J, \quad m \in \mathrm{M} \tag{6.9a}$$

Likewise, the capacity for the elevated tank ($T^{\max}$) must be greater than the one required in each period (T_m):

$$T_m \leq T^{\max}, \quad \forall m \in \mathrm{M} \tag{6.10}$$

Additionally, the freshwater sent to the elevated reservoir (f_m^{fresh}) must be lower than the capacity for this tank:

$$f_m^{fresh} \leq T^{\max}, \quad \forall m \in \mathrm{M} \tag{6.10a}$$

The maximum feeding stream to any storage tank ($f_{i,j,m}$) must be lower than the maximum capacity for the pipe of segment ($f_{i,j}^{cap}$):

$$f_{i,j,m} \leq f_{i,j}^{cap}, \quad \forall i \in I, \quad \forall j \in J, \quad \forall m \in \mathrm{M} \tag{6.11}$$

Additionally, the output stream from each storage tank (g_j^{cap}) must be lower than the capacity linked to the tank:

$$g_{j,m} \leq g_j^{cap}, \quad \forall j \in J, \quad \forall m \in \mathrm{M} \tag{6.12}$$

6.3.7 Costs

One objective function is the minimization for the total annual cost, which considers the capital costs for the collecting areas, storage devices, pumps, and pipes, and the freshwater cost and operating costs for pumping, treatment, and storage, which are calculated as follows.

6.3.8 Capital Cost for the Collecting Area

The capital cost for the collecting area is calculated considering the fixed part ($\mathrm{CF}_i^{\mathrm{captation}}$) that is independent of the size, which is multiplied by the binary variable y_i^{capt} to activate only when the collecting area is used, plus the variable charge that depends on the collecting area used and that is required to be conditioned (A_i):

$$CaptationCost = \sum_i \left(\mathrm{CF}_i^{\mathrm{captation}} y_i^{capt} + \mathrm{CV}_i^{\mathrm{captation}} (A_i)^{\delta} \right) \tag{6.13}$$

$\mathrm{CF}_{\mathrm{i}}^{\mathrm{captation}}$ and $\mathrm{CV}_{\mathrm{i}}^{\mathrm{captation}}$ are the fixed and variable charges for the capital cost linked to the location *i*, and δ is an exponent to account for the economies of scale. The unit costs depend on the type of area considered.

6.3.9 Capital Cost for the Storage Devices

The capital cost for the storage devices (*StorageCost*) involve the fixed and variable parts. The first part is activated when the storage exists (y_j^{stg} equals one) whereas the variable part considers the size of the unit:

$$StorageCost = \sum_j [\mathrm{CF}_{\mathrm{j}}^{\mathrm{storage}} y_j^{stg} + \mathrm{CV}_{\mathrm{j}}^{\mathrm{storage}} (S_j^{\mathrm{max}})^{\alpha}] \tag{6.14}$$

$\mathrm{CF}_{\mathrm{j}}^{\mathrm{storage}}$ and $\mathrm{CV}_{\mathrm{j}}^{\mathrm{storage}}$ are the unit fixed and variable part for the capital costs of the storage devices, and α is an exponent to account for the economies of scale. The size for the storage device (S_j^{max}) is the greatest required in all periods.

The capital cost for the elevated reservoir (*StorageCostT*) is determined as follows:

$$StorageCostT = [\mathrm{CF}^{\mathrm{ElevTank}} + \mathrm{CV}^{\mathrm{ElevTank}} (T^{max})^{\alpha}] \tag{6.15}$$

$\mathrm{CF}^{\mathrm{ElevTank}}$ and $\mathrm{CV}^{\mathrm{ElevTank}}$ are the unit fixed and variable capital costs for the elevated reservoir, and α is the exponent to account for the economies of scale. The elevated reservoir is always used, hence a binary variable is not needed to calculate its existence, but its cost has to be determined and included in the total cost.

6.3.10 Capital Costs Associated to the Pipes

There are two types of new pipes associated to the rainwater harvesting system for a residential development; one is from the collecting areas to the storages and the other one is from the storages to the elevated reservoir (It is considered that any other pipes required for the water distribution currently exist). To calculate the capital costs for the new pipes, it is needed to consider the fixed ($\mathrm{CF}_{\mathrm{i,j}}^{\mathrm{pipe}}$ and $\mathrm{CF}_{\mathrm{j}}^{\mathrm{pipe}}$) and variable ($\mathrm{CV}_{\mathrm{i,j}}^{\mathrm{pipe}}$ and $\mathrm{CV}_{\mathrm{j}}^{\mathrm{pipe}}$) parts and the exponent for the economies of scale (β) for the different segments considered:

$$PipeCost = \sum_i \sum_j \left[\mathrm{CF}_{\mathrm{i,j}}^{\mathrm{pipe}} z_{i,j} + \mathrm{CV}_{\mathrm{i,j}}^{\mathrm{pipe}} (f_{i,j}^{cap})^{\beta}\right] + \sum_j \left[\mathrm{CF}_{\mathrm{j}}^{\mathrm{pipe}} x_j + \mathrm{CV}_{\mathrm{j}}^{\mathrm{pipe}} (g_j^{cap})^{\beta}\right] \tag{6.16}$$

$f_{i,j}^{cap}$ and g_j^{cap} are the greatest flow rates for all time periods for the segments involved. Likewise, the unit costs consider the distance associated to each segment, and the fixed costs are activated through the corresponding binary variables.

6.3.11 Treatment Costs

The rainwater collected needs a treatment (usually chlorination) to be used in the residential development for domestic purposes:

$$TreatmentCost = \sum_i \sum_j \sum_m \mathrm{Cost}^{\mathrm{Treatment}} f_{i,j,m} \tag{6.17}$$

TreatmentCost is the annual treatment cost and $\mathrm{Cost}^{\mathrm{Treatment}}$ is the unit treatment cost.

6.3.12 Pumping Cost

The annual pumping cost (*PumpingCost*) for the rainwater harvested is:

$$PumpingCost = \sum_i \sum_j \sum_m \mathrm{Cost}_{\mathrm{i,j}}^{\mathrm{pumping}} f_{i,j,m} + \sum_j \sum_m \mathrm{Cost}_{\mathrm{j}}^{\mathrm{pumping}} g_{j,m} \tag{6.18}$$

$\mathrm{Cost}_{\mathrm{i,j}}^{\mathrm{pumping}}$ and $\mathrm{Cost}_{\mathrm{j}}^{\mathrm{pumping}}$ are the pumping unit costs associated to the segments from the collecting area to the storages and from the storages to the elevated reservoir, respectively.

6.3.13 Fresh Water Cost

The freshwater cost (*FreshCost*) is calculated considering the unit freshwater cost ($\mathrm{Cost}_{\mathrm{m}}^{\mathrm{fresh}}$):

$$FreshCost = \sum_m \mathrm{Cost}_{\mathrm{m}}^{\mathrm{fresh}} f_m^{fresh} \tag{6.19}$$

The possibility of having variations in the unit cost of fresh water is accounted in the formulation. This is because in some regions the water cost depends on the season of the year, during rainy seasons the cost is lower due to the availability of the resource, in the meantime during drought seasons the price of water rises as a measure to limit the consumption and avoid cutoffs in the supply.

6.3.14 Total Cost

One objective function is the minimization of the total annual cost for the system:

$$TAC = K_F(PipeCost + StorageCost + StorageCostT + CaptationCost) \\ + (PumpingCost + TreatmentCost + FreshCost) \quad (6.20)$$

K_F is the factor used to annualize the capital costs and this considers the years that will be used the system and the interest rate.

6.3.15 Total Fresh Water Consumed

The other objective function is to minimize the freshwater consumption (*FreshTotal*) during all periods:

$$FreshTotal = \sum_m f_m^{fresh} \quad (6.21)$$

6.3.16 Objective Function

The objective function is stated as the simultaneous minimization for the total annual cost and the freshwater consumption:

$$Min\{TAC; FreshTotal\} \quad (6.22)$$

The optimization model is a multi-objective mixed-integer nonlinear programming (MINLP) problem, where several scenarios are analyzed to yield a Pareto front that constitutes a set of optimal solutions that compensate both objectives. This Pareto front can be useful to choose the solution that best fits the investment and water reduction targets of the decision maker. To solve each MINLP problem, the model was coded in the software GAMS (Brooke et al. 2013), and the solvers DICOPT/CONOPT/CPLEX were used to solve the associated MINLP/NLP/LP problems.

6.4 Case Study

To verify the applicability of the presented optimization approach for the rainwater harvesting system for domestic use into a residential development; a case study for a residential development in Morelia, México was considered. This residential development corresponds to the one called "*Fraccionamiento Fresnos-Arboledas*". The city of Morelia has serious problems to meet the water demands, mainly because this city has a population of 729,279 inhabitants (INEGI 2010), which signify a consumption of 90,168,374.77 m^3/year (CONAGUA 2011). To fulfill these water demands, Morelia accounts for a dam and 105 deep wells, which are recharged by rainwater. Though, due to the population growth and the change in the precipitation patterns, the available water to meet the domestic demands has drastically diminished, which has represented serious problems and has produced the increment in the cost of freshwater. Consequently, this has resulted in a serious social problem that needs to be properly addressed. Furthermore, Morelia has two main seasons during the year, the rainy season goes from May to October, while the rest of the year (November to April) precipitation considerably decreases (Table 6.1). The city aquifer reservoir has a deficit due to the overutilization, mainly during the dry season. Although during the rainy season the excess water (water that cannot be used naturally to recharge the natural reservoirs) is not properly used because there is not an enough storage capacity. Hence, the optimal rainwater harvesting and distribution for a residential development of this city may help to meet the water demands, taking advantage of the excess water in the rainy season and palliating the situation during drought periods. Also, the problem consists in calculating the optimal rainwater harvesting system to satisfy the water demands in the mentioned residential development. The residential development was divided in nine sections to simplify the problem, and this is shown in Fig. 6.3. Sections 6–9 are gardens, and the rest are houses, the white-yellow area are terrain without construction and possible future expansions. Table 6.2 presents the potential area for each section that can be used to capture rainwater; whereas sections 6–9 are not able to be conditioned to capture rainwater. It is necessary to install eleven tanks presented in Fig. 6.3. Besides, there is required to install an elevated reservoir to store the water that is distributed to the residential development, and whose size must be optimized. The water demands for the nine areas during each month of the year are presented in Table 6.3. The water consumption is higher during the hottest months.

The problem then consists of 1266 continuous variables, 82 binary variables and 1547 constraints, and the solution for each point of the Pareto curve was solved in an average of 2.23 s of CPU time in a computer with an Intel i7 processor with 8 MB of RAM. The set of optimal solutions that compensates the *TAC* and freshwater consumption determined using the presented optimization approach are shown in Fig. 6.4, where solutions above the Pareto curve represents suboptimal solutions, whereas solutions below are infeasible solutions. Four scenarios are identified in this Pareto curve for further discussion. First, Scenario A is the solution for the minimum total annual cost (but also this represents the highest freshwater consumption in the

Table 6.1 Average rainfall in Morelia México

Month	Precipitation (mm)	Month	Precipitation (mm)
January	1.8	July	175
February	10	August	163
March	10	September	119
April	10	October	53
May	43	November	15
June	137	December	13

Printed with permission of "*Simultaneous design of water reusing and rainwater harvesting systems in a residential complex*", *Computers and Chemical Engineering*, Copyright (2015), Elsevier

Fig. 6.3 Schematic representation of the case study. Printed with permission of "*Simultaneous design of water reusing and rainwater harvesting systems in a residential complex*", *Computers and Chemical Engineering*, Copyright (2015), Elsevier

Pareto curve); Scenario B is the solution for the minimum freshwater consumption (but also this represents the solution with the highest cost). Two additional scenarios (C and D) that compensate the contradicting objectives are identified.

6.4.1 Scenario A

This solution gives the minimum cost, which uses five storage tanks of eleven potential options, the capacity of the used storage tanks is presented in Fig. 6.5 that presents

Table 6.2 Available area for catching rainwater in the residential development of the case study

Section (sinks/sources)	No. houses	No. inhabitants	Available harvesting area (m^2)
1	93	372	29,760
2	94	376	30,080
3	79	316	25,280
4	84	336	26,880
5	111	444	35,520
6	–	–	–
7	–	–	–
8	–	–	–
9	–	–	–

Table 6.3 Water demands for each section per day over the year in m^3

Section-Sinks	Jan	Feb	Mar	Apr	May	June	July	Aug	Sept	Oct	Nov	Dec
1	17	17	14	14	16	16	18	18	15	15	13	13
2	17	17	14	14	16	16	18	18	15	15	13	13
3	14	14	12	12	14	14	15	15	13	13	11	11
4	15	15	13	13	14	14	16	16	14	14	12	12
5	20	20	17	17	19	19	21	21	18	18	16	16
6	0	0	0.5	0.5	1	1	1	1	0.5	0.5	0	0
7	0	0	0.5	0.5	1	1	1	1	0.5	0.5	0	0
8	0	0	0.5	0.5	1	1	1	1	0.5	0.5	0	0
9	0	0	0.5	0.5	1	1	1	1	0.5	0.5	0	0

the collecting area used in this scenario, where only Sections 1 and 4 were considered for rainwater harvesting. The schematic representation for the solution of Scenario A is presented in Fig. 6.5, while Table 6.4 presents the costs. This solution has a reduction of 80.1% in the freshwater consumption for the use of harvested rainwater, furthermore this solution has a total annual cost 96.4% lower than the current case when only freshwater is used. The reduction in the cost is a consequence of the seasonal dependence of water cost (i.e. during dry months water is more expensive), hence rainwater is stored and used during these periods. This yields a nonlinear relation between the reduction of freshwater consumption and total savings.

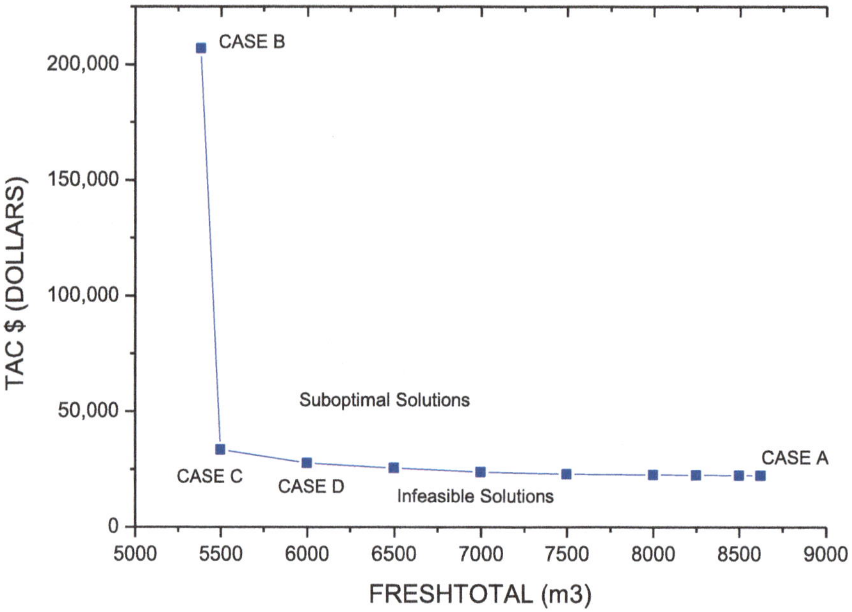

Fig. 6.4 Pareto optimal solutions for the case study. Printed with permission of "*Simultaneous design of water reusing and rainwater harvesting systems in a residential complex*", *Computers and Chemical Engineering*, Copyright (2015), Elsevier

Fig. 6.5 Solution for Scenario A of the case study. Printed with permission of "*Simultaneous design of water reusing and rainwater harvesting systems in a residential complex*", *Computers and Chemical Engineering*, Copyright (2015), Elsevier

Table 6.4 Costs associated to the solutions of Scenarios A, B, C and D

Concept	Scenario A	Scenario B	Scenario C	Scenario D
Treatment cost ($/year)	2189.4	5278.6	5064.8	4225.0
Elevated tank cost($/year)	14,476.3	13,359.2	94,905.5	57,683.3
Pipe cost($/year)	2643.0	166,915.0	7187.7	5371.2
Pumping cost ($/year)	1509.6	6249.7	3847.3	2874.5
Storage cost($/year)	2323.5	5985.5	5985.5	5985.5
Harvesting cost($/year)	22,850.9	58,132.3	53,260.2	44,261.7
Fresh water cost($/year)	14,807.6	9057.1	8800.0	9600.0
TAC($/year)	22,736.0	207,271.6	33,846.1	28,029.8

6.4.2 Scenario B

This solution corresponds to the minimum freshwater consumption (i.e. the maximum use of harvested rainwater), nevertheless it corresponds to the highest cost because the high capital costs required for conditioning the harvesting areas and installing the storage tanks. In this Scenario B, the eleven storage tanks (Table 6.5) were used, and all the available harvesting rainwater area was conditioned (Table 6.6). The schematic representation for the solution of this Scenario B is shown in Fig. 6.6 and the associated costs are given in Table 6.4. The cost for this Scenario is 87.6% greater than the one of the Scenario B; while the freshwater consumption is 37.6% lower than the one of Scenario A. This is the best environmental solution (due the reduction of the freshwater consumption), but this represents the highest inversion. Up to 87.6% of the water demand can be meet using rainwater. For this scenario, water is collected and used in consecutive periods, this means that despite that water could be stored and used during dry seasons to reduce the cost (as in scenario A) the preference is given to the freshwater reduction. This is achieved by using all the available rainwater and then recharging the storage devices. This operation requires less storage but in contrast requires a higher investment in pipe networks and pumping.

6.4.3 Scenario C

This scenario represents a solution that compensates both objectives (*TAC* and total freshwater consumption); in this scenario the associated *TAC* is $33,846.14 USD, which is 32.83% greater than the minimum one obtained in Scenario A; this solution has a total freshwater consumption of 5500 m^3, which is 2.20% greater than the one obtained for the minimum of Scenario B. This is an interesting solution because the investment required is not so significant like the one of Scenario B, but the reduction

Table 6.5 Used capacity for storage tanks in the different scenarios (m^3)

No. STG	Scenario A	Scenario B	Scenario C	Scenario D	
1	1000	1000	1000	1000	
2	700	700	700	700	
3	–	1000	1000	1000	
4	–	900	900	900	
5	1100	1100	1100	1100	
6	838	1000	1000	1000	
7	–	900	900	900	
8	800	800	800	800	
9	–	850	850	850	
10	–	950	950	950	
11	–	1500	1500	1500	

Table 6.6 Used harvesting area for the scenarios of the solution of the case study (m^2)

Harvesting area (i)	Scenario A	Scenario B	Scenario C	Scenario D
1	16,594.0	7088.7	28,678.9	29,760.0
2	–	9090.9	–	5704.0
3	–	22,596.6	–	–
4	10,303.0	20,617.7	11,532.8	12,142.8
5	–	5454.5	11,693.0	14,615.4

in the freshwater consumption is like the one of Scenario B. The solution of this Scenario C is shown in Fig. 6.7, where 11 storage reservoirs are required, whose capacity is presented in Table 6.5. Additionally, the total rainwater harvesting area required for this scenario is 51,904.75 m^2, which is 19.96% lower than the one of Scenario B. In this Scenario C, the main cost corresponds to the one associated to the elevated reservoir cost.

6.4.4 Scenario D

This Scenario D represents a solution that balances both objectives; therefore, the freshwater consumption limit is higher than the used in Scenario C. This way, the total annual cost is $28,029.82 and the associated freshwater consumption is 6000 m^3. Here, 11 storage reservoirs are needed, whose capacity is presented in Table 6.5.

No. STG (j)	CAPACITY
1	1000
2	700
3	1000
4	900
5	1100
6	1000
7	900
8	800
9	850
10	950
11	1500

Captation area (i)	CAPACITY
1	7088
2	9090.90
3	22596.57
4	20617
5	5454

Fig. 6.6 Solution for Scenario B of the case study. Printed with permission of "*Simultaneous design of water reusing and rainwater harvesting systems in a residential complex*", *Computers and Chemical Engineering*, Copyright (2015), Elsevier

No. STG (j)	CAPACITY
1	1000
2	700
3	1000
4	900
5	1100
6	1000
7	900
8	800
9	850
10	950
11	1500

Captation area (i)	CAPACITY
1	28678
2	
3	
4	11532
5	11693

Fig. 6.7 Solution for Scenario C of the case study. Printed with permission of "*Simultaneous design of water reusing and rainwater harvesting systems in a residential complex*", *Computers and Chemical Engineering*, Copyright (2015), Elsevier

Fig. 6.8 Solution for Scenario D of the case study. Printed with permission of "*Simultaneous design of water reusing and rainwater harvesting systems in a residential complex*", *Computers and Chemical Engineering*, Copyright (2015), Elsevier

While the total rainwater harvesting area is 62,222 m^2 and the schematic representation for this solution is presented in Fig. 6.8. This scenario involves an increment in the total annual cost of 18.89% concerning the minimum of Scenario A; while the total freshwater consumption has an increment of 10.30% concerning the minimum of Scenario B. Additionally, comparing Scenarios C and D, Scenario C represents a reduction in the freshwater consumption of 9.09% concerning Scenario D; while the total cost for Scenario C only increases 17.18% concerning Scenario D. For this Scenario D, the main cost is the one associated to the elevated reservoir.

6.5 Concluding Remarks

A multi-objective mixed-integer nonlinear programming model for designing the best rainwater harvesting system to meet residential development demands has been given in this chapter. The model that is being given suggests preparing the available space for rainwater collection and installing storage units so that the residential development can use this water when needed. The presented model takes into consideration the capital costs related to the harvesting area, storage facilities, pipes, and pumps. It also considers the expenses related to the freshwater needed, as well as the running costs for treatment and pumping. The multi-objective optimization methodology is designed to optimize both the freshwater usage and the overall annual cost at the same time. This research also proposed a methodical approach to illustrate the competing goals using Pareto curves.

Analysis was done on a case study of a residential development from the Mexican city of Morelia. In this instance, the possible benefits of putting such tactics into practice were illustrated, offering appealing outcomes from both an environmental and economic standpoint. The findings demonstrate that rainwater harvesting can cut freshwater usage by up to 87.6% when cutting freshwater consumption is the top priority and up to 80% when the least amount of expenditure is made. The model's ability to accommodate constraints, such as the maximum budget for investment or the amount of fresh water available, was demonstrated by the various alternative configurations that were obtained. In this manner, the planning process could be carried out using the previously mentioned analysis, which illustrates the trade-offs between the variables under consideration. The best solutions point to the necessity of storing water during wet seasons and using it during dry seasons when fresh water sources are scarce and more expensive. In general, this strategy provides a more favorable balance between water conservation and expenses than the freshwater reduction policy alone, which advocates for continuous use and collection of water over time. This is a crucial feature of the suggested multi-objective optimization strategy, which provides improved understanding of the system and its constituent parts, in contrast to single-objective approaches that could overlook appealing alternatives.

Lastly, the presented methodology is broad enough to be used in various contexts and case studies.

Nomenclatures

Acronyms

cap	Rainwater harvesting area
ElevTank	Elevated reservoir
fresh	Freshwater consumed
stg	Storage
Treatment	Relative to water treatment

Sets

i	Number of harvesting areas that are delineated on the roof of the buildings
j	Number of storage units for harvested rainwater where this can be treated and stored
l	Water users

m	Months of the year

Parameters

α, β, δ	Exponents for the capital cost functions
A_i^{max}	Maximum available harvesting area
$CF_i^{captation}$	Unit fixed cost for rainwater harvesting area
$CF^{ElevTank}$	Unit fixed cost of elevated reservoir
$CF_{i,j}^{pipe}$	Unit fixed cost for pipe network from the storages to the elevated reservoir
CF_j^{pipe}	Unit fixed cost for pipe network
$CF_j^{storage}$	Unit fixed cost for storage tanks
$Cost_m^{fresh}$	Unit freshwater cost
$Cost_j^{pumping}$	Unit pumping cost from the storage tanks to the elevated reservoir
$Cost^{Treatment}$	Unit water treatment cost
$CV_i^{captation}$	Unit variable cost for the rainwater harvesting area
$CV^{ElevTank}$	Unit variable cost for the elevated reservoir
$CV_{i,j}^{pipe}$	Unit variable cost for pipes from the harvesting area to the storage tank
CV_j^{pipe}	Unit variable cost for the pipe from the storage tanks to the elevated reservoir
$CV_j^{storage}$	Unit variable cost for the storage tanks
$D_{l,m}$	Required water for each sink during each period
F^{max}	Maximum allowed flow rate in pipes
G^{max}	Maximum allowed flow rate in pipes
K_F	Factor used to annualize the capital costs
P_m	Rainwater precipitation per month
STG_j^{max}	Maximum storage capacity for tanks

Variables

A_i	Used rainwater harvesting area
$CaptationCost$	Capital cost for harvesting area
$F_{i,m}$	Harvested rainwater for each location
$f_{i,j}^{cap}$	Maximum flow rate for the pipes from the harvesting area to the storages
f_m^{fresh}	Fresh water consumed during each period
F^{max}	Rainwater harvested during each period
$FreshCost$	Freshwater cost

$FreshTotal$	Total fresh water consumed from the municipal service
g_j^{cap}	Maximum flow rate from storages to the elevated reservoir
$G^{\max}$	Water distributed from the storages to the elevated tanks during each period
$PipeCost$	Capital cost for pipes
$PumpingCost$	Pumping cost
$S_{j,m}$	Accumulation for each tank per month
$S_j^{\max}$	Maximum capacity for storages
$StorageCost$	Capital cost for storages
$StorageCostT$	Capital cost for the elevated tank
T_m	Accumulation for the elevated tank per month
$T^{\max}$	Maximum capacity for the elevated tank
TAC	Total annual cost
$TreatmentCost$	Treatment cost for collected water

Binary Variables

x_j	Binary variable for the existence of pipe from storages to elevated tank
y_i^{capt}	Binary variable for the existence of rainwater harvesting area
y_j^{stg}	Binary variable for the existence of storages
$z_{i,j}$	Binary variable for the existence of pipes from the harvesting area to the storages

References

Abdulla FA, Al-Shareef AW (2009) Roof rainwater harvesting systems for household water supply in Jordan. Desalination 243:195–207

Appan A (2000) A dual-mode system for harnessing roofwater for non-potable uses. Urban Water 1:317–321

Atilhan S, Bin-Mahfouz AS, Batchelor B, Linke P, Abdel-Wahab A, Nápoles-Rivera F, Jiménez-Gutierrez A, El-Halwagi MM, (2012) A systems-integration approach to the optimization of macroscopic water desalination and distribution networks: a general framework applied to Qatar's water resources. Clean Technol Environ Policy 14:161–171

Brooke A, Kendrick D, Meereaus A, Raman R (2013) GAMS a user's guide. Washintong, D.C.

Burgara-Montero O, Ponce-Ortega JM, Serna-González M, El-Halwagi MM (2012) Optimal design of distributed treatment systems for the effluents discharged to the rivers. Clean Technol Environ Policy 14:925–942

Campisano A, Modica C (2012) Optimal sizing of storage tanks for domestic rainwater harvesting in Sicily. Resour Conserv Recycl 63:9–16

Chilton JC, Maidment GG, Marriott D, Francis A, Tobias G (2000) Case study of a rainwater recovery system in a commercial building with a large roof. Urban Water 1:345–354

CONAGUA (2011) National council of water, statistic of the water in México, chap 3 appendix. Available at: http://www.conagua.gob.mx/CONAGUA07/Contenido/Documentos/SINA/Capitulo3Usosdelagua.zip. Accessed Aug 2013
Deng C, Feng X (2011) Targeting for conventional and property-based water network with multiple resources. Ind Eng Chem Res 50:3722–3737
Deng C, Feng X, Wen Z (2013) Optimization of water network integrated with process models. Clean Technol Environ Policy 15:473–487
Domènech L, Saurí D (2011) A comparative appraisal of the use of rainwater harvesting in single and multi-family buildings of the Metropolitan Area of Barcelona (Spain): social experience, drinking water savings and economic costs. J Clean Prod 19:598–608
Domènech L, Heijnen H, Saurí D (2012) Rainwater harvesting for human consumption and livelihood improvement in rural Nepal: benefits and risks. Water Environ J 26:465–472
Fu J, Cai T, Xu Q (2012) Coupling multiple water-reuse network designs for agile manufacturing. Comput Chem Eng 45:62–71
Gikas P, Angelakis AN (2009) Water resources management in Crete and in the Aegean Islands, with emphasis on the utilization of non-conventional water sources. Desalination 248:1049–1064
Hadadin N, Qaqish M, Akawwi E, Bdour A (2010) Water shortage in Jordan—sustainable solutions. Desalination 250:197–202
Huang Y, Chen J, Zeng S, Sun F, Dong X (2013) A stochastic optimization approach for integrated urban water resource planning. Water Sci Technol 67:1634–1641
Imteaz MA, Shanableh A, Rahman A, Ahsan A (2011) Optimisation of rainwater tank design from large roofs: a case study in Melbourne, Australia. Resour Conserv Recycl 55:1022–1029
Imteaz MA, Adeboye OB, Rayburg S, Shanableh A (2012a) Rainwater harvesting potential for southwest Nigeria using daily water balance model. Resour Conserv Recycl 62:51–55
Imteaz MA, Rahman A, Ahsan A (2012b) Reliability analysis of rainwater tanks: a comparison between South-East and Central Melbourne. Resour Conserv Recycl 66:1–7
INEGI (2010) National census of population and housing. National Institute of Statistics and Geography. Basic questionnaire. Available at: http://www.inegi.org.mx/sistemas/mexicocifras/default.aspx?e¼16. Accessed Aug 2012
Khastagir A, Jayasuriya N (2010) Optimal sizing of rain water tanks for domestic water conservation. J Hydrol 381:181–188
Khor CS, Chachuat B, Shah N (2012) A superstructure optimization approach for water network synthesis with membrane separation-based regenerators. Comput Chem Eng 42:48–63
Li Z, Boyle F, Reynolds A (2010) Rainwater harvesting and greywater treatment systems for domestic application in Ireland. Desalination 260:1–8
Lü YP, Yang K, Che Y, Shang ZY, Zhu HF, Jian Y (2013) Cost-effectiveness-based multi-criteria optimization for sustainable rainwater utilization: a case study in Shanghai. Urban Water J 10:127–143
Morales-Pinzón T, Rieradevall J, Gasol CM, Gabarrell X (2012) Potential of rainwater resources based on urban and social aspects in Colombia. Water Environ J 26:550–559
Nápoles-Rivera F, Ponce-Ortega JM, El-Halwagi MM, Jiménez-Gutiérrez A (2010) Global optimization of mass and property integration networks with in-plant property interceptors. Chem Eng Sci 65:4363–4377
Nápoles-Rivera F, Ponce-Ortega JM, El-Halwagi MM, Jiménez-Gutiérrez A (2012) Global optimization of wastewater integration networks for processes with multiple contaminants. Environ Prog Sustain Energ 31:449–458
Nápoles-Rivera F, Serna-González M, El-Halwagi MM, Ponce-Ortega JM (2013) Sustainable water management for macroscopic systems. J Clean Prod 7:102–117
Ng DKS, Foo DCY, Tan RR, El-Halwagi M (2010) Automated targeting technique for concentration- and property-based total resource conservation network. Comput Chem Eng 34:825–845
Ponce-Ortega JM, Mosqueda-Jiménez FW, Serna-González M, Jiménez-Gutiérrez A, El-Halwagi MM (2011) A property-based approach to the synthesis of material conservation networks with economic and environmental objectives. AIChE J 57:2369–2387

Ponce-Ortega J, Nápoles-Rivera F, El-Halwagi M, Jiménez-Gutiérrez A (2012) An optimization approach for the synthesis of recycle and reuse water integration networks. Clean Technol Environ Policy 14:133–151

Rojas-Torres MG, Ponce-Ortega JM, Serna-González M, Nápoles-Rivera F, El-Halwagi MM (2012) Synthesis of water networks involving temperature-based property operators and thermal effects. Ind Eng Chem Res 52:442–461

Rozos E, Makropoulos C (2011) Assessing the combined benefits of water recycling technologies by modelling the total urban water cycle. Urban Water J 9:1–10

Rozos E, Makropoulos C, Butler D (2010) Design robustness of local water-recycling schemes. J Water Resour Plan Manag 136:531–538

Rubio-Castro E, Ponce-Ortega JM, Nápoles-Rivera F, El-Halwagi MM, Serna-González M, Jiménez-Gutiérrez A (2010) Water integration of eco-industrial parks using a global optimization approach. Ind Eng Chem Res 49:9945–9960

Rubio-Castro E, Ponce-Ortega JM, Serna-González M, Jiménez-Gutiérrez A, El-Halwagi MM (2011) A global optimal formulation for the water integration in eco-industrial parks considering multiple pollutants. Comput Chem Eng 35:1558–1574

Rubio-Castro E, Ponce-Ortega JM, Serna-González M, El-Halwagi MM (2012) Optimal reconfiguration of multi-plant water networks into an eco-industrial park. Comput Chem Eng 44:58–83

Rubio-Castro E, Ponce-Ortega JM, Serna-González M, El-Halwagi MM, Pham V (2013) Global optimization in property-based interplant water integration. AIChE J 59:813–833

Sotelo-Pichardo C, Ponce-Ortega JM, El-Halwagi MM, Frausto-Hernández S (2011) Optimal retrofit of water conservation networks. J Clean Prod 19:1560–1581

Sotelo-Pichardo C, Ponce-Ortega JM, Nápoles-Rivera F, Serna-González M, El-Halwagi MM, Frausto-Hernández S (2013) Optimal reconfiguration of water networks based on properties. Clean Technol Environ Policy 16:303–328

Vázquez-Castillo JA, Ponce-Ortega JM, Segovia-Hernández JG, El-Halwagi MM (2013) A multiobjective approach for property-based synthesis of batch water networks. Chem Eng Process 65:83–96

Yu J, Chen Y, Shao S, Zhang Y, Liu S, Zhang S (2013) A study on establishing an optimal water network in a dyeing and finishing industrial park. Clean Technol Environ Policy 16:45–57

Chapter 7
Mixed-Integer Nonlinear Programming (MINLP) Model for Determining the Optimal Site for a New Industrial Facility with Integrated Economic and Environmental Criteria

7.1 Introduction

Nowadays, only economic factors are typically considered when choosing a site for a new industrial plant; environmental harm caused by industrial effluents during the ultimate wastewater disposal process is not considered (El-Baz et al. 2005a). Once the new plants are installed, the problems associated with the pollution that the new plants produce are identified, and then is required to install new equipment to treat the industrial wastes to repair the damage and satisfy the environmental regulations, which represents an additional cost that may yield uneconomical processes.

Various approaches have been reported to address the problem of recycle/reuse water networks inside industrial facilities (Dunn and El-Halwagi 2003; El-Halwagi 1997, 2006; El-Halwagi and Spriggs 1998). These approaches allow yielding a proper use of resources and reducing waste streams discharged into the environment. These resource conservation approaches have been classified as pinch techniques and techniques formulated as mathematical programming models (Pillai and Bandyopadhyay 2007). Pinch-based approaches use heuristic rules based on targeting following by design conducting to sequential methods. Foo (2009) presented a review of the pinch techniques for synthesizing water networks (Wang and Smith 1994; Dhole et al. 1996; El-Halwagi and Spriggs 1998; Polley and Polley 2000; Hallale 2002; Manan et al. 2004; El-Halwagi et al. 2003; Feng et al. 2007). Other approaches are based on algebraic techniques (Sorin and Bedard 1999; Agrawal and Shenoy 2006; Gomes et al. 2007). Lastly, mathematical programming approaches have been reported to automate the synthesis of recycle and reuse water networks inside industrial plants (Savelski and Bagajewicz 2000, 2001, 2003; Alva-Argaez et al. 1998, 1999; Benko et al. 2000; Teles et al. 2008; Gabriel and El-Halwagi 2005; Kuo and Smith 1997; Doyle and Smith 1997; Galan and Grossmann 1998; Hernandez-Suarez et al. 2004; Gunaratnam et al. 2005; Karuppiah and Grossmann 2006; Putra and Amminudin 2008; Ponce-Ortega et al. 2009a, 2010; Nápoles-Rivera et al. 2010). Nevertheless, previous formulations considered exclusively the processes that happen inside the

J. M. Ponce-Ortega et al., *Sustainable Water Systems*, Water Resources Development and Management, https://doi.org/10.1007/978-3-031-72498-5_7

industrial facilities, ignoring the processes that occur outside the industrial plant. Wastewaters discharged by industrial plants are transported by the drainage systems to their final disposal, and these final disposals commonly are lakes or seas that are severely impacted by the polluted discharges from the industries.

On the other hand, watershed systems are how the effluents are transported to their final disposal. In the watersheds naturally exist mixing and splitting processes, which change naturally the properties of the streams, in addition to the natural degradation of the components due to the chemical, biochemical, and physical transformations (Brunner and Rechberg 2004). Watersheds involve several tributaries that feed into reaches, streams, or rivers that finally lead to catchment areas like lakes, seas, or oceans; and watersheds impact and are impacted radically by their surroundings. Baccini and Brunner (1991) reported a material flow analysis (MFA) model for analyzing ecosystems with human activities where mass, energy, and information are being exchanged with the surroundings. Lampert and Brunner (1999) reported an MFA model to track nutrient discharges into the Danube River basin. El-Baz et al. (2005a) developed an MFA model to track the nitrogenous compounds in the Bahr El-Baqar drainage, while El-Baz et al. (2005b) reported an MFA model to include mass integration in the Bahr El-Baqar drainage system in Egypt. Lovelady et al. (2009) introduced an approach to integrate the discharges of an industrial process with macroscopic watershed systems based on an MFA model that they called the reverse problem formulation. In the reverse problem formulations (Lovelady et al. 2009), the maximum concentrations for the pollutants discharged to the environment are fixed to avoid hazardous accumulation in the final disposal to ensure sustainable processes through the MFA technique; but this is a sequential approach that does not optimize simultaneously the model.

In addition, wastewater treatment costs for the new plant are strongly dependent on the new location. Thus, this chapter shows a model based on mathematical programming for the optimal location of a new industrial plant considering the minimization of the total cost and the treatment of the wastewater streams to meet the environmental regulations and the natural degradation of the involved contaminants to avoid pollution problems in the final disposal due to the accumulation of the contaminants, and thus to guarantee a sustainable environment. Furthermore, the model involves different and independent constraints for the water quality in the watershed regions (or sections) that are impacted by the new discharge because fresh water could be taken from some of these regions for human, agricultural, or other uses, or because there is a city or a place that must be protected to avoid hazardous situations. Then, the requirements for the water quality in these different locations of the watershed depend on the specific use. Lastly, the model can be used as a tool to make a sensitivity analysis for the case when the requirements for the water quality change. Also, the set of optimal solutions for different scenarios could be identified through a Pareto curve. Thus, the model could be used as a tool for decision-makers when a new industrial plant needs to be located (Lira-Barragán et al. 2011).

7.2 Problem Statement

The addressed problem in this chapter is stated as follows:

Given are a set of alternative locations to install the new industrial plant $P = \{p|p = 1, 2, \ldots, N_p\}$. Each allowed location p has linked an installation cost, $CLAND_p$, which involves the cost associated with the transportation of the raw materials and final products, and the land cost. Furthermore, multiple pollutants can be considered ($C = \{c|c = 1, 2, \ldots, N_c\}$). The problem consists in determining the optimal location for the new plant that satisfies the environmental constraints (y_c^{EnvReg}) for the industrial effluents and, at the same time, that does not exceed the accumulation of the pollutants in the final disposal ($y_c^{sustainability}$), and satisfies independent constraints for specific regions n through the watershed that depend on the type of the water use ($y_{c,n(r)}^{desired}$).

The discharge at the exit of the new plant is fixed by the process conditions (total flowrate discharged and its composition), but the quality of the discharge over the river near the new plant must be determined because this must be treated to satisfy the environmental regulations and to avoid the accumulation of pollutants at the final disposal, and this treatment depends on the location of the new plant. To satisfy this constraint, the behavior of the river is simulated by considering all inlet and outlet streams through the MFA technique.

The objective function consists of the minimization of the total annual cost (TAC) which is constituted of two components: the installation costs for the new plant and the treatment cost for the wastewater stream leaving the plant. The installation cost includes the cost of land and production, as well as the transportation costs for the raw materials and the products. The second component accounts for the treatment cost of wastewater streams using the available treatment technologies, this cost is split into two parts; one that is independent of the treated flowrate and the other one that depends on the treated flowrate in each treatment technology. The part of the cost for the treatment units that is independent of the treated flowrate is a very important factor to avoid infeasible networks with a very large number of treatment units.

Figure 7.1 presents a general watershed representation, where the main stream of the river is fed by several tributary streams ($FT_{r,t}$). The main stream of the river exchanges water during its route. According to its use, the following types of water exchanges can be recognized: water for agricultural use, wastewater discharged to the river with and without treatment, industrial and residential effluents, etc. In addition, the model considers phenomena like natural precipitation, filtration, and vaporization. These processes significantly modify the composition for the materials transported through the river; thus, to effectively track the mean composition of hazardous pollutants, it is required to segment the river considering the parts where the overall composition can be considered constant (these sections are represented in Fig. 7.1 with ellipses). These sections are called reaches, which identify sections where no bigger effluents are discharged and/or extracted. The flowrate and concentrations in each reach are different from other reaches because the inlet and extraction streams. Tributaries are channels or branches of the river, which may contain discharges with/without treatment, industrial discharges, etc., and their flowrates ($FT_{r,t}$) and concentrations ($CT_{c,r,t}$) influence the reaches where they are discharged.

The reactive term is considered in the model, which considers the possible chemical and biochemical reactions that can be carried out by the flora and fauna inside the rivers, which can decompose or produce the hazardous materials.

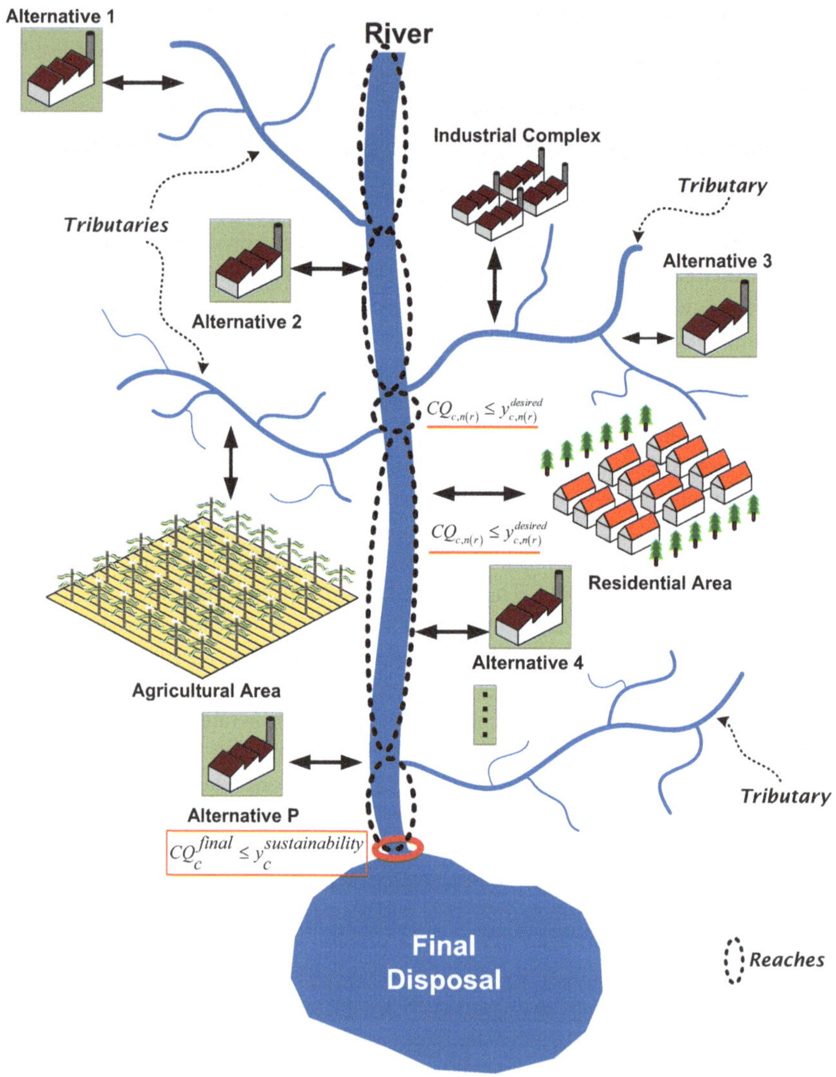

Fig. 7.1 Alternatives to locate the new industrial plant. Printed with permission of "*An MINLP model for the optimal location of a new industrial plant with simultaneous consideration of economic and environmental criteria*", *Industrial and Engineering Chemistry Research*, Copyright (2011), American Chemical Society

One of the objectives is to maintain clean the final disposal of the watershed considered the current and new discharges. The wastewater discharges for the new plant must yield a concentration lower than the one that ensures a sustainable process for the final disposal. Furthermore, the model must meet constraints for the water quality throughout the whole watershed, which are independent of each other but depend on the specific location for the water use (Fig. 7.1). It is common to extract water from a river to feed a city or agricultural areas; therefore, the water quality through the river is very important.

Another important aspect of the model formulation is that it can consider the trade-offs between the pollutant concentration discharged to the final disposal and the total associated cost, optimizing simultaneously the location and the treatment system. Thus, when there are stricter environmental regulations, the associated cost to treat the waste streams and the location cost are larger. On the other hand, for relaxed environmental regulations, the associated total cost is lower. Therefore, the final location for the new plant depends on the decision makers (investors), government, and society, and the approach presented in this chapter allows for optimization simultaneously of the total cost and the quality of the water discharged. Lastly, local governments usually offer incentives to bring industries to their areas because of the social benefits through the generation of jobs that the installation of a new plant yields; consequently, the proposed model can be a useful tool to determine the total expenses that a local government needs to pay to be an attractive location for the installation of the new industry respect to other locations.

7.3 Model Formulation

The used sets in the implanted model are defined firstly for a better understanding. c is a subindex used to indicate the component, N_c represents the total number of components and C is a set that contains all c. r is a subindex used to indicate a section of the river, and N_r indicates the total number of sections of the river used for a specific problem. R is a set that contains all r. t is a subindex for the effluents discharged to the river, N_t is the total number of effluents, and T their set. p is a subindex for the locations for the new plant, N_p indicates the total number of possible locations, and P is the corresponding set. j is a subindex used for the interceptors allowed to treat the composition of the effluent of the new plant, N_j indicates the total number of interceptors (including a fictitious interceptor used for modeling the bypassing stream) and J represents the corresponding set.

The equations for the model are described as follows:

7.3.1 Balance for Each Reach

The exit flowrate from each reach r (Q_r) is equal to the inlet flowrate to this reach (Q_{r-1}) plus the average precipitation (P_r), direct industrial discharges (D_r), residential discharges (H_r), the sum of all effluents inlet to the reach ($\sum_{t=1}^{N_{t(r)}} FT_{r,t}$), including the possible discharge of the new plant in this reach ($QPNEW_{r(p)}$), minus the extractions associated to filtration and evaporation (L_r) and the use of water in that section of the river (U_r).

$$Q_r = Q_{r-1} + P_r + D_r + H_r + \sum_{t=1}^{N_{t(r)}} FT_{r,t} + QPNEW_{r(p)} - L_r - U_r, \quad \forall r \in R \quad (7.1)$$

$N_{t(r)}$ corresponds to the total number of tributaries discharged to the reach r.

7.3.2 Component Balances for Each Reach

The exit mass from the reach for the toxic compound ($Q_r CQ_{c,r}$) is equal to the inlet mass to the reach ($Q_{r-1}CQ_{c,r-1}$) plus the precipitation ($P_r CP_{c,r}$), industrial discharges ($D_r CD_{c,r}$), residential discharges ($H_r CH_{c,r}$), tributaries ($\sum_{t=1}^{N_{t(r)}} FT_{r,t}CT_{c,r,t}$), and the mass that can be discharged by the new plant ($QPNEW_{r(p)}CPNEW_{c,r(p)}$), minus the lost ($L_r CL_{c,r}$) and use ($U_r CU_{c,r}$). The reaction term, $\int_{V=0}^{V_r} r_{c,r} dV_r$ is very important, which considers the chemical and biochemical reactions that take place in the river by the flora and fauna because of the interaction with the component inside the river, which can produce or eliminate the component. Finally, the component balances for each reach are stated as follows:

$$\begin{aligned} Q_r CQ_{c,r} &= Q_{r-1}CQ_{c,r-1} + P_r CP_{c,r} + D_r CD_{c,r} + H_r CH_{c,r} \\ &+ \sum_{t=1}^{N_{t(r)}} FT_{r,t}CT_{c,r,t} + QPNEW_{r(p)}CPNEW_{c,r(p)} \\ &- L_r CL_{c,r} - U_r CU_{c,r} - \int_{V=0}^{V_r} r_{c,r} dV_r, \quad \forall c \in C, \quad r \in R \end{aligned} \quad (7.2)$$

7.3.3 Balances for Tributaries

The balance for the tributary t that discharges to the reach r is described as follows:

$$FT_{r,t} = S_{r,t}^{untreated} + S_{r,t}^{treated} + I_{r,t} + P_{r,t} + D_{r,t} + QPNEW_{r,t(p)} - L_{r,t} - U_{r,t}, \quad \forall r \in R, \quad t \in T \tag{7.3}$$

The total flowrate that exits the tributary t and inlets to the reach r ($FT_{r,t}$) is equal to the sum of the discharges with/without treatment ($S_{r,t}^{treated}$, $S_{r,t}^{untreated}$), industrial discharges ($I_{r,t}$) and pluvial discharges ($P_{r,t}$), direct discharges ($D_{r,t}$) and the new discharge if the new plant is located on this tributary ($QPNEW_{r,t(p)}$) minus the losses ($L_{r,t}$) and use or extraction of water ($U_{r,t}$).

7.3.4 Component Balances for Each Tributary

The mass exits to each tributary t and directs to the reach r is calculated as follows:

$$FT_{r,t}CT_{c,r,t} = S_{r,t}^{untreated}CS_{c,r,t}^{untreated} + S_{r,t}^{treated}CS_{c,r,t}^{treated} + I_{r,t}CI_{c,r,t} + P_{r,t}CP_{c,r,t} + D_{r,t}CD_{c,r,t} + QPNEW_{r,t(p)}CPNEW_{c,r,t(p)} - L_{r,t}CL_{c,r,t} - U_{r,t}CU_{c,r,t} - \int_{V=0}^{V_{r,t}} r_{c,r,t}dV_{r,t}, \quad \forall c \in C, \quad r \in R, \quad t \in T \tag{7.4}$$

The component balances for each tributary include the reaction term $\int_{V=0}^{V_{r,t}} r_{c,r,t}dV_{r,t}$ to consider the natural degradation of the components inside the river.

7.3.5 Agricultural Discharges and Uses

The agricultural discharges $D_{r,t}$ and uses $U_{r,t}$ are related to the surrounding areas because they are proportional to the agricultural cultivated areas. These discharges and uses are calculated as follows:

$$D_{r,t} = \alpha_{r,t} * A_{r,t}, \quad \forall r \in R, \quad t \in T$$
$$U_{r,t} = \beta_{r,t} * A_{r,t}, \quad \forall r \in R, \quad t \in T$$

$\alpha_{r,t}$ is the flowrate per unit of area in m^3/acre s, $\beta_{r,t}$ is the agricultural use of water per unit of area for the tributary t in m^3/acre s and $A_{r,t}$ is the area that covers the tributary t in acres.

7.3.6 Location of the New Plant

The flowrate and the composition discharged from the new plant depend on the location of the plant. For example, if the new plant is in the location 1, the flowrate and composition discharged from the new plant in that position can be greater than zero; otherwise, if the new plant is not located in the position 1, the flowrate and composition discharged should be zero. For a set of possible locations p where the new plant can be placed, this is modeled through the following disjunction:

$$\begin{bmatrix} W_1 \\ CPNEW_{c,1} \geq 0, QPNEW_1 \geq 0 \\ CPNEW_{c,2} = 0, QPNEW_2 = 0 \\ \vdots \\ CPNEW_{c,p} = 0, QPNEW_p = 0 \end{bmatrix} \vee \begin{bmatrix} W_2 \\ CPNEW_{c,1} = 0, QPNEW_1 = 0 \\ CPNEW_{c,2} \geq 0, QPNEW_2 \geq 0 \\ \vdots \\ CPNEW_{c,p} = 0, QPNEW_p = 0 \end{bmatrix}$$

$$\vee \cdots \vee \begin{bmatrix} W_p \\ CPNEW_{c,1} = 0, QPNEW_1 = 0 \\ CPNEW_{c,2} = 0, QPNEW_2 = 0 \\ \vdots \\ CPNEW_{c,p} \geq 0, QPNEW_p \geq 0 \end{bmatrix}, \quad \forall c \in C$$

W_p is a Boolean variable associated to the location p of the new plant, $CPNEW_{c,p}$ is the concentration for the discharge of the new plant to the river in location p. Because the final discharge for the new plant to the river depends on the location of the new plant, when the optimal location for the new plant is section 1, the Boolean variable W_1 is set as true and $CPNEW_{c,1}$ and $QPNEW_{c,1}$ must be greater than zero, all the others Boolean variables must be false and $CPNEW_{c,p}$ and $QPNEW_p$ associated to their locations must be set as zero. Similar situation occurs for the case when the selected location is 2 or 3 or so on for the p possible locations. The value for the discharge concentration of the new plant is strongly influenced by the location through the material flow analysis model, satisfying the constraints given to maintain under control the pollutants in their final disposal as well as the maximum concentration allowable to be discharged. In this chapter, since the plant design is the same for any location, it is assumed that the flowrate discharged from the new plant is a constant value specified prior to the optimization for the location.

Lastly, previous disjunction is modeled using the convex hull reformulation (Raman and Grossmann 1994; Ponce-Ortega et al. 2009b) as follows:

The Boolean variables are transformed into a set of binary variables. When the Boolean variables are true, the associated binary variables are one; otherwise, when the Boolean variables are false, the associated binary variables are zero. To select only one location to place the new plant, the following relationship is used:

$$\sum_{p \in P} w_p = 1 \tag{7.5}$$

Then, upper limits are imposed to the variables $CPNEW_{c,p}$ and $QPNEW_p$ depending on the location of the new plant.

$$CPNEW_{c,p} \le CPNEW^{UP} w_p, \quad \forall c \in C, \quad p \in P \tag{7.6}$$

$$QPNEW_p \le QPNEW^{UP} w_p, \quad \forall p \in P \tag{7.7}$$

Equations (7.8) and (7.9) are used to ensure that $CPNEW_{c,p}$ and $QPNEW_p$ take values greater than zero only when the new plant is located in that position and these take values of zero when the location is not selected to place the new plant.

7.3.7 Wastewater Treatment System

Figure 7.2 shows a set of interceptors available to satisfy the discharge limits for the new plant. As the optimization process must determine the optimal selection of the treatment system, this must be in the place where the new plant is located. Thus, it is important to remark that the treatment requirement depends strongly on the location of the new plant to satisfy the sustainability of the system in the final disposal.

First, the concentration discharged for the new plant is calculated as the sum of the discharges for any allowable location because only one location can place the new plant, which is modeled as follows:

$$\sum_{p \in P} CPNEW_{c,p} = y_c^{disc}, \quad \forall c \in C \tag{7.8}$$

Also, for the flowrate discharged for the new plant:

$$\sum_{p \in P} QPNEW_p = QPNEWDATA \tag{7.9}$$

where $QPNEWDATA$ is a constant parameter known before the optimization process.

Figure 7.2 presents a representation of the interception network used to treat the wastewater for the new plant, where the treatment of the wastewater stream may or may not be required according to the location of the new plant and the MFA model. On the other hand, to model the performance of the interceptors, an efficiency factor for each component ($\gamma_{c,j}$) is used to simulate the unit j. This efficiency factor depends on the design and operating parameters for each unit, as it can be seen in the next equation:

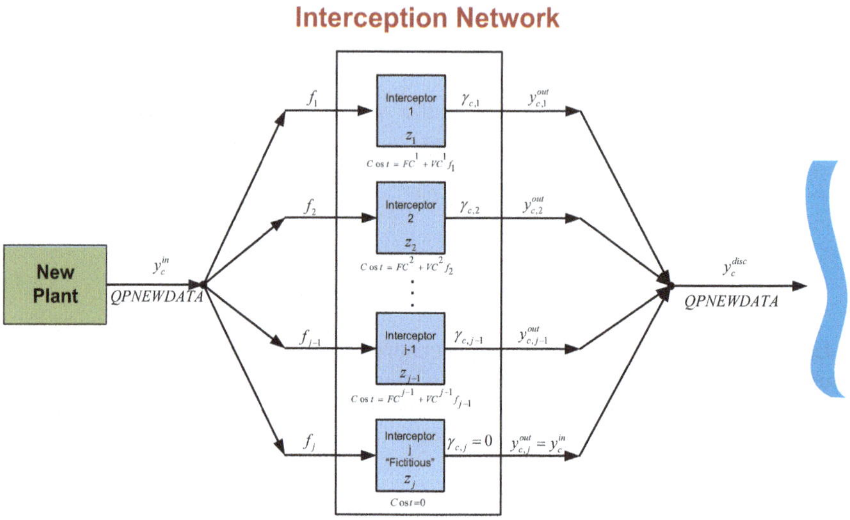

Fig. 7.2 Interceptors network for wastewater treatment. Printed with permission of "*An MINLP model for the optimal location of a new industrial plant with simultaneous consideration of economic and environmental criteria*", *Industrial and Engineering Chemistry Research*, Copyright (2011), American Chemical Society

$$\gamma_{c,j} = f\,(Design\ and\ Operating\ Parameters)$$

The available treatment units can be simulated (for example in ASPEN PLUS, using design equations, etc.) and the efficiency factors for each unit are determined before to the optimization process. While the final concentrations depend on the technologies selected and the inlet concentration, and the coefficients for the cost functions (FC^j and VC^j) for each treatment unit j depend on the type of unit (with their corresponding efficiencies), and the total cost for the treatment network accounts for the type of selected unit and the treated flowrate. Thus, the optimization consists in the selection of the treatment units with known characteristics and the treated flowrate. In this sense, the interceptor units to be used depend on each process, but they can be distillation columns, adsorption columns, stripping columns, liquid–liquid extraction columns, etc.

The following are important remarks for the treatment network:

- The wastewater stream can be treated by one or several technologies available to remove the pollutant.
- The flowrate at the exit of the plant is the same that the flowrate at the exit of the interceptor network, but the composition is not the same.
- Each interceptor has an efficiency factor to remove the pollutant, which is directly related to the amount of pollutant that it can remove. Furthermore, the cost for each interceptor is constituted by a fixed charge FC^j, and a variable charge VC^j that depends on the flowrate treated.

- The last interceptor in the interception network is a fictitious interceptor, which is included for modeling the bypass streams. This fictitious interceptor has a conversion factor of zero and an associated cost of zero.

The following equation is implemented to determine the concentration at the exit of each interceptor considered:

$$y_{c,j}^{out} = y_c^{in}\left(1 - \gamma_{c,j}\right), \quad \forall c \in C, \ j \in J$$

The exit concentration from each interceptor for each component is a given parameter because this depends on the parameter inlet concentration and the conversion factors; but the optimal decision is given in terms of the flowrates inlet to each interceptor.

The balances for the interceptor network are presented as follows:

$$QPNEWDATA = \sum_{j \in J} f_j \tag{7.10}$$

$$\sum_{j \in J} f_j \, y_{c,j}^{out} = QPNEWDATA \; y_c^{disc}, \quad \forall c \in C \tag{7.11}$$

f_j is the segregated wastewater flowrate to the interceptor j, $\gamma_{c,j}$ is the efficiency factor to remove the pollutant c for the interceptor j, y_c^{in} is the concentration of the pollutant c inlet to the interceptor network and outlet to the plant, $y_{c,j}^{out}$ is the concentration of the pollutants at the exit of the interceptor j.

The following constraints are needed to active the existence of the interceptors:

$$f_j \geq \delta z_j, \quad \forall j \in J \tag{7.12}$$

$$f_j \leq f^{UP} z_j, \quad \forall j \in J \tag{7.13}$$

δ corresponds to a small parameter used for modeling purposes, z_j is a binary variable associated with the existence of the interceptor j and f^{UP} is an upper limit for f_j. f^{UP} is defined as $f^{UP} = QPNEWDATA$.

7.3.8 Constraints for the Quality of Specific Reaches

Some reaches through the watershed require specific constraints for the water quality because they need to provide water for specific uses. The presented model can consider these constraints through the following relationship:

$$CQ_{c,n(r)} \leq y_{c,n(r)}^{desired}, \quad \forall c \in C, \ \ n(r) \in N(R) \tag{7.14}$$

$N(R)$ corresponds to a subset of reaches n from all the reaches r that requires a specific water quality.

7.3.9 Constraint for the Final Disposal

The concentration of the waste discharged to the river by the new industrial plant must be restricted by the environmental regulations as follows:

$$y_c^{disc} \leq y_c^{EnvReg}, \quad \forall c \in C$$

Nevertheless, the environmental regulation does not ensure that the concentration discharged to the final disposal avoids its accumulation in the lake or sea. To get a sustainable process, the following constraint must be added to the model:

$$CQ_c^{final} \leq y_c^{sustainability}, \quad \forall c \in C \tag{7.15}$$

7.3.10 Objective Function

The objective function consists in minimizing the total annual cost considering the installation of the new plant as well as the treatment costs. Therefore, the objective function is stated as follows:

$$Min\ TAC = \sum_{p \in P} CLAND_p w_p + k_f \sum_{j \in J} FC^j z_j + H_Y \sum_{j \in J} VC^j \left(f_j\right)^{c_{\exp}} \tag{7.16}$$

TAC is the total annual cost, $CLAND_p$ is the annualized installation cost for the new plant in the site p, k_f is an annualization factor, FC^j is the fixed charge for the interceptor j, VC^j is the variable charge for the interceptor j, H_Y is the hour per year that operates the new plant and $c_{\exp}$ is an exponential factor to consider the economies of scale.

7.3.11 Remarks About the Presented Model

- The chemical and biochemical reactions can be represented by a simple kinetic model, although several chemical and biochemical reactions can occur.
- The installation costs ($CLAND_p$) include three different terms: transportation of raw materials, transportation of products and the land. All these costs depend strongly on the location.

- The flowrate discharged for the new plant, *QPNEWDATA*, is considered as a constant. The same case applies for the pollutant concentration exits to the plant, y_c^{in}, while the pollutant concentration for the final discharge of the new plant to the river y_c^{disc} is an optimization variable.

7.4 Case Studies

Two case studies are presented to show the applicability of the proposed methodology, which are described as follows.

7.4.1 *Example 7.1*

The data for this example correspond to the Bahr El-Baqar drain, which is one of the largest drains on the Nile Delta in Egypt and ends in the lake Manzala (Lovelady et al. 2009). The system has a total length of 106.5 km and an approximated width of 23 m at the beginning and 70 m at the end, whereas the final disposal has a total area of 1000 km^2 and an average depth of 1 m. The drain receives several types of discharges including agricultural drainage, treated and untreated domestic wastewater, and industrial wastewater. Lovelady et al. (2009) reported detailed information with respect to this drainage system. A new fertilizing plant must be installed discharging phosphorous as a major contaminant (in this case only one key component was selected). The phosphorous discharged by the new plant must not exceed a final concentration in the lake of 1.3 ppm to maintain the system.

Figure 7.3 presents a schematic representation of the watershed system considered in this example. In this system, there are residential, agricultural, and industrial discharges that significantly change the composition of the phosphorous in the river. Figure 7.3 presents four sites where it is possible to locate the new industrial plant; each one of these sites has associated different installation costs as can be seen in Table 7.1 (land cost, costs for transportation of raw materials and products, etc.).

The flowrate for the new plant (*QPNEWDATA*) is 2 m^3/s and the concentration for the discharge of the new plant before treatment (y^{in}) is 12.5 ppm of phosphorous. Noticeably, the installation of the new plant will alter and impact the environment through the new discharges to the watershed, increasing the composition of the contaminants in the surrounding rivers, as well as contaminating the final disposal in the lake (CQ^{final}). Furthermore, water is taken from the zone corresponding to reach 6; this water is used to feed a town, which requires that the maximum concentration of phosphorous be lower than 2.1 ppm. In reach 11, some farmers take water to their crops; then, the water must contain a concentration lower than 3.15 ppm. Also, the maximum concentration allowed to be discharged into the lake is of 1.3 ppm to allow the natural degradation of the phosphorous in the lake. This upper value was determined from a sustainability analysis (Lovelady et al. 2009). It should be

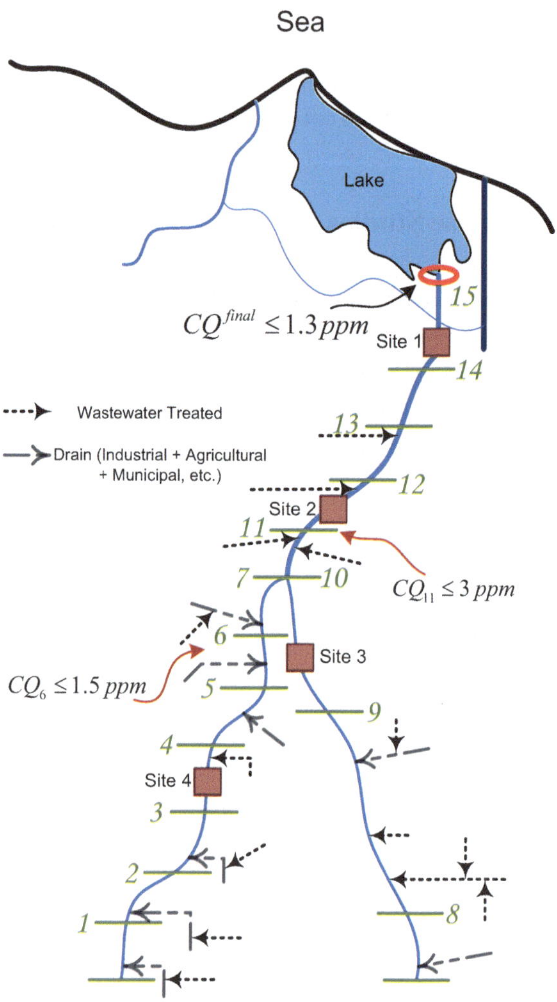

Fig. 7.3 Candidate sites for the location of the new industrial plant for Example 7.1. Printed with permission of "*An MINLP model for the optimal location of a new industrial plant with simultaneous consideration of economic and environmental criteria*", *Industrial and Engineering Chemistry Research*, Copyright (2011), American Chemical Society

Table 7.1 Installation cost for the different sites to locate the new plant for Example 7.1

Site	Annualized installation cost, \$MM/year
1	10
2	17
3	35
4	18

Printed with permission of "*An MINLP model for the optimal location of a new industrial plant with simultaneous consideration of economic and environmental criteria*", *Industrial and Engineering Chemistry Research*, Copyright (2011), American Chemical Society

noted that the system discharges water to the final disposal with a concentration of 1.17 ppm of phosphorous before the installation of the new industrial plant.

To model the system, the following assumptions were used in this example problem (Lovelady et al. 2009):

1. Insignificant precipitation. The flowrate due to precipitation was eliminated according to the weather in Egypt.
2. Insignificant losses. Water filtration and vaporization are too small compared to the convective flows.
3. The concentration of phosphorus at the exit condition to the wastewater treatment plants (WWTP) was taken as 9 mg/L. This assumption is based on the laboratory analysis of different treatment plants.
4. The concentration of the phosphorus for the wastewater without treatment is 15 mg/L.
5. The concentration of the phosphorous for the water treated is 9.75 mg/L.
6. The concentration of the phosphorous for the agricultural wastewater is 1.5 mg/L.

The reaction kinetics for the degradation of the phosphorous in the reaches takes the following form:

$$\int_{V=0}^{V_r} r_r dV_r = k * CQ_r * V_r, \quad \forall r \in R$$

k is the kinetics constant for the reaction measured experimentally, CT_r is the concentration of the phosphorus in the reach and V_r is the volume of the reach.

Whereas for the tributaries, the reaction kinetics takes the following form:

$$\int_{V=0}^{V_{r,t}} r_{r,t} dV_{r,t} = k * CT_{r,t} * V_{r,t}, \quad \forall r \in R, \quad t \in T$$

$k = 9.041909 \times 10^{-6}$/s for reaches and tributaries.

Furthermore, values of $\alpha_{r,t}$ are 0.000066 m^3/acre * s for all reaches. β = 0.000023 m^3/acre * s for reaches 1–6, 8, and 12–15; β = 0.000011 m^3/acre * s for reaches 7, and 9–12. The operation time for the plant is 8600 h/year, the annualization factor is 0.1 and the exponential factor that considers the economies of scale is 1.

Finally, the interceptors available with their unitary costs and efficiencies are shown in Table 7.2.

The model was executed in the General Algebraic Modeling System, GAMS (Brooke et al. 2006), and the solver DICOPT (Viswanathan and Grossmann 1990) was used to solve it. The optimal solution for this example corresponds to locating the new plant in site 1, including a treatment for 1.453 m^3/s of the wastewater discharged

Table 7.2 Treatment cost for the available interceptors for Example 7.1

Interceptor	Fixed cost, $	Variable cost, $ h/m^3	Efficiency, γ_j
1	568,000	0.235	0.95
2	446,000	0.193	0.84
3	353,000	0.168	0.76
4	0	0	0

from the new plant using interceptor 3, whereas 0.547 m^3/s bypass the treatment units. For the optimal solution of this example, the wastewater discharged from the new plant located in site 1 must be 5.597 ppm to ensure that the natural degradation satisfies the constraint for the discharge to the final disposal of 1.3 ppm, and to obtain concentrations for the reaches 6 and 11 lower than 1.313 ppm and 2.842 ppm, respectively. The total annual cost to install the new plant in site 1 is $\$17.6 \times 10^6$/year. Table 7.3 presents a result comparison for the different locations available to install the new plant. Sites 1 and 2 require treatment to satisfy the sustainability of the system, and sites 3 and 4 do not require treating the wastewater discharged for the new plant because they are far away from the final disposal. Thus, in these cases, the pollutant can be disintegrated naturally through the river. Nevertheless, although sites 3 and 4 do not require treatment of the new wastewater, the installation cost for these places is more expensive than for site 1, including the cost to treat the wastewater required for this last location. Additionally, if the problem is solved without considering the environmental aspects (i.e., considering exclusively the cost), the economic solution is given for the site with the smallest installation cost (smallest $CLAND_p$), that is site 1. In this case, however, the final discharge to the lake has a concentration of 1.5 ppm, which is 15.38% bigger than the maximum concentration required to maintain the system without dangerous accumulations.

Table 7.3 allows to identify how much the other locations differ from the optimal one must pay as incentives to be attractive economically for the installation of the new plant. For example, site 2 must pay 1.95×10^6 \$/year, whereas site 4 must pay

Table 7.3 Results comparison for a constraint of 1.3 ppm in Example 7.1

Site	Installation cost, $/year	Treatment cost, $/year	Total annual cost, $/year
1	10×10^6	7.6×10^6	17.59×10^6
2	17×10^6	2.54×10^6	19.54×10^6
3	35×10^6	0	35×10^{76}
4	18×10^6	0	18×10^6

Table 7.4 Total cost for several sustainability constraints in each site for Example 7.1

Concentration discharged to the final disposal lower than	Total annual cost, \$/year for each site			
	1	2	3	4
1.2	21.73×10^6	28.02×10^6	46.63×10^6	27.55×10^6
1.25	19.46×10^6	23.56×10^6	39.86×10^6	19.39×10^6
1.3	17.59×10^6	19.54×10^6	35×10^6	18×10^6
1.4	13.86×10^6	17×10^6	35×10^6	18×10^6
1.503	10×10^6	17×10^6	35×10^6	18×10^6

0.41×10^6 \$/year to be attractive economically while satisfying the environmental constraints. Consequently, the proposed model is a useful tool for governments to determine the incentives that must be paid to make the installation of a new industrial plant satisfying the environmental constraints for the new pollutants discharged.

Table 7.4 presents the optimal costs associated with the installation of the new plant for different upper limits of the pollutant concentration discharged to the final disposal for all possible locations. The information reported in Table 7.4 allows to make a sensibility analysis for the case when the pollutant concentration discharged to the final disposal changes, and this could be a useful tool for the decision maker and the governments interested in analyzing several scenarios prior to installing the new plant. For example, site 1 is the one that is less sensitive to changes for the constraint in the final disposal, whereas site 3 is always the most expensive.

Furthermore, using the proposed model it is also possible to develop a multiobjetive optimization approach to determine the optimal trade-offs between the total annual cost and the concentration of the pollutant discharged to the final disposal, considering simultaneously the optimal location and the treatment requirements. The optimal Pareto curve can be seen in Fig. 7.4. The solutions above the Pareto curve represent suboptimal solutions, whereas the solutions below the Pareto curve are infeasible.

7.4.2 Example 7.2

The data for the Balsas watershed system located in Mexico was considered in this example (Conagua 2009a, b). This river is one of the biggest in Mexico, where several industrial zones discharge their effluents. Figure 7.5 shows the main streams of the river and 20 possible locations to place the new industrial plant that discharges a hazardous pollutant (in this case to key compound is arsenic, which is very toxic).

The following assumptions are made for the pollutant considered in this example:

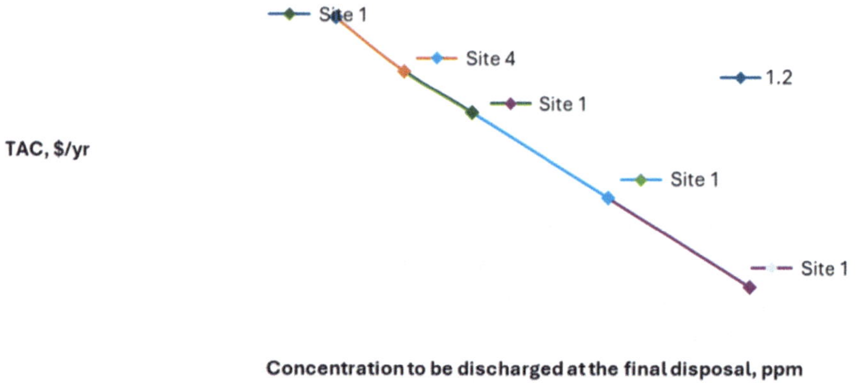

Fig. 7.4 Pareto curve for Example 7.1. Printed with permission of "*An MINLP model for the optimal location of a new industrial plant with simultaneous consideration of economic and environmental criteria*", *Industrial and Engineering Chemistry Research*, Copyright (2011), American Chemical Society

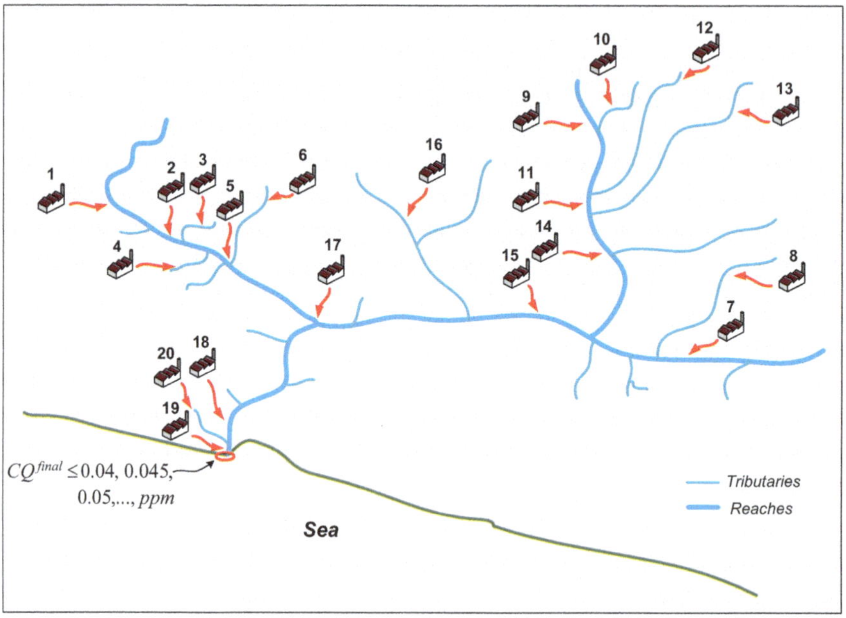

Fig. 7.5 Available locations for the new industrial plant for Example 7.2. Printed with permission of "*An MINLP model for the optimal location of a new industrial plant with simultaneous consideration of economic and environmental criteria*", *Industrial and Engineering Chemistry Research*, Copyright (2011), American Chemical Society

1. The concentration of the sanitary discharges without treatment ($CS_{r,t}^{untreated}$) is 0.5 ppm.
2. The concentration of the sanitary discharges with treatment ($CS_{r,t}^{treated}$) is 0.03 ppm.
3. The concentration of the industrial discharges ($CI_{r,t}$) is 0.06 ppm.
4. The concentration of the precipitation ($CP_{r,t}$) is 0 ppm.
5. The concentration of the agricultural discharges ($CD_{r,t}$) is 0.055 ppm.
6. The concentration of the industrial and sanitary discharges (CH_r) is 0.07 ppm.
7. The concentration of the direct discharges (CD_r) is 0.075 ppm.

The flowrate for the new plant ($QPNEWDATA$) is 5 m^3/s and the concentration of its hazardous pollutant discharged is 20 ppm.

Table 7.5 presents the data for the installation costs for the different location alternatives of the new industrial plant (these installation costs involve the transportation cost for the raw materials and products and the land cost). Table 7.5 indicates that sites 18, 19 and 20 are the cheapest ones because these sites are the closers to the sea and therefore the transportation costs are the smallest because that zone counts with the port facilities.

On the other hand, the chemical interaction between the pollutant and the environment is represented by a first order reaction with a kinetic constant of $k = 4.52095 \times 10^{-7}/s$.

Furthermore, in this case $\alpha_{r,t} = 0.000148\,m^3/ha\,s$ and $\beta_{r,t} = 0.000296\,m^3/ha\,s$, the operation time for the new plant is 8400 h/year, the annualization factor is 0.1, and the exponential factor that considers the economies of scale is 1. The following constraints are required to maintain under optimal conditions the water quality through the watershed: (1) In reach 5, some farmers take water from the river; therefore, the composition must be lower than 2.5 ppm; (2) there is a city in the area corresponding to site 15; therefore, the pollutant concentration needs to be lower than 0.3 ppm; (3) another city is located near to site 18 and this city requires a concentration for the water river lower than 0.15 ppm. The data for the available interceptors to treat the wastewater discharged from the new plant for this example are shown in Table 7.6.

To consider adequately the environmental aspects, it is required to include a constraint for the concentration of the pollutant in the final disposal once the new plant was installed (i.e., the concentration of the final disposal must be less than the one that naturally the system is able to degrade avoiding the accumulation of this pollutant). The concentration discharged to the final disposal without the discharge of the new plant is 0.037 ppm for the pollutant considered.

First, the solution of the problem considering exclusively economic aspects was obtained, which states that the new plant must be in site 19 with a total annual cost of $\$4.8 \times 10^6$/year; in this case, the concentration for the pollutant in the final disposal is 0.206 ppm. Then, several scenarios have been considered for the solution of this example.

Case A. The concentration allowable to be discharged to the final disposal must be lower than 0.04 ppm. The optimal solution for case A is to install the new plant

Table 7.5 Installation costs for Example 7.2

Site	Annualized installation cost, \$/year
1	12×10^6
2	8×10^6
3	9.5×10^6
4	8×10^6
5	9.5×10^6
6	8×10^6
7	12.5×10^6
8	10.5×10^6
9	7.7×10^6
10	7×10^6
11	7.8×10^6
12	7.5×10^6
13	8.2×10^6
14	8.5×10^6
15	7.6×10^6
16	9.5×10^6
17	9×10^6
18	5×10^6
19	4.8×10^6
20	5.5×10^6

in site 2, which represents a total annual cost of \$11.85 $\times$ 10^6/year with a treatment of 0.471 m^3/s in the interceptor 9 and 4.529 m^3/s in the interceptor 10 to yield a concentration for the final disposal that is naturally degraded. Therefore, the treated stream discharged for the new plant has a concentration of 7.125 ppm. Case A requires to invest 247% more than the economic solution to obtain a reduction of 80% of the concentration in the stream discharged to the final disposal.

Case B. The concentration for the final disposal must be lower than 0.045 ppm for the pollutant considered. In this case, the optimal location for the new plant is also in site 2, but the wastewater stream does not require any treatment and the total annual cost is \$8 $\times$ 10^6/year. Here, the comparison with the economic solution shows that it is possible to have a concentration for the final disposal less than 78% (0.045 ppm) investing \$8 $\times$ 10^6/year, which is approximately 166% of the economic solution.

Case C. The concentration in the final disposal must be lower than 0.05 ppm. In this case, the minimum total annual cost is \$7.35 $\times$ 10^6/year and the optimal location

Table 7.6 Data for the available interceptors for Example 7.2

Interceptor	Fixed cost, $	Variable cost, $ h/m^3	Efficiency factor, γ_j
1	195,000	0.058	0.95
2	186,000	0.052	0.88
3	179,500	0.049	0.85
4	175,400	0.047	0.82
5	170,000	0.043	0.79
6	162,000	0.039	0.75
7	158,000	0.035	0.73
8	144,000	0.031	0.71
9	137,500	0.028	0.68
10	126,000	0.025	0.64
11	0	0	0

for the new plant is in site 10 using interceptor 10 to treat 0.447 m^3/s of the total wastewater discharged for the new plant. The pollutant concentration for the treated discharge of the new plant is 18.855 ppm. Consequently, investing 153% more than the economic solution yields a reduction of 75% in the concentration for the final disposal concerning the economic solution.

Case D. The concentration allowable to be discharged to the final disposal must be lower than 0.06 ppm. The optimal solution for this case is to install the new plant in site 10, which represents a total annual cost of $\$7 \times 10^6$/year without any treatment because the pollutant is naturally degraded through the watershed. Compared with the economic solution, case D requires to invest 146% more to obtain a reduction of 70% for the concentration discharged to the final disposal.

Case E. The concentration for the final disposal must be lower than 0.14 ppm. Because in this case, the constraint is less severe than in the other cases, the optimal location is in site 18, a site closer to the final disposal with a lower installation cost of $\$6.17 \times 10^6$/year. It is required to treat 2.235 m^3/s using interceptor 10 to satisfy the environmental constraints and to yield a treated discharge from the new plant with a concentration of 14.279 ppm. In this case, it is shown that by investing 1.28% more than the economic solution, the concentration discharged to the final disposal increases 32%.

Table 7.7 presents a result comparison for the cases previously analyzed (including the solution without considering the environmental constraint in the last row). Table 7.7 presents the concentrations for the specific locations of the river that are restricted, and one can see that all these constraints are satisfied. Table 7.7 shows that the total annual cost increases proportionally to the regulation for the final disposal of the pollutant.

Table 7.7 Result comparison for several sustainability constraints in Example 7.2

Constraint	Site	y^{disc} (ppm)	Concentration for specific reaches, ppm			Installation cost, \$/year	Treatment cost, \$/year	Total annual cost, \$/year
			1	2	3			
0.04 ppm X	2	7.125	0.943	0.139	0.032	8×10^6	3.85×10^6	11.85×10^6
0.045 ppm X	2	20	2.445	0.139	0.037	8×10^6	0	8×10^6
0.05 ppm X	10	18.855	0.123	0.277	0.042	7×10^6	3.50×10^5	7.35×10^6
0.06 ppm X	10	20	0.123	0.285	0.043	7×10^6	0	7×10^6
0.14 ppm X	18	14.279	0.123	0.139	0.137	5×10^6	1.7×10^6	6.7×10^6
0.206 ppm X	19	20	0.123	0.139	0.029	4.8×10^6	0	4.8×10^6

Furthermore, Table 7.8 presents different scenarios to install the new plant in different locations and for different concentrations in the final disposal. All options considered must satisfy the constraints for the concentration through the watershed. For instance, if an upper limit for the concentration in the final disposal of 0.05 ppm was fixed, then the optimal solution is in site 10. Nevertheless, if the local governments for sites 15 and 17 were interested in catching this new industry, they would require incentives to compensate the cost of the optimal solution by \$2.23 $\times$ 10^6 and \$3.74 $\times$ 10^6, respectively.

Figure 7.6 presents the Pareto curve for this case. There is a trade-off between the total annual cost and the concentration discharged to the final disposal. In addition, the optimal location depends on the specific concentration in the final disposal. This Pareto curve shows a set of optimal solutions, and it is very useful to analyze different scenarios.

Finally, shows the size and CPU time for both examples in a Core 2 processor with 3 MB of RAM. It is noteworthy that the CPU time is small (Table 7.9).

7.5 Concluding Remarks

The optimal location of a new industrial plant that satisfies environmental regulations for wastewater streams discharged to rivers and constraints imposed to ensure the sustainability for the watershed—including local constraints for specific sites

Table 7.8 Total cost for several sustainability constraints and several locations in Example 7.2

Upper limit for the concentration in the final disposal, ppm	Total annual cost for the site, \$/year					
	2	10	15	17	18	19
0.04	11.85×10^6	13.25×10^6	14.84×10^6	16.30×10^6	Infeasible	Infeasible
0.045	8×10^6	9.54×10^6	11.13×10^6	12.60×10^6	13.73×10^6	Infeasible
0.05	8×10^6	7.35×10^6	9.58×10^6	11.09×10^6	13.14×10^6	13.16×10^6
0.06	8×10^6	7×10^6	7.6×10^6	9×10^6	11.96×10^6	12.15×10^6
0.14	8×10^6	7×10^6	7.6×10^6	9×10^6	6.70×10^6	7.13×10^6
0.206	8×10^6	7×10^6	7.6×10^6	9×10^6	5×10^6	4.8×10^6

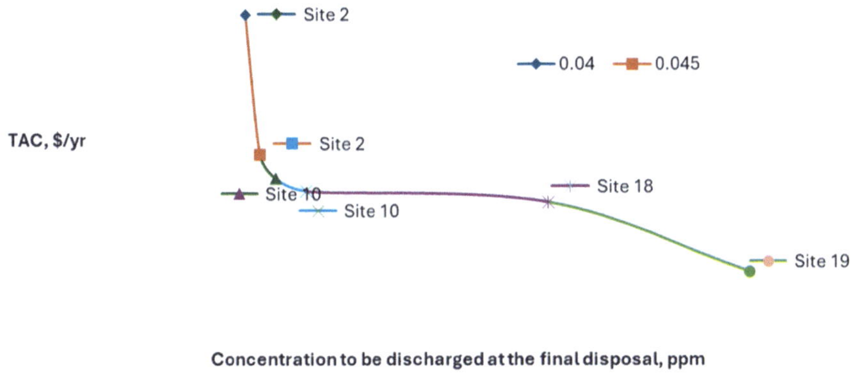

Fig. 7.6 Pareto curve for Example 7.2. Printed with permission of "*An MINLP model for the optimal location of a new industrial plant with simultaneous consideration of economic and environmental criteria*", *Industrial and Engineering Chemistry Research*, Copyright (2011), American Chemical Society

where the water can be used—is presented in this chapter using a new mixed integer nonlinear programming model. Both the mass integration with the watershed and the mass integration for the wastewater stream can be taken into consideration in the suggested model. The purpose of the model's formulation was to forecast how the newly polluted flows will affect the watershed. The material flow analysis model was utilized to create this challenge. The MFA model considers all extractions and discharges from industrial, residential, agricultural, and other sources. It also

Table 7.9 Size and computation time for the problems considered

Concept	Example 7.1	Example 7.2
Number of constraints	81	1479
Number of continuous variables	82	1476
Number of binary variables	8	31
CPU time (s)	0.37	1.68

considers the chemical processes that occur in the rivers. To determine the best location for a new industrial plant that satisfies environmental standards and integrates waste, the watershed model is integrated with a disjunctive model. The objective function of the model is to minimize the overall yearly cost, which is made up of the costs associated with the site of the new plant and the wastewater treatment costs. This allows for the best possible choice of treatment. The case studies' resolution demonstrates how a new industrial plant's location greatly affects the amount of treatment needed to produce a sustainable process. The Pareto curve displays a set of ideal solutions for the scenario where the concentration in the final disposal changes, but the suggested model can be used as a tool by governments and society to determine the incentives needed to make their locations attractive for the installation of the new plant to increase the offer of jobs in their location. Ultimately, the model can detect several contaminants; nevertheless, the primary goal of this work is to analyze the impact of the major component and follow it as it moves through the watershed.

Nomenclature

$A_{r,t}$	Area cover by effluent j in reach r, acre or ha
C	Set for the components, $\{c \mid c = 1, \ldots, N_c\}$
$c_{\exp}$	Exponential factor useful to consider the economies of scale.
$CD_{c,r}$	Concentration of direct discharges to the reach r, ppm
$CD_{c,r,t}$	Concentration of agricultural discharges to the tributary t to the reach r, ppm
$CH_{c,r}$	Concentration of total discharge (i.e., industrial + sanitary) to the reach r, ppm
$CI_{c,r,t}$	Concentration of industrial discharge from the tributary t to the reach r, ppm
$CL_{c,r}$	Concentration of total losses (filtration and evaporation) from the reach r, ppm

$CL_{c,r,t}$	Concentration of total losses (filtration and evaporation) from tributary t, ppm
$CLAND_p$	Annualized installation cost for the new plant in site p, \$/year
$CP_{c,r}$	Concentration of precipitation discharged to the reach r, ppm
$CP_{c,r,t}$	Concentration of precipitation discharged to the tributary t to the reach r, ppm
$CPNEW_{c,p}$	Concentration for the contaminant discharged for the new plant installed in site p, ppm
$CQ_{c,r}$	Concentration of flowrate exit to the reach r, ppm
$CQ_{c,r-1}$	Concentration of flowrate inlet to the reach r, ppm
CQ_c^{final}	Pollutant concentration discharged to the final disposal, ppm
$CT_{c,r,t}$	Concentration of discharge for the tributary t to the reach r, ppm
$CU_{c,r}$	Concentration of water used from reach r, ppm
$CU_{c,r,t}$	Concentration of water used from tributary t discharge to reach r, ppm
$CS_{c,r,t}^{treated}$	Concentration of residual wastewater discharged without treatment to the reach r for tributary t, ppm
$CS_{c,r,t}^{untreated}$	Concentration of residual treated wastewater discharged to the reach r for tributary t, ppm
$D_{r,t}$	Agricultural discharges to the tributary t to the reach r, m^3/s
D_r	Direct discharges to the reach r, m^3/s
FC^j	Fixed cost for interceptor j, \$/year
$FT_{r,t}$	Discharge for the tributary t to the reach r, m^3/s
f_j	Segregated flowrate from the wastewater of the new plant to the interceptor j, m^3/s
f^{UP}	Upper limit for f_j
H_Y	Operation time per year, h/year
H_r	Total discharge (i.e., industrial + sanitary) to the reach r, m^3/s
$I_{r,t}$	Industrial discharge from the tributary t to the reach r, m^3/s
J	Set for the interceptors
k	Kinetic constant for the degradation of the pollutant in the system
k_f	Annualization factor
$L_{r,t}$	Total losses (filtration and evaporation) from tributary t, m^3/s
L_r	Total losses (filtration and evaporation) from the reach r, m^3/s
N_c	Total number of components
N_j	Total number of interceptors
N_p	Total number of allowable sites to locate the new plant
N_r	Total number of reaches
N_t	Total number of tributaries
$N(R)$	Subset for specific reaches that requires composition constraints
P	Set for the allowable sites to locate the new plant
$P_{r,t}$	Precipitation discharged for the tributary t to the reach r, m^3/s
P_r	Precipitation discharged to the reach r, m^3/s
Q_r	Flowrate exit to the reach r, m^3/s

Q_{r-1}	Flowrate inlet to the reach r, m^3/s
$QPNEW_p$	Flowrate discharged for the new plant to the location p
$QPNEWDATA$	Constant flowrate discharged for the new plant
R	Set for the reaches
r_r	Reaction carries out in the reach r, g/s
$r_{r,t}$	Reaction carries out in the tributary t that discharges to the reach r, g/s
$S_{r,t}^{untreated}$	Residual wastewater discharged without treatment to the reach r for tributary t, m^3/s
$S_{r,t}^{treated}$	Residual treated wastewater discharged to the reach r for tributary t, m^3/s
T	Set for the tributaries
TAC	Total annual cost, $/year
$U_{r,t}$	Water used from tributary t discharged to reach r, m^3/s
U_r	Water used from reach r, m^3/s
VC^j	Variable cost for interceptor j, $/ m^3 year
$V_{r,t}$	Volume for tributary t discharged to reach r, m^3
V_r	Volume for reach r, m^3
W_p	Boolean variable for the location of the new plant
w_p	Binary variable for the location of the new plant
$y_{c,n(r)}^{desired}$	Limit for the desired concentration in some reaches, ppm
y_c^{disc}	Pollutant concentration for the final discharge of the new plant to the river, ppm
y_c^{EnvReg}	Environmental regulation for the pollutant concentration to be discharged by the new plant, ppm
y_c^{in}	Pollutant concentration exits to the plant and inlets to the treatment system, ppm
$y_{c,j}^{out}$	Pollutant concentration at the exit of interceptor j, ppm
$y_c^{sustainability}$	Sustainability constraint at the final disposal, ppm
z_j	Binary variable associated to the existence of the interceptor j

Greek Symbols

$\alpha_{r,t}$	Agricultural flowrate per area, m^3/ha s
$\beta_{r,t}$	Agricultural use of water from tributary t, m^3/ha s
γ_j	Efficiency factor to remove the pollutant for the interceptor j
δ	Small number

Indexes

c	Component
j	Interceptor
$n(r)$	Reaches that requires a composition constraint
p	Site to locate the new plant
r	Reach
t	Tributary
UP	Upper limit

References

Agrawal V, Shenoy UV (2006) Unified conceptual approach to targeting and design of water and hydrogen networks. AIChE J 52:1071–1082

Alva-Argaez A, Kokossis AC, Smith R (1998) Wastewater minimization of industrial systems using an integrated approach. Comput Chem Eng 22:S741–S744

Alva-Argaez A, Vallianatos A, Kokossis A (1999) A multi-contaminant transhipment model for mass exchange networks and wastewater minimization problems. Comput Chem Eng 23:1439–1453

Baccini P, Brunner P (1991) Metabolism of the anthroposphere. Springer, Berlin, Germany

Benko N, Rev E, Fonyo Z (2000) The use of nonlinear programming to optimal water allocation. Chem Eng Commun 178:67–101

Brooke A, Kendrick D, Meeruas A, Raman R (2006) GAMS-language guide. GAMS Development Corp. Washington, D.C.

Brunner PH, Rechberg H (2004) Practical handbook of material flow analysis. CRC

CONAGUA (2009a) Mexican National Water Commission. Water statistics 2008. http://www.conagua.gob.mx/OCB07/Temas/EstadisticasBALSAS-FINALago09.pdf

CONAGUA (2009b) Mexican National Water Commission. Water atlas in Mexico 2009

Dhole VR, Ramchandani N, Tainsh RA, Wasilewski M (1996) Make your process water pay for itself. Chem Eng 103:100–103

Doyle SJ, Smith R (1997) Targeting water reuse with multiple contaminants. Process Saf Environ Prot 75:181–189

Dunn RF, El-Halwagi MM (2003) Process integration technology review: background and application in chemical process industry. J Chem Technol Biotechnol 78:1011–1021

El-Baz AA, Ewida KT, Shouman MA, El-Halwagi MM (2005a) Material flow analysis and integration of watersheds and drainage systems: I. Simulation and application to ammonium management in Bahr El-Baqar drainage system. Clean Technol Environ Policy 7:51–61

El-Baz AA, Ewida KT, Shouman MA, El-Halwagi MM (2005b) Material flow analysis and integration of watersheds and drain systems: II. Integration and solution strategies with application to ammonium management in Bahr El-Baqar drain system. Clean Technol Environ Policy 7:78–86

El-Halwagi MM (1997) Pollution prevention through process integration: systematic design tools. Academic Press, San Diego

El-Halwagi MM (2006) Process integration. Academic Press, New York

El-Halwagi MM, Spriggs HD (1998) Solve design puzzles with mass integration. Chem Eng Prog 94:25–44

El-Halwagi MM, Gabriel F, Harell D (2003) Rigorous graphical targeting for resource conservation via material recycle/reuse networks. Ind Eng Chem Res 42:4319–4328

Feng X, Bai J, Zheng XS (2007) On the use of graphical method to determine the targets of single-contaminant regeneration recycling water systems. Chem Eng Sci 62:2127–2138
Foo CY (2009) State-of-the-art review of pinch analysis techniques for water network synthesis. Ind Eng Chem Res 48:5125–5159
Gabriel FB, El-Halwagi MM (2005) Simultaneous synthesis of waste interception and material reuse networks: Problem reformulation for global optimization. Environ Prog 24:171–180
Galan B, Grossmann IE (1998) Optimal design of distributed wastewater treatment networks. Ind Eng Chem Res 37:4036–4048
Gomes JFS, Queiroz EM, Pessoa FLP (2007) Design procedure for water/wastewater minimization: single contaminant. J Clean Prod 15:474–485
Gunaratnam M, Alva-Argaez A, Kokossis A, Kim JK, Smith R (2005) Automated design of total water systems. Ind Eng Chem Res 44:588–599
Hallale N (2002) A new graphical targeting method for water minimization. Adv Environ Res 6:377–390
Hernandez-Suarez R, Castellanos-Fernandez J, Zamora JM (2004) Superstructure decomposition and parametric optimization approach for the synthesis of distributed wastewater treatment networks. Ind Eng Chem Res 43:2175–2191
Karuppiah R, Grossmann IE (2006) Global optimization for the synthesis of integrated water systems in chemical processes. Comput Chem Eng 30:650–673
Kuo WCJ, Smith R (1997) Effluent treatment system design. Chem Eng Sci 52:4273–4290
Lampert C, Brunner PH (1999) Material accounting as a policy tool for nutrient management in the Danube Basin. Water Sci Technol 40:43–49
Lira-Barragán LF, Ponce-Ortega JM, Serna-González M, El-Halwagi MM (2011) An MINLP model for the optimal location of a new industrial plant with simultaneous consideration of economic and environmental criteria. Ind Eng Chem Res 50:953–964
Lovelady EM, El-Baz AA, El-Monayeri D, EL-Halwagi MM (2009) Reverse problem formulation for integrated process discharges with wastewater and drainage systems: managing phosphorus in lake Manzala. J Ind Ecol 13:914–927
Manan ZA, Tan YL, Foo DCY (2004) Targeting the minimum water rate using water cascade analysis technique. AIChE J 50:3169–3183
Nápoles-Rivera F, Ponce-Ortega JM, El-Halwagi MM, Jiménez-Gutiérrez A (2010) Global optimization of mass and property integration networks with in-plant property interceptors. Chem Eng Sci 65:4363–4377
Pillai HK, Bandyopadhyay S (2007) A rigorous targeting algorithm for resource allocation networks. Chem Eng Sci 62:6212–6221
Polley GT, Polley HL (2000) Design better water networks. Chem Eng Prog 96:47–52
Ponce-Ortega JM, Hourtua AC, El-Halwagi MM, Jiménez-Gutiérrez A (2009a) A property-based optimization of direct recycle networks and wastewater treatment processes. AIChE J 55:2329–2344
Ponce-Ortega JM, Serna-González M, Jiménez-Gutiérrez A (2009b) A disjunctive programming model for simultaneous synthesis and detailed design of cooling networks. Ind Eng Chem Res 48:2991–3003
Ponce-Ortega JM, El-Halwagi MM, Jiménez-Gutiérrez A (2010) Global optimization of property-based recycle and reuse networks including environmental constraints. Comput Chem Eng 34:318–330
Putra ZA, Amminudin KA (2008) Two-step optimization approach for design a total water system. Ind Eng Chem Res 47:6045–6057
Raman R, Grossmann IE (1994) Modeling and computational techniques for logic based integer programming. Comput Chem Eng 18:563–578
Savelski MJ, Bagajewicz MJ (2000) On the optimality conditions of water utilization systems in process plants with single contaminants. Chem Eng Sci 55:5035–5048
Savelski MJ, Bagajewicz MJ (2001) Algorithmic procedure to design water utilization systems featuring a single contaminant in process plants. Chem Eng Sci 56:1897–1911

Savelski M, Bagajewicz M (2003) On the necessary conditions of optimality of water utilization systems in process plants with multiple contaminants. Chem Eng Sci 58:5349–5362
Sorin M, Bedard S (1999) The global pinch point in water reuse networks. Process Saf Environ Prot 77:305–308
Teles J, Castro PM, Novals AQ (2008) LP-based solution strategies for the optimal design of industrial water networks with multiple contaminants. Chem Eng Sci 63:376–394
Viswanathan J, Grossmann IE (1990) A combined penalty function and outer approximation method for MINLP optimization. Comput Chem Eng 14:769–782
Wang YP, Smith R (1994) Wastewater minimization. Chem Eng Sci 49:981–1006

Chapter 8
Designing Water Networks with a Focus on the Sustainability of the Adjacent Watersheds

8.1 Introduction

Chemical, petrochemical, petroleum refining, and other process industries release a lot of wastewater into the environment in addition to needing a lot of freshwater for stripping, liquid–liquid extraction, steam generation, reaction medium, and other washing operations. The process industries are now driven to maximize water recycling and reuse to minimize water consumption due to the growing shortage of freshwater resources. Furthermore, before being released into various water disposal locations, industrial wastewater must have its chemical and thermal pollution reduced to an acceptable level due to the environment's ever-tougher restrictions.

In the process industries usually, there is a set of process equipment (sinks) that require water with specific characteristics. Part of this water becomes wastewater after they are used because the equipment does not operate at 100% efficiency for the conversion and separation of the species. Freshwater consumption is one of the key elements in the processes because the large consumption of this resource impacts economically and environmentally (for the costs associated with the fresh sources and the production of waste streams discharged to the environment). The recycle and reuse water networks are very convenient for decreasing the consumption of freshwater using the segregation and mixing of different waste streams to recycle them directly to the process sinks or to include a treatment and then reuse these streams in the industrial plant. The recycle and reuse strategy aids in reducing the total wastewater discharged to the environment and at the same time it helps to reduce the consumption of freshwater (with the corresponding reduction in the total annual cost for the overall process). Several techniques have been considered for the synthesis of recycle and reuse networks owing to the important environmental and economic savings produced. These techniques can be classified as heuristic approaches and techniques based on mathematical programming. The heuristic approaches are the pinch point-based methodologies (Wang and Smith 1994; Dhole et al. 1996; El-Halwagi and Spriggs 1998; Polley and Polley 2000; Hallale 2002; Manan et al.

J. M. Ponce-Ortega et al., *Sustainable Water Systems*, Water Resources Development and Management, https://doi.org/10.1007/978-3-031-72498-5_8

2004; El-Halwagi et al. 2003; Feng et al. 2007; Foo 2009) and the approaches based on algebraic techniques (Sorin and Bedard 1999; Agrawal and Shenoy 2006; Gomes et al. 2007; Dunn and El-Halwagi 2003). Furthermore, recent methodologies are formulated as mathematical programming models (Pillai and Bandyopadhyay 2007; Savelski and Bagajewicz 2000, 2001, 2003; Alva-Argaez et al. 1998, 1999; Benko et al. 2000; Teles et al. 2008; Gabriel and El-Halwagi 2005; Kuo and Smith 1997; Doyle and Smith 1997; Galan and Grossmann 1998; Hernandez-Suarez et al. 2004; Gunaratnam et al. 2005; Karuppiah and Grossmann 2006; Putra and Amminudin 2008; Ponce-Ortega et al. 2009a, 2010a; Nápoles-Rivera et al. 2010). Nevertheless, the main limitation of these approaches is that they only consider the processes that happen inside the industrial facilities, without considering the processes that occur outside the plant for the waste streams discharged to the environment. Even recently some papers have been reported to account for environmental aspects of the synthesis of mass integration networks (see for example Ku-Pineda and Tan 2006; Tan et al. 2007; Lim and Park 2007, 2008a, b; Ponce-Ortega et al. 2010b), these formulations do not track the pollutants discharged to the surrounding watershed and their effects in the surrounding areas cannot be considered adequately.

This way, watersheds are the systems where the wastewater discharged by the industrial plants is transported to their final disposal, and these final disposals frequently are lakes or seas that are severely impacted by the pollutants discharged from the industries. In addition to the industrial discharges, other types of streams like municipal, agricultural, treated and untreated wastewater, precipitation, etc. are discharged in the watershed All these discharges change extremely the compositions and properties of the streams in the rivers; consequently, the watershed systems impact and are strongly impacted by their surroundings, but at the same time the natural degradation of the pollutants is carried out through the watershed by the chemical, biochemical and physical transformations of the flora and fauna (Brunner and Rechberg 2004). During the route, watersheds are fed by tributaries, and finally, the watershed systems transport the pollutants to catchment areas. It is important to track the pollutants discharged as wastewater for the industries to avoid the hazardous accumulation of these pollutants in the final disposal and to impose specific constraints for the areas where the water is used for human and agricultural uses.

Regarding the modeling of the pollutant behavior in the watershed system, Baccini and Brunner (1991) developed a material flow analysis (MFA) model for analyzing ecosystems with human activities, where mass, energy, and information are exchanged with the surroundings. Furthermore, Lampert and Brunner (1999) reported an MFA model to track nutrient discharges into the Danube river basin. El-Baz et al. (2005a) proposed an MFA model to track the nitrogenous compounds in the Bahr El-Baqar drainage system, and El-Baz et al. (2005b) reported an MFA model to include a mass integration for the Bahr El-Baqar drainage system in Egypt. Lovelady et al. (2009) reported a reverse problem formulation to integrate the discharges of an industrial process with macroscopic watershed systems based on an MFA model, where the key idea consists of fixing the maximum concentration for the pollutants discharged to the environment to avoid hazardous accumulation in the final disposal to ensure sustainable processes; but, the main disadvantage of this model is that it does not optimize simultaneously the process and it uses a set of heuristic rules. Lira-Barragán et al. (2011b) reported an approach to locate a new industrial plant considering all discharges and extractions over the watershed; nevertheless, this approach does not include the mass integration inside of the plant, in addition, the wastewater discharged to the environment is a parameter given only by economic aspects (Fig. 8.1) and this work only contemplates the associated costs for the installation of the new plant (fixed for each location) and the costs for the central treatment technology required to satisfy the environmental constraints. Therefore, it is not possible to consider the reduction of the costs associated with fresh sources, distributed treatment, and pipes because mass integration inside the industrial plant was not taken into consideration.

Hence, this chapter reports a model to integrate the synthesis of recycle and reuse water networks for new plants together with the watershed system where the pollutants from the new plant are discharged considering the possibility of installing the new plant in different allowable alternatives (Lira-Barragán et al. 2011a). The optimal quantity and quality for the wastewater discharged from the new plant depend on the location selected to install the plant because the impact on the watershed depends significantly on the location selected (considering the integration with other streams and the chemical and biochemical reactions that occur in the watershed system); and so, the recycle and reuse water network is highly influenced by the location of the new plant. The presented model is a disjunctive programming formulation that is reformulated as a MINLP problem considering the minimization of the total annual cost for the recycle and reuse water network and the installation of the new plant to satisfy the process and environmental constraints as well as the sustainability of the watershed. The model can consider multiple pollutants in the waste streams and the selection of several technologies to treat them.

8.2 Problem Statement

The problem addressed in this chapter is expressed as follows.

A set of alternatives to locate a new industrial facility are given ($P = \{p|p = 1, 2, \ldots, N_p\}$). Each allowed location p has associated an installation cost, ($Cland_p$), which involves the cost associated with the transportation of raw materials, products, and services, as well as the land cost. Also is given a set of process sources (process streams), ($I = \{i|i = 1, 2, \ldots, N_i\}$). Each process source has several components ($C = \{c|c = 1, 2, \ldots, N_c\}$), and these streams can be segregated to be recycled and/or reused in the process sinks. The flowrate and composition for each component are fixed by the process and given by W_i and $y_{i,c}^{InSource}$, respectively. Also, given is a set of process sinks (process units), $J = \{j|j = 1, 2, \ldots, N_j\}$ and each process sink requires a given flowrate (G_j), with specific limits for the inlet composition ($y_{j,c}^{InSink}$). Also, there is available a set of fresh sources ($K = \{k|k = 1, 2, \ldots, N_k\}$) with flowrates ($F_k$) to be determined,

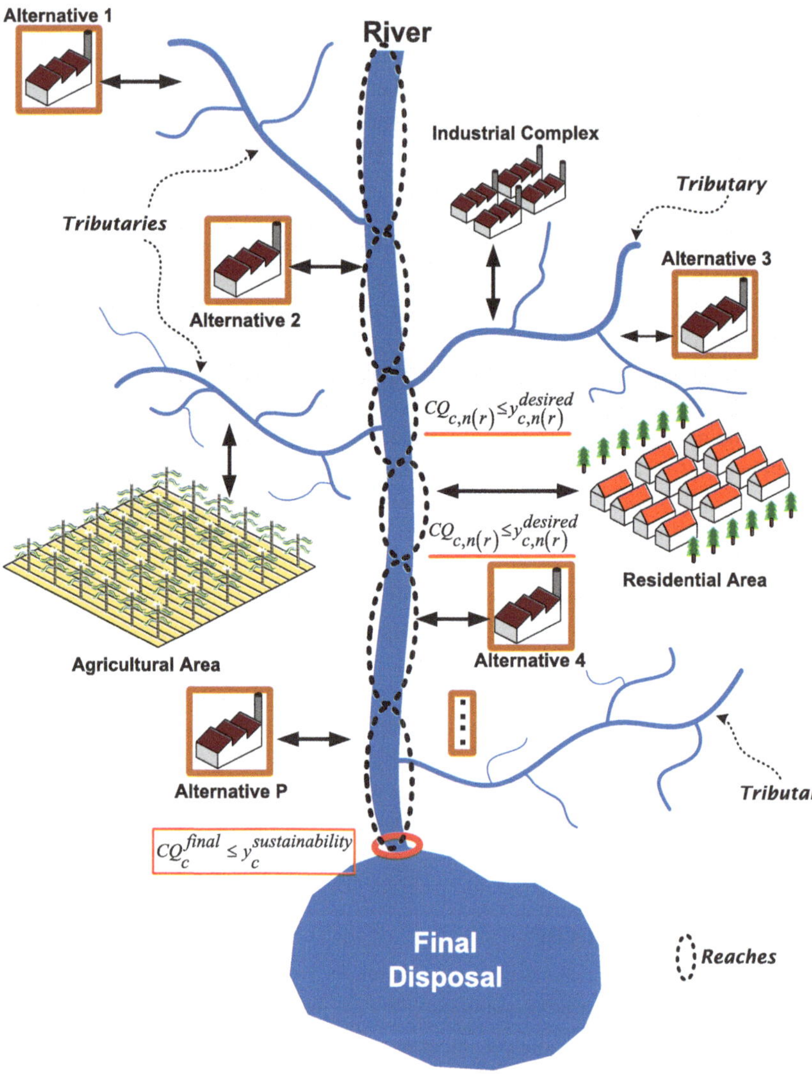

Fig. 8.1 Alternatives to locate the new industrial plant in a watershed. Printed with permission of "*Synthesis of water networks considering the sustainability of the surrounding watershed*", *Computers and Chemical Engineering*, Copyright (2011), Elsevier

and each one with composition given ($y^{Fresh}_{k,c}$). There is available a set of interceptors to treat the pollutants (INT). Lastly, there is a set of environmental regulations ($y^{disc,\max}_{c}$) for the wastewater discharged into the environment ($Waste$). This new stream will be discharged to the surrounding river or watershed, which is divided into several reaches ($R = \{r | r = 1, 2, \ldots, N_r\}$) and each reach can receive several effluents ($T = \{t | t = 1, 2, \ldots, N_t\}$). Hence, there is imposed a set of sustainability constraints for the composition to be discharged to the final disposal. Then, the problem consists of determining simultaneously the synthesis of the recycle and reuse water network and the optimal location for the new plant that discharges wastewater to the surrounding watershed that avoid the accumulation of the pollutants in the final disposal and at the same time meets the environmental constraints for the pollutants through the watershed. The objective function is to minimize the total annual cost (TAC) including the installation costs for the new plant, the treatment costs for the process sources, the costs of fresh sources, and the piping costs.

The total wastewater discharged for the new plant and its concentration are optimization variables, which depend strongly on its optimal position to be determined by the model; this means that the required treatment depends on the location of the new plant. Furthermore, this concentration must meet the environmental constraints as well as the sustainability constraints to avoid the accumulation of pollutants at the final disposal. To meet this last objective, the behavior of the river is simulated by considering all inlet and outlet streams through the MFA technique. On the other hand, the required treatment also depends on the composition constraints at the inlet of process sinks. Therefore, the treatment is carried out by segregating the process sources to treat the pollutants in the interceptors, where each interceptor can treat all pollutants; the efficiencies to remove each component depend on the type of unit used.

Figure 8.2 shows that the main stream of the river is fed by several streams or tributaries ($FT_{r,t}$). The watershed exchanges water during its trajectory; according to their use, the following types of water exchanges can be identified: water for agricultural use, wastewater discharged to the river with and without treatment, industrial and residential effluents, etc., and the watershed discharges to the final disposal. Besides, natural phenomena like precipitation, filtration, and evaporation are considered in the model. These processes alter the composition of the materials transported in the river; consequently, to track adequately the mean composition for the hazardous compounds, it becomes essential that the river is sectioned into parts where the composition can be considered constant (these sections are represented in Fig. 8.2 with ellipses). These parts o sections are called reaches and these identify zones where no substantial intermediate effluents are discharged and/or extracted. Therefore, the model identifies the main effluents discharged in each reach, where these effluents are called tributaries. Tributaries are channels or branches of the river, which may contain discharges with/without treatment, industrial discharges, etc., their flowrates ($FT_{r,t}$) and concentrations ($CT_{c,r,t}$) impact the reaches where they are discharged. The discharges to the watershed contain pollutants with specific concentrations, and they interact chemically and biochemically with the system through the flora and fauna living in the watershed.

One of the goals of the model of this chapter is to ensure the sustainability of the surrounding watershed. Thus, it is required to implement an analysis before the optimization of the capacity of the final disposal to decompose chemically and biochemically each pollutant. This analysis yields the sustainability constraints which ensure that the pollutants will not be accumulated in the system.

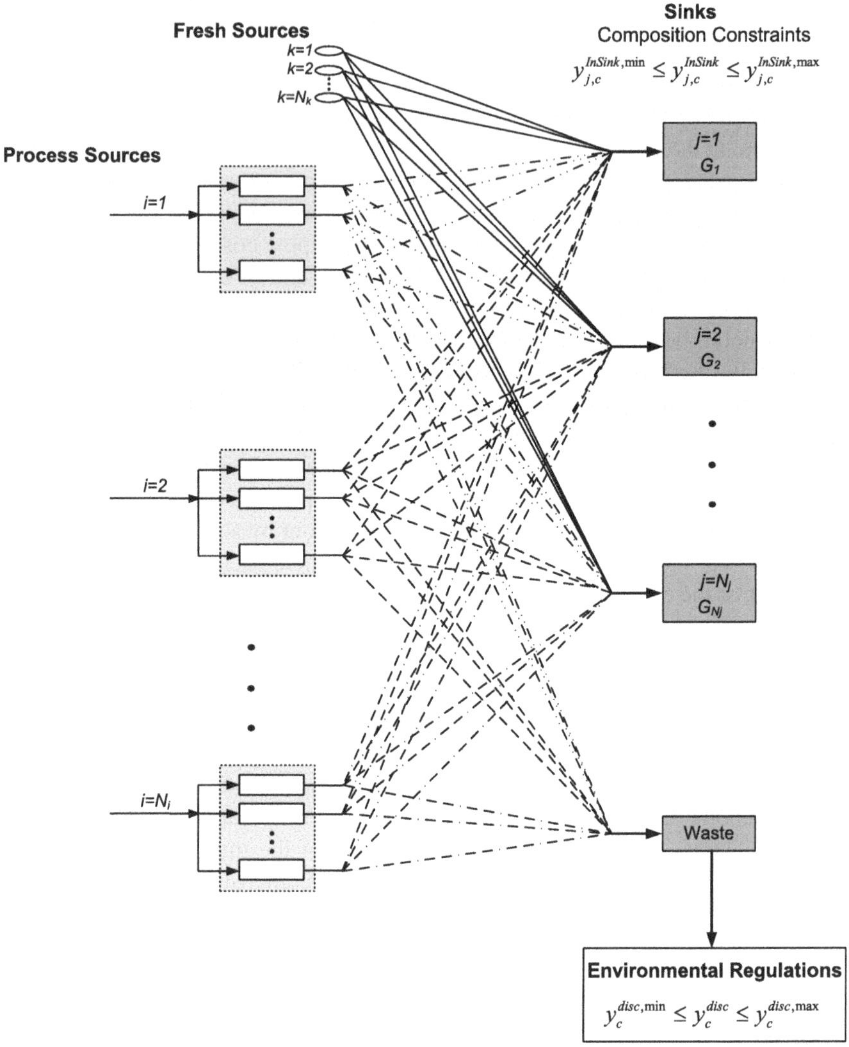

Fig. 8.2 Superstructure for the recycle and reuse water network. Printed with permission of "*Synthesis of water networks considering the sustainability of the surrounding watershed*", *Computers and Chemical Engineering*, Copyright (2011), Elsevier

Figure 8.3 presents the considered superstructure (based on the formulation previously reported by Gabriel and El-Halwagi 2005 and then taken by Ponce-Ortega et al. 2010a), where the mass integration is carried out. As can be seen in this superstructure, process sources are segregated to treat all the required pollutants, avoiding the mixing of different process sources. At the exit of the treatment systems, each stream is split to be sent to each process sink or the wastewater. The available fresh water can be used in the process sinks when they are required. Each interceptor has associated an efficiency factor (γ_c^{int}) for each required pollutant to be removed. These efficiencies are related to the amount of pollutants that can be removed and they are calculated before the optimization process. Then, the composition at the exit of each treatment unit for each component is calculated as follows:

$$y_{i,c}^{Out,\mathrm{int}} = \left(1 - \gamma_c^{\mathrm{int}}\right) y_{i,c}^{InSource}, \quad \forall i \in I, \quad c \in C, \quad \mathrm{int} \in INT$$

It is worth noticing that to model the direct recycle streams, the model includes a set of fictitious interceptors for each compound with conversion factors equal to zero.

8.3 Implemented Mathematical Model

The implemented mathematical model is based on the material flow analysis technique to predict the watershed performance, as well as in a disjunctive model to locate the new industrial plant and the required equations to model the superstructure for the recycle and reuse water network inside the plant. The model generated is explained as follows.

8.3.1 *Material Flow Analysis for Multiple Pollutants*

The MFA technique is utilized to model the watershed and to calculate the trajectory for the pollutants discharged from the new industrial plant and their interactions with the environment. Thus, the watershed is split into tributaries and reaches, and the following material and component balances are required.

8.3.1.1 Overall Balance for Each Reach

This balance determines that the exit flowrate (Q_r) from each reach r is equal to the inlet flowrate (Q_{r-1}), plus precipitation (P_r), direct industrial discharges (D_r), residential discharges (H_r), the sum of all effluents entering the reach ($FT_{r,t}$), and the possible discharge for the new industrial plant when this discharges in the reach r of the p allowable ($QPNEW_{r(p)}$), minus the extractions for filtration and evaporation (L_r), and the used of water (U_r) in the r section of the river. This material balance can be stated as follows:

$$Q_r = Q_{r-1} + P_r + D_r + H_r + \sum_{t=1}^{N_{t(r)}} FT_{r,t} + QPNEW_{r(p)} - L_r - U_r, \quad \forall r \in R \quad (8.1)$$

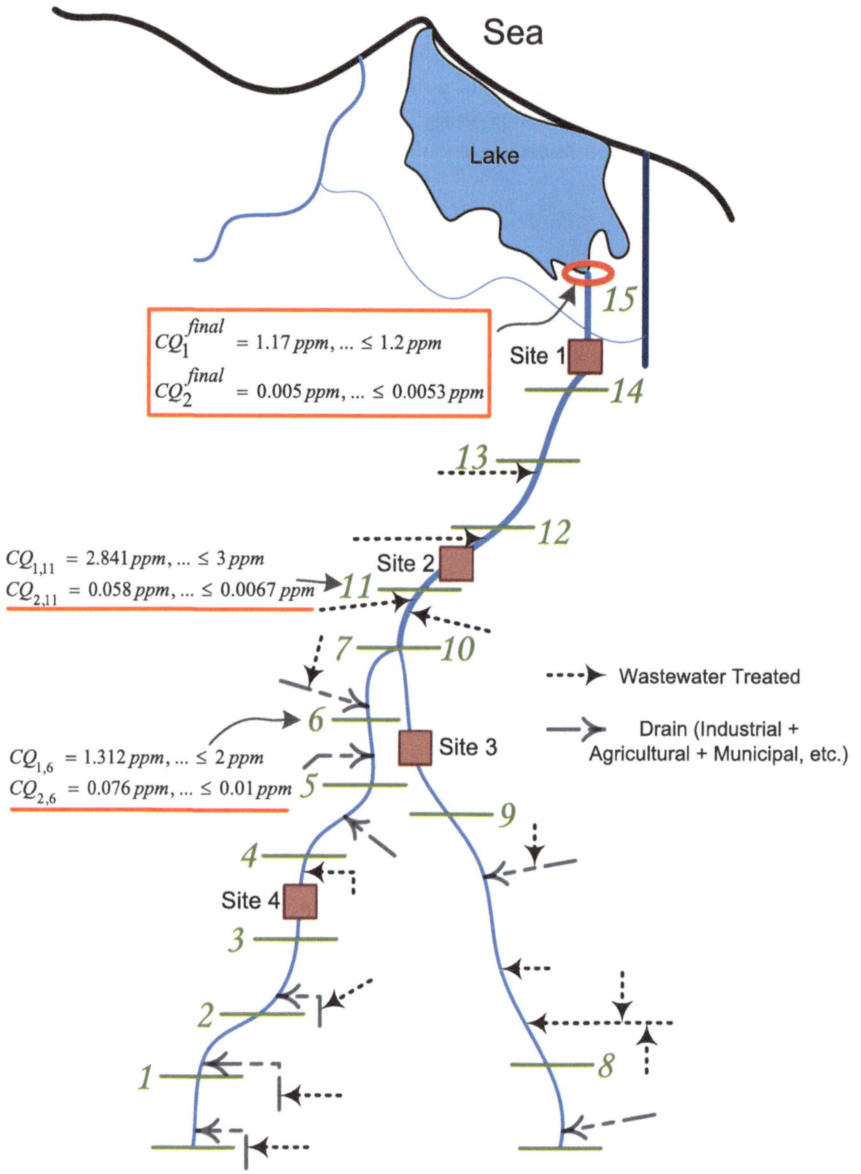

Fig. 8.3 Candidate sites for the location of the new industrial plant and constraints for Example 8.1. Printed with permission of "*Synthesis of water networks considering the sustainability of the surrounding watershed*", *Computers and Chemical Engineering*, Copyright (2011), Elsevier

$N_{t(r)}$ is the total number of tributaries that are discharged to reach r.

8.3.1.2 Component Balances for Each Reach

The mass for each hazardous compound c at the exit of each reach r ($Q_r CQ_{c,r}$) is equal to the mass at the inlet ($Q_{r-1} CQ_{c,r-1}$), plus the mass contained in the precipitation ($P_r CP_{c,r}$), industrial discharges ($D_r CD_{c,r}$), residential discharges ($H_r CH_{c,r}$), the sum of tributaries ($\sum_{t=1}^{N_{t(r)}} FT_{r,t} CT_{c,r,t}$), and the potential discharge from the new plant ($QPNEW_{r(p)} CPNEW_{c,r(p)}$), minus loses ($L_r CL_{c,r}$), uses ($U_r CU_{c,r}$) and the mass decomposed by the chemical reactions that are involved in that section of the river ($\int_{V=0}^{V_r} r_{c,r} dV$). The reactive term counts the chemical and biochemical reactions that take place in the river by the flora and fauna due to the interaction between the components and the system, which may produce or eliminate pollutants. Then, the component balance for each compound c in each section of the river r is stated as follows:

$$
\begin{aligned}
Q_r CQ_{c,r} = {} & Q_{r-1} CQ_{c,r-1} + P_r CP_{c,r} + D_r CD_{c,r} + H_r CH_{c,r} \\
& + \sum_{t=1}^{N_{t(r)}} FT_{r,t} CT_{c,r,t} + QPNEW_{r(p)} CPNEW_{c,r(p)} \\
& - L_r CL_{c,r} - U_r CU_{c,r} - \int_{V=0}^{V_r} r_{c,r} dV_r, \quad \forall c \in C, \quad r \in R
\end{aligned} \tag{8.2}
$$

8.3.1.3 Overall Balance for Each Tributary

The total balance for the tributary t that discharges to the reach r is explained as follows:

$$
\begin{aligned}
FT_{r,t} = {} & S_{r,t}^{untreated} + S_{r,t}^{treated} + I_{r,t} + P_{r,t} + D_{r,t} + QPNEW_{r,t(p)} \\
& - L_{r,t} - U_{r,t}, \quad \forall r \in R, \quad t \in T
\end{aligned} \tag{8.3}
$$

The total flowrate discharged from the tributary t to the reach r ($FT_{r,t}$) is equal to the sum of the following types of discharges: discharges without treatment ($S_{r,t}^{untreated}$), discharges with treatment ($S_{r,t}^{treated}$), industrial discharges ($I_{r,t}$), pluvial discharges ($P_{r,t}$), direct discharges ($D_{r,t}$), and the new discharge if the new facility is located on this tributary ($QPNEW_{r,t(p)}$), minus losses ($L_{r,t}$) and use or extraction ($U_{r,t}$) of water.

8.3.1.4 Component Balances for Each Tributary

The mass for the component c discharged from the tributary t to the reach r is estimated as follows:

$$\begin{aligned} FT_{r,t}CT_{c,r,t} = {} & S_{r,t}^{untreated}CS_{c,r,t}^{untreated} + S_{r,t}^{treated}CS_{c,r,t}^{treated} + I_{r,t}CI_{c,r,t} + P_{r,t}CP_{c,r,t} \\ & + D_{r,t}CD_{c,r,t} + QPNEW_{r,t(p)}CPNEW_{c,r,t(p)} - L_{r,t}CL_{c,r,t} \\ & - U_{r,t}CU_{c,r,t} - \int_{V=0}^{V_{r,t}} r_{c,r,t}dV_{r,t}, \quad \forall c \in C, \quad r \in R, \quad t \in T \end{aligned} \tag{8.4}$$

Here appears the term to account for the reactions that can be carried out in the tributary.

8.3.1.5 Reactive Terms

The kinetic reaction for the degradation of the pollutants in the reaches can be represented by the following equation:

$$\int_{V=0}^{V_r} r_{c,r}dV_r = k_c\left(CQ_{c,r}\right)^{\sigma_c}V_r, \quad \forall c \in C, \quad r \in R \tag{8.5}$$

k_c is the pollutant kinetic constants for the reactions determined experimentally, σ_c is the reaction order for each compound, $CQ_{c,r}$ is the concentration of the pollutants in the reach r, and V_r is the volume of the reach.

Whereas for the tributaries, the kinetic reaction takes the following form:

$$\int_{V=0}^{V_{r,t}} r_{c,r,t}dV_{r,t} = k_c\left(CT_{c,r,t}\right)^{\sigma_c}V_{r,t}, \quad \forall c \in C, \quad r \in R, \quad t \in T \tag{8.6}$$

$CT_{c,r,t}$ is the pollutant concentration c in the tributary t that discharges in the reach r and $V_{r,t}$ is the volume of the tributary.

8.3.1.6 Agricultural Discharges and Uses

The agricultural discharges ($D_{r,t}$) and uses ($U_{r,t}$) are determined by the following equations:

$$D_{r,t} = \alpha_{r,t}A_{r,t}, \quad \forall r \in R, \quad t \in T$$

$$U_{r,t} = \beta_{r,t} A_{r,t}, \quad \forall r \in R, \quad t \in T$$

$\alpha_{r,t}$ is the water needed per area for agricultural purposes and its units are $m^3/acre * s$, $\beta_{r,t}$ is the discharged flowrate per area and its units also are $m^3/acre * s$. $A_{r,t}$ is the area used for agricultural purposes near the tributary t that discharges to the reach r.

8.3.2 Location of the New Industrial Plant

There are p alternatives to locating the new industrial plant; besides, for each possible location, it is possible to include the new wastewater from the new industrial plant. Consequently, if the new industrial plant is in the allowable position 1, then the wastewater and its composition associated with this site must be bigger than zero; otherwise, if the new plant is not located in position 1, the wastewater and its compositions discharged would be zero. This is modeled through the following disjunction:

$$\begin{bmatrix} Y_1 \\ CPNEW_{c,1} \geq 0, QPNEW_1 \geq 0 \\ CPNEW_{c,2} = 0, QPNEW_2 = 0 \\ \vdots \\ CPNEW_{c,p} = 0, QPNEW_p = 0 \end{bmatrix} \vee \begin{bmatrix} Y_2 \\ CPNEW_{c,1} = 0, QPNEW_1 = 0 \\ CPNEW_{c,2} \geq 0, QPNEW_2 \geq 0 \\ \vdots \\ CPNEW_{c,p} = 0, QPNEW_p = 0 \end{bmatrix}$$

$$\vee \cdots \vee \begin{bmatrix} Y_p \\ CPNEW_{c,1} = 0, QPNEW_1 = 0 \\ CPNEW_{c,2} = 0, QPNEW_2 = 0 \\ \vdots \\ CPNEW_{c,p} \geq 0, QPNEW_p \geq 0 \end{bmatrix}, \quad \forall c \in C$$

Y_p is the Boolean variable associated with the location p of the new plant, $QPNEW_p$ is the flowrate of the new wastewater stream that can be generated by the new industrial plant in location p, $CPNEW_{c,p}$ is the concentration for the compound c from the same stream. Therefore, the disjunctive model states that if the optimal location for the new plant is on site 1, then thc Boolean variable Y_1 is set as true and $CPNEW_{c,1}$ and $QPNEW_1$ must be greater than zero, while all the other Boolean variables must be false and the $CPNEW_{c,p}$ and $QPNEW_p$ associated with their locations must be set as zero. Similar situations occur for the case when the selected location is 2, 3 or so on for the p possible locations. It should be noted that the value for the flowrate of the wastewater and the concentrations of the new plant are strongly influenced by the location through the material flow analysis model, satisfying the environmental regulations given to maintain under control the pollutants that include the sustainability constraints at the final disposal. Noticed that all these variables are influenced by the mass integration inside the plant. Lastly, previous disjunction is reformulated using the convex hull technique (Raman and Grossmann 1994; Ponce-Ortega et al. 2009b) as follows.

The Boolean variables (Y_p) are transformed into a set of binary variables (y_p), when the Boolean variables are true, the associated binary variables are one; otherwise, when the Boolean variables are false, the associated binary variables are zero. Therefore, the following relationship is used to select only one location to place the new industrial plant:

$$\sum_{p \in P} y_p = 1 \tag{8.7}$$

The next upper limits for the variables $CPNEW_{c,p}$ and $QPNEW_p$ must be satisfied in any p position and for any compound.

$$QPNEW_p \leq QPNEW^{UP} y_p, \quad \forall p \in P \tag{8.8}$$

$$CPNEW_{c,p} \leq CPNEW^{UP} y_p, \quad \forall c \in C, \ \ p \in P \tag{8.9}$$

$QPNEW^{UP}$ and $CPNEW^{UP}$ are the upper limits for $QPNEW_p$ and $CPNEW_{c,p}$, respectively. $CPNEW^{UP}$ is an upper bound for the concentration of any compound. These constraints are used to ensure that $QPNEW_p$ and $CPNEW_{c,p}$ take values greater than zero only when the new plant is located in that position; on the other hand, $QPNEW_p$ and $CPNEW_{c,p}$ take values equal to zero when the location is not selected to place the new plant.

8.3.3 Superstructure Modeling

The implemented superstructure was based on the one reported by Ponce-Ortega et al. (2010a). The size for this formulation increases exponentially according to the number of treated compounds; but the model of the superstructure for the treatment system inside the industrial plant is linear because it avoids the mixing of different streams and it is very useful for the optimization process. The complete model is nonlinear because of the bilinear terms in the MFA model ($Q_r CQ_{c,r}$ and $FT_{r,t} CT_{c,r,t}$). Furthermore, there are constraints for the composition of the streams recycled to the sinks and for the waste stream discharged to the environment to satisfy the environmental regulations.

When the superstructure model for mass integration inside the plant is combined with the material flow analysis model to consider the processes outside the plant, including the disjunctive model for the location, it is required to make a connection between the variables used in each one of these models. Consequently, the next equations are needed to link these models:

$$\sum_{p \in P} QPNEW_p = Waste \tag{8.10}$$

$$\sum_{p \in P} CPNEW_{c,p} = y_c^{disc}, \quad \forall c \in C \tag{8.11}$$

Waste is the flowrate discharged by the new industrial plant and y_c^{disc} corresponds to the composition of each pollutant. Equation (8.7) guarantees that only one location is selected for the optimal solution, then only one of the p variables for $QPNEW_p$ and $CPNEW_{c,p}$ can be greater than zero, and Eqs. (8.10) and (8.11) are fulfilled.

8.3.3.1 Splitting Fresh Sources

Each fresh source can be split and sent to any process sink.

$$F_k = \sum_{j \in J} f_{k,j}, \quad \forall k \in K \tag{8.12}$$

F_k is the total flowrate for the fresh source k and $f_{k,j}$ is the split flowrate for the fresh source k directed to process sink j. Fresh sources only can be sent to the process sinks.

8.3.3.2 Process Sources Splitting to be Sent to the Interceptors

The wastewater interception network is segregated avoiding mixing of different wastewater streams; mixing is only allowed after treatment (Fig. 8.3). When a process source does not require any treatment, a false unit is used (placed at the end of the superstructure) with efficiency and cost equal to zero.

The next equation denotes the segregation of every process source to treat the pollutants in the available interceptors, each one with a given efficiency and cost.

$$W_i = \sum_{int \in INT} w_i^{int}, \quad \forall i \in I \tag{8.13}$$

8.3.3.3 Splitting Sources at the Exit of the Water Interception Network

Once the pollutants of the process sources are treated, they are sent to process sinks and to the wastewater stream.

$$w_i^{int} = \sum_{j \in J} g_{i,j}^{int} + g_{i,Waste}^{int}, \quad \forall i \in I, \quad int \in INT \tag{8.14}$$

8.3.3.4 Overall Balance at the Mixing Point Before any Sink

The total inlet flowrate to any process sink is given by the sum of the flowrates from any process source treated and the flowrates from the fresh sources.

$$G_j = \sum_{i \in I} \sum_{\mathrm{int} \in INT} g_{i,j}^{int} + \sum_{k \in K} f_{k,j}, \quad \forall j \in J \tag{8.15}$$

where the total flowrate at the inlet of any sink is given by the process.

8.3.3.5 Component Balances at the Mixing Point Before any Sink

Component balance balances are needed to determine the composition of the stream at the inlet point to any process sink.

$$G_j y_{j,c}^{InSink} = \sum_{i \in I} \sum_{\mathrm{int} \in INT} g_{i,j}^{int} y_{i,c}^{Out,int} + \sum_{k \in K} f_{k,j} y_{k,c}^{InFresh}, \quad \forall j \in J, \quad c \in C \tag{8.16}$$

8.3.3.6 Whole Balance for the Waste Streams

The total flowrate of the wastewater discharged to the environment is determined by the sum of the treated process sources.

$$Waste = \sum_{i \in I} \sum_{\mathrm{int} \in INT} g_{i,Waste}^{int} \tag{8.17}$$

8.3.3.7 Component Balances for the Wastewater Stream

Component balances are required to determine the compositions of the wastewater stream discharged to the environment.

$$Waste \; y_c^{disc} = \sum_{i \in I} \sum_{\mathrm{int} \in INT} g_{i,Waste}^{int} \; y_{i,c}^{Out,int}, \quad \forall c \in C \tag{8.18}$$

These balances provide the fulfillment of the environmental regulations imposed on the wastewater stream discharged to the surrounding watershed.

8.3.3.8 Relationships Needed to Activate the Fixed Costs for the Interceptors

These relationships are needed to activate the fixed part for the capital costs for the interceptors; without them, the model just should consider the variable part for the capital costs of the treatment units. Thus, the relationships to activate the fixed part of the capital cost for the interceptors are the next:

$$w_i^{int} \geq w^{LO} z_i^{\text{int}}, \quad \forall i \in I, \quad \text{int} \in INT \tag{8.19}$$

$$w_i^{int} \leq w^{UP} z_i^{\text{int}}, \quad \forall i \in I, \quad \text{int} \in INT \tag{8.20}$$

w^{LO} is the minimum flowrate that can be treated in any interceptor. w^{UP} is an upper limit for w_i^{int}, and z_i^{int} is a binary variable associated with the existence of the interceptor *int* to treat the process source *i*.

8.3.3.9 Constraints for the Sink

Besides the environmental constraints, there are additional types of constraints for the composition of the inlet streams to the process sinks. Each process sink incorporates a set of process constraints for several compounds. Constraints per compound are given in terms of the minimum compositions and maximum compositions as follows:

$$y_{j,c}^{InSink,\min} \leq y_{j,c}^{InSink} \leq y_{j,c}^{InSink,\max}, \quad \forall j \in J, \quad c \in C \tag{8.21}$$

8.3.3.10 Environmental Constraints

The environmental rules impose a set of constraints in terms of compositions for each component as follows:

$$y_c^{disc,\min} \leq y_c^{disc} \leq y_c^{disc,\max}, \quad \forall c \in C \tag{8.22}$$

8.3.4 Constraints for the Quality of Specific Reaches

Selected reaches in the watershed need specific constraints for the water quality since they use water for specific targets. The presented model can consider these constraints through the following relationship:

$$CQ_{c,n(r)} \leq y_{c,n(r)}^{desired}, \quad \forall c \in C, \quad n(r) \in N(R) \tag{8.23}$$

N(R) corresponds to a subset of reaches n from all the reaches r that require a specific water quality, and $y_{c,n(r)}^{desired}$ are the requested concentrations in some reaches to preserve the water quality.

8.3.5 Sustainability Constraints

Even though the environmental regulations are fulfilled in the wastewater stream, it is possible that this is not enough to avoid the accumulation of pollutants in the final disposal (lake, sea, or ocean), generating unsustainable processes. Then, to guarantee the sustainability of the process, the following constraints are needed:

$$CQ_c^{final} \leq y_c^{sustainable}, \quad \forall c \in C \tag{8.24}$$

CQ_c^{final} is the concentration of each component in the final or last reach; thus, this is the concentration of the pollutants that the final disposal receives after installing the new industrial plant and $y_c^{sustainable}$ is the set of sustainability constraints required for each compound. It should be noticed that this last set of constraints is generated by analyzing the final disposal before the optimization process.

8.3.6 Objective Function

The objective function consists of the minimization of the total annual cost (TAC), which is established by five components: the installation costs for the new industrial plant (where $Cland_p$ is the installation cost associated with the location p), the fresh sources costs (where $FreC_k$ is the unit costs for the fresh source k), the capital cost for the treatment units (where $VarC_{cap}^{int}$ and $FixC_{cap}^{int}$ are the unit coefficients for the variable and fixed capital cost for the treatment unit int), the operational costs for the treatment units (where $VarC_{op}^{int}$ is the unit operational costs for unit int) and the piping costs for the different segments of pipes ($TrPipC_i^{int}$, $WaPipC_i^{int}$, $EqPipC_{i,j}^{int}$, and $FrPipC_{k,j}$). The installation cost incorporates the cost of land and the transportation costs for the raw materials, products, and services. The second term indicates the fresh sources costs; the problem gives the opportunity to select the type of fresh source from the ones available and the optimal fresh flowrate sent to the process sinks. Therefore, the fresh source costs depend on the used flowrate. For the third and fourth component costs, process sources need to be treated before being sent to the process sinks, as well as before being discharged to the watershed. Process sources can be treated through the available treatment technologies; the treatment cost is divided into capital cost and operating cost. The piping costs for the different segments considered depend on the flowrates. The objective function also counts the piping cost for the segment of the treatment units ($TrPipC_i^{int}$). Furthermore, the costs for the pipeline to send a process source to the waste are included ($WaPipC_i^{int}$), and the required pipeline costs to connect the process sources and fresh sources with process sinks ($EqPipC_{i,j}^{int}$ and $FrPipC_{k,j}$) are also included in the objective function, which is described as follows:

$$\begin{aligned}
Min\ TAC = & \sum_{p \in P} Cland_p y_p + H_Y \sum_{k \in K} FreC_k F_k \\
& + k_f \left[\sum_{i \in I} \sum_{int \in INT} VarC_{cap}^{int} \left(w_i^{int} \right)^{c_{\exp}} + \sum_{i \in I} \sum_{int \in INT} FixC_{cap}^{int} z_i^{int} \right] \\
& + H_Y \sum_{i \in I} \sum_{int \in INT} VarC_{op}^{int} \left(w_i^{int} \right) \\
& + H_Y \left[\sum_{i \in I} \sum_{\mathrm{int} \in INT} \left(TrPipC_i^{int} w_i^{int} + WaPipC_i^{int} g_{i,Waste}^{int} \right) \right. \\
& \left. + \sum_{i \in I} \sum_{\mathrm{int} \in INT} \sum_{j \in J} EqPipC_{i,j}^{int} g_{i,j}^{int} + \sum_{k \in K} \sum_{j \in J} FrPipC_{k,j} f_{k,j} \right]
\end{aligned} \tag{8.25}$$

H_Y is the hours per year that operates the new industrial plant, k_f is an annualization factor, and $c_{\exp}$ is an exponential factor to consider the economies of scale.

8.3.7 Remarks for the Optimization Model

- The chemical and biochemical reactions can be represented by a simple kinetic model, even that may occur several chemical and biochemical reactions.
- The installation costs ($Cland_p$) involve three different terms: transportation cost of raw materials, transportation cost of products and services, and the land cost for the alternative p. These costs depend directly on the location selected for the new plant.
- Efficiency factors (γ_c^{int}) are parameters known for each unit to treat each pollutant, these are determined through experimental data, simulation, design equations like Kremser's equation, etc. (depending on the treatment units). Because the concentrations at the exit of the interception network depend on the efficiency factors and the inlet concentrations (data given by the process); next, the component balances in the superstructure for the treatment systems are not required in the optimization model, and to calculate the outlet concentration from the interception network only the following relationship is used $y_{i,c}^{Out,\mathrm{int}} = \left(1 - \gamma_c^{\mathrm{int}}\right) y_{i,c}^{InSource}$.

- The model of this chapter is based on the composition of main toxic or hazardous materials; but, for cases with many compounds that yield hazardous properties that are regulated (i.e., chemical oxygen demand, pH, toxicity, color, and/or odor), the environmental constraints must be given in terms of these properties. Consequently, an extension of the present work is to include property tracking relationships in addition to the component balances for the hazardous materials to track the properties through the watershed. Several papers have included the property balance for the synthesis of water networks (Ponce-Ortega et al. 2009a, 2010a; Nápoles-Rivera et al. 2010); nevertheless, these approaches are limited to processes that happen inside the plant and to include these balances in the watershed, several processes must be included in the MFA model to consider the effect of the kinetic reactions carried out by the flora and fauna for the modification of the properties of the streams in the rivers and the effects of external factors such as temperature, pressure, humidity, etc.

8.4 Results and Discussion

Two example problems are presented to show the applicability of the presented approach.

8.4.1 *Example 8.1*

The data for this example match the Bahr El-Baqar drain addressed by Lovelady et al. (2009). The considered pollutants are phosphorous and phenol. Figure 8.4 presents a graphical representation of the considered watershed. Here, there are residential, agricultural, and industrial discharges that significantly alter the compositions of phosphorous and phenol in the river. Figure 8.4 also shows four places where it is allowed to locate the new industrial plant; each one of these places has associated different installation costs as Table 8.1 shows.

Figure 8.4 shows the characteristics for reaches 6, 11, and the last one; so, reach 6 has 1.312 ppm of phosphorous and 0.0076 ppm of phenol, and reach 11 contains 2.841 ppm and 0.0058 ppm of phosphorous and phenol, respectively; whereas the final disposal receives 1.17 ppm of phosphorous and 0.005 ppm of phenol. Nevertheless, at the same time, each one of the reaches above-mentioned has its set of constraints to maintain in good conditions the water quality. From the zone corresponding to reach 6, is obtained water by farmers, and the water in reach 11 is near one city, so, in reach 6, the concentration must not be greater than 2 ppm and 0.01 ppm for phosphorous and phenol, respectively; in reach 11, the upper limit for concentrations of phosphorous and phenol are 3 ppm and 0.0067 ppm, respectively, and finally, the phosphorous discharged by the new industrial plant must not exceed a final concentration in the lake of 1.2 ppm and the phenol concentration must not

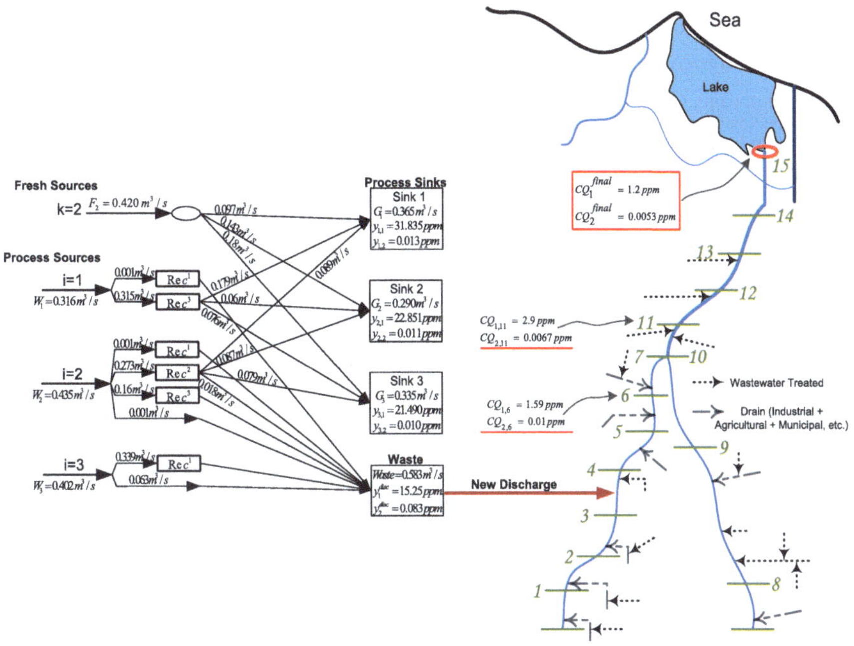

Fig. 8.4 Optimal solution for Example 8.1. Printed with permission of "*Synthesis of water networks considering the sustainability of the surrounding watershed*", *Computers and Chemical Engineering*, Copyright (2011), Elsevier

Table 8.1 Installation cost for the different sites to locate the new plant for Example 8.1

Site	Annualized installation cost, \$/year
1	18×10^6
2	17×10^6
3	35×10^6
4	10×10^6

Printed with permission of "*Synthesis of water networks considering the sustainability of the surrounding watershed*", *Computers and Chemical Engineering*, Copyright (2011), Elsevier

be greater than 0.0053 ppm (Fig. 8.4). For functional purposes in this example, the phosphorous is called component 1 and phenol is component 2.

The installation of the new industrial plant will modify and affect the environment through the new discharges to the watershed, increasing the composition of the pollutants in the surrounding rivers, and contaminating the final disposal in the lake through the concentration of the discharge that the final disposal will receive (CQ_{15}). Table 8.2 presents the considered assumptions in this example problem.

The degradation of the phosphorous and phenol is represented by first-order reactions with kinetic constants of $k_1 = 9.041909 \times 10^{-6}$/s for phosphorous and

Table 8.2 Data used for Example 8.1

Type of discharge/compound	Phosphorous, ppm	Phenol, ppm
Untreated sewage, $CS_{c,r,t}^{untreated}$	12.5	0.5
Treated sewage, $CS_{c,r,t}^{treated}$	9.5	0.005
Industrial discharges, $CI_{c,r,t}$	10	0.1
Precipitation, $CP_{c,r,t}$, $CP_{c,r}$	0	0
Agricultural drainage, $CD_{c,r,t}$	1.5	0.0025
Loses, $CL_{c,r,t}$, $CL_{c,r}$	0	0
Agricultural wastewater, $CD_{c,r}$	1.5	0.0045

Printed with permission of "*Synthesis of water networks considering the sustainability of the surrounding watershed*", *Computers and Chemical Engineering*, Copyright (2011), Elsevier

$k_2 = 5.20847 \times 10^{-7}$/s for phenol. The degradation of phosphorous is greater than the one of phenol, due to the interaction with the flora and fauna established in the river is easier for phosphorous. This way, phosphorous is a natural nutrient that can be found in most watersheds; nevertheless, the excess of phosphorous in the rivers is a risk for certain species altering the quantities of flora and fauna usually recommended, whereas the phenol is a component very toxic for the aquatic species. The operation time for the new industrial plant is 8000 h/year, $k_f = 0.1$, and the exponential factor $c_{exp} = 1$. Furthermore, the data of the fresh sources and process sources are presented in Table 8.3. The costs for the fresh sources 1 and 2 are \$0.194/m^3 and \$0.142/m^3, respectively. Table 8.4 presents the constraints for the process sinks.

In addition, to consider the environment, there is a set of environmental constraints, which are described as follows: the maximum allowable concentrations for industrial wastewater discharges are 50 ppm for phosphorous and 0.1 ppm for phenol.

Generally, an interceptor (treatment unit) can effectively separate some specific compounds; but it is evident that while a compound is separated with good efficiency, this cannot be the case for another compound treated in the same interceptor. For this reason and since the chosen compounds are very different, each interceptor

Table 8.3 Fresh and process sources for Example 8.1

Sources	Flowrate, m^3/s	Composition 1, ppm	Composition 2, ppm
Process			
1	0.316	88.7	0.14
2	0.435	77.2	0.22
3	0.402	52.4	0.16
Fresh			
1	–	0	0
2	–	10	0.005

Printed with permission of "*Synthesis of water networks considering the sustainability of the surrounding watershed*", *Computers and Chemical Engineering*, Copyright (2011), Elsevier

Table 8.4 Constraints for the process sinks for Example 8.1

Process sink	Flowrate, m^3/s	$y_{j,1}^{InSink,\max}$, ppm	$y_{j,2}^{InSink,\max}$, ppm
1	0.365	55.5	0.013
2	0.290	36.2	0.011
3	0.335	30	0.010

Printed with permission of "*Synthesis of water networks considering the sustainability of the surrounding watershed*", *Computers and Chemical Engineering*, Copyright (2011), Elsevier

functions with different efficiencies factors to remove each pollutant (Table 8.5). In this case, the capital and operational costs for the interceptors are described by $TreatC = H_Y \sum_{i \in I} \sum_{int \in INT} a^{int} \left(w_i^{int}\right)^{C_{\exp}} + \sum_{i \in I} \sum_{int \in INT} b^{int} z_i^{int}$, where a^{int} is a parameter that depends on the flowrate segregated to the treatment unit (it can be obtained by $VarC_{cap}^{int} + VarC_{op}^{int}$), b^{int} is the parameter corresponding to the fixed part of the capital cost (which is equal to $FixC_{cap}^{int}$). The most expensive interceptor is Rec3, which can get a better separation of phenol. Piping costs are divided into four categories. The piping cost to segregate process sources is \$0.01/m^3; the piping cost to send the process sources to the waste is \$0.006/m^3; while the cost to send the process sources and the fresh sources to the process sink is \$0.013/m^3.

The data and model for this case study were implemented in the software GAMS (Brooke et al. 2006) and the solver DICOPT (Viswanathan and Grossmann 1990) was used to resolve it. The optimal solution that satisfies all imposed constraints corresponds to installing the new industrial plant in site 4 with a total annual cost of \$15.583 $\times$ 10^6/year, where \$10 $\times$ 10^6/year represents the installation cost, \$3.06 $\times$ 10^6/year corresponds to the treatment cost, and \$1.72 $\times$ 10^6/year to the fresh source cost, whereas the piping cost is \$803,000/year. Here, all the process and environmental constraints are fulfilled as Fig. 8.5 shows.

Figure 8.5 includes water integration inside the plant. The wastewater stream presents pollutant concentrations lower than the maximum given by the environmental regulations; then, the environmental regulations do not ensure the sustainability of the system for this case study. Furthermore, the total annual cost for the solution without considering the sustainability constraints for the final disposal is \$14.93 $\times$ 10^6/year; but the concentrations to be discharged to the final disposal are 1.254 ppm of phosphorous and 0.0057 ppm of phenol. This put at risk the system

Table 8.5 Data of the available interceptors for Example 8.1

Interceptor	Efficiency factor 1, γ_1	Efficiency factor 2, γ_2	Fixed cost, \$	Variable cost, \$/m^3
Rec1	0.95	0.54	123,000	0.09
Rec2	0.68	0.65	112,000	0.063
Rec3	0.51	0.90	168,000	0.107

Printed with permission of "*Synthesis of water networks considering the sustainability of the surrounding watershed*", *Computers and Chemical Engineering*, Copyright (2011), Elsevier

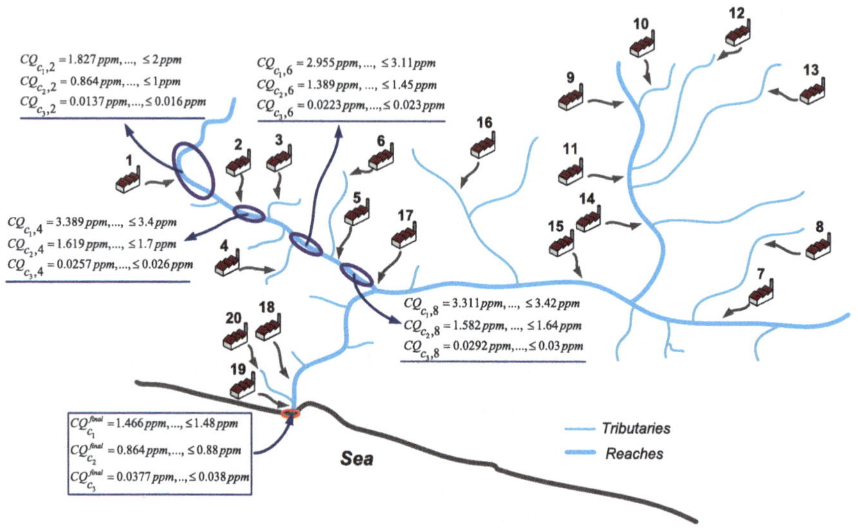

Fig. 8.5 Available locations for the new industrial plant, including pollutants concentrations and their constraints for Example 8.2. Printed with permission of "*Synthesis of water networks considering the sustainability of the surrounding watershed*", *Computers and Chemical Engineering*, Copyright (2011), Elsevier

since it would generate an excess of 4.5% for phosphorous and 7.55% for phenol that could not be naturally degraded.

Lastly, using the approach reported by Lira-Barragán et al. (2011a) in this case, under the same conditions and constraints for water quality in specific reaches and the same installation costs (Lira-Barragán et al. (2011a) only considers the treatment for phosphorous by a central system), the solution reports a total annual cost of $\$19.56 \times 10^6$/year, where $\$10 \times 10^6$/year corresponds to installation costs, and the cost for the central system is $\$9.56 \times 10^6$/year, locating the new plant in site 4; but this formulation does not take into account the fresh sources costs, piping costs and distributed treatment costs. It should be noted that the model of this chapter allows better results with a reduction of 25.5% in the total annual cost concerning the previous method by Lira-Barragán et al. (2011a, b) since of the consideration of the water integration inside the industrial plant.

8.4.2 *Example 8.2*

This example is based on the Balsas watershed, which is one of the biggest watersheds in Mexico, where several cities, and industrial and agricultural zones discharge their effluents. The data for the Balsas watershed system were taken from CONAGUA (2009a, b). For this example, three pollutants were called c_1, c_2 and c_3, which were considered for the analysis. c_1 is the lowest toxic compound, while c_3 is the most toxic. Figure 8.6 presents the system considered and the 20 alternatives to locate the new industrial plant. Additionally, Fig. 8.6 presents the pollutant concentrations before installing the new industrial plant and the constraints for some specific reaches.

For example, in reach 2 the pollutant concentration without to install the new industrial plant for c_1 is 1.827 ppm and the constraint after to install the new plant is 2 ppm, for component c_2 the concentration before the installation of the new plant is 0.864 ppm and it is requested a concentration lower than 1.1 ppm after the installation of the new plant.

There are some water extractions from reaches 2 and 6 to feed two different cities, while other water extractions for agricultural purposes exit from reaches 4 and 8. Even though reaches 2 and 6 feed two cities, the constraints for reach 2 are stricter than the one for reach 6, due to the city located near reach 2 is greater than the other one and the local government for reach 2 is stricter. Furthermore, the differences in the constraints for reaches 4 and 8 are to consider the type of crops. Table 8.6 presents the annualized installation costs for each available site to locate the new industrial plant.

Table 8.7 presents the assumptions used for the concentrations of the discharges to the watershed. The composition of pollutant c_1 is usually greater than the other pollutants.

Furthermore, there is a degradation for each compound in every section of the river with first-order kinetic reactions for the three compounds using the next kinetic constants: $k_{c_1} = 3.23764 \times 10^{-6}/\text{s}$, $k_{c_2} = 1.492 \times 10^{-6}/\text{s}$ and $k_{c_3} = 4.52095 \times 10^{-7}/\text{s}$. Also, for this case $\alpha_{r,t} = 0.000148\ \text{m}^3/\text{ha*s}$ and $\beta_{r,t} = 0.000296\ \text{m}^3/\text{ha*s}$, the operation time for the new plant is 8000 h/year, $k_f = 0.1\ \text{year}^{-1}$, and the exponential factor $c_{exp} = 1$.

Table 8.8 presents the data for the fresh sources and the process sources, as well as the costs for fresh sources 1, 2, 3, and 4 are \$0.094/m^3, \$0.085/m^3, \$0.072/m^3 and \$0.065/m^3, respectively. Table 8.9 presents the requirements for the process sinks.

Here, the environmental regulations are very important ($y_c^{disc,\max}$); thus, the wastewater stream for the new industrial plant considers the next maximum allowable concentration to be discharged to the surrounding river: $y_{c_1}^{disc,\max} = 50ppm$, $y_{c_2}^{disc,\max} = 35ppm$ and $y_{c_3}^{disc,\max} = 0.5ppm$. Table 8.10 presents the operational efficiency of removing each pollutant in each available interceptor as well as their unit costs. Like the previous example, for simplicity, the capital and operational costs for the interceptors are described by $TreatC = H_Y \sum_{i \in I} \sum_{int \in INT} a^{int} \left(w_i^{int}\right)^{c_{\exp}} + k_f \sum_{i \in I} \sum_{int \in INT} b^{int} z_i^{int}$. The most expensive interceptor (Rec3) is the one that can remove with better efficiency the most toxic compound (i.e., c_3).

Lastly, the unit piping costs are the following: $TrPipC = \$0.0025/\text{m}^3$, $WaPipC = \$0.001/\text{m}^3$, $EqPipC = \$0.006/\text{m}^3$ and $FrPipC = \$0.004/\text{m}^3$. All these costs are referenced to the flowrate transported by the pipeline.

Figure 8.6 presents the solution to this example, which was solved in the software GAMS with the solver DICOPT. The optimal solution indicates that the plant must

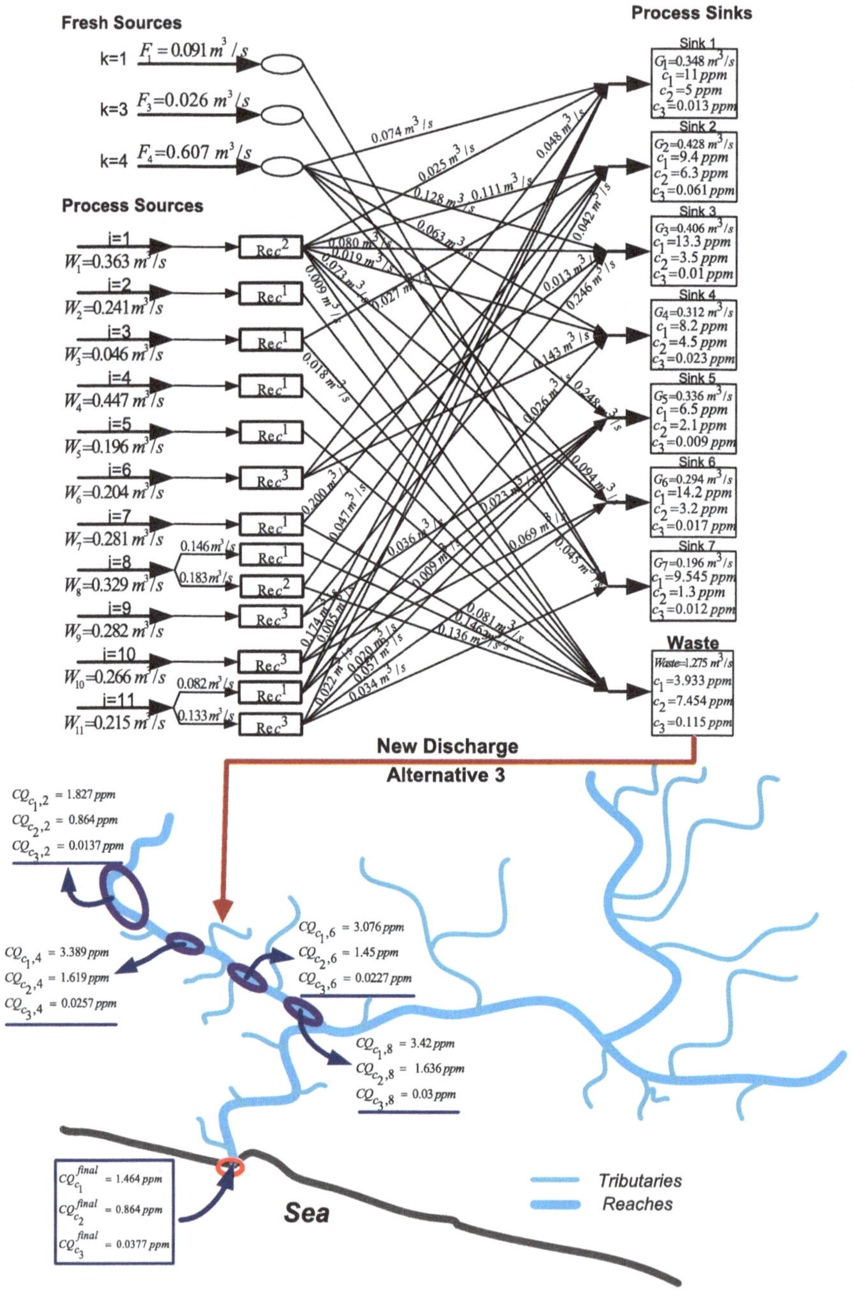

Fig. 8.6 Optimal solution for Example 8.2. Printed with permission of "*Synthesis of water networks considering the sustainability of the surrounding watershed*", *Computers and Chemical Engineering*, Copyright (2011), Elsevier

Table 8.6 Installation costs for Example 8.2

Site	Annualized installation cost, \$/year
1	8×10^6
2	8.5×10^6
3	9×10^6
4	10.5×10^6
5	9.5×10^6
6	9×10^6
7	15.5×10^6
8	14.5×10^6
9	17.7×10^6
10	22×10^6
11	18×10^6
12	17.5×10^6
13	24×10^6
14	28×10^6
15	23.5×10^6
16	17×10^6
17	20×10^6
18	19×10^6
19	16×10^6
20	17×10^6

Printed with permission of "*Synthesis of water networks considering the sustainability of the surrounding watershed*", *Computers and Chemical Engineering*, Copyright (2011), Elsevier

Table 8.7 Constraints used for Example 8.2

Type of discharge/component	c_1, ppm	c_2, ppm	c_3, ppm
Untreated sewage, $CS_{r,t}^{untreated}$	24.5	13	0.5
Treated sewage, $CS_{r,t}^{treated}$	10.5	5	0.03
Industrial discharges, $CI_{r,t}$	25	10	0.06
Agricultural drainage, $CD_{r,t}$	12.8	5.6	0.055
Total discharges, CH_r	21.8	8.5	0.07
Direct discharges, CD_r	18.5	9.5	0.075

Printed with permission of "*Syntehsis of water netwworks considering the sustainability of the surrounding watershed*", *Computers and Chemical Engineering*, Copyright (2011), Elsevier

Table 8.8 Fresh and process sources for Example 8.2

Sources	Flowrate, m^3/s	Composition c_1, ppm	Composition c_2, ppm	Composition c_3, ppm
Process				
1	0.363	64.6	22.4	0.11
2	0.241	38	21.6	0.24
3	0.046	55.3	10	0.22
4	0.447	30	32.5	0.3
5	0.196	49.8	16.5	0.31
6	0.204	12	28.7	0.15
7	0.281	74.2	36.6	0.14
8	0.329	57.8	42	0.3
9	0.282	46	18.2	0.09
10	0.266	32.1	25.1	0.1
11	0.215	66.4	17.8	0.13
Fresh				
1	–	0	0	0
2	–	1	0.5	0.002
3	–	2	0.7	0.005
4	–	2.5	1	0.007

Printed with permission of "*Synthesis of water networks considering the sustainability of the surrounding watershed*", *Computers and Chemical Engineering*, Copyright (2011), Elsevier

Table 8.9 Constraints for the process sinks for Example 8.2

Process sink	Flowrate, m^3/s	$y_{j,c_1}^{InSink,max}$, ppm	$y_{j,c_2}^{InSink,max}$, ppm	$y_{j,c_3}^{InSink,max}$, ppm
1	0.348	11	5	0.013
2	0.428	9.4	6.3	0.61
3	0.406	13.3	3.5	0.1
4	0.312	8.2	4.5	0.023
5	0.336	6.5	2.1	0.009
6	0.294	14.2	3.2	0.017
7	0.196	9.6	1.3	0.012

Printed with permission of "*Synthesis of water networks considering the sustainability of the surrounding watershed*", *Computers and Chemical Engineering*, Copyright (2011), Elsevier

be installed in site 3 with a total annual cost of \$12.69 $\times$ 10^6/year, which involves \$9 $\times$ 10^6/year for the installation cost, \$1.65 $\times$ 10^6/year for the treatment cost, \$1.44 $\times$ 10^6/year for the fresh sources cost, and \$6 $\times$ 10^5/year for the piping cost. Furthermore, in this case, the sustainability constraints were not active because the mean watershed flowrate is too much greater than the wastewater stream discharged

Table 8.10 Interceptors for Example 8.2

Interceptor	Efficiency factor for c_1	Efficiency factor for c_2	Efficiency factor for c_3	Fixed cost, $	Variable cost, $/m^3
Rec1	0.95	0.72	0.58	307,000	0.007
Rec2	0.68	0.92	0.64	353,000	0.013
Rec3	0.59	0.74	0.9	458,000	0.027
Rec4	0.73	0.8	0.52	270,000	0.018
Rec5	0.61	0.57	0.7	320,000	0.01

by the new industrial plant; nevertheless, the imposed constraints on reaches 2, 4, 6 and 8 were activated. Then, the total flowrate on the left side of the watershed is very small concerning the right side. It is worth noting that only if the new industrial plant is installed on the positions near the final disposal (i.e., sites 18, 19, and 20), the wastewater discharged by the new plant has a lower impact over the final reach concerning the optimal solution, but the total annual cost is not competitive because of the higher installation cost.

This example indicates also that the environmental regulations for the wastewater discharged from the industrial facilities are not enough to ensure the sustainability of the systems, and it is required to model the watershed to ensure the sustainability of the system like the proposed methodology, which includes a set of sustainability constraints.

Furthermore, applying the methodology previously proposed by Lira-Barragán et al. (2011a) to this example, considering the constraints for water quality in specific reaches, the sustainability constraints for the final disposal, the same installation costs, taking into account only one component (the same that considers the work above mentioned) and the treatment units used by Lira-Barragán et al. (2011a), the solution is infeasible because the available central treatment system is not enough to satisfy the environmental constraints imposed to maintain the water quality on intermediate reaches and at the final disposal. It should be noted that the constraints are stricter in present work concerning the constraints used by Lira-Baragán et al. (2011a) for this example, it is noteworthy that the water integration inside of the new industrial plant allows for satisfying these stricter constraints, and this is the main advantage of the presented methodology in this chapter.

Ultimately, Table 8.11 presents the size and the computation time for the problems addressed in this chapter using a computer with an i5 processor at 2.3 GHz with 4 MB of RAM.

Table 8.11 Size and computation time for the example problems considered

Concept	Example 8.1	Example 8.2
Number of constraints	195	4,062
Number of continuous variables	212	4,486
Number of binary variables	16	86
CPU time (s)	0.37	1.68

8.5 Concluding Remarks

This chapter presents a mathematical programming approach for the simultaneous synthesis of water networks inside new industrial plants and the consideration of the surrounding watershed to satisfy environmental and sustainability constraints. Furthermore, the material flow analysis technique was employed to create the model, which was designed to forecast the behavior of the watershed affected by the additional contaminated discharges. This work considers not only the chemical reactions occurring in the rivers, but also the discharges from industrial, residential, agricultural, and other sources, as well as all extractions. To determine the best location for a new industrial plant that satisfies environmental regulations and integrates waste, the watershed model is integrated with a disjunctive model. The objective function of the model is to minimize the overall yearly cost, which is made up of the costs associated with wastewater treatment, fresh source costs, piping costs, and the placement of the new plant. The model provides for the best choice of treatment. The resolution of the examined situations demonstrates that a new industrial plant's location greatly affects the amount of treatment needed to produce a sustainable process. Furthermore, the results indicate that to maintain the sustainability of the system, the sustainability constraints for the final disposal and the constraints specified for particular river reaches can occasionally become active. For this reason, all these constraints need to be considered when solving this problem.

The restrictions provided for the primary hazardous substances form the basis of this chapter's model. Nonetheless, further development of the current technique entails incorporating restrictions expressed in terms of attributes with composition limits for scenarios involving several compounds exhibiting hazardous properties.

Nomenclatures

Indexes

c Component

i	Process source
int	Interceptor
In	Inlet
j	Sink
k	Fresh source
LO	Lower value
max	Maximum
min	Minimum
$n(r)$	Specific reaches that require a composition constraint
Out	Out
p	Site to locate the new industrial plant
r	Reach
$Sink$	Sink
$Source$	Source
t	Tributary
UP	Upper value
$Waste$	Waste discharged to the environment

Sets

C	Set for the components, $\{c\vert c=1,\ldots,N_c\}$
I	Set for the process sources, $\{i\vert i=1,\ldots,N_i\}$
INT	Set for the interceptor, $\{\text{int}\vert\text{int}=1,\ldots,N_{\text{int}}\}$
J	Set for the sinks, $\{j\vert j=1,\ldots,N_j\}$
K	Set for the fresh sources, $\{k\vert k=1,\ldots,N_k\}$
$N(R)$	Subset for the specific reaches that require composition constraints
P	Set for the sites to locate the new plant, $\{p\vert p=1,\ldots,N_p\}$
R	Set for the reaches, $\{r\vert r=1,\ldots,N_r\}$
T	Set for the tributaries, $\{t\vert t=1,\ldots,N_t\}$

Parameters

$A_{r,t}$	Area cover by effluent t in reach r, acre or ha
a^{int}	Unitary cost that depends on the flowrate treated for interceptor int, \$/m^3
b^{int}	Parameter corresponding to the fixed part of the capital cost for interceptor int, \$
c_{exp}	Exponential factor useful to consider the economies of scale.
$Cland_p$	Annualized installation cost for the new plant in site p, \$/year
$CP_{c,r}$	Pollutants concentration of precipitation discharged to the reach r, ppm
$CD_{c,r}$	Pollutants concentration of direct discharges to the reach r, ppm

$CH_{c,r}$	Pollutants concentration of total discharge (i.e., industrial + sanitary) to the reach r, ppm
$CL_{c,r}$	Pollutants concentration of total losses (filtration and evaporation) from the reach r, ppm
$CU_{c,r}$	Pollutants concentration of water used from reach r, ppm
$CS_{c,r,t}^{treated}$	Pollutants concentration of residual wastewater discharged without treatment to the reach r for tributary t, ppm
$CS_{c,r,t}^{untreated}$	Pollutants concentration of residual treated wastewater discharged to the reach r for tributary t, ppm
$CI_{c,r,t}$	Pollutants concentration of industrial discharge from the tributary t to the reach r, ppm
$CP_{c,r,t}$	Pollutants concentration of precipitation discharged to the tributary t to the reach r, ppm
$CD_{c,r,t}$	Pollutants concentration of agricultural discharges to the tributary t to the reach r, ppm
$CL_{c,r,t}$	Pollutants concentration of total losses (filtration and evaporation) from tributary t, ppm
$CU_{c,r,t}$	Pollutants concentration of water used from tributary t discharge to reach r, ppm
$D_{r,t}$	Agricultural discharges to the tributary t to the reach r, m^3/s
D_r	Direct discharges to the reach r, m^3/s
$EqPipC_i^{int}$	Piping cost to send process sources to process sinks, $\$/m^3$
$FixC_{cap}^{int}$	Fixed part of the capital cost for interceptor *int*, \$/year
$FreC_k$	Unit cost for fresh utility k, $\$/m^3$
$FrPipC_{k,j}$	Piping cost to send fresh sources to process sinks, $\$/m^3$
G_j	Total flowrate for process sink j, m^3/s
H_Y	Operation time per year, h/year
H_r	Total discharge (i.e., industrial + sanitary) to the reach r, m^3/s
$I_{r,t}$	Industrial discharge from the tributary t to the reach r, m^3/s
k_c	Kinetic constant for the degradation for pollutant c
k_f	Annualization factor
$L_{r,t}$	Total losses (filtration and evaporation) from tributary t of the reach r, m^3/s
L_r	Total losses (filtration and evaporation) from the reach r, m^3/s
N_c	Total number of components
N_i	Total number of process sources
N_{int}	Total number of interceptors
N_j	Total number of sinks
N_k	Total number of fresh sources
N_p	Total number of sites allowable to locate the new plant
N_r	Total number of reaches
N_t	Total number of tributaries
$P_{r,t}$	Precipitation discharged for the tributary t to the reach r, m^3/s
P_r	Precipitation discharged to the reach r, m^3/s

$S_{r,t}^{untreated}$	Residual wastewater discharged without treatment to the reach r for tributary t, m^3/s
$S_{r,t}^{treated}$	Residual treated wastewater discharged to the reach r for tributary t, m^3/s
$TrPipC_i^{int}$	Treatment piping cost, $\$/m^3$
$U_{r,t}$	Water used from tributary t discharge to reach r, m^3/s
U_r	Water used from reach r, m^3/s
$VarC_{cap}^{int}$	Variable part of the capital cost for interceptor *int*, $\$/m^3$
$VarC_{op}^{int}$	Variable cost associated to operation of the interceptor *int*, $\$/m^3$
$V_{r,t}$	Volume for tributary t from reach r, m^3
V_r	Volume for reach r, m^3
W_i	Total flowrate for the process sources i, m^3/s
$WaPipC_i^{int}$	Waste piping cost, $\$/m^3$
$y_{c,n(r)}^{desired}$	Desired concentrations in some reaches, ppm
$y_c^{disc,\max}$	Environmental regulation for the pollutant concentration to be discharged by the new industrial plant, ppm
$y_{i,c}^{InSource}$	Concentration of the component c in the process source i, ppm
$y_c^{sustainable}$	Sustainability constraints at the final disposal, ppm
$\alpha_{r,t}$	Agricultural flowrate per area, m^3/ha s
$\beta_{r,t}$	Agricultural use of water from tributary t, m^3/ha s
γ_c^{int}	Efficiency factor to remove the pollutant c for the interceptor *int*
δ	Small number, $1e10^{-3}$
σ_c	Reaction order for pollutant c

Variables

$CPNEW_{c,p}$	Concentration for the contaminants discharged for the new plant installed in site p, ppm
CQ_c^{final}	Pollutants concentration discharged to the final disposal, ppm
$CQ_{c,r}$	Pollutants concentration of flowrate exit to the reach r, ppm
$CQ_{c,r-1}$	Pollutants concentration of flowrate inlet to the reach r, ppm
$CT_{c,r,t}$	Pollutants concentration of the tributary t to the reach r, ppm
F_k	Total flowrate for fresh source k, m^3/s
$f_{k,j}$	Segregated flowrate from fresh source k to sink j, m^3/s
$FT_{r,t}$	Discharge for the tributary t to the reach r, m^3/s
$g_{i,j}^{int}$	Segregated flowrate from interceptors *int* to process sink j for process source i, m^3/s
$g_{i,Waste}^{int}$	Segregated flowrate from interceptors *int* to waste for process source i, m^3/s
Q_r	Flowrate exit to the reach r, m^3/s
Q_{r-1}	Flowrate inlet to the reach r, m^3/s
$QPNEW_p$	Flowrate discharged for the new plant for the location p

$r_{c,r}$	Reaction carry out in the reach r for pollutant c, g/s
$r_{c,r,t}$	Reaction carry out in the tributary t that discharges to the reach r for pollutant c, g/s
$TreatC$	Treatment cost, $
TAC	Total annual cost, $/year
w_i^{int}	Segregated flowrate from process source i to interceptor int, m^3/s
$Waste$	Total flowrate for the waste stream discharged to the environment, m^3/s
Y_p	Boolean variable for the location of the new plant
y_p	Binary variable for the location of the new plant
y_c^{disc}	Pollutant concentration for the final discharge of the new plant to the river, ppm
z_i^{int}	Binary variable for the interceptor int from the process source i

References

Agrawal V, Shenoy UV (2006) Unified conceptual approach to targeting and design of water and hydrogen networks. AIChE J 52(3):1071–1082

Alva-Argaez A, Kokossis AC, Smith R (1998) Wastewater minimization of industrial systems using an integrated approach. Comput Chem Eng 22(Suppl):S741–S744

Alva-Argaez A, Vallianatos A, Kokossis A (1999) A multi-contaminant transhipment model for mass exchange networks and wastewater minimization problems. Comput Chem Eng 23(10):1439–1453

Baccini P, Brunner P (1991) Metabolism of the anthroposphere. Springer, Berlin

Benko N, Rev E, Fonyo Z (2000) The use of nonlinear programming to optimal water allocation. Chem Eng Commun 178:67–101

Brooke A, Kendrick D, Meeruas A, Raman R (2006) GAMS-language guide. GAMS Development Corporation, Washington, D.C.

Brunner PH, Rechberg H (2004) Practical handbook of material flow analysis. CRC

CONAGUA (2009a) Mexican National Water Commission. Water statistics 2008. http://www.conagua.gob.mx/OCB07/Temas/EstadisticasBALSAS-FINALago09.pdf

CONAGUA (2009b) Mexican National Water Commission. Water atlas in Mexico 2009

Dhole VR, Ramchandani N, Tainsh RA, Wasilewski M (1996) Make your process water pay for itself. Chem Eng 103(1):100–103

Doyle SJ, Smith R (1997) Targeting water reuse with multiple contaminants. Process Saf Environ Prot 75(B3):181–189

Dunn RF, El-Halwagi MM (2003) Process integration technology review: background and application in chemical process industry. J Chem Technol Biotechnol 78(9):1011–1021

El-Baz AA, Ewida KT, Shouman MA, El-Halwagi MM (2005a) Material flow analysis and integration of watersheds and drainage systems: I. Simulation and application to ammonium management in Bahr El-Baqar drainage system. Clean Technol Environ Policy 7(1):51–61

El-Baz AA, Ewida KT, Shouman MA, El-Halwagi MM (2005b) Material flow analysis and integration of watersheds and drain systems: II. Integration and solution strategies with application to ammonium management in Bahr El-Baqar drain system. Clean Technol Environ Policy 7(1):78–86

El-Halwagi MM, Spriggs HD (1998) Solve design puzzles with mass integration. Chem Eng Prog 94(8):25–44

El-Halwagi MM, Gabriel F, Harell D (2003) Rigorous graphical targeting for resource conservation via material recycle/reuse networks. Ind Eng Chem Res 42(19):4319–4328

Feng X, Bai J, Zheng XS (2007) On the use of graphical method to determine the targets of single-contaminant regeneration recycling water systems. Chem Eng Sci 62(8):2127–2138

Foo CY (2009) State-of-the-art review of pinch analysis techniques for water network synthesis. Ind Eng Chem Res 48:5125–5159

Gabriel FB, El-Halwagi MM (2005) Simultaneous synthesis of waste interception and material reuse networks: problem reformulation for global optimization. Environ Prog 24(2):171–180

Galan B, Grossmann IE (1998) Optimal design of distributed wastewater treatment networks. Ind Eng Chem Res 37(10):4036–4048

Gomes JFS, Queiroz EM, Pessoa FLP (2007) Design procedure for water/wastewater minimization: single contaminant. J Clean Prod 15(5):474–485

Gunaratnam M, Alva-Argaez A, Kokossis A, Kim JK, Smith R (2005) Automated design of total water systems. Ind Eng Chem Res 44(3):588–599

Hallale N (2002) A new graphical targeting method for water minimization. Adv Environ Res 6(3):377–390

Hernandez-Suarez R, Castellanos-Fernandez J, Zamora JM (2004) Superstructure decomposition and parametric optimization approach for the synthesis of distributed wastewater treatment networks. Ind Eng Chem Res 43(9):2175–2191

Karuppiah R, Grossmann IE (2006) Global optimization for the synthesis of integrated water systems in chemical processes. Comput Chem Eng 30(4):650–673

Kuo WCJ, Smith R (1997) Effluent treatment system design. Chem Eng Sci 52(23):4273–4290

Ku-Pineda V, Tan RR (2006) Environmental performance optimization using process water integration and sustainable process index. J Clean Prod 14:1586–1592

Lampert C, Brunner PH (1999) Material accounting as a policy tool for nutrient management in the Danube Basin. Water Sci Technol 40(10):43–49

Lim SR, Park JM (2008a) Cooperative water networks system to reduce carbon footprints. Environ Sci Technol 42:6230–6236

Lim SR, Park JM (2007) Environmental and economic analysis of a water networks system using LCA and LCC. AIChE J 53:3253–3262

Lim SR, Park JM (2008b) Synthesis of an environmental friendly water network system. Ind Eng Chem Res 47:1988–1994

Lira-Barragán LF, Ponce-Ortega JM, Serna-González M, El-Halwagi MM (2011a) An MINLP model for the optimal location of a new industrial plant with simultaneous consideration of economic and environmental criteria. Ind Eng Chem Res 50(2):953–964

Lira-Barragán LF, Ponce-Ortega JM, Serna-González M, El-Halwagi MM (2011b) Comput Chem Eng 35:2837–2852

Lovelady EM, El-Baz AA, El-Monayeri D, EL-Halwagi MM (2009) Reverse problem formulation for integrated process discharges with wastewater and drainage systems: Managing phosphorus in lake Manzala. J Ind Ecol 13(6):914–927

Manan ZA, Tan YL, Foo DCY (2004) Targeting the minimum water rate using water cascade analysis technique. AIChE J 50(12):3169–3183

Nápoles-Rivera F, Ponce-Ortega JM, El-Halwagi MM, Jiménez-Gutiérrez A (2010) Global optimization of mass and property integration networks with in-plant property interceptors. Chem Eng Sci 65(15):4363–4377

Pillai HK, Bandyopadhyay S (2007) A rigorous targeting algorithm for resource allocation networks. Chem Eng Sci 62:6212–6221

Polley GT, Polley HL (2000) Design better water networks. Chem Eng Prog 96(2):47–52

Ponce-Ortega JM, Hortua AC, El-Halwagi MM, Jiménez-Gutiérrez A (2009a) A property-based optimization of direct recycle networks and wastewater treatment processes. AIChE J 55(9):2329–2344

Ponce-Ortega JM, Serna-González M, Jiménez-Gutiérrez A (2009b) A disjunctive programming model for simultaneous synthesis and detailed design of cooling networks. Ind Eng Chem Res 48(6):2991–3003

Ponce-Ortega JM, El-Halwagi MM, Jiménez-Gutiérrez A (2010a) Global optimization of property-based recycle and reuse networks including environmental constraints. Comput Chem Eng 34(3):318–330

Ponce-Ortega JM, Mosqueda-Jiménez FW, Serna-González M, Jiménez-Gutiérrez A, El-Halwagi A (2010b) A property-based approach to the synthesis of material conservation networks with economic and environmental objectives. AIChE J 57(9):2369–2387. https://doi.org/10.1002/aic.12444

Putra ZA, Amminudin KA (2008) Two-step optimization approach for design a total water system. Ind Eng Chem Res 47(16):6045–6057

Raman R, Grossmann IE (1994) Modeling and computational techniques for logic based integer programming. Comput Chem Eng 18(7):563–578

Savelski MJ, Bagajewicz MJ (2000) On the optimality conditions of water utilization systems in process plants with single contaminants. Chem Eng Sci 55(21):5035–5048

Savelski MJ, Bagajewicz MJ (2001) Algorithmic procedure to design water utilization systems featuring a single contaminant in process plants. Chem Eng Sci 56(5):1897–1911

Savelski M, Bagajewicz M (2003) On the necessary conditions of optimality of water utilization systems in process plants with multiple contaminants. Chem Eng Sci 58(23–24):5349–5362

Sorin M, Bedard S (1999) The global pinch point in water reuse networks. Process Saf Environ Prot 77(B5):305–308

Tan RR, Foo DCY, Ng DKS, Chiang CL, Hul S, Ku-Pineda V (2007) An approximated mixed integer linear programming (MILP) model for the design of water reuse/recycle networks with minimum energy. Asia-Pac J Chem Eng 42:566–574

Teles J, Castro PM, Novals AQ (2008) LP-based solution strategies for the optimal design of industrial water networks with multiple contaminants. Chem Eng Sci 63(2):376–394

Viswanathan J, Grossmann IE (1990) A Combined penalty function and outer approximation method for MINLP optimization. Comput Chem Eng 14(7):769–782

Wang YP, Smith R (1994) Wastewater minimization. Chem Eng Sci 49(7):981–1006

Chapter 9
Integrating Water Networks Characteristics and Watershed Sustainability into Industrial Facility Site Selection

9.1 Introduction

The chemical and process industries are among the largest consumers of freshwater and are responsible for tremendous wastewater discharges. Thus, numerous techniques have been proposed to economically reduce freshwater consumption in the process industry considering the possibility to recycle, reuse, and regenerate the wastewater produced in the industrial facilities (Fig. 9.1). These approaches accounted for different objectives (e.g., minimizing freshwater consumption, minimizing wastewater discharge, minimizing regeneration costs, etc.) and using different techniques (e.g., graphical, algorithmic, and mathematical programming-based approaches). Heuristic approaches permit the identification of targets ahead of the design (Wang and Smith 1994; Dhole et al. 1996; El-Halwagi and Spriggs 1998; Polley and Polley 2000; Hallale 2002; Manan et al. 2004; El-Halwagi et al. 2003; Feng et al. 2007; Foo 2009) or to algebraically resolve this problem (Sorin and Bedard 1999; Gomes et al. 2007; Foo et al. 2006). Moreover, different water integration techniques based on mathematical programming approaches have been proposed to get the optimal solution for the water integration problem inside industrial facilities (see for example Takama et al. 1980; El-Halwagi et al. 1996; Savelski and Bagajewicz 2000, 2001, 2003; Alva-Argaez et al. 2003, 1999; Benko et al. 2000; Teles et al. 2008; Gabriel and El-Halwagi 2005; Kuo and Smith 1997; Doyle and Smith 1997; Galan and Grossmann 1998; Hernandez-Suarez et al. 2004; Gunaratman et al. 2005; Karuppiah and Grossmann 2006; Putra and Amminudin 2008; Ponce-Ortega et al. 2008; Nápoles-Rivera et al. 2011; Faria and Bagajewicz 2011a, b, c, d, e). For addressing problems with several pollutants that are based on stream properties other than contaminant concentration, Ng et al. (2008), Ponce-Ortega et al. (2009a, b, 2010, 2011), Nápoles-Rivera et al. (2010),[45] Kheireddine et al. (2011), Deng and Feng (2011) developed approaches for water integration inside industrial facilities by using the property integration framework proposed by Shelley and El-Halwagi (2000) and El-Halwagi et al. (2004).

J. M. Ponce-Ortega et al., *Sustainable Water Systems*, Water Resources Development and Management, https://doi.org/10.1007/978-3-031-72498-5_9

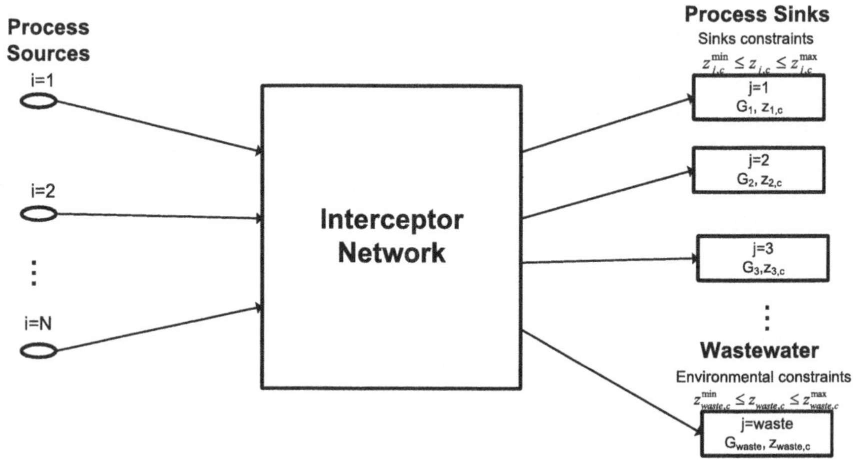

Fig. 9.1 Recycle and reuse networks. Printed with permission of "*Incorporating property-based water networks and surrounding watersheds in site selection of industrial facilities*", *Industrial and Engineering Chemistry Research*, Copyright (2012), American Chemical Society

Previous approaches have considered the process activities carried out inside the industrial facilities, without considering the activities occurring outside the industrial plant for the wastewater streams discharged to the environment. Furthermore, despite the increasing number of works that consider environmental aspects in the synthesis of water networks (Ku-Pineda and Tan 2006; Tan et al. 2007; Lim and Park 2007, 2008a, b), these approaches do not consider the interrelations between the wastewater streams discharged to the environment and the surrounding watershed. Consequently, the physical, chemical, and biochemical relations between the industrial discharges and other discharges and uses such as agricultural, industrial, residential, precipitation, filtration, and others are not considered.

Meanwhile, the watersheds are imperative ecological elements through which the discharged industrial, residential, and agricultural effluents are moved to their final disposal while undergoing physical, chemical, and biological changes, the watersheds, and their surroundings are interrelated as Fig. 9.2 presents. The discharges received by a watershed alter the chemical composition and the properties of the receiving water bodies. The discharged compounds undergo chemical, biochemical, and physical transformations that are produced by the flora and fauna of the watersheds (Brunner and Rechberg 2004; Baccini and Brunner 1991; Lampert and Brunner 1999; El-Baz et al. 2005a, b; Lovelady et al. 2009). In addition, Lira-Barragán et al. (2011a, b) reported two mathematical programming approaches for optimal water use inside industrial facilities accounting for the surrounding watershed through an MFA model. Nevertheless, these studies were based on the composition of the wastewater streams without considering tracking stream properties for multiple pollutants involved.

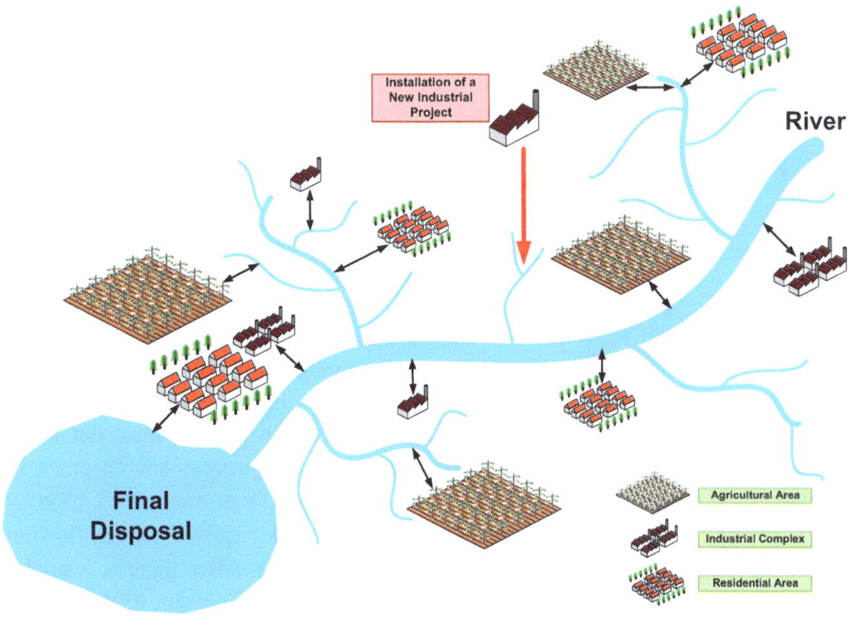

Fig. 9.2 Impact of a new industrial plant in a watershed system. Printed with permission of "*Incorporating property-based water networks and surrounding watersheds in site selection of industrial facilities*", *Industrial and Engineering Chemistry Research*, Copyright (2012), American Chemical Society

This chapter shows a mathematical programming approach that incorporates property tracking and constraints in addition to other environmental, technical, and economic considerations (Lira-Barragán et al. 2012). The focus is to involve the design of water networks with multiple pollutants and properties (rather than just contaminant concentrations) in the site selection of new industrial facilities while incorporating modeling of the surrounding watershed. A set of sites are considered as candidates for locating a new industrial plant in a given watershed. Consequently, the optimum water network and the optimum flow and quality of the wastewater discharged in a receiving water body depend on the location selected to install the plant, which depends on the interactions between the new industrial plant and the watershed. Therefore, the new industrial plant and all sources and uses of the watershed are seen as interacting systems rather than comprising isolated components in an overall system. A material flow analysis (MFA) approach is incorporated for the integrated assessment of water use options to ensure the sustainability of the watershed in the synthesis of property-based water networks. A superstructure is developed to consider several alternatives for in-plant water integration and the plant location. The mathematical model corresponds to a disjunctive programming formulation that is reformulated as an MINLP problem considering the minimization of the total annual cost for the recycle and reuse network and the installation of the new plant to meet the process and environmental constraints and the sustainability of the watershed in terms of properties. The model can consider several properties and determine the selection of several technologies to treat the wastewater streams.

9.2 Problem Statement

Given is a set of alternatives to locate a new industrial facility ($L = \{l|l = 1, 2, ..., N_l\}$). Each place has accompanying an installation cost, ($Cland_l$). Moreover, it is given a set of process sources (wastewater streams), ($I = \{i|i = 1, 2, \ldots, N_i\}$). Each process source has a set of given properties ($P = \{p|p = 1, 2, \ldots, N_p\}$), which may involve composition, density, pH, toxicity, viscosity, between others. These wastewater streams can be split to be recycled and/or reused in the process sinks and the optimization process must determine the values for the flowrate and the properties for each process source given by W_i and $P_{i,p}^{InSource}$, respectively. Likewise, given is a set of process sinks (process units), $J = \{j|j = 1, 2, \ldots, N_j\}$, were each process sink requires a given flowrate (G_j), with specific limits for the inlet properties ($P_{j,p}^{InSink}$). Also, there is available a set of fresh sources ($K = \{k|k = 1, 2, \ldots, N_k\}$) that can be sent to the process sinks (F_k), each one with given properties ($P_{k,p}^{Fresh}$). Figure 9.4 shows that each property can be individually treated in a set of interceptors. Thus, there is a set of interceptors to treat the first property (INT^1), and for the second property exists a second set of interceptors (INT^2), while for the N property to be treated there is the last set of interceptors (INT^N). Lastly, there is a set of environmental constraints ($P_p^{Waste,\max}$) for the wastewater discharged into the environment (*Waste*). This new stream will be discharged to a river or watershed, which is segregated into several reaches ($R = \{r|r = 1, 2, \ldots, N_r\}$) and each reach can receive several effluents ($T = \{t|t = 1, 2, \ldots, N_t\}$). Thus, a set of sustainability constraints is imposed for the properties of the stream discharged to the final disposal. The problem consists of determining simultaneously the recycle and reuse water network based on properties and the optimal position for the new plant that discharges wastewater to the surrounding watershed avoiding the accumulation of pollutants and maintaining control of the water discharged at the final disposal (considering the properties of the streams in the watershed) and, at the same time, satisfying the environmental constraints for the pollutants through the watershed system. The objective function considers minimizing the total annual cost (TAC). Which includes the installation costs for the new plant, the treatment costs for the process sources, the costs of fresh sources, and the piping costs.

The total flowrate discharged from the new industrial plant and its properties are optimization variables, which depend effectively on its optimal location to be decided by the optimization process. This means that the required treatment depends on the location of the new industrial plant. To replicate the behavior of the river, the model uses the MFA technique considering all inlet and outlet streams. Then, the required treatment also depends on the constraints at the inlet of process sinks, and

Table 9.1 Mixing property operators for wastewater streams

Property	Operator
Composition	$\psi_z(z) = z$
Toxicity	$\psi_{Tox}(Tox) = Tox$
Chemical Oxygen Demand	$\psi_{COD}(COD) = COD$
pH	$\psi_{pH}(pH) = 10^{pH}$
Density	$\psi_{\rho}(\rho) = 1/\rho$
Viscosity	$\psi_{\mu}(\mu) = \log(\mu)$
Reid vapor pressure	$\psi_{RVP}(RVP) = RVP^{1.44}$
Electric resistivity	$\psi_R(R) = 1/R$
Paper reflectivity	$\psi_{R_\infty}(R_\infty) = R_\infty^{5.92}$
Color	$\psi_{Color}(Color) = Color^{0.606}$
Odor	$\psi_{Odor}(Odor) = Odor$

the treatment is carried out by segregating the process sources to treat them in the interceptors.

A key point included in this chapter is the tracking of properties instead of compositions. Thus, likewise the composition balances for key pollutants, the model includes property balances for pH, toxicity, COD, color, viscosity, etc., which are based on the property operator functions contained in Table 9.1 (Shelley and El-Halwagi 2000).

Furthermore, the main stream of the river is fed by several streams called tributaries ($FT_{r,t}$). Figure 9.3 shows that the watershed system exchanges water during its path (i.e., water for agricultural use, wastewater discharged to the river with and without treatment, industrial and residential effluents, etc.) and, ultimately, discharges a total stream to the final disposal. In addition, natural phenomena like precipitation, filtration, and vaporization are considered in the model. Since all these processes can significantly modify the characteristics of the river, for tracking adequately the mean properties, the river is sectioned in parts where the properties can be considered constant (these sections are denoted in Fig. 9.3 with ovals). These parts or sections are called reaches, representing zones where no significant intermediate effluents are discharged and/or extracted. Similarly, the model accounts for the main effluents discharged in each reach, which are called tributaries. The tributaries can be channels or branches of the river, which may contain discharges with/without treatment, industrial discharges, etc. The flowrates ($FT_{r,t}$) and properties ($P_{p,r,t}^{Trib}$) of the tributaries impact the reaches where they are discharged. The discharges to the watershed contain compounds that interact chemically and biochemically with the system through the flora and fauna living in the watershed.

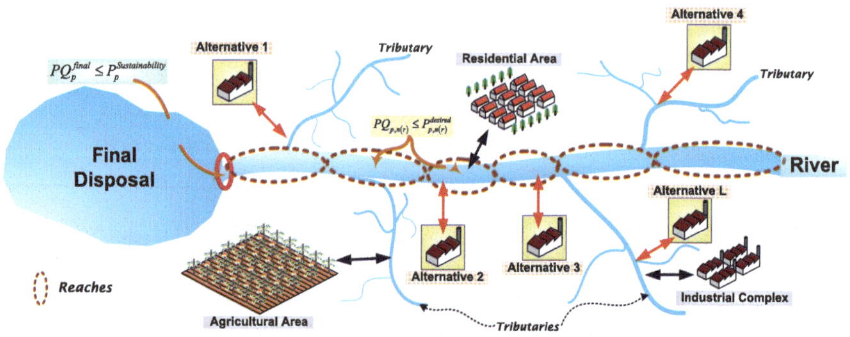

Fig. 9.3 Alternatives to locate the new industrial plant in a watershed system. Printed with permission of "*Incorporating property-based water networks and surrounding watersheds in site selection of industrial facilities*", *Industrial and Engineering Chemistry Research*, Copyright (2012), American Chemical Society

To guarantee the sustainability of the surrounding watershed, it is needed to carry out an analysis based on properties before the optimization to determine the capacity of the final disposal to decompose chemically and biochemically the pollutants. The analysis yields the sustainability constraints, which ensure that the pollutants will not accumulate in the system and maintain the properties of the watershed under adequate limits.

Figure 9.4 presents the implemented superstructure to treat the properties of the wastewater process streams. In this superstructure, process sources are segregated to treat the first property, without mixing different process sources. At the exit of the first treatment, process sources are segregated again and sent to the following interceptors to treat the second property. The same sequence is used to treat all required properties in the property interception network, where the mixing of different process sources is avoided in all the steps of the treatment. Each process source that leaves the treatment units is segregated to be sent to each process sink or the wastewater stream. The available fresh sources can be used in the process sinks when they are necessary.

A set of interceptors can be used to adjust the stream properties. Several treatment units can be used for these purposes, including separation units to recover some hazardous materials, neutralization units to modify the pH, and aeration units to modify the chemical oxygen demand, among others. Consequently, a given adjustment factor is used to distinguish the execution of each available interceptor, and these adjustment factors ($\gamma_p^{int^1,\ldots,\mathrm{int}^N}$) can be calculated before the optimization process. This approach models linear models for the interceptors. The adjustment factor depends on several variables such as the value of the property at the inlet conditions of the interceptor, design, and operating parameters, and the flowrate. The offered treatment technologies considered in the property interception network can be simulated (i.e., using process simulators) to get their adjustment factors. The property operator values for the inlet condition of the process sources are also known before

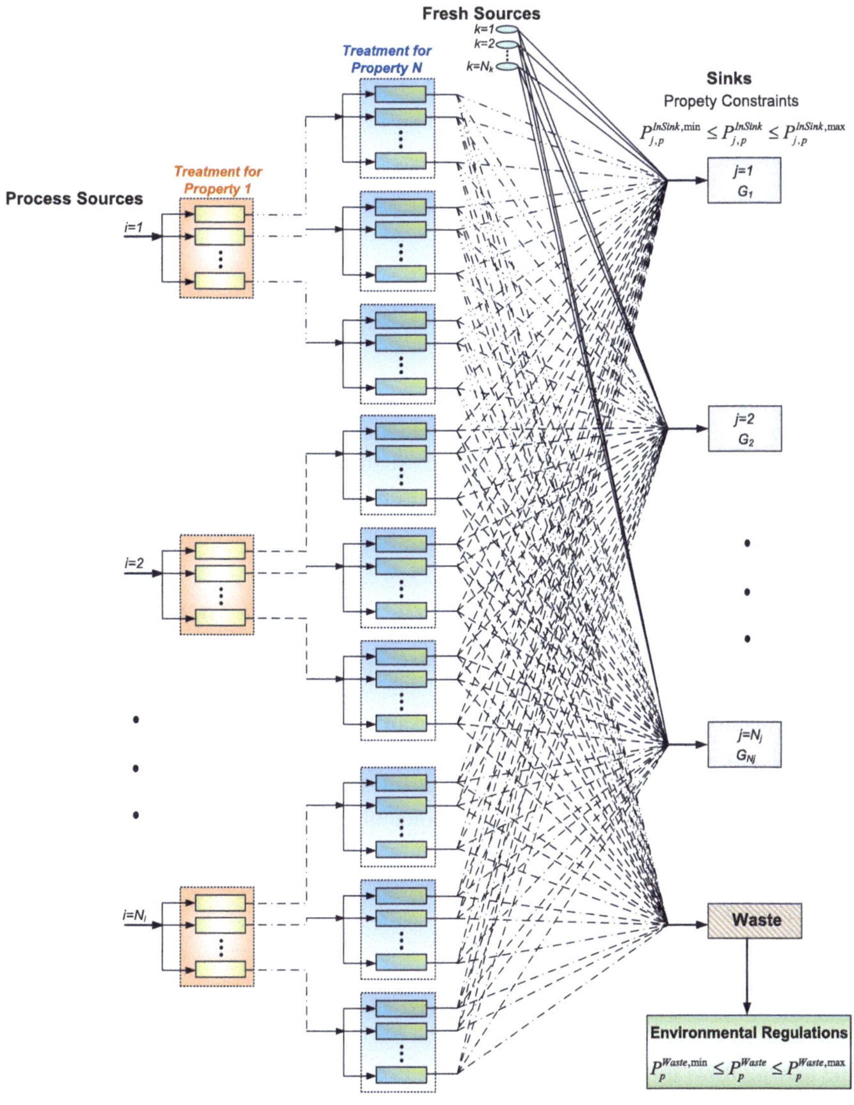

Fig. 9.4 Superstructure for the recycle and reuse network. Printed with permission of "*Incorporating property-based water networks and surrounding watersheds in site selection of industrial facilities*", *Industrial and Engineering Chemistry Research*, Copyright (2012), American Chemical Society

the optimization process, $\psi_p\left(P_{i,p}^{InSource}\right)$. Hence, the property values at the exit conditions of the property interception network can be calculated before optimization as follows:

$$\psi_p\left(P_{i,p}^{OutPIN,\text{int}^1,\ldots,\text{int}^N}\right) = \gamma_p^{\text{int}^1,\ldots,\text{int}^N}\psi_p\left(P_{i,p}^{InSource}\right),\quad i \in I, p \in P, \text{int}^1 \in INT^1,\ldots,\text{int}^N \in INT^N$$

9.3 Model Formulation

The implemented mathematical model is a combination of the Material Flow Analysis technique, a disjunctive programming model to determine the optimal location of the new industrial facility, and the superstructure model for the water integration inside the industrial plant. The next sections describe these models.

9.3.1 *Material Flow Analysis for the Watershed Based on Properties*

Lira-Barragán et al. (2011a, b) implemented an MFA model for watersheds based on the composition of the streams, and this model is extended in this chapter to consider property tracking balances that are suited for wastewater streams constituted by numerous compounds. Also, the environmental constraints are given in terms of limits for specific properties that affect the environment. The balances required to model the watersheds in terms of properties are based on the representation of Fig. 9.3, which are described as follows.

9.3.1.1 Overall Balance for Each Reach

The leaving flowrate (Q_r) from each reach r is equal to the inlet flowrate (Q_{r-1}) plus precipitation (P_r), direct industrial discharges (D_r), residential discharges (H_r), the sum of all effluents arriving to the reach ($FT_{r,t}$) and the potential discharge from the new industrial plant if this is installed near the reach r of the l allowable ($QPNEW_{r(l)}$), minus the extractions due to natural phenomena such as filtration and evaporation (L_r) as well as used water (U_r) in the r section of the river. Consequently, the whole balance for each reach can be written as follows:

$$Q_r = Q_{r-1} + P_r + D_r + H_r + \sum_{t=1}^{N_{t(r)}} FT_{r,t} + QPNEW_{r(l)} - L_r - U_r,\quad \forall r \in R \quad (9.1)$$

$N_{t(r)}$ indicates the total number of tributaries that are discharged to the reach r.

9.3.1.2 Property Tracking Balances for Each Reach

The product of the property operator times the flowrate ($\psi_p\left(P_{p,r}^{Reach}\right)Q_r$) for each reach r is equal to the property operator times the flowrate at the inlet ($\psi_p\left(P_{p,r-1}^{InReach}\right) Q_{r-1}$), adding the property operator contained in the precipitation ($\psi_p\left(P_{p,r}^{Prec}\right)P_r$), industrial discharges ($\psi_p\left(P_{p,r}^{Ind}\right)D_r$), residential discharges ($\psi_p\left(P_{p,r}^{Resid}\right)H_r$), the sum of tributaries ($\sum_{t=1}^{N_{t(r)}} \psi_p\left(P_{p,r,t}^{Trib}\right)FT_{r,t}$) and the property operator from the new industrial plant is this is located in this reach ($\psi_p\left(P_{p,r(l)}^{PNEW}\right)QPNEW_{r(l)}$), deducting the losses ($\psi_p\left(P_{p,r}^{Loses}\right)L_r$), uses ($\psi_p\left(P_{p,r}^{Uses}\right)U_r$), and considering the adjust in the property operator associated to the chemical reactions that are carried out in that section of the river ($\int_{V=0}^{V_r} r_{p,r}dV$). The reactive term considers the chemical and biochemical reactions that take place in the river due to the interaction between the pollutants and the system. Consequently, the property tracking balance for each section of the river r is stated as follows:

$$\begin{aligned}\psi_p\left(P_{p,r}^{Reach}\right)Q_r = {} & \psi_p\left(P_{p,r-1}^{InReach}\right)Q_{r-1} + \psi_p\left(P_{p,r}^{Prec}\right)P_r + \psi_p\left(P_{p,r}^{Ind}\right)D_r + \psi_p\left(P_{p,r}^{Resid}\right)H_r \\ & + \sum_{t=1}^{N_{t(r)}} \psi_p\left(P_{p,r,t}^{Trib}\right)FT_{r,t} + \psi_p\left(P_{p,r(l)}^{PNEW}\right)QPNEW_{r(l)} - \psi_p\left(P_{p,r}^{Loses}\right)L_r \\ & - \psi_p\left(P_{p,r}^{Uses}\right)U_r - \int_{V=0}^{V_r} r_{p,r}dV_r, \quad \forall p \in P, \quad r \in R \end{aligned} \tag{9.2}$$

9.3.1.3 Overall Balance for Each Tributary

For the tributary t that discharges to the reach r, the overall balance is stated as follows:

$$\begin{aligned} FT_{r,t} = {} & S_{r,t}^{Untreat} + S_{r,t}^{Treat} + I_{r,t} + P_{r,t} + D_{r,t} \\ & + QPNEW_{r,t(l)} - L_{r,t} - U_{r,t}, \quad \forall r \in R, \quad t \in T \end{aligned} \tag{9.3}$$

The total flowrate discharged from the tributary t to the reach r ($FT_{r,t}$) is the sum of discharges without treatment ($S_{r,t}^{Untreated}$), discharges with treatment ($S_{r,t}^{Treated}$), industrial discharges ($I_{r,t}$), pluvial discharges ($P_{r,t}$), direct discharges ($D_{r,t}$), and the new discharge if the new facility is located on this tributary ($QPNEW_{r,t(p)}$), subtracting the losses ($L_{r,t}$) and use ($U_{r,t}$) of water.

9.3.1.4 Property Balances for Each Tributary

The following tracking balance that includes the reaction term is used to calculate the property operator discharged from the tributary t to the reach r.

$$\begin{aligned}\psi_p\left(P_{p,r,t}^{Trib}\right)FT_{r,t} = {} & \psi_p\left(P_{p,r,t}^{Untreat}\right)S_{r,t}^{Untreat} + \psi_p\left(P_{p,r,t}^{Treat}\right)S_{r,t}^{Treat} \\ & + \psi_p\left(P_{p,r,t}^{Ind}\right)I_{r,t} + \psi_p\left(P_{p,r,t}^{Prec}\right)P_{r,t} \\ & + \psi_p\left(P_{p,r,t}^{Direct}\right)D_{r,t} + \psi_p\left(P_{p,r,t(l)}^{PNEW}\right)QPNEW_{r,t(l)} \\ & - \psi_p\left(P_{p,r,t}^{Loses}\right)L_{r,t} - \psi_p\left(P_{p,r,t}^{Uses}\right)U_{r,t} \\ & - \int_{V=0}^{V_{r,t}} r_{p,r,t}dV_{r,t}, \quad \forall p \in P, \quad r \in R, \quad t \in T \end{aligned} \tag{9.4}$$

9.3.1.5 Modelling Reactive Terms

The chemical reactions that are experimented inside the rivers have strong effects on the properties of the streams in the watershed. Numerous components can be damaged affecting the properties of the streams. The result of the chemical or biochemical reactions over the properties for reaches (Eq. 9.5) and tributaries (Eq. 9.6) can be presented as follows:

$$\int_{V=0}^{V_r} r_{p,r}dV_r = k_p\left[\psi_p\left(P_{p,r}^{Reach}\right)\right]^{\sigma_p} V_r, \quad \forall p \in P, \quad r \in R \tag{9.5}$$

$$\int_{V=0}^{V_{r,t}} r_{p,r,t}dV_{r,t} = k_p\left[\psi_p\left(P_{p,r,t}^{Trib}\right)\right]^{\sigma_p} V_{r,t}, \quad \forall p \in P, \quad r \in R, \quad t \in T \tag{9.6}$$

k_p is the kinetic constant of degradation for each property p determined experimentally, σ_p is the reaction order for each property, $\psi_p\left(P_{p,r}^{Reach}\right)$ is the property operator at the reach r, while $\psi_p\left(P_{p,r,t}^{Trib}\right)$ is the property operator at the tributary t, and V_r and $V_{r,t}$ are the volumes for the reach and tributary, respectively.

9.3.1.6 Agricultural Discharges and Uses

Information is used to calculate the agricultural discharges ($D_{r,t}$) as well as the agricultural uses ($U_{r,t}$) through a pair of experimental parameters that depend on the agricultural area ($A_{r,t}$).

$$D_{r,t} = \alpha_{r,t} A_{r,t}, \quad \forall r \in R, \quad t \in T$$
$$U_{r,t} = \beta_{r,t} A_{r,t}, \quad \forall r \in R, \quad t \in T$$

$\alpha_{r,t}$ is the water required per area and its units are m^3/ha s, while $\beta_{r,t}$ is the discharged flowrate per area and its units also are m^3/ha s.

9.3.2 Location of the New Industrial Plant

This chapter considers a set of alternatives to locate the new industrial facility (L), which are given before the optimization process. Also, for each alternative, it is needed to consider the potential discharge of the new industrial wastewater flowrate ($QPNEW_l$) and its property operators ($\psi_p\left(P_{p,l}^{PNEW}\right)$). If the new industrial plant is located on alternative 1, the wastewater flowrate and the property operators associated with this location must be greater than zero; else, if the new industrial plant is not located in position 1, the wastewater flowrate and the property operators discharged must be zero. This is included in the following disjunction:

$$\begin{bmatrix} Y_1 \\ \psi_p\left(P_{p,1}^{PNEW}\right) \geq 0, QPNEW_1 \geq 0 \\ \psi_p\left(P_{p,2}^{PNEW}\right) = 0, QPNEW_2 = 0 \\ \vdots \\ \psi_p\left(P_{p,l}^{PNEW}\right) = 0, QPNEW_l = 0 \end{bmatrix} \vee \begin{bmatrix} Y_2 \\ \psi_p\left(P_{p,1}^{PNEW}\right) = 0, QPNEW_1 = 0 \\ \psi_p\left(P_{p,2}^{PNEW}\right) \geq 0, QPNEW_2 \geq 0 \\ \vdots \\ \psi_p\left(P_{p,l}^{PNEW}\right) = 0, QPNEW_l = 0 \end{bmatrix}$$
$$\vee \cdots \vee \begin{bmatrix} Y_l \\ \psi_p\left(P_{p,1}^{PNEW}\right) = 0, QPNEW_1 = 0 \\ \psi_p\left(P_{p,2}^{PNEW}\right) = 0, QPNEW_2 = 0 \\ \vdots \\ \psi_p\left(P_{p,l}^{PNEW}\right) \geq 0, QPNEW_l \geq 0 \end{bmatrix}, \quad \forall p \in P$$

In this disjunction, the Boolean variable associated with site l of the new industrial plant is Y_l, whereas $\psi_p\left(P_{p,l}^{PNEW}\right)$ corresponds to the property operator for property p from the wastewater stream of the new industrial plant ($QPNEW_l$). As can be seen, these variables depend on the final location of the new industrial plant. Therefore, the disjunctive model states that if the optimal location for the new industrial plant is on alternative 1, then the Boolean variable Y_1 is set as true; hence, $\psi_p\left(P_{p,1}^{PNEW}\right)$ and $QPNEW_1$ must be greater than zero, but all the others Boolean variables (Y_l) must be false and $\psi_p\left(P_{p,l}^{PNEW}\right)$ and $QPNEW_l$ associated to such locations must be zero.

Similarly for the case when the selected location is 2, 3, and so on for the l potential places. It is important to remark that the value for the flowrate of the wastewater and the property operators of the new industrial plant are powerfully influenced by the location through the material flow analysis model, satisfying the environmental constraints given to maintain control of the properties and the desired constraints in intermediate reaches, as well as the sustainability constraints at the final disposal. It should be noted that these variables also depend on the water integration inside the industrial plant where there are sink constraints to be satisfied at the inlet to process units. The previous disjunction is reformulated using the convex hull technique (Raman and Grossmann, 1994; Ponce-Ortega et al. 2009a, b).

The algebraic reformulation involves the conversion of the Boolean variables (Y_l) into a set of binary variables (y_l). When the Boolean variables are true, the associated binary variables must one; or else, when the Boolean variables are false, the associated binary variables are zero. Then, the following relationship is used to select only one site to install the new industrial plant:

$$\sum_{l \in L} y_l = 1 \tag{9.7}$$

The following upper limits for the variables $QPNEW_l$ and $\psi_p\left(P_{p,l}^{PNEW}\right)$ are needed in the algebraic reformulation.

$$QPNEW_l \leq QPNEW^{Up} y_l, \quad \forall l \in L \tag{9.8}$$

$$\psi_p\left(P_{p,l}^{PNEW}\right) \leq \psi_p\left(P^{PNEW}\right)^{Up} y_l, \quad \forall p \in P, \ l \in L \tag{9.9}$$

$QPNEW^{Up}$ and $\psi_p\left(P_{p,l}^{PNEW}\right)^{Up}$ are the upper limits for $QPNEW_l$ and $\psi_p\left(P_{p,l}^{PNEW}\right)$, respectively. It should be noted that $\psi_p\left(P_{p,l}^{PNEW}\right)^{Up}$ is an upper bound for all properties p. These constraints are required to ensure that $QPNEW_l$ and $\psi_p\left(P_{p,l}^{PNEW}\right)$ take values greater than zero only when the new plant is in a particular position l.

9.3.3 Additional Equations to Interconnect the Models

The following equations are needed to interconnect the Material Flow Analysis model with the superstructure model for the mass integration inside the plant:

$$\sum_{l \in L} QPNEW_l = Waste \tag{9.10}$$

$$\sum_{l \in L} \psi_p\left(P_{p,l}^{PNEW}\right) = \psi_p\left(P_p^{disc}\right), \quad \forall l \in L \tag{9.11}$$

Waste is the wastewater discharged to the environment by the new industrial industrial plant and $\psi_p\left(P_p^{disc}\right)$ is the property operator for each property of such stream. This way, Eq. (9.7) guarantees that only one site is selected, then only one of the l variables for $QPNEW_l$ and $\psi_p\left(P_{p,l}^{PNEW}\right)$ can be greater than zero, and Eqs. (9.8) and (9.9) are fulfilled.

9.3.4 Superstructure Modeling for In-Plant Treatment

The implemented superstructure for the in-plant wastewater treatment is presented in Fig. 9.4, which is based on the ones previously reported by Ponce-Ortega et al. (2010) and Gabriel and El-Halwagi (2005). The size for this formulation increases exponentially as a function of the number of treated properties (N); but the model obtained for this superstructure has a linear behavior, which is the main advantage of this approach because it avoids the mixing of different streams and, consequently, it is very useful for the optimization process. Figure 9.4 shows that for each property to be treated there is a set of treatment units; consequently, it generates N treatment units for each stream.

9.3.4.1 Splitting Freshwater Sources

Each fresh water source is split and sent to any process sink as follows:

$$F_k = \sum_{j \in J} f_{k,j}, \quad \forall k \in K \tag{9.12}$$

F_k corresponds to the total flowrate for the fresh source k and $f_{k,j}$ is the segregated flowrate from the fresh source k sent to process sink j. It should be noticed in Fig. 9.4 that the fresh sources just are directed to the process sinks.

9.3.4.2 Splitting of Process Sources to the Property Interceptors

The interception network is segregated evading the mixing of different process sources. If a process source does not need any treatment, it is sent to a fictitious unit (located at the end of the set of property interceptors) with effectiveness and cost equal to zero. The next equation represents the segregation of each process source to treat the first property on the first set of available interceptors.

$$W_i = \sum_{int^1 \in INT^1} w_i^{int^1}, \quad \forall i \in I \tag{9.13}$$

Later, to treat the second property, the flowrates at the exit of the first line of interceptors are segregated evading the mixing of streams, and again these are fed to the second set of interceptors.

$$w_i^{int^1} = \sum_{int^2 \in INT^2} w_i^{int^1, int^2}, \quad \forall i \in I, \quad int^1 \in INT^1 \tag{9.14}$$

It should be noted that it does not allow mixing of different process streams. This approach removes the nonlinearities involved in the property balances, even if the problem size increases exponentially.

Following the same sequence for N properties to be intercepted, the next equations model the segregations.

$$w_i^{int^1, \mathrm{int}^2, \ldots, \mathrm{int}^{N-1}} = \sum_{int^N \in INT^N} w_i^{int^1, int^2, \ldots, \mathrm{int}^N}, \quad \forall i \in I, \quad int^1 \in INT^1, \ldots, int^{N-1} \in INT^{N-1} \tag{9.15}$$

9.3.4.3 Splitting of Sources at the Exit of the Interception Network

Next the treatment network, the process sources are segregated and sent to the process sinks and to the wastewater stream discharged to the environment.

$$w_i^{int^1, \ldots, \mathrm{int}^N} = \sum\nolimits_{j \in J} g_{i,j}^{int^1, \ldots, \mathrm{int}^N} + g_{i, Waste}^{int^1, \ldots, \mathrm{int}^N}, \quad \forall i \in I, \quad int^1 \in INT^1, \ldots, int^N \in INT^N \tag{9.16}$$

9.3.4.4 Total Balance at the Mixing Point Before Any Sink

The total flowrate at the inlet of any sink (G_j) must be completed by the sum of intercepted flowrates from process sources, as well as the flowrates from fresh sources.

$$G_j = \sum_{i \in I} \sum_{int^1 \in INT^1} \cdots \sum_{\mathrm{int}^N \in INT^N} g_{i,j}^{int^1, \ldots, \mathrm{int}^N} + \sum_{k \in K} f_{k,j}, \quad \forall j \in J \tag{9.17}$$

9.3.4.5 Property Tracking Balances at the Mixing Point Before Any Sink

Property tracking balances are needed to calculate the properties at the inlet to any process sink, which is stated as follows:

$$\psi_p\left(P_{j,p}^{InSink}\right)G_j = \sum_{i\in I}\sum_{\text{int}^1\in INT^1}\cdots\sum_{\text{int}^N\in INT^N}\left[\psi_p\left(P_{i,p}^{OutPIN,int^1,\dots,\text{int}^N}\right)g_{i,j}^{int^1,\dots,\text{int}^N}\right] + \sum_{k\in K}\left[\psi_p\left(P_{k,p}^{Fresh}\right)f_{k,j}\right],\quad \forall j\in J,\ \ p\in P \tag{9.18}$$

9.3.4.6 Overall Balance for the Waste Stream

The total flowrate of the wastewater stream discharged to the environment is provided by the sum of the process sources after the treatment units that are sent to the waste.

$$Waste = \sum_{i\in I}\sum_{\text{int}^1\in INT^1}\cdots\sum_{\text{int}^N\in INT^N} g_{i,Waste}^{int^1,\dots,\text{int}^N} \tag{9.19}$$

9.3.4.7 Property Tracking Balances for the Wastewater Stream

The following equation is used to calculate the properties of the wastewater stream discharged to the environment.

$$\psi_p\left(P_p^{Waste}\right)Waste = \sum_{i\in I}\sum_{\text{int}^1\in INT^1}\cdots\sum_{\text{int}^N\in INT^N}\left(\psi_p\left(P_{i,p}^{OutPIN,int^1,\dots,\text{int}^N}\right)g_{i,Waste}^{int^1,\dots,\text{int}^N}\right),\quad \forall p\in P \tag{9.20}$$

9.3.4.8 Sink Constraints Based on Properties

Process sinks require a set of constraints to properly work, which regularly are given in terms of limits for specific properties such as the composition of specific compounds, density, pH, viscosity, thermal conductivity, etc. In this chapter, these constraints are stated in terms of the property operators as follows:

$$\psi_p\left(P_{j,p}^{InSink,min}\right)\le\psi_p\left(P_{j,p}^{InSink}\right)\le\psi_p\left(P_{j,p}^{InSink,max}\right),\quad j\in J,\ \ P\in P \tag{9.21}$$

9.3.4.9 Environmental Constraints in Terms of Properties

The environmental regulations require a set of constraints on the wastewater from industrial plants, which are given in terms of limits for properties as follows:

$$\psi_p\left(P_p^{Waste,\min}\right) \le \psi_p\left(P_p^{Waste}\right) \le \psi_p\left(P_p^{Waste,\max}\right), \quad \forall p \in P \tag{9.22}$$

These limited constraints include upper and lower bounds for specific properties in the wastewater discharged to the environment like pH, COD, toxicity, color, etc.

9.3.4.10 Constraints for the Quality of Some Reaches Based on Properties

Since the water in some specific places of the rivers can be used for different purposes, it becomes mandatory to impose a set of property constraints on such reaches to guarantee the quality of the water extracted. The presented model can consider these constraints through the following relationship:

$$\psi_p\left(P_{p,m(r)}^{DesiredMIN}\right) \le \psi_p\left(P_{p,m(r)}^{Reach}\right) \le \psi_p\left(P_{p,m(r)}^{DesiredMAX}\right), \quad \forall p \in P, \quad m(r) \in M(R) \tag{9.23}$$

$M(R)$ corresponds to a subset of reaches of all the reaches r that require a specific water quality, whereas $\psi_p\left(P_{p,m(r)}^{DesiredMAX}\right)$ and $\psi_p\left(P_{p,m(r)}^{DesiredMIN}\right)$ are the limits for the property operators with required properties in specified reaches to maintain the water quality in them.

9.3.4.11 Sustainability Constraints

To ensure the sustainability of the watershed and its environment, and to avoid putting at risk the final disposal (lake, sea, or ocean), the next constraints must be satisfied due to the environmental regulations imposed on the industrial plant discharge could not be enough:

$$\psi_p\left(P_p^{SustainableMIN}\right) \le \psi_p\left(P_{p,final}^{Reach}\right) \le \psi_p\left(P_p^{SustainableMAX}\right), \quad \forall p \in P \tag{9.24}$$

$\psi_p\left(P_{p,final}^{Reach}\right)$ is the property operator in the final disposal after installing the new industrial plant, while $\psi_p\left(P_p^{SustainableMIN}\right)$ and $\psi_p\left(P_p^{SustainableMAX}\right)$ are the set of sustainability constraints for property p. It should be noted that the sustainability constraints are generated by making a sustainability analysis for the final disposal before the optimization process. Satisfying these constraints guarantees that the natural degradation of the pollutants in the final disposal will maintain the properties under sustainable limits.

9.3.5 Objective Function

Lastly, the objective function consists of minimizing the total annual cost (*TAC*), which is established by the annualized installation costs for the new industrial plant, the fresh sources costs, the treatment cost (including operational cost, variable and fixed capital cost), and the piping costs.

$$\begin{aligned}
Min\ TAC = {} & \sum_{l \in L} Cland_l y_l + H_Y \sum_{k \in K} FreC_k F_k \\
& + H_Y \sum_{i \in I} \left[\sum_{int^1 \in INT^1} VarC_{op}^{int^1} w_i^{int^1} + \cdots \sum_{\mathrm{int}^N \in INT^N} \left(VarC_{op}^{\mathrm{int}^N} w_i^{\mathrm{int}^1,\ldots,\mathrm{int}^N} \right) \right] \\
& + H_Y \sum_{i \in I} \left[\sum_{int^1 \in INT^1} VarC_{cap}^{int^1} w_i^{int^1} + \cdots \sum_{\mathrm{int}^N \in INT^N} \left(VarC_{cap}^{\mathrm{int}^N} w_i^{\mathrm{int}^1,\ldots,\mathrm{int}^N} \right) \right] \\
& + k_f \sum_{i \in I} \left[\sum_{int^1 \in INT^1} FixC_{cap}^{int^1} z_i^{int^1} + \cdots \sum_{\mathrm{int}^N \in INT^N} \left(FixC_{cap}^{\mathrm{int}^N} z_i^{\mathrm{int}^1,\ldots,\mathrm{int}^N} \right) \right] \\
& + H_Y \left[\begin{array}{l}
TrPipC \sum_{i \in I} \left(\sum_{\mathrm{int} \in INT} w_i^{int^1} + \cdots \sum_{\mathrm{int}^N \in INT^N} w_i^{\mathrm{int}^1,\ldots,\mathrm{int}^N} \right) \\
+ WaPipC \sum_{i \in I} \sum_{int^1 \in INT^1} \cdots \sum_{\mathrm{int}^N \in INT^N} g_{i,Waste}^{\mathrm{int}^1,\ldots,\mathrm{int}^N} \\
+ EqPipC \sum_{i \in I} \sum_{int^1 \in INT^1} \cdots \sum_{\mathrm{int}^N \in INT^N} \sum_{j \in J} g_{i,j}^{\mathrm{int}^1,\ldots,\mathrm{int}^N} \\
+ FrPipC \sum_{k \in K} \sum_{j \in J} f_{k,j}
\end{array} \right]
\end{aligned} \tag{9.25}$$

H_Y corresponds to the hours per year that operates the new industrial facility and k_f represents the factor used to annualize the inversion. The annualized installation cost ($Cland_l$) considers the land cost and the transportation costs for raw materials, products, and services. It should be noted that this annualized installation cost considers simultaneously capital and operational costs and that the factor used to annualize the capital costs is used to provide the associated cost per year. Fresh sources cost ($FreC_k$) is associated with the required amount of fresh sources to meet the water requirements of the process sinks as well as the property constraints imposed on them. Likewise, process sources need to be treated before being sent to the process sinks, as well as before being discharged into the environment. A set of technologies must be selected to treat the process sources yielding capital and operational costs for each one of the interceptors required. The operational cost ($VarC_{op}^{\text{int}^N}$) for the interceptors depends on the treated flowrate, but the capital cost contains a variable cost ($VarC_{cap}^{\text{int}^N}$) that also depends on the treated flowrate and a fixed component ($FixC_{cap}^{\text{int}^N}$). Lastly, the piping costs for the four sections presented in the superstructure are considered. The first section involves the pipes required when the process sources are sent to treatment ($TrPipC$); afterwards, once the process sources are treated they can be sent to the wastewater stream ($WaPipC$) or process sinks ($EqPipC$). The final section covers the pipes to connect the fresh sources with process sinks ($FrPipC$).

9.3.6 Remarks for the Optimization Model

- The presented model considers simultaneously the water integration inside the industrial facility and the sustainability of the surrounding watershed.
- The presented model is based on the properties of the streams, considering property balances, as well as process and environmental constraints given in terms of limits for the properties. This condition is very useful for streams formed by several pollutants, where the component balances are not suited to characterize them.
- The effect of the chemical and biochemical reactions on the properties can be represented by a kinetic model. The kinetic constants can be measured experimentally.
- The annualized installation costs ($Cland_l$) include different costs: transportation cost of raw materials, transportation cost of products, services, and the land cost for the alternative l. These costs depend directly on the location selected for the new industrial plant.
- Conversion factors ($\gamma_p^{\text{int}^1,\ldots,\text{int}^N}$) are parameters known for each unit to adjust each property, which can be determined experimentally or estimated by simulation. Consequently, the property balances in the superstructure for the treatment systems are not required in the optimization model. Furthermore, to determine the outlet property from the interception network only the following relationship is used: $\psi_p\left(P_{i,p}^{OutPIN,\text{int}^1,\ldots,\text{int}^N}\right) = \gamma_p^{\text{int}^1,\ldots,\text{int}^N}\psi_p\left(P_{i,p}^{InSource}\right)$.

9.4 Results and Discussion

The presented model was coded in the General Algebraic Modeling System (GAMS) (Brook et al. 2011). Two examples are presented to show the implementation of the proposed formulation, one from Egypt and the other one from Mexico. $FixC_{cap}^{\text{int}^N}$ are fixed as zero for both examples.

9.4.1 Example 9.1

The Bahr El-Baqar system is in Egypt and its report was provided by Lovelady et al. (2009). The system receives numerous types of discharges (i.e., agricultural, industrial, municipal effluents) and the properties restricted in the environment are the toxicity, COD, odor, color, and composition of a dangerous pollutant. The properties restricted in the process sinks are the composition and the toxicity. Four candidate sites are identified for the installation of a new industrial plant, where each one has associated an installation cost (Table 9.2). Figure 9.5 presents the Bahr El-Baqar system, the four available sites, the properties considered, and the constraints imposed on the system. The set of constraints is explained by considering the reach 6. Before installing the new industrial plant, it had a composition and toxicity of 1.312 ppm and 0.0076, and the maximum composition and toxicity allowed after the installation of the plant were 1.34 ppm and 0.0078, respectively. A similar explanation applies to the constraints imposed on the remaining reaches of Fig. 9.5.

The natural degradation of pollutants along the system helps to improve properties like toxicity, COD, color, etc. Particularly, in this system, the kinetic constants in terms of properties are: $k_{Comp} = 9.041909 \times 10^{-6}/\text{s}$, $k_{Tox} = 5.208473 \times 10^{-7}/\text{s}$, $k_{COD} = 4.08396 \times 10^{-6}/\text{s}$, $k_{Odor} = 2.083389 \times 10^{-6}/\text{s}$ and $k_{Color} = 3.645931 \times 10^{-6}/\text{s}$, which correspond to a kinetic of the first order. Furthermore, the new industrial plant operates 8000 h/year and $k_f = 0.1$. Table 9.3 presents the characteristics of the fresh sources and the process sources; it should be noted that the flowrates for the fresh sources are optimization variables, and the specifications of COD, Odor, and Color are not presented because fresh sources can only be sent to process sinks, which do not contain constraints in terms of these properties for this example.

The flowrate required for each process sink as well as the maximum composition and toxicity allowed are shown in Table 9.4, whereas Table 9.5 shows the available interceptors to treat the properties considered.

The costs for the fresh sources 1 and 2 are \$0.190/m^3 and \$0.142/m^3, respectively. In addition, the environmental constraints for the wastewater stream in terms of properties are: $P_{Comp}^{Sustainable} = 50$, $P_{Tox}^{Sustainable} = 0.1$, $P_{COD}^{Sustainable} = 75$, $P_{Odor}^{Sustainable} = 2.7$, and $P_{Color}^{Sustainable} = 100$. Therefore, to meet these constraints the model considers a set

Table 9.2 Installation cost for the different sites to locate the new plant for Example 9.1

Site	Annualized installation cost, \$/year
1	14,000,000
2	12,500,000
3	11,000,000
4	10,000,000

Printed with permission of "*Incorporating property-based water networks and surrounding watersheds in site selection of industrial facilities*", *Industrial and Engineering Chemistry Research*, Copyright (2012), American Chemical Society

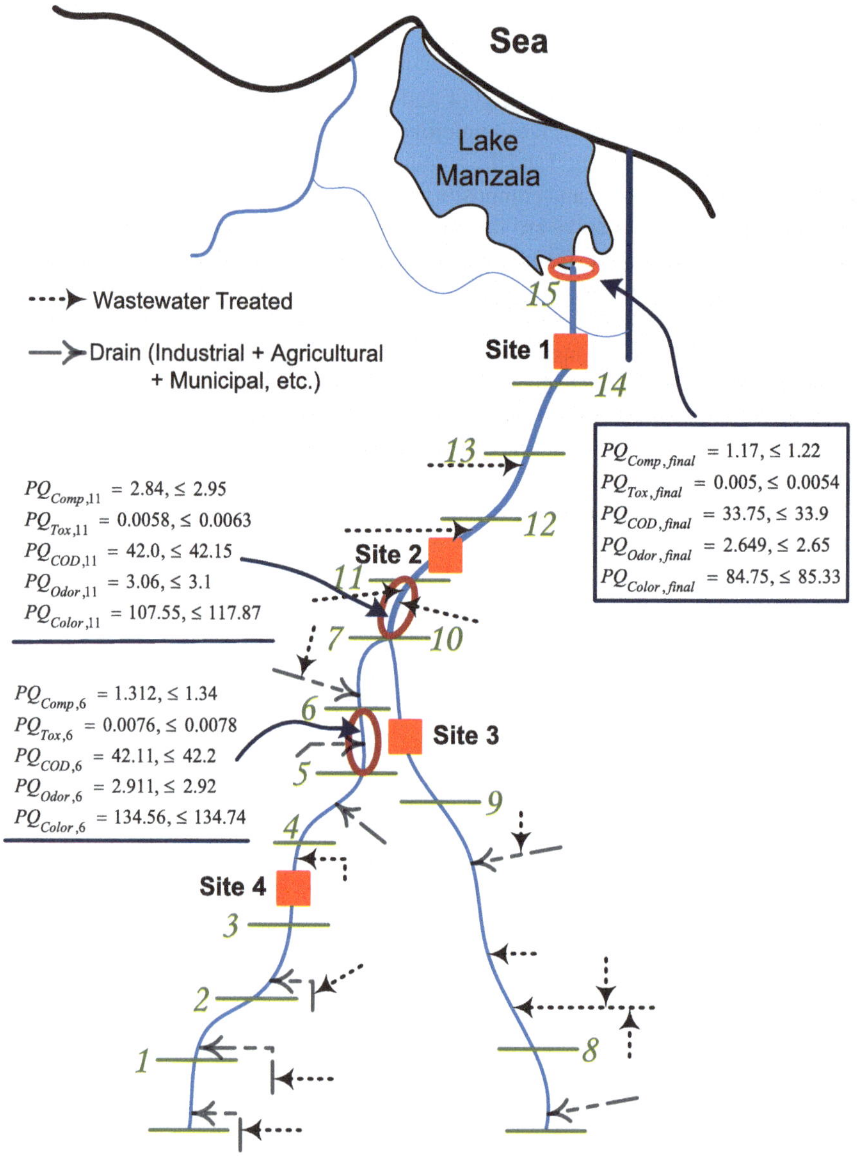

Fig. 9.5 Bahr El-Baqar system. Printed with permission of "*Incorporating property-based water networks and surrounding watersheds in site selection of industrial facilities*", *Industrial and Engineering Chemistry Research*, Copyright (2012), American Chemical Society

Table 9.3 Fresh and process sources for Example 9.1

Sources	Flowrate, m^3/s	Composition, ppm	Toxicity, %	COD, mgO_2/lt	Odor	Color
Process						
1	0.316	88.7	0.14	139	6	150
2	0.435	77.2	0.22	93	7	262
3	0.402	52.4	0.16	128	5.5	290
Fresh						
1	–	0	0	–	–	–
2	–	10	0.005	–	–	–

Table 9.4 Data for the process sinks for Example 9.1

Process sink	Flowrate, m^3/s	Composition^{max}	Toxicity^{max}
1	0.365	55.5	0.013
2	0.290	36.2	0.011
3	0.335	30	0.010

of interceptors with different efficiencies to remove the pollutants and to improve the properties. Lastly, the piping costs for each section in the superstructure are: $TrPipC = 0.003\ \$/m^3$, $WaPipC = 0.006\ \$/m^3$, $EqPipC = 0.010\ \$/m^3$, and $FrPipC = 0.008\ \$/m^3$.

The problem for this case study was implemented in the software GAMS (Brooke et al. 2011), and it was solved using the solver DICOPT (Viswanathan and Grossmann 1990). The optimal solution involves a total annual cost of $\$17.317 \times 10^6$/year, which is constituted by the next cost parts: $\$11.1 \times 10^6$/year of installation cost, $\$4.816 \times 10^6$/year of treatment cost, \$573,330/year of fresh source cost, and \$827,532/year of piping cost. The results obtained for the solution of the proposed model show that the new industrial plant must be installed in site 3 with water integration based on properties shown in Fig. 9.6. It should be noted that some properties of the stream discharged to the environment are limited by environmental constraints, while others are lower than the environmental constraint. Particularly, composition, toxicity, and color require a larger decrease to meet the constraints imposed on the river or the final disposal to maintain the quality of the water around the watershed (Fig. 9.7). Consequently, the results indicate that if the new industrial plant just satisfies the environmental regulations imposed by the legislation on the wastewater, the sustainability requirements of the system are not satisfied because the associated

Table 9.5 Data of the available interceptors for Example 9.1

Interceptor	Adjustment factor	Cost, $/m^3$
Composition		
Rec^1	0.05	0.090
Rec^2	0.22	0.072
Toxicity		
Tox^1	0.08	0.105
Tox^2	0.17	0.075
COD		
Aer^1	0.06	0.055
Aer^2	0.23	0.032
Odor		
Ads^1	0.02	0.065
Ads^2	0.16	0.047
Color		
Dec^1	0.10	0.070
Dec^2	0.20	0.051

Printed with permission of "*Incorporating property-based water networks and surrounding watersheds in site selection of industrial facilities*", *Industrial and Engineering Chemistry Research*, Copyright (2012), American Chemical Society

environmental impact of the wastewater discharged from the new industrial plant is not in line with the estimated degradation capacity of the watershed.

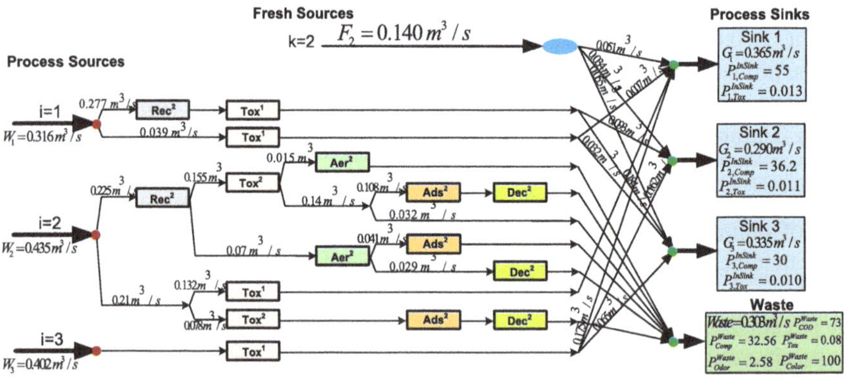

Fig. 9.6 Mass integration based on properties for Example 9.1. Printed with permission of "*Incorporating property-based water networks and surrounding watersheds in site selection of industrial facilities*", *Industrial and Engineering Chemistry Research*, Copyright (2012), American Chemical Society

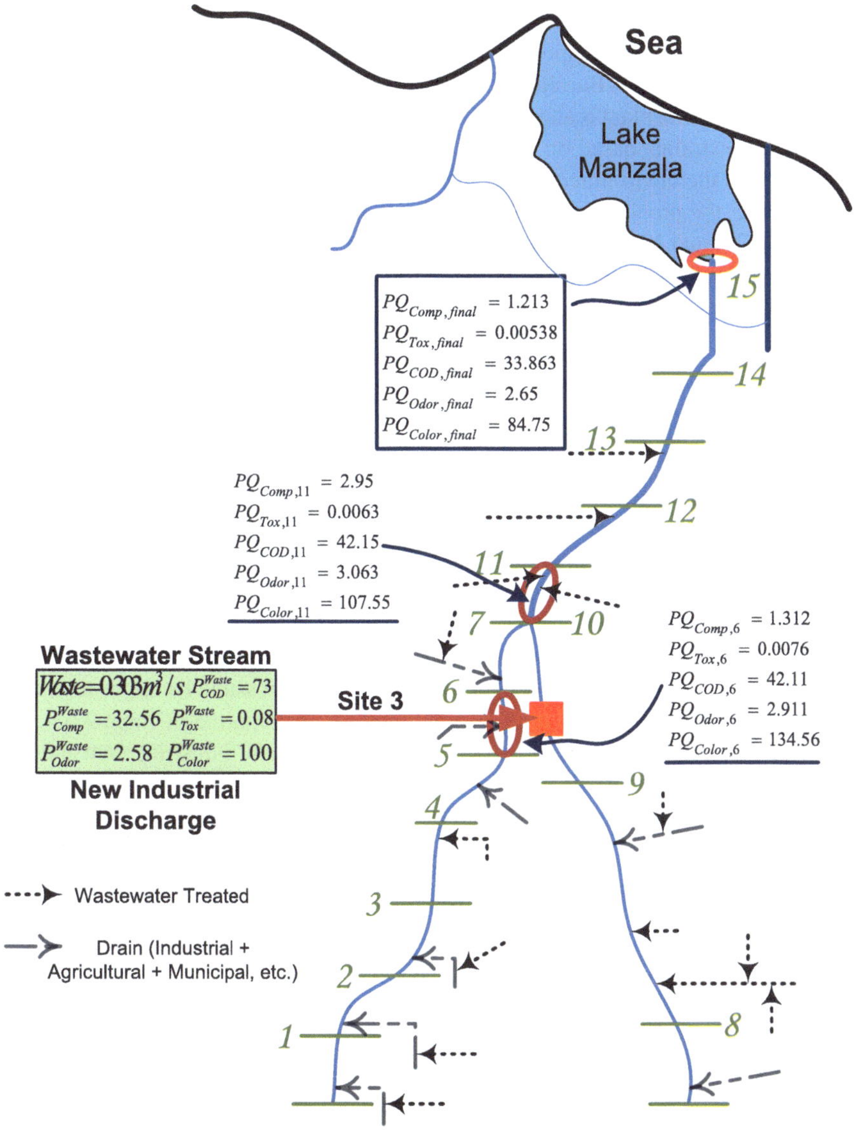

Fig. 9.7 Impact of the new wastewater stream over Bahr El-Baqar system. Printed with permission of "*Incorporating property-based water networks and surrounding watersheds in site selection of industrial facilities*", *Industrial and Engineering Chemistry Research*, Copyright (2012), American Chemical Society

Furthermore, this case was addressed previously by Lira-Barragán et al. (2011b) but just considering the compositions of chemical compounds and without considering the properties. Lira-Barragán et al. (2011b) determined that the optimal location of the new industrial plant is site 4 with a total annualized cost of $\$12.121 \times 10^6$/year. When the constraints based on properties for the in-plant sinks and the wastewater discharged to the environment are not included, some constraints may be violated. For example, the properties in reach 6 (toxicity of 0.0088, COD, odor, and color of 42.808, 2.936, and 134.69, respectively) represent a risk for the water extracted for human activities, and this problem is evaded in the solution given by the formulation proposed in this chapter.

9.4.2 *Example 9.2*

This case involves the Balsas watershed, which is one of the largest in Mexico. This watershed transports numerous types of discharges until the Pacific Ocean and CONAGUA (2010a, b) reported the data for it, which is presented in Fig. 9.8. Here, the properties considered were the composition, toxicity, density, COD, odor, and color, and the properties restricted for the stream discharged to the environment are composition, toxicity, COD, odor, and color, while the process sinks limit just composition and density. The watershed presents twenty possible locations to install the new industrial facility and their installation costs are reported in Table 9.6.

Figure 9.8 presents the properties before installing the new industrial plant and the set of constraints imposed on the system. For example, the composition discharged on the Pacific Ocean before the installation of the new industrial plant is 1.466 ppm and it is required that this concentration does not exceed 1.47 ppm when the new industrial plant is working. The property odor discharged to the Pacific Ocean is 0.654 and once the new industrial plant is operating must be lower or equal to 0.68. This explanation must be followed to understand all the constraints imposed on the river. It is noteworthy that in the limited reaches, toxicity is strictly regulated.

To model the reactive effects, the kinetic constants in terms of properties for first order kinetics are: $k_{Comp} = 3.2376 \times 10^{-6}/\text{s}$, $k_{Tox} = 4.5205 \times 10^{-7}/\text{s}$, $k_{COD} = 1.9273 \times 10^{-6}/\text{s}$, $k_{Odor} = 8.315 \times 10^{-7}/\text{s}$ and $k_{Color} = 1.1423 \times 10^{-6}/\text{s}$. Remember that the density is not restricted in the environment. Additionally, the new industrial plant will operate 8000 h/year with a $k_f = 0.1$. Table 9.7 reports the data for the fresh and process sources; it should be noted that for fresh sources only the specifications of composition and density are presented because they cannot be sent to the wastewater stream discharged to the environment where the other properties are restricted.

Table 9.8 presents the conditions for the flowrate, maximum composition, as well as the minimum and maximum density for each process sink. The unit cost for fresh source 1 is $\$0.128/\text{m}^3$, for fresh source 2 is $\$0.113/\text{m}^3$, and for fresh source 3 is $\$0.097/\text{m}^3$. Table 9.9 presents the characteristics (efficiency and variable costs) of the property interceptors. In this case, the variable costs are included in only one term.

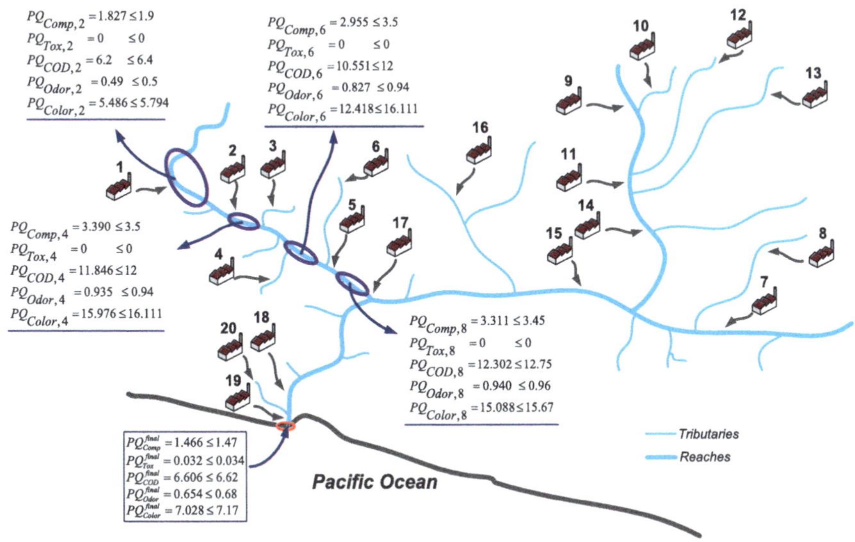

Fig. 9.8 Balsas watershed system (possible locations and the constraints on the river). Printed with permission of "*Incorporating property-based water networks and surrounding watersheds in site selection of industrial facilities*", *Industrial and Engineering Chemistry Research*, Copyright (2012), American Chemical Society

The environmental regulations impose the following constraints on the wastewater stream: $P^{Sustainable}_{Comp} = 50$, $P^{Sustainable}_{Tox} = 0.1$, $P^{Sustainable}_{COD} = 75$, $P^{Sustainable}_{Odor} = 2.7$, and $P^{Sustainable}_{Color} = 100$. Whilst the piping costs for each section in the superstructure are: $TrPipC = 0.001$ \$/m^3, $WaPipC = 0.002$ \$/m^3, $EqPipC = 0.005$ \$/m^3, and $FrPipC = 0.003$ \$/m^3.

Next, the implemented model was applied for this case study yielding an optimal solution where the new industrial facility must be located in alternative 6 with a total annual cost of \$14.9 × 10^6/year, composed of \$9 × 10^6/year of installation cost, \$3.6 × 10^6/year of treatment cost, \$1.5203 × 10^6/year of fresh water cost and \$779,040/year of piping cost. Figure 9.9 presents the water network based on properties for this example; it should be noted that the value of toxicity in the wastewater stream is zero, satisfying the constraint for reach 8, where it must be zero (Fig. 9.10). Similarly, it is important to see that most of the properties in the wastewater streams are almost at the upper limits of the environmental constraints. Nevertheless, the composition requires a larger reduction to meet the constraints imposed on reach 8 to satisfy the quality of the water extracted for human activities. Furthermore, if the implemented model by Lira-Barragán et al. (2011b) is used to solve this example (i.e., if the properties are not considered), the optimal solution to install the new industrial plant is the alternative 1 obtaining a total annualized cost of \$10.214 × 10^6/year. However, in that solution, some of the properties have values greater than the maximum allowed yielding not sustainable solutions.

Table 9.6 Installation costs for Example 9.2

Site	Annualized installation cost, $/year
1	8,000,000
2	8,500,000
3	9,000,000
4	10,500,000
5	9,500,000
6	9,000,000
7	15,500,000
8	14,500,000
9	17,700,000
10	22,000,000
11	18,000,000
12	17,500,000
13	24,000,000
14	28,000,000
15	23,500,000
16	17,000,000
17	20,000,000
18	19,000,000
19	16,000,000
20	17,000,000

Ultimately, Table 9.10 presents the computation time for the problems addressed in this chapter, using a computer with an i5 processor at 2.3 GHz with 4 GB of RAM.

9.5 Concluding Remarks

This chapter presented a water integration approach that considers both the sustainability of the surrounding watershed and the integration within the plant when choosing a location for a new industrial complex. To track, manipulate, and assign properties and compositions while considering water integration possibilities and the impact on the surrounding watershed, a material flow analysis is devised and combined with an optimization formulation. The model also considers the combination of the industrial facility's wastewater outflow with other regional discharges that affect the watershed (e.g., evaporation, filtration, household, agricultural, and

Table 9.7 Fresh and process sources for Example 9.2

Sources	Flowrate, m^3/s	Composition, ppm	Toxicity, %	COD, mgO_2/lt	Odor	Color	Density, g/ml
Process							
1	0.363	64.6	0.11	149	6.3	250	0.94
2	0.241	58	0.04	88	7.7	234	2.41
3	0.046	45.3	0	108	5.1	298	1.14
4	0.447	60	0.13	185	4.8	202	1.20
5	0.196	52	0.21	120	7.6	194	2.70
6	0.204	62	0.25	76	8.2	164	1.64
7	0.281	54.5	0	243	3.7	179	1.85
8	0.329	47.8	0.13	164	7.4	346	1.75
9	0.282	46	0.09	297	4.8	264	2.45
10	0.266	82.1	0.7	326	8.5	394	0.89
11	0.215	26.4	0.3	146	5.7	195	2.15
Fresh							
1	–	0	–	–	–	–	2.3
2	–	1	–	–	–	–	2.45
3	–	2	–	–	–	–	1.8

Table 9.8 Requirements for the process sinks for Example 9.2

Process sink	Flowrate, m^3/s	$Composition^{max}$	$Density^{min}$	$Density^{max}$
1	0.348	16	2.2	2.4
2	0.428	19.4	1.95	2.1
3	0.406	23.3	2.35	2.55
4	0.312	28.5	1.7	1.9
5	0.336	18	2.2	2.3
6	0.294	14	1.8	2
7	0.196	22.5	2.45	2.55

other industrial discharges). The natural degradation of pollutants within the watershed and their impact on the properties that are restricted in the system are taken into consideration by the phenomena that occur within it. The optimization formulation considers both environmental and economic factors. Two case studies, one from

Table 9.9 Interceptors for Example 9.2

Interceptor	Adjustment factor	Variable cost, $/m^3
Composition		
Rec^1	0.11	0.026
Rec^2	0.23	0.021
Toxicity		
Tox^1	0.00	0.054
Tox^2	0.08	0.049
COD		
Aer^1	0.18	0.022
Aer^2	0.43	0.018
Odor		
Ads^1	0.14	0.013
Ads^2	0.39	0.011
Color		
Dec^1	0.10	0.019
Dec^2	0.20	0.015

Egypt and the other from Mexico, have been used to apply the presented model. The findings indicate that the conventional approaches for integrating water, which disregard the surrounding watershed, would not be adequate to guarantee the watershed's survival. The interaction of the wastewater streams released into the environment with other outputs and uses through the watershed, as well as the physical, chemical, and biological phenomena taking place in the watershed, must be considered to meet sustainability standards. Finally, a few property requirements might not be met if one looks solely at the composition.

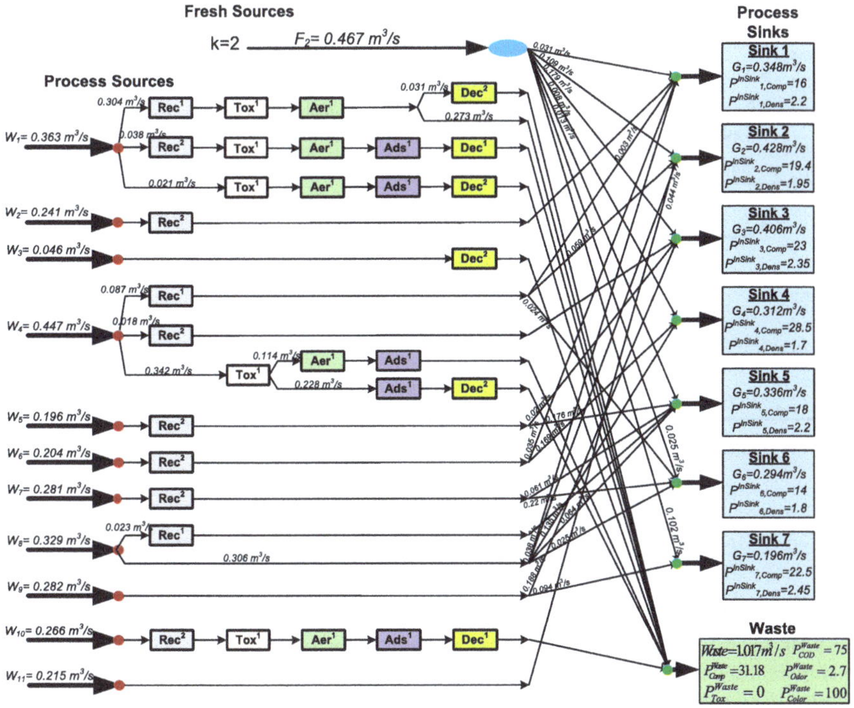

Fig. 9.9 Mass integration for Example 9.2. Printed with permission of "*Incorporating property-based water networks and surrounding watersheds in site selection of industrial facilities*", *Industrial and Engineering Chemistry Research*, Copyright (2012), American Chemical Society

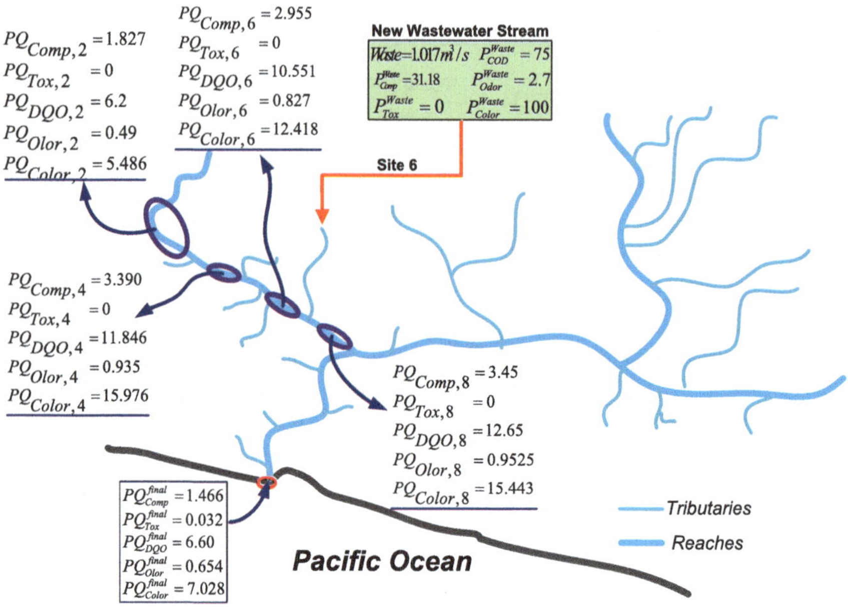

Fig. 9.10 Impact of the new wastewater stream over Balsas system. Printed with permission of "*Incorporating property-based water networks and surrounding watersheds in site selection of industrial facilities*", *Industrial and Engineering Chemistry Research*, Copyright (2012), American Chemical Society

Table 9.10 Size and computation time for the addressed problems

Concept	Example 9.1	Example 2
Number of constraints	1415	9527
Number of continuous variables	5030	30,876
Number of binary variables	4	20
CPU time (s)	4.36	19.96

Printed with permission of "*Incorporating property-based water networks and surrounding watersheds in site selection of industrial facilities*", *Industrial and Engineering Chemistry Research*, Copyright (2012), American Chemical Society

Nomenclatures

Indexes

Direct	Direct
Ind	Industrial
i	Process source
int	Interceptor

In	Inlet
j	Sink
k	Fresh source
l	Site to locate the new industrial facility
$m(r)$	Specific reaches that require property constraints
max	Maximum
min	Minimum
N	Number of properties to be treated
Out	Out
PNEW	New industrial plant
Pr *ec*	Precipitation
p	Property
Resid	Residential
r	Reach
Sink	Sink
Source	Source
t	Tributary
Treat	Treated
Trib	Tributary
Untreat	Untreated
Waste	Waste discharged to the environment

Sets

I	Set for the process sources, $\{i \mid i = 1, \ldots, N_i\}$
INT^N	Set for property interceptor N, $\{\text{int} \mid \text{int} = 1, \ldots, N_{\text{int}^N}\}$
J	Set for the sinks, $\{j \mid j = 1, \ldots, N_j\}$
K	Set for the fresh sources, $\{k \mid k = 1, \ldots, N_k\}$
$M(R)$	Subset for the specific reaches that require properties constraints
L	Set for the sites to locate the new plant, $\{l \mid l = 1, \ldots, N_l\}$
P	Set for the properties, $\{p \mid p = 1, \ldots, N_p\}$
R	Set for the reaches, $\{r \mid r = 1, \ldots, N_r\}$
T	Set for the tributaries, $\{t \mid t = 1, \ldots, N_t\}$

Parameters

$A_{r,t}$	Area cover by effluent t in reach r, acre or ha
$Cland_p$	Annualized installation cost for the new plant in site p, \$/year
$D_{r,t}$	Agricultural discharges from tributary t to reach r, m^3/s
D_r	Direct discharges to the reach r, m^3/s
$EqPipC$	Piping cost to send process sources to process sinks, \$/m^3

$FixC_{cap}^{int^N}$	Fixed part of the capital cost for interceptor int^N, \$/year
$FreC_k$	Unit cost for fresh utility k, \$/m^3
$FrPipC$	Piping cost to send fresh sources to process sinks, \$/m^3
G_j	Total flowrate for process sink j, m^3/s
H_Y	Operation time per year, h/year
H_r	Total discharge (i.e., industrial + sanitary) to the reach r, m^3/s
$I_{r,t}$	Industrial discharge from the tributary t to the reach r, m^3/s
k_f	Factor used to annualize the capital costs
k_p	Kinetic constant for the degradation for property p
$L_{r,t}$	Total losses (filtration and evaporation) from tributary t of the reach r, m^3/s
L_r	Total losses (filtration and evaporation) from the reach r, m^3/s
N_i	Total number of process sources
N_{int^N}	Total number of interceptors
N_j	Total number of sinks
N_k	Total number of fresh sources
N_l	Total number of sites allowable to locate the new plant
N_p	Total number of properties
N_r	Total number of reaches
N_t	Total number of tributaries
$P_{r,t}$	Precipitation discharged for the tributary t to the reach r, m^3/s
P_r	Precipitation discharged to the reach r, m^3/s
$P_{p,m(r)}^{Desired}$	Desired properties in some reaches
$P_p^{Waste,\max}$	Environmental regulation for the property to be discharged by the new plant
$P_p^{Sustainable}$	Sustainability constraints at the final disposal
$S_{r,t}^{untreated}$	Residual wastewater discharged without treatment to the reach r for tributary t, m^3/s
$S_{r,t}^{treated}$	Residual treated wastewater discharged to the reach r for tributary t, m^3/s
$TrPipC$	Treatment piping cost, \$/m^3
$U_{r,t}$	Water used from tributary t discharged to reach r, m^3/s
U_r	Water used from reach r, m^3/s
$VarC_{cap}^{int^N}$	Variable part for the capital cost for interceptor int^N, \$/m^3
$VarC_{op}^{int^N}$	Variable cost associated to operation of the interceptor int^N, \$/m^3
$V_{r,t}$	Volume for tributary t from reach r, m^3
V_r	Volume for reach r, m^3
W_i	Total flowrate for the process sources i, m^3/s
$WaPipC$	Waste piping cost, \$/m^3
$\alpha_{r,t}$	Agricultural flowrate per area, m^3/ha * s
$\beta_{r,t}$	Agricultural use of water from tributary t, m^3/ha s
$\gamma_p^{int^1,\ldots,int^N}$	Efficiency factor to improve the property p for the interceptor int^N
ρ	Density
σ_p	Reaction order for property p

ψ_p Property operator for the mixing rule for property p

Variables

F_k	Total flowrate for fresh source k, m^3/s
$f_{k,j}$	Segregated flowrate from fresh source k to sink j, m^3/s
$FT_{r,t}$	Flowrate discharged by the tributary t to the reach r, m^3/s
$g_{i,j}^{int^1,\ldots,int^N}$	Segregated flowrate from interceptors int^N to process sink j for process source i, m^3/s
$g_{i,Waste}^{int^1,\ldots,int^N}$	Segregated flowrate from interceptors int^N to waste for process source i, m^3/s
P_p^{disc}	Property for the final discharge of the new plant to the river
$P_{p,r(l)}^{PNEW}$	Property of the waste stream, to be discharged for the new plant installed on site l
$P_{p,final}^{Reach}$	Property discharged to the final disposal
Q_r	Flowrate exit from the reach r, m^3/s
Q_{r-1}	Flowrate inlet to the reach r, m^3/s
$QPNEW_l$	Flowrate discharged for the new plant for the location p
$r_{p,r}$	Reaction carried out in the reach r in terms of property p
$r_{p,r,t}$	Reaction carried out in the tributary t that discharges to the reach r in terms of property p
TAC	Total annual cost, \$/year
$w_i^{int^1,\ldots,int^N}$	Segregated flowrate from process source i to interceptor int^N, m^3/s
$Waste$	Total flowrate for the waste stream discharged to the environment, m^3/s
Y_l	Boolean variable for the location of the new plant
y_l	Binary variable for the location of the new plant

References

Alva-Argaez A, Vallianatos A, Kokossis A (1999) A multi-contaminant transshipment model for mass exchange networks and wastewater minimization problems. Comput Chem Eng 23(10):1439–1453

Alva-Argaez A, Kokossis AC, Smith R (2003) Wastewater minimization of industrial systems using an integrated approach. Comput Chem Eng 22(Suppl):S741–S744

Baccini P, Brunner P (1991) Metabolism of the anthroposphere. Springer, Berlin

Benko N, Rev E, Fonyo Z (2000) The use of nonlinear programming to optimal water allocation. Chem Eng Commun 178:67–101

Brooke A, Kendrick D, Meeruas A, Raman R (2011) GAMS-Language guide. GAMS Development Corporation, Washington, D.C.

Brunner PH, Rechberg H (2004) Practical handbook of material flow analysis. CRC, Boca Raton, Florida
CONAGUA (2010a) Mexican National Water Commission. Water statistics. http://www.conagua.gob.mx/OCB07/Contenido/Documentos/EstadisticasBALSAS.pdf
CONAGUA (2010b) Mexican National Water Commission. Water statistics in Mexico edition
Deng C, Feng X (2011) Targeting for conventional and property-based water network with multiple resources. Ind Eng Chem Res 50(7):3722–3737
Dhole VR, Ramchandani N, Tainsh RA, Wasilewski M (1996) Make your process water pay for itself. Chem Eng 103(1):100–103
Doyle SJ, Smith R (1997) Targeting water reuse with multiple contaminants. Process Saf Environ Prot 75(B3):181–189
El-Baz AA, Ewida KT, Shouman MA, El-Halwagi MM (2005a) Material flow analysis and integration of watersheds and drainage systems: I. Simulation and application to ammonium management in Bahr El-Baqar drainage system. Clean Technol Environ Policy 7(1):51–61
El-Baz AA, Ewida KT, Shouman MA, El-Halwagi MM (2005b) Material flow analysis and integration of watersheds and drain systems: II. Integration and solution strategies with application to ammonium management in Bahr El-Baqar drain system. Clean Technol Environ Policy 7(1):78–86
El-Halwagi MM, Spriggs HD (1998) Solve design puzzles with mass integration. Chem Eng Prog 94(8):25–44
El-Halwagi MM, Hamad AA, Garrison GW (1996) Synthesis of waste interception and allocation networks. AIChE J 42(11):3087–3101
El-Halwagi MM, Gabriel F, Harell D (2003) Rigorous graphical targeting for resource conservation via material recycle/reuse networks. Ind Eng Chem Res 42(19):4319–4328
El-Halwagi MM, Glasgow IM, Eden MR, Qin X (2004) Property integration: componentless design techniques and visualization tools. AIChE J 50(8):1854–1869
Faria DC, Bagajewicz MJ (2011a) A new approach for global optimization of a class of MINLP problems with applications to water management and pooling problems. AIChE J 58(8):2320–2335. https://doi.org/10.1002/aic.12754
Faria DC, Bagajewicz MJ (2011b) Global optimization based on subspaces elimination: application to generalized pooling and water management problems. AIChE J 58(8):2336–2345. https://doi.org/10.1002/aic.12738
Faria DC, Bagajewicz MJ (2011c) Planning model for the design and/or retrofit of industrial water systems. Ind Eng Chem Res 50(7):3788–3797
Faria DC, Bagajewicz MJ (2011d) Global optimization of water management problems using linear relaxation and bound contraction methods. Ind Eng Chem Res 50(7):3738–3753
Faria DC, Bagajewicz MJ (2011e) Novel bound contraction procedure for global optimization of bilinear MINLP problems with application to water management problems. Comput Chem Eng 35(3):446–455
Feng X, Bai J, Zheng XS (2007) On the use of graphical method to determine the targets of single-contaminant regeneration recycling water systems. Chem Eng Sci 62(8):2127–2138
Foo CY (2009) State-of-the-art review of pinch analysis techniques for water network synthesis. Ind Eng Chem Res 48:5125–5159
Foo DCY, Kazantzi V, El-Halwagi MM, Manan ZA (2006) Surplus diagram and cascade analysis techniques for targeting property-based material reuse network. Chem Eng Sci 61(8):2626–2642
Gabriel FB, El-Halwagi MM (2005) Simultaneous synthesis of waste interception and material reuse networks: problem reformulation for global optimization. Environ Prog 24(2):171–180
Galan B, Grossmann IE (1998) Optimal design of distributed wastewater treatment networks. Ind Eng Chem Res 37(10):4036–4048
Gomes JFS, Queiroz EM, Pessoa FLP (2007) Design procedure for water/wastewater minimization: single contaminant. J Clean Prod 15(5):474–485
Gunaratnam M, Alva-Argaez A, Kokossis A, Kim JK, Smith R (2005) Automated design of total water systems. Ind Eng Chem Res 44(3):588–599

Hallale N (2002) A new graphical targeting method for water minimization. Adv Environ Res 6(3):377–390
Hernandez-Suarez R, Castellanos-Fernandez J, Zamora JM (2004) Superstructure decomposition and parametric optimization approach for the synthesis of distributed wastewater treatment networks. Ind Eng Chem Res 43(9):2175–2191
Karuppiah R, Grossmann IE (2006) Global optimization for the synthesis of integrated water systems in chemical processes. Comput Chem Eng 30(4):650–673
Kheireddine H, Dadmhammadi Y, Deng C, Feng X, El-Halwagi MM (2011) Optimization of direct recycle networks with the simultaneous consideration of property, mass, and thermal effects. Ind Eng Chem Res 50(7):3754–3762
Kuo WCJ, Smith R (1997) Effluent treatment system design. Chem Eng Sci 52(23):4273–4290
Ku-Pineda V, Tan RR (2006) Environmental performance optimization using process water integration and sustainable process index. J Clean Prod 14(18):1586–1592
Lampert C, Brunner PH (1999) Material accounting as a policy tool for nutrient management in the Danube Basin. Water Sci Technol 40(10):43–49
Lim SR, Park JM (2007) Environmental and economic analysis of a water networks system using LCA and LCC. AIChE J 53(12):3253–3262
Lim SR, Park JM (2008) Cooperative water networks system to reduce carbon footprints. Environ Sci Technol 42(16):6230–6236
Lim SR, Park JM (2008b) Synthesis of an environmentally friendly water network system. Ind Eng Chem Res 47(6):1988–1994
Lira-Barragán LF, Ponce-Ortega JM, Serna-González M, El-Halwagi MM (2011a) An MINLP model for the optimal location of a new industrial plant with simultaneous consideration of economic and environmental criteria. Ind Eng Chem Res 50(2):953–964
Lira-Barragán LF, Ponce-Ortega JM, Serna-González M, El-Halwagi MM (2011b) Synthesis of water networks considering the sustainability of the surrounding watershed. Comput Chem Eng 35(12):2837–2852
Lira-Barragán LF, Ponce-Ortega JM, Nápoles-Rivera F, Serna-González M, El-Halwagi MM (2012) Ind Eng Chem Res 52:91–107
Lovelady EM, El-Baz AA, El-Monayeri D, El-Halwagi MM (2009) Reverse problem formulation for integrated process discharges with wastewater and drainage systems: managing phosphorus in lake Manzala. J Ind Ecol 13(6):914–927
Manan ZA, Tan YL, Foo DCY (2004) Targeting the minimum water rate using water cascade analysis technique. AIChE J 50(12):3169–3183
Nápoles-Rivera F, Ponce-Ortega JM, El-Halwagi MM, Jiménez-Gutiérrez A (2010) Global optimization of mass and property integration networks with in-plant property interceptors. Chem Eng Sci 65(15):4363–4377
Nápoles-Rivera F, Ponce-Ortega JM, El-Halwagi MM, Jiménez-Gutiérrez A (2011) Global optimization of recycle-reuse mass integration networks for processes with multiples contaminants. Environ Prog Sustain Energ 31(3):449–458
Ng DKS, Foo DCY, Rabie A, El-Halwagi MM (2008) Simultaneous synthesis of property-based water reuse/recycle and interception networks for batch processes. AIChE J 54(10):2634–2632
Polley GT, Polley HL (2000) Design better water networks. Chem Eng Prog 96(2):47–52
Ponce-Ortega JM, Nápoles-Rivera F, El-Halwagi MM, Jiménez-Gutiérrez A (2008) An optimization approach for the synthesis of recycle and reuse water integration networks. Clean Technol Environ Policy 14(1):133–151
Ponce-Ortega JM, Serna-González M, Jiménez-Gutiérrez A (2009a) A disjunctive programming model for simultaneous synthesis and detailed design of cooling networks. Ind Eng Chem Res 48(6):2991–3003
Ponce-Ortega JM, Hortua AC, El-Halwagi MM, Jiménez-Gutiérrez A (2009b) A property-based optimization of direct recycle networks and wastewater treatment processes. AIChE J 55(9):2329–2344

Ponce-Ortega JM, El-Halwagi MM, Jiménez-Gutiérrez A (2010) Global optimization of property-based recycle and reuse networks including environmental constraints. Comput Chem Eng 34(3):318–330
Ponce-Ortega JM, Mosqueda-Jiménez FW, Serna-González M, Jiménez-Gutiérrez A, El-Halwagi A (2011) A property-based approach to the synthesis of material conservation networks with economic and environmental objectives. AIChE J 57(9):2369–2387
Putra ZA, Amminudin KA (2008) Two-step optimization approach for design a total water system. Ind Eng Chem Res 47(16):6045–6057
Raman R, Grossmann IE (1994) Modeling and computational techniques for logic based integer programming. Comput Chem Eng 18(7):563–578
Savelski MJ, Bagajewicz MJ (2000) On the optimality conditions of water utilization systems in process plants with single contaminants. Chem Eng Sci 55(21):5035–5048
Savelski MJ, Bagajewicz MJ (2001) Algorithmic procedure to design water utilization systems featuring a single contaminant in process plants. Chem Eng Sci 56(5):1897–1911
Savelski M, Bagajewicz M (2003) On the necessary conditions of optimality of water utilization systems in process plants with multiple contaminants. Chem Eng Sci 58(23–24):5349–5362
Shelley MD, El-Halwagi MM (2000) Componentless design of recovery and allocation systems: a functionality-based clustering approach. Comput Chem Eng 24(9–10):2081–2091
Sorin M, Bedard S (1999) The global pinch point in water reuse networks. Process Saf Environ Prot 77(B5):305–308
Takama N, Kuriyama T, Shiroko K, Umeda T (1980) Optimal water allocation in a petroleum refinery. Comput Chem Eng 4(4):251–258
Tan RR, Foo DCY, Ng DKS, Chiang CL, Hul S, Ku-Pineda V (2007) An approximated mixed integer linear programming (MILP) model for the design of water reuse/recycle networks with minimum energy. Asia-Pac J Chem Eng 42:566–574
Teles J, Castro PM, Novals AQ (2008) LP-based solution strategies for the optimal design of industrial water networks with multiple contaminants. Chem Eng Sci 63(2):376–394
Viswanathan J, Grossmann IE (1990) A Combined penalty function and outer approximation method for MINLP optimization. Comput Chem Eng 14(7):769–782
Wang YP, Smith R (1994) Wastewater minimization. Chem Eng Sci 49(7):981–1006

Chapter 10
Development of Eco-industrial Parks in Interaction with Adjacent Watershed

10.1 Introduction

Considerable amounts of wastewater are discharged from industrial facilities, and tremendous amounts of freshwater are consumed in industrial plants. The discharged effluents are usually loaded with several pollutants and may lead to a major impact on the surrounding watersheds. An effective strategy for reducing water usage and discharge for industrial facilities is to synthesize recycle and reuse water networks within the industrial facilities (El-Halwagi et al. 2003; Hortua and El-Halwagi 2013; Foo 2009). The synthesis of water networks has been extended from intra-plant integration to inter-plant integration using the concept of eco-industrial parks (EIP). EIP consider industrial symbiosis to integrate various forms of materials and energy as part of the emerging field of industrial ecology (Noureldin and El-Halwagi 2015; Acobsen 2006; Lombardi and Laybourn 2012; Chertow 2000). In water integration in EIP, neighboring industries can exchange resources (for example water streams) and common infrastructure (for example treatment units) to decrease the consumption of freshwater and the wastewater discharges to the environment (Ehrenfeld and Gertler 1997; Lamberta and Boonsb 2002; Yu et al. 2015; Wang et al. 2014). Figure 10.1 presents a schematic representation of an EIP constituted of several industrial facilities where it is possible to recycle wastewater streams to the same plant or other plants. Furthermore, a central treatment facility can be used to receive and treat wastewater streams to enable recycling to the participating plants. Subsequently, water integration is improved in the inter-plant integration compared to the single-plant integration. This integration decreases the overall need for freshwater and the total wastewater discharged to the environment. Moreover, the inter-plant integration allows for decreasing the capital and operating costs of the possibility of using a shared treatment facility and/or distributed treatment units in each plant.

Numerous approaches for synthesizing inter-plant water networks based on heuristic rules have been presented. This way, Foo (2008) included threshold problems, Bandyopadhyay et al. (2010) implemented a decomposition algorithm, and

J. M. Ponce-Ortega et al., *Sustainable Water Systems*, Water Resources Development and Management, https://doi.org/10.1007/978-3-031-72498-5_10

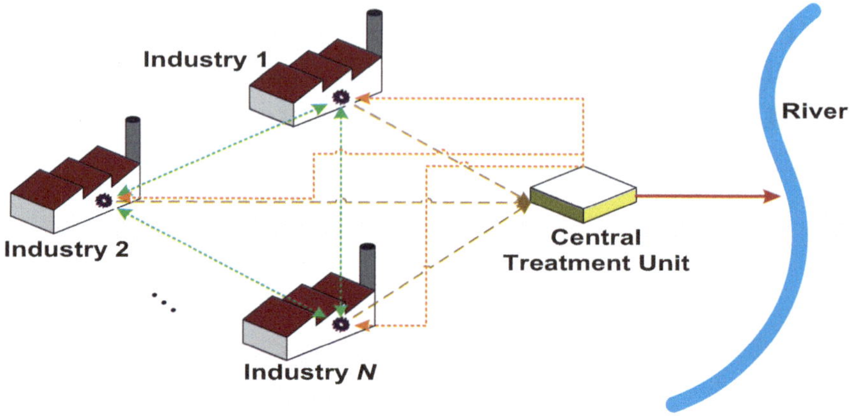

Fig. 10.1 General configuration for an eco-industrial park. Printed with permission of "*Synthesis of eco-industrial parks interacting with a surrounding watershed*", *ACS Sustainable Chemistry and Engineering Research*, Copyright (2015), American Chemical Society

Chew et al. (2010) addressed batch processes. To improve the results of heuristic approaches for synthesizing inter-plant water networks, mathematical programming approaches have been employed, including targeting approaches (2009), direct recycle networks (Chew et al. 2008; Chew and Foo 2009), involving multiple interceptors or treatment units (Lim and Park 2010; Rubio-Castro et al. 2010, 2011; Chew et al. 2011; Tan et al. 2011a, b; Alnouri et al. 2014), and incorporating retrofitting options (Rubio-Castro et al. 2012), similarly property-based constraints have been involved (Lovelady et al. 2009a, b; Rubio-Castro et al. 2013). Additionally, different flexibility opportunities (Liao et al. 2007; Cheng-Liang et al. 2011; Montastruc et al. 2013; Bishnu et al. 2014), multi-objective optimization methods (Boix et al. 2012), and fuzzy mathematical programming approaches (Aviso et al. 2010, 2011; Chen et al. 2010; Tan et al. 2011a, b; Tashkhiri et al. 2011) have been reported, and such approaches have been implemented in different types of industries (Lovelady et al. 2007; Noureldin and El-Halwagi 2015; Boix et al. 2015). In addition to the integration of water usage and discharge among several plants, it is important to track the impact of these discharges on the surrounding environment. Numerous approaches have included material flow analysis (MFA) to involve the impact of water discharges on the surroundings (El-Baz et al. 2005; Lira-Barragán et al. 2011a, b, 2013; Burgara-Montero et al. 2012, 2013a, b; Martinez-Gómez et al. 2013; López-Villarreal et al. 2014).

Despite the valuable contributions made by the previous approaches for synthesizing and retrofitting inter-plant water networks, they have at least one of the following weaknesses:

- Just a single eco-EIP has been considered. Since the industrial facilities may be clustered into several groups based on their proximity to a certain geographical location, the installation of multiple EIP can present improved results.

- The effects of the wastewater discharges on the surrounding watershed have not been considered.
- The optimal location of the retrofitted eco-industrial plant has not been included.
- Just economic objectives were considered without considering the environmental objectives incorporating the implications of the discharged wastewater on the surrounding watershed.
- The interaction among the various discharges has not been considered.
- The impact of physical, chemical, and biological natural phenomena that occur in the watershed has not been determined.

To overcome these limitations, this chapter presents a multi-objective optimization approach for designing a set of EIP for water integration considering the industries located in different clusters and the interaction with the surrounding watershed (López-Díaz et al. 2015). The objectives are to minimize the total annualized cost and the environmental impact on the surrounding watershed while allowing retrofitting and recycling strategies while satisfying process and environmental constraints. The presented optimization formulation is based on a superstructure that implies the optimal selection of the industries and location for installing the eco-industrial parks. The formulation similarly accounts for the collective impact on water recycling, reusing, and discharging in the surrounding watershed.

10.2 Problem Statement

The referred problem is described as follows: Given is a set of industrial facilities that are installed surrounding a watershed. These industrial plants can be grouped into a set of clusters: $\text{CLUSTER} = \{p \mid p = 1, 2, \ldots, N_{\text{Clusters}}\}$. Each cluster p corresponds to a group of industrial plants that are in the same geographical place and can be integrated into an EIP. The industrial plants belonging to each group p offer several process sources (effluents) that are discharged into the environment. Each source, i, has a known flowrate and composition of key pollutants c. Given is similarly a set of fresh sources of water. Each fresh source w is available to be sent directly to the process sinks as well as a set of available interception technologies in place p to remove the considered pollutants. For each treatment technology, *int*, the fixed and operating costs, and the pollutant-removal efficiencies are known. The treatment wastewater network allows for reusing the process sources in a set of process sinks located in the same place. For each sink, j, the flowrate and upper limits for the inlet composition are known. The watershed receiving the industrial effluents can receive other discharges (e.g., agricultural, residential, pluvial, etc.). For modeling the watershed, this is split into a set of reaches r and tributaries t. The existing conditions for the watershed are given, including the flowrates and compositions for reaches and tributaries and all the effluents and natural phenomena interacting with the watershed. Lastly, the unit costs for pipelines are given.

The problem is intended to determine the optimal selection of industries to yield EIP and their corresponding configurations and determine the effects of the wastewater discharges in the surrounding watershed whereas considering as objective functions the minimization of the total annualized costs (*TAC*) and the minimization of the environmental impact on the surrounding watershed. The *TAC* is composed of freshwater costs, treatment costs, and piping costs. Furthermore, to get a decrease in the pollutant concentrations discharged to the final disposal, the model of this chapter considers designing EIP interacting with the watershed and thus, this process will reduce the industrial effluents coming from these industrial plants. To implement this task, the presented superstructure for water integration is shown in Fig. 10.2, where each process source can be sent to the process sinks in the same plant or even to other industries located in the same place, to each interceptor of the first column in the central treatment unit or lastly to the wastewater stream. The fresh sources are just sent to the process sinks. Once the treatment is completed, the outlet streams are sent to the process sinks of all the industrial facilities and they can be sent to the discharge stream. Figure 10.2 presents an example of a recycle/reuse superstructure for two industrial plants, where each facility has two process sources and two process sinks, fresh water, and three interception units for treating one pollutant (for each component considered, there is a column of interceptors).

The model of this chapter uses the MFA technique to represent the watershed and to track the pollutants during their path in the watershed until reach the final disposal (Fig. 10.3). The MFA model requires the segregation of the main river into several sections that are called reaches. The main branches or channels that feed to the main waterway (e.g., river) are called tributaries. Usually, a watershed interacts with several users placed in the surrounding areas such as industrial, agricultural, and residential areas. Several natural phenomena change the pollutant compositions such as evaporation, filtration, and the natural degradation of the compounds through biochemical reactions (Lovelady et al. 2009a, b). In addition to constraints throughout the watershed, it is also important to limit the maximum amount of pollutants that are discharged into the final disposal of the watershed (e.g., outfall or catchment area).

10.3 Mathematical Model Formulation

Before presenting the proposed approach, the subscripts, superscripts, and sets used in the mathematical model formulation are defined. The index r refers to a reach, t is a tributary, and p is the group of industrial plants in the same geographical location. The industries in group p may be integrated within the same EIP to be in that geographical area. Hence, the sources and sinks that can be integrated are the ones located in the same place p and this way there can be $N_{Clusters}$ EIPs. The index c corresponds to a component or pollutant, i denotes a process source, j is the process sinks, w corresponds to the type of fresh water and int indicates an interceptor. The superscripts in, out, m, and max are used to indicate inlet, outlet, and upper limit, respectively. Lastly, the sets R, T, P, and C are utilized to indicate reaches, tributaries,

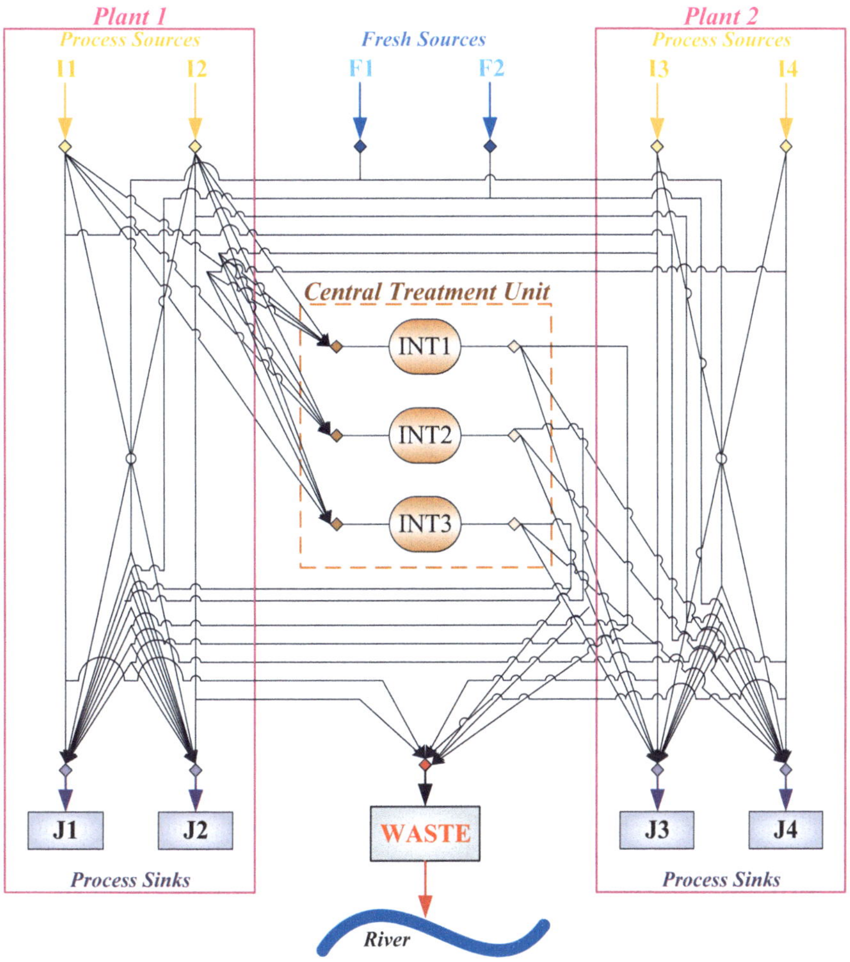

Fig. 10.2 Superstructure for the proposed water integration in each eco-industrial park. Printed with permission of "*Synthesis of eco-industrial parks interacting with a surrounding watershed*", *ACS Sustainable Chemistry and Engineering Research*, Copyright (2015), American Chemical Society

possible industrial parks available to be reconfigured, and compounds, respectively; whereas the sets for process sources, process sinks, and the types of fresh waters are denoted by I, J, and W, respectively, and INT indicates the interceptors.

10.3.1 Material Flow Analysis Approach

The material flow analysis (MFA) approach is utilized to track the quality of the water in the watershed previously and after the retrofitting modifications. The MFA

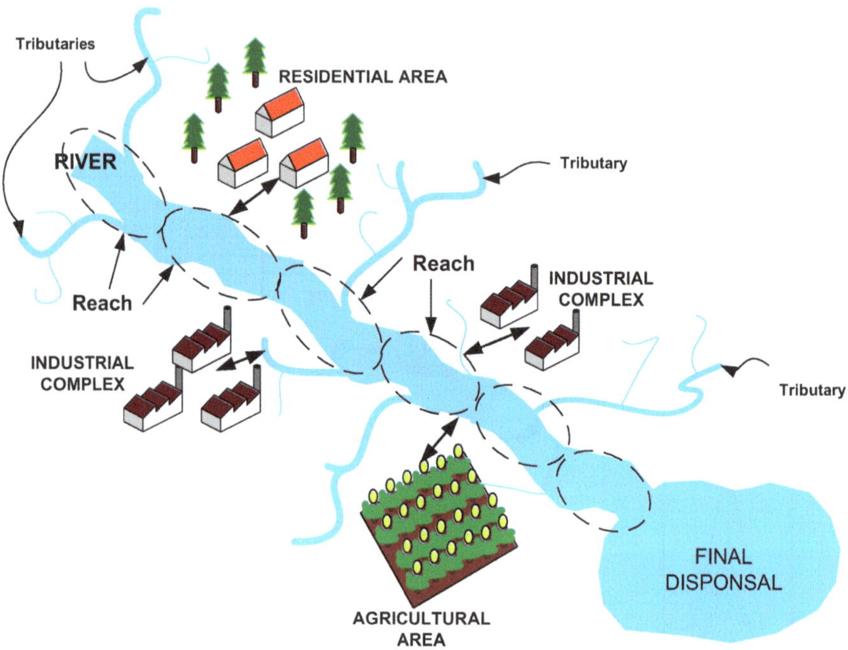

Fig. 10.3 A watershed system interacting with several users. Printed with permission of "*Synthesis of eco-industrial parks interacting with a surrounding watershed*", *ACS Sustainable Chemistry and Engineering Research*, Copyright (2015), American Chemical Society

technique involves chemical and biochemical interactions of the natural degradation that is caused by the flora and fauna of the watersheds. Figure 10.3 shows that the MFA model makes a subdivision of the watershed in reaches and tributaries. Therefore, the mathematical relationships needed by this approach correspond to water and pollutant balances in each section of the system and these are described in the next sections.

10.3.1.1 Overall Balance for Each Reach

The flowrate leaving reach r (Q_r) should be equal to the flowrate entering in the same reach r (Q_{r-1}), plus precipitation (P_r), direct industrial discharges (D_r), residential discharges (H_r), the sum of all effluents entering the reach ($FT_{r,t}$), industrial discharges for the industries belonging to the industrial park available to be reconfigured in an eco-industrial park p associated to the reach r ($IP_{p(r)}$), and the eco-industrial discharges from park p discharging at reach r after the reconfiguration ($EP_{p(r)}$), minus the terms of losses (owing to filtration and evaporation) (L_r) and uses (U_r) in the r section of the river, which is stated as follows:

$$Q_r = Q_{r-1} + P_r + D_r + H_r + \sum_{t=1}^{N_{t(r)}} FT_{r,t} + IP_{p(r)} + EP_{p(r)} - L_r - U_r, \quad \forall r \in R \tag{10.1}$$

$N_{t(r)}$ indicates the total number of tributaries that are discharged to the reach r. In the previous relationship, the flowrate leaving the reach r-1 is the same as the flowrate at the inlet for the reach r, and it requires making an association between each industrial park available to be reconfigured as an eco-industrial park and the reach where the park discharges its waste stream.

10.3.1.2 Component Balance for Each Reach

For the component balance in the reaches, each term of the previous equation is multiplied by the pollutant concentrations to obtain a balance in terms of mass instead of flowrate and incorporating the reactive term ($\int_{V=0}^{V_r} r_{c,r} dV$), which involves the chemical reactions that are carried out in that section of the river; generally, the chemical reactions decompose the components in the system. Thus, the component balance for each reach can be written as follows:

$$\begin{aligned} Q_r CQ_{c,r} = {} & Q_{r-1} CQ_{c,r-1} + P_r CP_{c,r} + D_r CD_{c,r} + H_r CH_{c,r} + \sum_{t=1}^{N_{t(r)}} FT_{r,t} CT_{c,r,t} \\ & + IP_{p(r)} CIP_{c,p(r)} + EP_{p(r)} CEP_{c,p(r)} - L_r CL_{c,r} - U_r CU_{c,r} \\ & - \int_{V=0}^{V_r} r_{c,r} dV_r, \quad \forall c \in C, \; r \in R \end{aligned} \tag{10.2}$$

10.3.1.3 Overall Balance for Each Tributary

According to the MFA approach, each reach can receive flowrates from various tributaries; these tributaries are channels or river branches that drag the chemical compounds from different polluted sources to the reaches or the central river. Next, the total flowrate discharged from the tributary t to the reach r ($FT_{r,t}$) is equal to the sum of the following types of discharges: residual discharges without treatment ($S_{r,t}^{untreated}$), with treatment ($S_{r,t}^{treated}$), industrial discharges ($I_{r,t}$), pluvial discharges ($P_{r,t}$), direct discharges ($D_{r,t}$), discharges coming from industries inside the industrial parks selected to be readjusted as an eco-industrial park p discharging at tributary t of the reach r ($IP_{p(r)}$) and the wastewater stream of the eco-industrial park p discharged at tributary t of the reach r ($EP_{p(r)}$), minus the losses ($L_{r,t}$) and use or extraction ($U_{r,t}$) of water. Therefore, this balance is established as follows:

$$FT_{r,t} = S_{r,t}^{untreated} + S_{r,t}^{treated} + I_{r,t} + P_{r,t} + D_{r,t} + IP_{p(r,t)} + EP_{p(r,t)} - L_{r,t} - U_{r,t}, \quad \forall r \in R, \ t \in T \tag{10.3}$$

10.3.1.4 Component Balance for Each Tributary

Likewise in the case of the reaches, the following relationship is obtained by multiplying each term of the previous equation by the pollutant concentrations and it is considered the chemical reaction term:

$$\begin{aligned} FT_{r,t}CT_{c,r,t} = {} & S_{r,t}^{untreated}CS_{c,r,t}^{untreated} + S_{r,t}^{treated}CS_{c,r,t}^{treated} + I_{r,t}CI_{c,r,t} \\ & + P_{r,t}CP_{c,r,t} + D_{r,t}CD_{c,r,t} + IP_{p(r,t)}CIP_{c,p(r,t)} \\ & + EP_{p(r,t)}CEP_{c,p(r,t)} - L_{r,t}CL_{c,r,t} - U_{r,t}CU_{c,r,t} \\ & - \int_{V=0}^{V_{r,t}} r_{c,r,t}dV_{r,t}, \quad \forall c \in C, \ r \in R, \ t \in T \end{aligned} \tag{10.4}$$

To determine the actual conditions (before the implementation of the proposed approach), in the previous equations the values of $IP_{p(r)}$ and $CIP_{p(r)}$ associated with each industry available for the retrofitting in the eco-industrial parks have values greater than zero and the terms related to the eco-industrial discharges once the reconfigurations are done ($EP_{p(r)}$ and $CEP_{p(r)}$) must be equal to zero; but the roles are inverted after the reconfiguration in eco-industrial parks.

10.3.1.5 Reactive Terms

The terms to model the chemical and biochemical reactions that occur in the watersheds included in Eqs. (10.2) and (10.4) are described as follows:

$$\int_{V=0}^{V_r} r_{c,r}dV_r = k_c\left(CQ_{c,r}\right)^{\sigma_c}V_r, \quad \forall c \in C, \ r \in R \tag{10.5}$$

$$\int_{V=0}^{V_{r,t}} r_{c,r,t}dV_{r,t} = k_c\left(CT_{c,r,t}\right)^{\sigma_c}V_{r,t}, \quad \forall c \in C, \ r \in R, \ t \in T \tag{10.6}$$

k_c represents the kinetic constant for compound c, σ_c represents the reaction order for each component, $CQ_{c,r}$ corresponds to the concentration of the pollutants in the reach r and V_r refers to the volume of the reach; while $CT_{c,r,t}$ corresponds to the concentration for pollutant c in the tributary t that discharges in the reach r and $V_{r,t}$

represents the volume of the tributary. The parameters k_c, σ_c, V_r and $V_{r,t}$ should be experimentally determined.

10.3.1.6 Agricultural Discharges and Uses

The agricultural sector demands huge amounts of water for its proper operation. Water is taken from watersheds. Likewise, a portion of this water is given back to the watershed mainly through filtration. Therefore, this chapter considers a simplified way to determine the agricultural discharges and uses, which is represented as follows:

$$D_{r,t} = \alpha_{r,t} A_{r,t}, \quad \forall r \in R, \quad t \in T \tag{10.7}$$

$$U_{r,t} = \beta_{r,t} A_{r,t}, \quad \forall r \in R, \quad t \in T \tag{10.8}$$

$\alpha_{r,t}$ represents a known parameter for the water required per area cultivated and its units are m^3/acre s, $\beta_{r,t}$ represents the discharged flowrate per area cultivated and its units also are m^3/acre s. $A_{r,t}$ corresponds to the cultivated area related to the tributary t that discharges to the reach r and its units are m^3.

10.3.2 *Reconfiguration to Eco-industrial Parks*

The proposed approach to synthesizing EIP that integrates water networks of grouped industrial plants is based on the superstructure presented in Fig. 10.2. Each process source in each plant can be sent directly to the process sinks, to the central treatment unit, or to the wastewater stream; whereas the fresh water only can be received by the process sinks, and finally the streams at the exit of the treatment are segregated to be directed to the process sinks and to the wastewater stream. Each treatment stage only can treat a single pollutant and the implemented approach does not require a special index for each plant inside the industrial park (afterward, the process sources and process sinks are consecutively located, and it is identified previously to the optimization), but it is required the index p for the whole industrial park available to be retrofitted into an eco-industrial park.

10.3.2.1 Whole Balance for Each Process Source

The flowrate of each process source i for each industrial park available for the retrofitting p ($FS_{i,p}$) is split and directed to the process sinks ($fss_{i,j,p}$), to the first treatment stage ($fsi_{i,\text{int}=1,p}$) and to the waste stream ($fse_{i,p}$). This is stated as follows:

$$FS_{i,p} = \sum_{j=1}^{J} fss_{i,j,p} + \sum_{\text{int}=1}^{INT} fsi_{i,\text{int}=1,p} + fse_{i,p}, \quad \forall i \in I, \quad p \in P \tag{10.9}$$

10.3.2.2 Whole Balance for Each Process Sink

The flowrate at the inlet of each process sink j in each industrial park p ($FU_{j,p}$) is given and it must be provided by the flowrates coming from process sources ($fss_{i,j,p}$), the exit of the interception network ($fis_{\text{int=NT},j,p}$) and fresh water ($fws_{w,j,p}$), which is stated as follows:

$$FU_{j,p} = \sum_{i=1}^{I} fss_{i,j,p} + \sum_{\text{int}=NT}^{INT} fis_{\text{int}=NT,j,p} + \sum_{w=1}^{W} fws_{w,j,p}, \quad \forall j \in J, \quad p \in P \tag{10.10}$$

10.3.2.3 Component Balance for Each Process Sink

Each term of the previous balance is multiplied by its concentration for each considered pollutant as follows:

$$FU_{j,p} cu_{c,j,p} \geq \sum_{i=1}^{I} fss_{i,j,p} cs_{c,i,p} + \sum_{\text{int}=NT}^{INT} fis_{\text{int}=NT,j,p} ci^{out}_{c,\text{int}=NT,p} + \sum_{w=1}^{W} fws_{w,j,p} cw_{c,w,p}, \quad \forall c \in C, \quad j \in J, \quad p \in P \tag{10.11}$$

$cu_{c,j,p}$ represents the pollutant concentration in each process sink, $cs_{c,i,p}$ corresponds to the component concentration in each process source, $ci^{out}_{c,\text{int}=NT,p}$ is the pollutant concentration at the outlet of the interception network, and lastly the pollutant concentration in each fresh water is represented by $cw_{c,w,p}$.

10.3.2.4 Whole Balance at the Inlet of the Interceptors

The flowrate entering the first stage of the interception network ($FI_{int=1,p}$) is given by the process sources ($fsi_{i,\text{int},p}$):

$$FI_{\text{int},p} = \sum_{i=1}^{I} fsi_{i,\text{int},p}, \quad \forall \text{int} = 1, \quad p \in P \tag{10.12}$$

10.3.2.5 Component Balance at the Inlet of the Interceptors

A component balance at the inlet of the interception network is required to establish the pollutant concentration at this point ($ci^{in}_{c,\mathrm{int},p}$). This relationship is stated as follows:

$$FI_{\mathrm{int},p} ci^{in}_{c,\mathrm{int},p} = \sum_{i=1}^{I} fsi_{i,\mathrm{int},p} cs_{c,i,p}, \quad \forall c \in C, \quad \mathrm{int} \in INT; \quad p \in P \tag{10.13}$$

10.3.2.6 Whole Mass Balances in the Interception Network After Stage 1

According to the presented mathematical model, just a pollutant is treated in each stage of the interception network, afterwards, the number of stages required is the same as the number of pollutants. Therefore, the inlet flowrate to any stage of the interception network, excluding the first one ($FI_{int \neq 1,p}$), is provided by the sum of the outlet flows from the previous stage of the interception network ($fii_{int\,=\,int-1,p}$). This is represented as follows:

$$FI_{\mathrm{int},p} = \sum_{\mathrm{int}=\mathrm{int}-1}^{\mathrm{int}=NT} fii_{\mathrm{int},p}, \quad \forall \mathrm{int} \neq 1, \quad p \in P \tag{10.14}$$

It is noteworthy that the flowrate at the exit of the last stage is segregated and sent to the process sinks ($fis_{int=NT,p}$) and the environment ($fie_{int=NT,p}$):

$$FI_{\mathrm{int},p} = \sum_{j=1}^{J} fis_{\mathrm{int},j,p} + fie_{\mathrm{int},p}, \quad \forall \mathrm{int} = NT, \quad p \in P \tag{10.15}$$

Likewise, the following equation is needed to determine the concentration in the inner stages:

$$FI_{\mathrm{int},p} ci^{in}_{c,\mathrm{int},p} = \sum_{\mathrm{int}=\mathrm{int}-1}^{\mathrm{int}=NT} fii_{\mathrm{int},p} ci^{out}_{c,\mathrm{int},p}, \quad \forall c \in C, \quad \mathrm{int} \neq 1, \quad p \in P \tag{10.16}$$

$ci^{in}_{c,\mathrm{int},p}$ is the pollutant concentration at the inlet of the interceptor *int* and $ci^{out}_{c,\mathrm{int},p}$ represents the outlet pollutant concentration in the interceptor *int*.

It should be noted that a fictitious interceptor is placed at the end of each treatment stage with the purpose of implementing a bypass; this interceptor has a cost equal to zero and it cannot remove any pollutant.

10.3.2.7 Interception Balances

This chapter considers a given conversion factor ($RR_{c,int}$) that models the efficiency of each available treatment technology *int* to remove the pollutant *c*. This factor can be fixed before the optimization process, and it is valuable to determine the pollutant concentration at the outlet of each interceptor ($ci^{out}_{c,\mathrm{int},p}$):

$$ci^{out}_{c,\mathrm{int},p} = ci^{in}_{c,\mathrm{int},p}\left(1 - RR_{c,\mathrm{int}}\right), \quad \forall c \in C, \quad \mathrm{int} \in INT, \quad p \in P \tag{10.17}$$

Furthermore, to determine the operating cost of the system, it is required to know the removed pollutant load ($cim_{c,int,p}$), which is implemented by the following equation:

$$cim_{c,\mathrm{int},p} = FI_{\mathrm{int},p}\left(ci^{in}_{c,\mathrm{int},p} - ci^{out}_{c,\mathrm{int},p}\right), \quad \forall c \in C, \quad \mathrm{int} \in INT, \quad p \in P \tag{10.18}$$

10.3.2.8 Whole Balance at the Mixer Before the Wastewater Stream

The flowrate discharged to the environment by the retrofitted eco-industrial park (FE_p) is provided by the portions of flowrates coming from the process sources ($fse_{i,p}$) and from the exit of the interception network ($fie_{int,p}$), which is stated as follows:

$$EP_p = \sum_{i=1}^{I} fse_{i,p} + \sum_{\mathrm{int}=1}^{INT} fie_{\mathrm{int},p}, \quad \forall p \in P \tag{10.19}$$

10.3.2.9 Component Balance at the Mixer Before the Wastewater Stream

This relationship is obtained by multiplying each term of the previous balance times the concentration of the involved pollutants:

$$EP_p CEP_{c,p} = \sum_{i=1}^{I} fse_{i,p} cs_{c,i,p} + \sum_{\mathrm{int}=NT}^{NT} fie_{\mathrm{int},p} ci^{out}_{c,\mathrm{int},p}, \quad \forall c \in C, \quad p \in P \tag{10.20}$$

It is noteworthy that the flowrate (EP_p) and the pollutant concentration ($CEP_{c,p}$) discharged to the environment are related to the MFA approach because this wastewater stream impacts the surrounding watershed.

10.3.2.10 Environmental Constraints Through the Watershed

There are also needed environmental constraints for the pollutant concentrations through the watershed subject on the required use in each reach:

$$CEP_{c,p} \leq CEP_{c,p}^{\max} \tag{10.21}$$

10.3.2.11 Pipelines

The model of this chapter considers the pipeline costs. It should be noted that this is an important factor for eco-industrial parks (where the involved distances to transport the flowrates can be considerable). Therefore, the first step consists of determining the existence of the required pipeline; thus, the flowrate in the pipeline must be greater than a minimum value and lower than a maximum value. Therefore, the following relationships determine the existence of the required pipelines between process sources and sinks:

$$M_{fss_{i,j,p}}^{\min} x_{i,j,p}^{1} \leq fss_{i,j,p} \leq M_{fss_{i,j,p}}^{\max} x_{i,j,p}^{1}, \quad \forall i \in I, \ \ j \in J, \ \ p \in P \tag{10.22}$$

$x_{i,j,p}^{1}$ represents a binary variable used to model the existence of the pipeline between process sources and sinks in the eco-industrial park p.

For the pipeline between the process sources and the interception network in each eco-industrial park:

$$M_{fsi_{i,\text{int},p}}^{\min} x_{i,\text{int},p}^{2} \leq fsi_{i,\text{int},p} \leq M_{fsi_{i,\text{int},p}}^{\max} x_{i,\text{int},p}^{2}, \quad \forall i \in I, \ \ \text{int} = 1, \ \ p \in P \tag{10.23}$$

$x_{i,\text{int},p}^{2}$ represents the associated binary variable.

Once the treatment is implemented in the central unit, it must consider the pipeline at the exit of the interception network to the process sinks:

$$M_{fis_{\text{int},j,p}}^{\min} x_{\text{int},j,p}^{3} \leq fis_{\text{int},j,p} M_{fis_{\text{int},j,p}}^{\max} x_{\text{int},j,p}^{3}, \quad \forall \text{int} = NT, \ \ j \in J, \ \ p \in P \tag{10.24}$$

$x_{\text{int},j,p}^{3}$ represents the binary variable associated to this case.

Additionally, the pipeline between the outlet of the treatment unit and the mixer before to the wastewater stream is considered:

$$M_{fie_{\text{int},p}}^{\min} x_{\text{int},p}^{4} \leq fie_{\text{int},p} \leq M_{fie_{\text{int},p}}^{\max} x_{\text{int},p}^{4}, \quad \forall \text{int} = NT, \ \ p \in P \tag{10.25}$$

$x_{\text{int},p}^{4}$ represents a binary variable used to determine the existence of a pipeline between the interception network and the wastewater stream. Here, $M^{\min}$ corresponds to the minimum value of the flowrate required to consider its existence, $M^{\max}$ represents an upper value for the flowrate associated with each section of the pipes.

10.3.3 Economic Objective Function

The economic objective function consists of minimizing the total annual cost (TAC), which is constituted by the freshwater cost (FWC), treatment costs (TC), and pipeline costs (PC); therefore, each component of the TAC is described as follows.

10.3.3.1 Fresh Water Cost

The cost for the fresh water provided to the system can be obtained through the following equation:

$$FWC = H_Y Dsh \sum_{p=1}^{P} \sum_{j=1}^{J} \sum_{w=1}^{W} CU_w fws_{w,j,p} \tag{10.26}$$

H_Y represents the operating hours for the eco-industrial parks, Dsh is the conversion of seconds into hours and CU_w corresponds to the cost for the fresh water w.

10.3.3.2 Treatment Cost

The cost function for interceptors includes fixed and variable costs as follows:

$$TC = K_F \sum_{p=1}^{P} \sum_{\text{int}=1}^{INT} CU_{\text{int}} z_{\text{int},p}^{treat} + H_Y \sum_{p=1}^{P} \sum_{\text{int}=1}^{INT} \sum_{c=1}^{C} CUM_{c,\text{int}} cim_{c,\text{int},p} \tag{10.27}$$

K_F represents a factor used to annualize the investment, CU_{int} corresponds to the fixed charge for the interceptor *int*, $z_{\text{int},p}^{treat}$ represents the binary variable for the interceptors, and $CUM_{c,\text{int}}$ is the unit cost for the mass removed of the pollutant c for each interceptor.

It is noteworthy that a set of restrictions are required to activate the binary variables for the interceptors:

$$M_{FI_{\text{int},p}}^{\min} z_{\text{int},p}^{treat} \leq FI_{\text{int},p} \leq M_{FI_{\text{int},p}}^{\max} z_{\text{int},p}^{treat}, \quad \forall \text{int} \in INT, \; p \in P \tag{10.28}$$

10.3.3.3 Piping Cost

The cost for each pipeline segment is determined as follows (Chew et al. 2008):

$$PC = K_F \left[\begin{array}{l} cpp \sum_{i=1}^{I} \sum_{j=1}^{J} \frac{D^1_{i,j,p} fss_{i,j,p}}{3600 \rho v} + x^1_{i,j} D^1_{i,j,p} CUP \\ + cpp \sum_{int=1}^{INT=1} \sum_{i=1}^{I} \frac{D^2_{i,\text{int},p} fsi_{i,\text{int},p}}{3600 \rho v} + x^2_{i,\text{int}} D^2_{i,\text{int},p} CUP \\ + cpp \sum_{int=NT}^{INT=NT} \sum_{j=1}^{J} \frac{D^3_{\text{int},j,p} fis_{\text{int},j,p}}{3600 \rho v} + x^3_{\text{int},j} D^3_{\text{int},j,p} CUP \\ + cpp \sum_{int=NT}^{INT=NT} \frac{D^4_{\text{int},p} fie_{\text{int},p}}{3600 \rho v} + x^4_{\text{int}} D^4_{\text{int},p} CUP \end{array} \right] \quad (10.29)$$

D is the length of each pipe segment, ρ corresponds to the density, υ represents the velocity, *cpp* corresponds to a parameter for cross-plant pipeline cost and lastly, *CUP* is the unit cost.

Thus, the *TAC* is determined as the sum of the previous terms as follows:

$$TAC = FWC + TC + PC \quad (10.30)$$

10.3.4 Environmental Objective Function

The environmental objective function involves the sustainability of the final disposal at the end of the watershed, which is addressed by the minimization of the pollutants discharged to catchment area to guarantee the natural degradation of the pollutants:

$$EOF = CQ_c^{final} \quad (10.31)$$

CQ_c^{final} corresponds to the target values for the concentration of the pollutants discharged to the final disposal. In this chapter, these values must be determined before the optimization process. Therefore, if it is desired to generate sustainable systems, the maximum concentration for each pollutant that can be decomposed at the final disposal through natural interaction has to be quantified and set as a target.

10.3.5 Implemented Multi-objective Optimization Approach

The problem is formulated as a multi-objective optimization model, where one objective corresponds to minimizing the *TAC* and the other one to minimizing the *EOF*, which is stated as follows:

$$\text{Objective Function} = \text{Min}\, TAC; \text{Min}\, EOF \quad (10.32)$$

It is noteworthy that these objective functions contradict each other. To obtain a proper environmental performance (EOF), in other words, low values for the pollutant compositions, it is necessary to increase the treatment in the EIPs to decrease their polluted discharges; therefore, the costs increase. Whereas low values for the total annual cost are associated with lower levels for treatment of discharges, which yields discharge with higher values for the concentrations of the pollutants; this scheme yields low costs and high levels for the pollutants discharged to the final disposal.

It should be noted that most of the nonlinearities included in the mathematical model formulation are bilinear terms. Therefore, Eq. (10.2) presents the product $Q_r CQ_{c,r}$, Eq. (10.4) has $FT_{r,t}CT_{c,r,t}$, Eqs. (10.5) and (10.6) involve the reaction order σ_c for the unknown concentrations $CQ_{c,r}$ and $CT_{c,r,t}$, respectively. While, Eq. (10.11) involves the term $fis_{\text{int}=NT,j,p}ci^{out}_{c,\text{int}=NT,p}$, relationships (10.13) and (10.16) have the product $FI_{\text{int},p}ci^{in}_{c,\text{int},p}$, Eq. (10.18) has several unknown variables ($cim_{c,\text{int},p}$, $FI_{\text{int},p}$, $ci^{in}_{c,\text{int},p}$ and $ci^{out}_{c,\text{int},p}$), and Eq. (10.20) has the term $EP_p CEP_{c,p}$. Noticed that all the previous symbols are optimization variables (except the exponent σ_c).

To resolve this multi-objective optimization formulation, the epsilon constraint optimization approach was implemented to obtain Pareto solutions that compensate for the contradicting objectives (Haimes 1970; Haimes et al. 1971; Bixby et al. 2000; Diwekar 2020). The mathematical model corresponds to a multi-objective mixed-integer nonlinear programming problem. The software GAMS (Brooke et al. 2015) was used to resolve this optimization problem using the algorithm DICOPT in a computer with an intel® core™ i7-3612QM CPU processor at 2.10 GHz and 6.00 GB of RAM. Lastly, the problem is resolved by using a deterministic mathematical programming method because of the large number of degrees of freedom associated with the model, which complicates the use of multi-objective stochastic optimization algorithms.

10.4 Results and Discussion

A case study is used to show the capabilities of the presented optimization approach. This example considers the Balsas watershed, which is one of the most important watersheds in Mexico and covers 117,305 km^2. This watershed is impacted by several wastewater effluents (such as industrial, agricultural, and residential discharges and wastewater streams from power plants); however, the most polluted discharges are the industrial effluents. Furthermore, Fig. 10.4 shows that the Balsas watershed has been segregated into 23 reaches, where there are four industrial parks available to be reconfigured as eco-industrial parks to reduce the pollutant concentrations at the final disposal. Parks 1 and 2 are integrated by three industrial plants, while parks 3 and 4 include two industrial plants each one; providing a total of ten industries available to participate in the implementation of the presented approach. This way, this example considers two pollutants where pollutant 1 is the most toxic one.

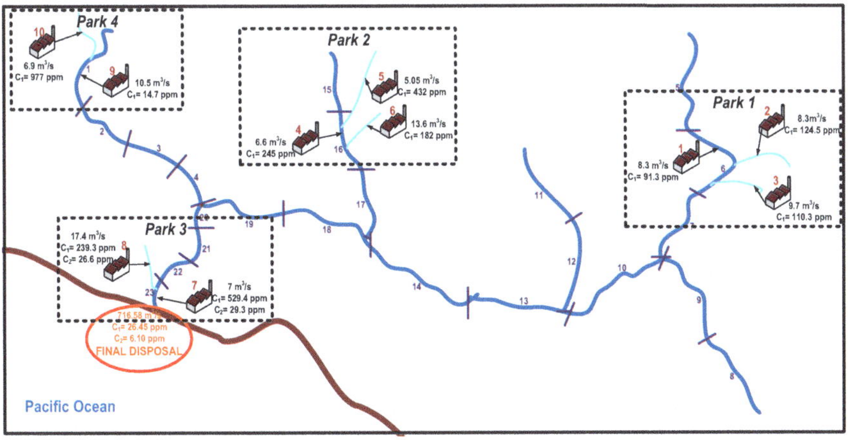

Fig. 10.4 Industrial parks located at Balsas watershed. Printed with permission of "*Synthesis of eco-industrial parks interacting with a surrounding watershed*", *ACS Sustainable Chemistry and Engineering Research*, Copyright (2015), American Chemical Society

Similarly, Fig. 10.4 presents the flowrates and the pollutant concentrations discharged by each individual plant (this information is the result of the sum of all the process sources for each industry, which are discharged to the environment without any treatment) and the actual conditions for the final disposal (before the implementation of the methodology), which receives 716.58 m^3/s with a concentration of 26.45 ppm and 6.10 ppm for pollutants 1 and 2, respectively. The complete data for the sources and sinks for the considered plants are reported in Table 10.1. To model this watershed, the information of Conagua (2013) was used (this is an annual report published by the Mexican Commission of water). Additionally, making an experimental study, the pollutant degradation follows a kinetic of first order with the next constants: $k_1 = 4.0284 \times 10^{-6}\ s^{-1}$ and $k_2 = 1.492 \times 10^{-6}\ s^{-1}$. Furthermore, the parameters D, K_F, H_Y, CU_{int}, CUP, ρ and υ are 100 m, 0.231/year, 8000 h/year, 12,600 US$, 250 US$, 1000 kg/m^3 and 1 m/s, respectively. The interceptors considered can operate with efficiencies (RR) of 0.6 and 0.8 and their costs (CUM) are 1.46 and 2.06 $/kg removed, respectively. The fresh water has a unit cost of 0.13 US$/ton.

The pollutant concentrations at the final disposal have high values that put at risk the aquatic life in the Pacific Ocean near the coast. Therefore, it is needed to implement a strategy to reduce the pollutant concentrations at this point. The discussion of the results includes several cases to solve the above-described problem. Case 1 analyzes the solution when only one industrial park is reconfigured as an eco-industrial park, Case 2 presents the possible solution involving two readjusted eco-industrial parks, Case 3 allows three reconfigurations, and Case 4 shows the solution when all the industrial parks can be retrofitted. According to each case, the generated solutions are different; nonetheless, if a greater number of industrial parks are reconfigured, the investment costs increase although the values for the environmental

Table 10.1 Sources and sinks for each plant

Sinks					Sources			
Plant	Process sink	Flowrate (m^3/s)	Pollutant composition 1 (ppm)	Pollutant composition 2 (ppm)	Process source	Flowrate (m^3/s)	Pollutant composition 1 (ppm)	Pollutant composition 2 (ppm)
1	1	2.22	70		1	3.33	80	
	2	3.33	80		2	2.22	110	
	3	2.78	100		3	2.78	90	
2	4	3.61	90		4	3.61	125	
	5	2.50	110		5	1.67	95	
	6	4.17	120		6	3.06	140	
3	7	3.06	115		7	2.64	120	
	8	4.44	85		8	3.19	115	
	9	2.22	125		9	3.89	100	
4	10	0.56	0		10	0.56	100	
	11	1.85	50		11	1.85	80	
	12	2.78	50		12	2.78	100	
	13	1.16	80		13	1.16	800	
	14	0.28	400		14	0.28	800	
5	15	0.56	0		15	0.56	100	
	16	1.85	50		16	1.85	80	
	17	0.43	80		17	0.43	400	
	18	1.19	100		18	2.02	800	
	19	0.19	400		19	0.19	1000	

(continued)

Table 10.1 (continued)

Sinks					Sources			
Plant	Process sink	Flowrate (m^3/s)	Pollutant composition 1 (ppm)	Pollutant composition 2 (ppm)	Process source	Flowrate (m^3/s)	Pollutant composition 1 (ppm)	Pollutant composition 2 (ppm)
6	20	0.56	0		20	0.56	100	
	21	2.22	25		21	2.22	50	
	22	1.39	25		22	1.39	125	
	23	1.11	50		23	1.11	800	
	24	8.33	100		24	8.33	150	
7	25	2.78	3	20	25	2.78	35	600
	26	3.61	1	80	26	3.61	30	500
	27	0.69	3	0	27	0.69	3	400
8	28	8.33	2	15	28	8.33	30	0
	29	5.56	8	70	29	5.56	35	400
	30	3.47	2	30	30	3.47	5	550
9	31	3.75	5		31	2.78	10	
	32	4.33	12		32	4.17	15	
	33	5.42	9		33	3.61	18	
10	34	4.17	100		34	2.22	800	
	35	5.83	50		35	2.08	1.2	
	36	3.94	130		36	2.64	950	

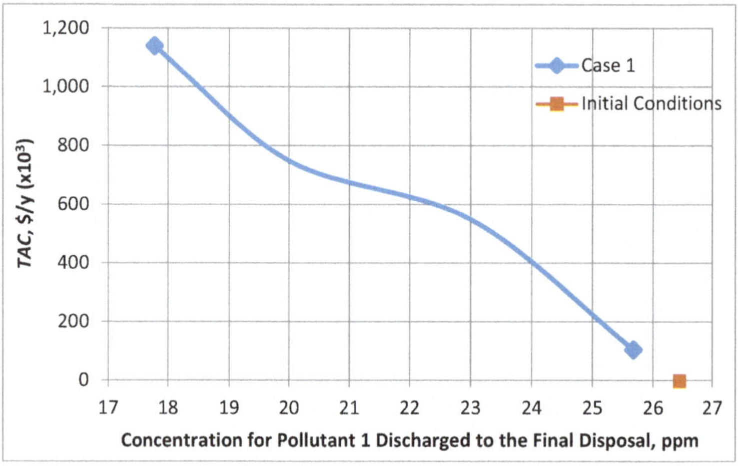

Fig. 10.5 Pareto curve for case 1. Printed with permission of "*Synthesis of eco-industrial parks interacting with a surrounding watershed*", *ACS Sustainable Chemistry and Engineering Research*, Copyright (2015), American Chemical Society

objective are reduced. In this context, Figs. 10.5, 10.6, and 10.7 present the Pareto curves for the set of solutions for Cases 1, 2, and 3, respectively. Case 4 corresponds to the specific solution where all the industrial parks are retrofitted, additionally, there are different scales and values in Pareto curves. The following discussion considers the extreme cases for each Pareto curve; nonetheless, intermediate solutions can be selected by the decision-makers.

10.4.1 Case 1

Case 1 includes constraints to just allow the installation of one EIP (of the four available parks). There are two aspects to be accounted for in each solution; the economic objective function and the pollutant concentration at the final disposal (environmental issue). Hence, the best economic solution in this section is obtained when Park 1 is retrofitted, yielding a *TAC* of $\$106.6 \times 10^3$/year and the pollutants 1 and 2 have concentrations at the final disposal of 25.68 ppm and 5.87 ppm, respectively. In this solution, it can be obtained a decrease of 2.9% for pollutant 1 and 3.8% for pollutant 2. Else, when the environmental concern is prioritized, the selected industrial park for retrofitting is Park 3, which achieves concentrations for pollutants 1 and 2 of 17.78 ppm and 5.48, respectively. This way, it can reduce 32.78% of the pollutant 1 discharged to the Pacific Ocean and 10.16% of the pollutant 2; nevertheless, the cost to reconfigure Park 3 is $\$1141.6 \times 10^3$/year (Fig. 10.5).

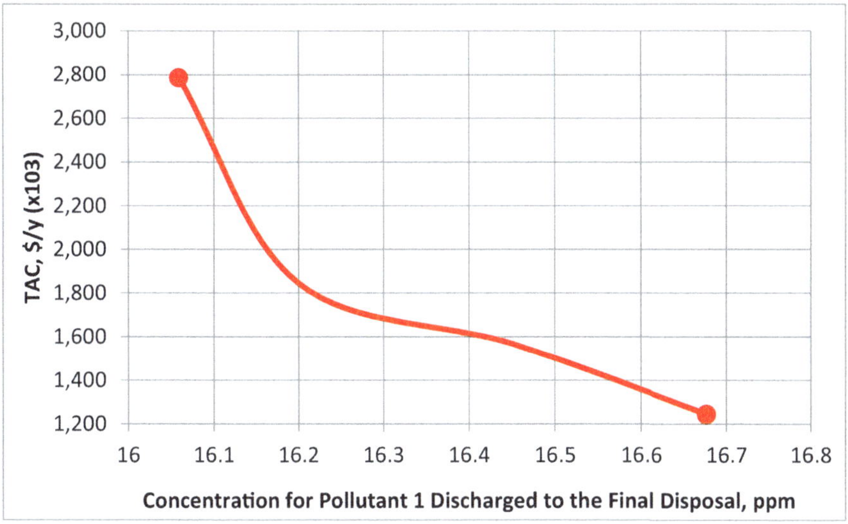

Fig. 10.6 Pareto curve for case 2. Printed with permission of "*Synthesis of eco-industrial parks interacting with a surrounding watershed*", *ACS Sustainable Chemistry and Engineering Research*, Copyright (2015), American Chemical Society

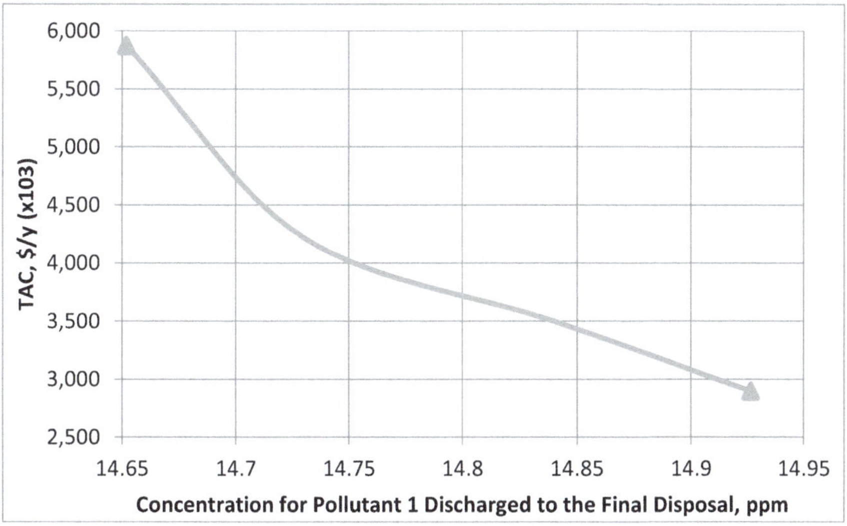

Fig. 10.7 Pareto curve for case 3. Printed with permission of "*Synthesis of eco-industrial parks interacting with a surrounding watershed*", *ACS Sustainable Chemistry and Engineering Research*, Copyright (2015), American Chemical Society

10.4.2 Case 2

Case 2 considers the possibility of retrofitting up to two EIPs to get the lowest cost. The solution selects Parks 1 and 3 to be reconfigured and yields a *TAC* of $\$1248.3 \times 10^3$/year, however, the condition for the discharge to the final disposal is a concentration for pollutants 1 and 2 of 16.68 ppm and 5.22 ppm, respectively (Fig. 10.6). If we consider in this case the minimization of the concentration at the final disposal for pollutant 1, the concentrations for pollutants 1 and 2 are 16.06 ppm and 5.74 ppm, respectively, yielding a *TAC* of $\$2788.8 \times 10^3$/year (representing a reduction in the concentration of pollutant 1 of 3.17% but with an augmentation in the *TAC* of 124%).

10.4.3 Case 3

This case considers the selection of up to three EIPs to be installed in Balsas watershed. The best economic solution considers the retrofitting of Parks 1, 2, and 3 and yields a *TAC* of $\$2895.4 \times 10^3$/year with 14.93 ppm and 5.48 ppm of pollutants 1 and 2 discharged to the Pacific Ocean, respectively (Fig. 10.7). Considering as optimization objective function the minimization of the concentration of pollutant 1 discharged to the Pacific Ocean, the solution considers the retrofitting of parks 2, 3, and 4; this combination has a *TAC* of $\$5873.8 \times 10^3$/year and discharges 14.65 ppm of pollutant 1 and 5.76 ppm of pollutant 2 to the final disposal. In addition, for the case when the optimal solution involves the minimization of the concentration of pollutant 2 discharged to the Pacific Ocean, it is possible to get a concentration for pollutant 2 discharged to the ocean of 5.24 ppm (i.e., 4.38% lower than the best economic solution); but the concentration of pollutant 1 has a value of 15.24 ppm and the *TAC* augments to $\$4333.4 \times 10^3$/year, this solution is obtained when the parks 1, 3 and 4 are retrofitted as eco-industrial parks.

10.4.4 Case 4

Lastly, in Case 4 there is no constraint on the number of EIPs to be reconfigured. The optimal economic solution is presented in Fig. 10.8, but Figs. 10.9, 10.10, 10.11 and 10.12 show the configuration for the EIP found in Fig. 10.8. This solution presents the wastewater stream and its concentration for each park once the reconfiguration is carried out. Comparing Figs. 10.4 and 10.8, there are significant reductions in the flowrates discharged to the watershed and the fulfillment of the environmental regulations, which yields significant environmental benefits, specifically the concentration for pollutant 1 discharged to the ocean is 13.48 ppm; whereas the pollutant 2 has a concentration of 5.51 ppm (in the original case the concentrations for pollutants

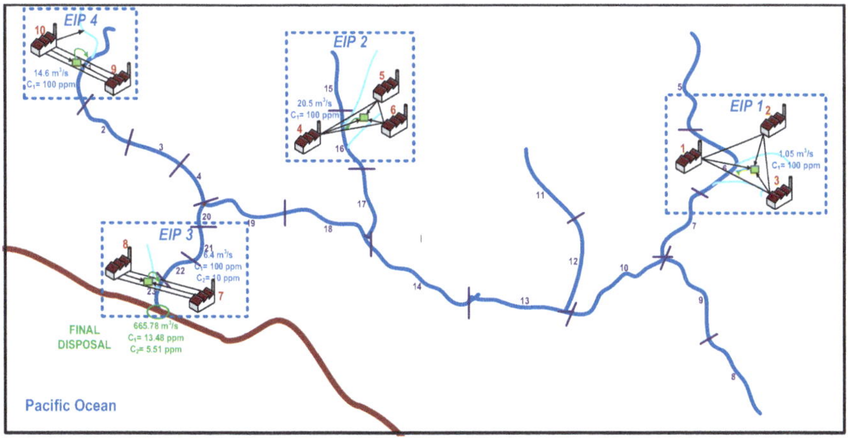

Fig. 10.8 Industrial parks reconfigured into eco-industrial parks for the Balsas watershed. Printed with permission of "*Synthesis of eco-industrial parks interacting with a surrounding watershed*", *ACS Sustainable Chemistry and Engineering Research*, Copyright (2015), American Chemical Society

1 and 2 are 26.45 ppm and 6.10 ppm, respectively). In this case, if the minimization of the concentration of pollutant 1 is considered, it is possible to reduce pollutant 1 49% concerning the initial conditions (remember that pollutant 1 is more dangerous than pollutant 2), while pollutant 2 can be decreased by 9.67%; but the *TAC* generated by the transformation of the four industrial parks is \$5980.5 $\times$ 10^3/year, which corresponds to a reduction of 65.43% concerning the original process (\$17,298.0 $\times$ 10^3/year). Therefore, the implementation of the proposed approach does not only include environmental benefits, but it also improves the economic aspects.

Lastly, this problem involves 2481 continuous variables, 1049 binary variables, and 2537 constraints; while the CPU time varied depending on the Pareto point, a mean time of 9.7 s was consumed in the used computer.

10.5 Concluding Remarks

An optimization approach for the retrofitting of several EIP and their integration with the nearby watersheds has been provided in this chapter. The goals are to reduce, to the lowest possible cost, the environmental impact on the nearby watershed. The best location, size, and purpose of an EIP are provided by the presented model, which also establishes the set of industrial facilities that must be chosen to be a member of an EIP. The optimization approach considers how the industrial effluents interact with other discharges and uses (domestic, agricultural, and a few natural phenomena) in the surrounding watershed. The influence of the natural events on the flows and concentrations of the pollutants is included in the geographical depiction of the

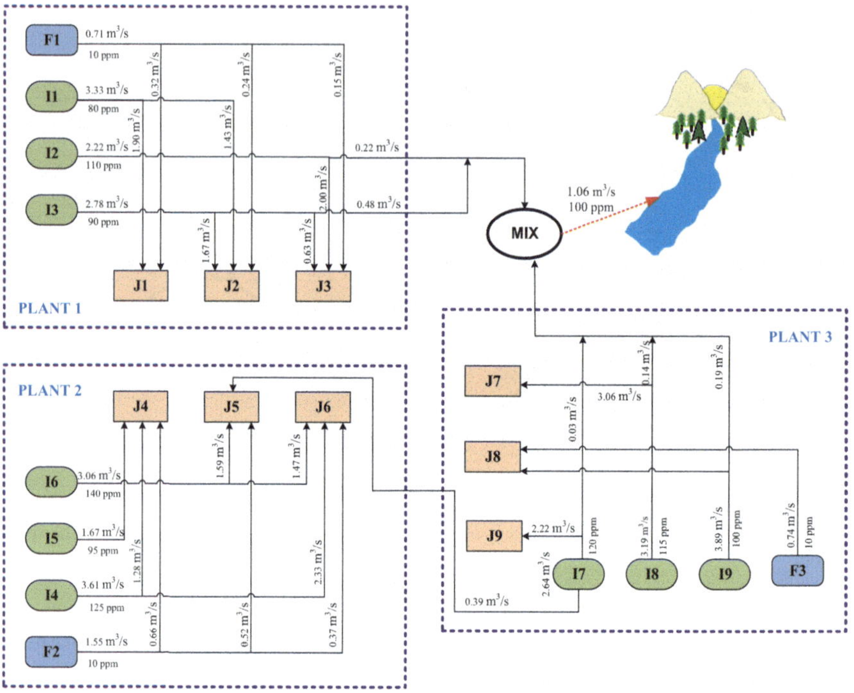

Fig. 10.9 Optimal configuration for the eco-industrial park (EIP) 1. Printed with permission of "*Synthesis of eco-industrial parks interacting with a surrounding watershed*", *ACS Sustainable Chemistry and Engineering Research*, Copyright (2015), American Chemical Society

watershed. A case study from Mexico is provided to demonstrate how the presented optimization formulation might be used. To increase the sustainability of the Balsas watershed, several nearby industrial facilities have been considered to integrate EIP. The findings demonstrate that it is feasible to meet water demands and enhance the watershed's water quality at a comparatively modest cost.

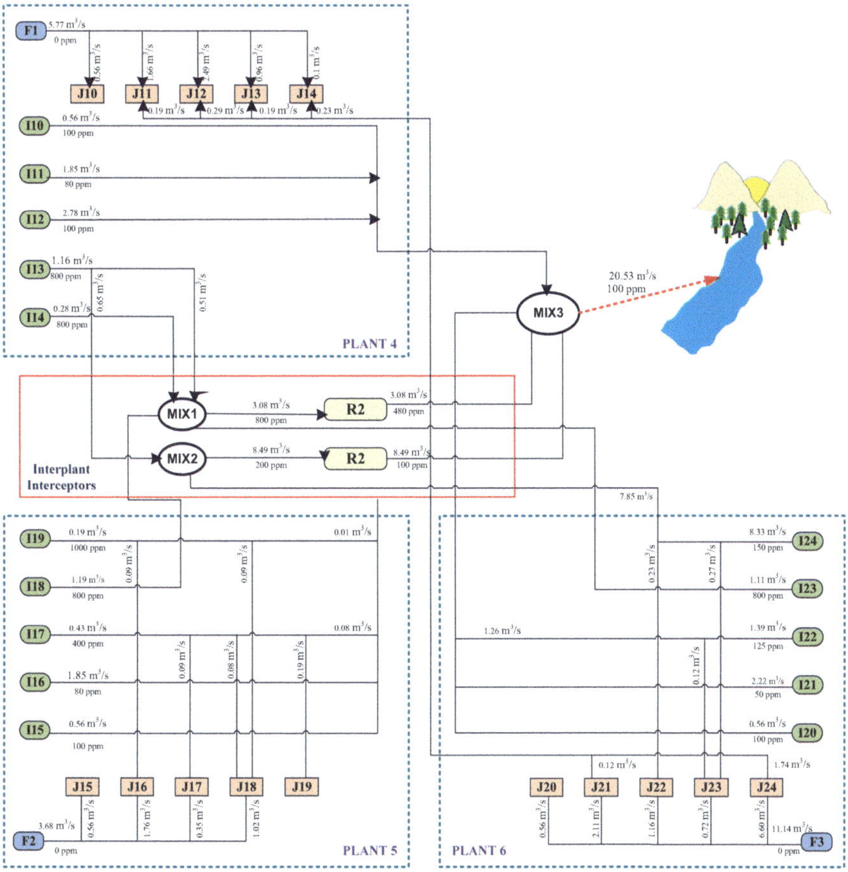

Fig. 10.10 Optimal configuration for the eco-industrial park (EIP) 2. Printed with permission of "*Synthesis of eco-industrial parks interacting with a surrounding watershed*", *ACS Sustainable Chemistry and Engineering Research*, Copyright (2015), American Chemical Society

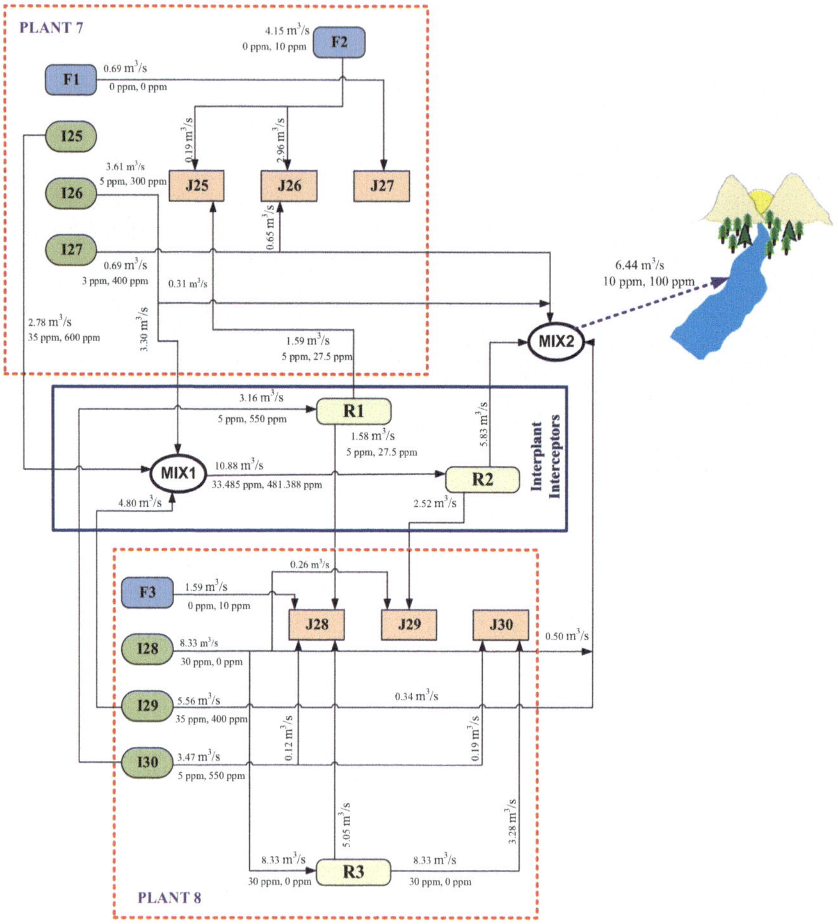

Fig. 10.11 Optimal configuration for the eco-industrial park (EIP) 3. Printed with permission of "*Synthesis of eco-industrial parks interacting with a surrounding watershed*", *ACS Sustainable Chemistry and Engineering Research*, Copyright (2015), American Chemical Society

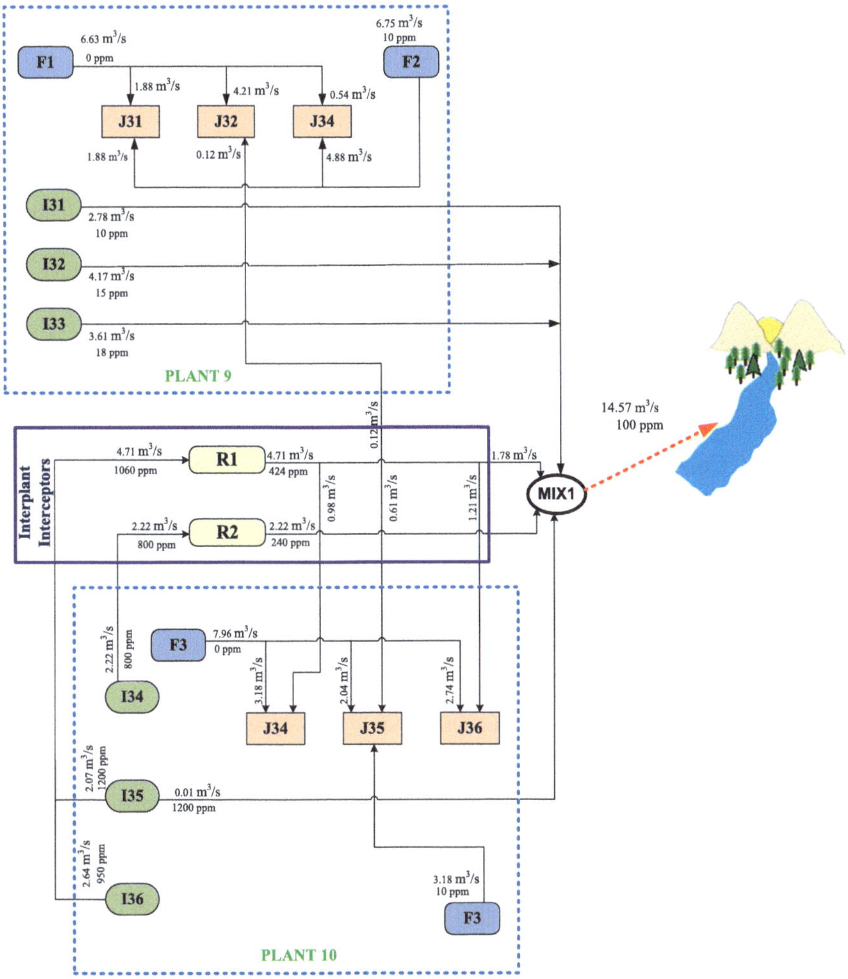

Fig. 10.12 Optimal configuration for eco-industrial park (EIP) 4. Printed with permission of "*Synthesis of eco-industrial parks interacting with a surrounding watershed*", *ACS Sustainable Chemistry and Engineering Research*, Copyright (2015), American Chemical Society

Nomenclatures

$A_{r,t}$	Cultivated area associated with the tributary t in the reach r, ha
C	Set for compounds $\{c \mid c = 1, \ldots, Nc\}$
$CD_{c,r}$	Concentration of the component c for direct discharges to the reach r, ppm
$CD_{c,r,t}$	Concentration of the component c for agricultural discharges in the tributary t of the reach r, ppm

$CEP_{c,p}$	Concentration of component c for the discharge from the eco-industrial park p after the reconfiguration, ppm
$CH_{c,r}$	Concentration of the component c for total discharges to the reach r, ppm
$CI_{c,r,t}$	Concentration of the component c for industrial discharges in the tributary t of the reach r, ppm
$CIP_{c,p}$	Concentration of the component c for the discharge of the industrial park p before the reconfiguration, ppm
$ci_{c,\text{int},p}^{in}$	Inlet concentration of the component c for the interceptor *int* in the park p, ppm
$ci_{c,\text{int},p}^{out}$	Outlet concentration of the compound c for the interceptor *int* in the park p, ppm
$cim_{c,\text{int},p}$	Removed load for the compound c in the interceptor *int* for the park p, g/s
$CL_{c,r}$	Concentration of the component c for total losses in the reach r, ppm
$CL_{c,r,t}$	Concentration of the component c for total losses in the tributary t of the reach r, ppm
$CP_{c,r}$	Concentration of the component c for the precipitation discharged to the reach r, ppm
$CP_{c,r,t}$	Concentration of the component c for precipitation discharges in the tributary t of the reach r, ppm
cpp	Parameter for capital cost for cross-plant piping
$CQ_{c,r}$	Concentration of the component c in the reach r, ppm
$cs_{c,i,p}$	Concentration of the component c for the source i in the park p, ppm
$CS_{l,r,t}^{treated}$	Concentration of the component c for residual treated wastewater discharged to the tributary t of the reach r, ppm
$CS_{c,r,t}^{untreated}$	Concentration of the component c for residual wastewater discharged without treatment to the tributary t of the reach r, ppm
$CT_{c,r,t}$	Concentration of the component c for flowrate discharged from tributary t to the reach r, ppm
$CU_{c,r}$	Concentration of the component c for water used from the reach r, ppm
$CU_{c,r,t}$	Concentration of the component c for water used in the tributary t of the reach r, ppm
$cu_{c,j,p}$	Concentration of the component c for the sink j in the park p, ppm
$CUM_{c,\text{int}}$	Unit cost for the mass removed of the pollutant c in the interceptor *int*, US$/Kg
CU	Unit cost of the pipe, US$
CU_{int}	Unit cost for interceptor *int*, US$
CU_w	Unit cost for fresh water w, US$/m^3
$cw_{c,w,p}$	Concentration of the component c for the fresh source w in the park p, ppm
D	Length of pipe segment, m
D_r	Direct discharges to the reach r, m^3/s
$D_{r,t}$	Agricultural discharges in the tributary t of the reach r, m^3/s
Dsh	Conversion factor to change seconds into hours, 3600 s/h

EP_p	Discharges coming from the eco-industrial park p after the reconfiguration, m^3/s
EOF	Environmental objective function, ppm
$FT_{r,t}$	Discharge from the tributary t to the reach r, m^3/s
$fie_{\text{int},p}$	Flowrate sent from the interceptor int to the environment in the park p, m^3/s
$FI_{\text{int},p}$	Flowrate in the interceptor int in the park p, m^3/s
$fii_{\text{int},p}$	Internal flowrates for the interceptor int in the park p, m^3/s
$fis_{\text{int},j,p}$	Flowrates sent from the interceptor int to the sink j in the park p, m^3/s
$FS_{i,p}$	Flowrate of the source i in the park p, m^3/s
FU_j	Flowrate entering to the sink j in the park p, ton/h
$fse_{i,p}$	Flowrate of the source i discharged to the environment in the park p, m^3/s
$fsi_{i,\text{int},p}$	Flowrate sent from the source i to the interceptors int in the park p, m^3/s
$fss_{i,j,p}$	Flowrate sent directly from the source i to the sink j in the park p, m^3/s
FWC	Fresh water cost, US$/year
$fws_{w,j,p}$	Flowrate of the fresh water w sent to the sink j in the park p, m^3/s
H_r	Total discharge (i.e., industrial, sanitary) to the reach r, m^3/s
H_Y	Annual working hours, h/year
I	Set for process sources $\{i \mid i = 1, \ldots, Ni\}$
$I_{r,t}$	Industrial stream discharged to the tributary t of the reach r, m^3/s
INT	Set for interceptors $\{int \mid int = 1, \ldots, Nint\}$
IP_p	Discharges coming from the industrial park p before the reconfiguration, m^3/s
J	Set for process sinks $\{j \mid j = 1, \ldots, Nj\}$
K_F	Factor used to annualize the inversion, year^{-1}
k_c	Kinetic constant for the degradation for compound c
L_r	Total losses (filtration and evaporation) from the reach r, m^3/s
$L_{r,t}$	Total losses (filtration and evaporation) in the tributary t to the reach r, m^3/s
M	Lower or upper values respect a specific variable
P	Set for parks $\{p \mid p = 1, \ldots, Np\}$
P_r	Precipitation discharged to the reach r, m^3/s
$P_{r,t}$	Precipitation discharged in the tributary t of the reach r, m^3/s
PC	Cross plant pipeline capital cost, US$/year
Q_r	Flowrate leaving the reach r, m^3/s
$r_{c,r}$	Reaction carried out in the reach r for compound c, g/s
$r_{c,r,t}$	Reaction carried out in the tributary t of the reach r for compound c, g/s
R	Set for reaches $\{r \mid r = 1, \ldots, Nr\}$
$RR_{c,\text{int}}$	Conversion factor for the compound c in the interceptor int, dimensionless
$S_{r,t}^{untreated}$	Residual wastewater without treatment discharged to the tributary t of the reach r, m^3/s

$S_{r,t}^{treated}$	Residual treated wastewater discharged to the tributary t of the reach r, m^3/s
T	Set for tributaries $\{t \mid t = 1, \ldots, Nt\}$
TAC	Total annual cost, US$/year
TC	Treatment cost, US$/year
U_r	Water used from reach r, m^3/s
$U_{r,t}$	Water used from tributary t that discharges to reach r, m^3/s
V_r	Volume for reach r, m^3
$V_{r,t}$	Volume for tributary t from reach r, m^3
W	Set for types of fresh water $\{w \mid w = 1, \ldots, Nw$

Greek Symbols

$\alpha_{r,t}$	Required agricultural flowrate related to the tributary t that discharges to the reach r, m^3/ha s
$\beta_{r,t}$	Discharged agricultural flowrate related to the tributary t that discharges to the reach r, m^3/ha s
σ_c	Reaction order for compound c
ρ	Density, kg/m^3
v	Velocity, m/s

Binary Variables

$x_{i,j,p}^1$	Used to determine the existence of pipe from source i to process sink j in the park p
$x_{i,\text{int},p}^2$	Used to determine the existence of pipe from source i to interceptor *int* in the park p
$x_{\text{int},j,p}^3$	Used to determine the existence of pipe from interceptor *int* to process sink j in the park p
$x_{\text{int},p}^4$	Used to determine the existence of pipe from interceptor *int* to environment in the park p
$z_{\text{int},p}^{treat}$	Used to determine the existence of the interceptor *int* in the park p

Subscripts

c	Compound
i	Source
j	Sink
int	Interceptor

p	Park
r	Reach
t	Tributary
w	Type of fresh water

Superscripts

in	Inlet
max	Upper limit
min	Lower limit
out	Outlet

References

Acobsen NB (2006) Industrial symbiosis in Kalundborg, Denmark: a quantitative assessment of economic and environmental aspects. J Ind Ecol 10:239–255

Alnouri SY, Linke P, El-Halwagi MM (2014) Optimal interplant water networks for industrial zones: addressing interconnectivity options through pipeline merging. AIChE J 60(8):2853–2874

Aviso KB, Tan RR, Culaba AB (2010) Designing eco-industrial water exchange networks using fuzzy mathematical programming. Clean Technol Environ Policy 12(4):353–363

Aviso KB, Tan RR, Culaba AB, Foo DCY, Hallale H (2011) Fuzzy optimization of topologically constrained eco-industrial resource conservation networks with incomplete information. Eng Optim 43(3):257–279

Bandyopadhyay S, Sahu GC, Foo DCY, Tan RR (2010) Segregated targeting for multiple resource networks using decomposition algorithm. AIChE J 56(5):1235–1248

Bishnu SK, Linke P, Alnouri SY, El-Halwagi MM (2014) Multiperiod planning of optimal industrial city direct water reuse networks. Ind Eng Chem Res 53(21):8844–8865

Bixby RE, Fenelon M, Gu Z, Rothberg E, Wunderling R (2000) MIP: theory and practice—closing the gap. In: Powell MJD, Scholtes S (eds) System modelling and optimization: methods, theory and applications, pp 19–50

Boix M, Montastruc L, Pibouleau L, Azzaro-Pantel C, Domenech S (2012) Industrial water management by multiobjective optimization: from individual to collective solution through eco-industrial parks. J Clean Prod 22:85–97

Boix M, Montastruc L, Azzaro-Pantel C, Domenech S (2015) Optimization methods applied to the design of eco-industrial parks: a literature review. J Clean Prod 87(1):303–317

Brooke A, Kendrick D, Meeruas A, Raman R (2015) GAMS-Language guide. GAMS Development Corp, Washington, D.C.

Burgara-Montero O, Ponce-Ortega JM, Serna-González M, El-Halwagi MM (2012) Optimal design of distributed treatment system for the effluents discharged to the rivers. Clean Technol Environ Policy 14:925–942

Burgara-Montero O, El-Baz AA, Ponce-Ortega JM, El-Halwagi MM (2013a) Optimal design of a distributed treatment system for increasing dissolved oxygen in watersheds through self-rotating discs. ACS Sustain Chem Eng 1(10):1267–1279

Burgara-Montero O, Ponce-Ortega JM, Serna-González M, El-Halwagi MM (2013b) Incorporation of the seasonal variations in the optimal treatment of industrial effluents discharged to watersheds. Ind Eng Chem Res 52(14):5145–5160

Chen CL, Hung SW, Lee JY (2010) Design of inter-plant water network with central and decentralized water mains. Comput Chem Eng 34(9):1522–1531

Cheng-Liang C, Chung-Yen L, Foo DCY (2011) Synthesis of inter-plant water networks involving batch and continuous processes. Chem Eng Trans 25:587–592

Chertow MR (2000) Industrial symbiosis: literature and taxonomy. Annu Rev Energy Env 25:313–337

Chew IML, Foo DCY (2009) Automated targeting for inter-plant water integration. Chem Eng J 153(1–3):23–36

Chew IML, Ng DKS, Foo DCY, Majozi T, Gouws J (2008) Synthesis of direct and indirect interplant water network. Ind Eng Chem Res 47:9485–9496

Chew IML, Thukkaivarrna SL, Tan RR, Foo DCY (2011) Analysis of inter-plant water integration with direct integration schemes through game theory approach: Pareto optimal solution with interventions. Clean Technol Environ Policy 13(1):49–62

Chew IML, Foo DCY, Tan RR (2010) Flowrate targeting for interplant resource conservation network. Part 2: unassisted integration scheme, Ind Eng Chem Res 49(14): 6456–6468

Conagua (2013) Mexican National Water Commission. Water atlas in Mexico 2013. Mexico City, Mexico

Diwekar UM (2020) Introduction to applied optimization, vol 22. Springer Nature, USA

Ehrenfeld J, Gertler N (1997) Industrial ecology in practice: the evolution of interdependence at Kalundborg. J Ind Ecol 1:67–79

El-Baz AA, Ewida KT, Shouman MA, El-Halwagi MM (2005) Material flow analysis and integration of watersheds and drain systems II: integration and solution strategies with application to ammonium management in Bahr El-Baqar drain system. Clean Technol Environ Policy 7(1):78–86

El-Halwagi MM, Gabriel F, Harrell D (2003) Rigorous graphical targeting for resource conservation via material recycle/reuse networks. Ind Eng Chem Res 42:4319–4328

Foo DCY (2008) Flowrate targeting for threshold problems and plant-wide integration for water network synthesis. J Environ Manage 88(2):253–274

Foo DCY (2009) State-of-the-art review of pinch analysis techniques for water network synthesis. Ind Eng Chem Res 48:5125–5159

Haimes YY, Lasdon LS, Wismer DA (1971) On a bicriterion formulation of the problems of integrated system identification and system optimization. IEEE Trans Syst Man Cybern 1:296–297

Haimes YY (1970) The integration of system identification and system optimization. Ph.D. dissertation, Engineering System Department, School of Engineering and Applied Science, University of California, Los Angeles, California

Hortua AC, El-Halwagi M (2013) Integrated approach for simultaneous mass and property integration for resource conservation. ACS Sustain Chem Eng 1(1):29–38

Lamberta AJD, Boonsb FA (2002) Eco-industrial parks: stimulating sustainable development in mixed industrial parks. Technovation 22(8):471–484

Liao ZW, Wu JT, Jian BB, Wang JD, Yang YR (2007) Design methodology for flexible multiple plant water networks. Ind Eng Chem Res 46:4954–4963

Lim SR, Park JM (2010) Interfactory and intrafactory water network system to remodel a conventional industrial park to a green eco-industrial park. Ind Eng Chem Res 49:1351–1358

Lira-Barragán LF, Ponce-Ortega JM, Serna-González M, El-Halwagi MM (2011a) An MINLP model for the optimal location of the new industrial plant with simultaneous consideration of economic and environmental criteria. Ind Eng Chem Res 50(2):953–964

Lira-Barragán LF, Ponce-Ortega JM, Serna-González M, El-Halwagi MM (2011b) Synthesis of water networks considering the sustainability of the surrounding watershed. Comput Chem Eng 35:2837–2852

Lira-Barragán LF, Ponce-Ortega JM, Nápoles-Rivera F, Serna-González M, El-Halwagi MM (2013) Incorporating property-based water networks and surrounding watersheds in site selection of industrial facilities. Ind Eng Chem Res 52(1):91–107

Lombardi DR, Laybourn P (2012) Redefining industrial symbiosis. J Ind Ecol 16:28–37

López-Díaz DC, Lira-Barragán LF, Rubio-Castro E, Ponce-Ortega JM, El-Halwagi MM (2015) ACS Sustain Chem Eng 3:1564–1578

López-Villarreal F, Lira-Barragan LF, Rico-Ramirez V, Ponce-Ortega JM, El-Halwagi MM (2014) An MFA optimization approach for pollution treading considering the sustainability of the surrounding watershed. Comput Chem Eng 63:140–151

Lovelady EM, El-Halwagi MM, Krishnagopalan GA (2007) An integrated approach to the optimization of water usage and discharge in pulp and paper plants. Int J Environ Pollut 29(1–3):274–307

Lovelady EM, El-Baz AA, El-Monayeri D, El-Halwagi MM (2009a) Reverse problem formulation for integrating process discharges with watersheds and drainage systems. J Ind Ecol 13:914–927

Lovelady EM, El-Halwagi MM, Chew IMI, Ng DKS, Foo DCY, Tan RR (2009b) A property-integration approach to the design and integration of eco-industrial parks. In: El-Halwagi MM, Linninger AA (eds) Design for energy and the environment. Proceedings of the 7th international conference on the foundations of computer-aided process design (FOCAPD). CRC Press/Taylor & Francis, pp 559–568

Martinez-Gomez J, Burgara-Montero O, Ponce-Ortega JM, Nápoles-Rivera F, Serna-González M, El-Halwagi MM (2013) On the environmental, economic and safety optimization of distributed treatment systems for industrial effluents discharged to watersheds. J Loss Prev Process Ind 26(5):908–923

Montastruc L, Boix M, Pibouleau L, Azzaro-Pantel C, Domenech S (2013) On the flexibility of an eco-industrial park (EIP) for managing industrial water. J Clean Prod 43:1–11

Noureldin MMB, El-Halwagi MM (2015) Synthesis of C–H–O symbiosis networks. AIChE J 61(4):1242–1262

Rubio-Castro E, Ponce-Ortega JM, Nápoles-Rivera F, El-Halwagi MM, Serna-González M, Jimenez-Gutiérrez A (2010) Water integration of eco-industrial parks using a global optimization approach. AIChE J 40:9945–9960

Rubio-Castro E, Ponce-Ortega JM, Serna-Gonzalez M, Jimenez-Gutierrez A, El-Halwagi MM (2011) A global optimal formulation for the water integration in eco-industrial parks considering multiple pollutants. Comput Chem Eng 35:1558–1574

Rubio-Castro E, Ponce-Ortega JM, Serna-González M, El-Halwagi MM (2012) Optimal reconfiguration of multi-plant water networks into an eco-industrial park. Comput Chem Eng 44:58–83

Rubio-Castro E, Ponce-Ortega JM, El-Halwagi MM, Pham V (2013) Global optimization in property-based interplant water integration. AIChE J 59(3):813–833

Tan RR, Aviso KB, Cruz JB Jr, Culaba AB (2011a) A note on an extended fuzzy bi-level optimization approach for water exchange in eco-industrial parks with hub topology. Process Saf Environ Prot 89(2):106–111

Tan RR, Taskhiri MS, Chiu ASF (2011b) MILP model for energy optimization in EIP water networks. Clean Technol Environ Policy 13(5):703–712

Tashkhiri MS, Tan RR, Chiu ASF (2011) Energy-based fuzzy optimization approach for water reuse in an industrial park. Resour Conserv Recycl 55(7):730–737

Wang G, Feng X, Hoonga CK (2014) Symbiosis analysis on industrial ecological system. Chin J Chem Eng 22(6):690–698

Yu F, Han F, Cui Z (2015) Evolution of industrial symbiosis in an eco-industrial park in China. J Clean Prod 87:339–347

Chapter 11
Selecting Optimal Site for Biorefineries: A Focus on Environmental Sustainability Integration

11.1 Introduction

With the growing interest in sustainable development, it is required to consider the effective usage of renewable resources. Therefore, biorefineries (Fig. 11.1) offer interesting options for the conversion of biomass into fuels and chemicals and the reduction of greenhouse gas emissions through the sequestration of carbon dioxide during photosynthesis (Sandun et al. 2006; Kent 2009). There are numerous challenges related to the viable operation of a biorefinery. Biomass availability depends on location and seasonal variability within these locations, there is needed to plan the optimal location of these facilities in a distributed scheme (Kelloway and Daoutidis 2014; Stuart and El-Halwagi 2012). The implementation of a system of distributed biorefineries will certainly raise the demand for biomass and increase the cultivation areas for the growth of the needed crops. In some regions, it is relatively straightforward to determine the lands to be used for cultivation (Okwo and Thomas 2014). Nevertheless, frequently numerous factors must be considered and reconciled to determine the cultivation areas. Key factors involve the type of bioresources to be cultivated, the available land, water resources, energy, infrastructure, and socio-economic issues (Kokossis et al. 2015). The location of cultivation areas is linked with the location of end-use biorefineries. Numerous contributions have been reported around site location for biorefining supply chains while considering the available biomass resources and the product (eg., biofuel) demands. Bowling et al. (2011), De Meyer et al. (2015), Marvin et al. (2013), Santibañez-Aguilar et al. (2015a, b), and Zhang et al. (2012) reported approaches for optimizing the facility location of biorefineries as part of considering the involved supply chain. Murillo-Alvarado et al. (2014), and Yeh et al. (2015) optimized supply chains for biorefineries considering the available feedstocks and harvest management in a location. Other works have considered environmental and economic issues as objective functions while optimizing biorefinery supply chains (e.g., Osmani and Zhang 2014; Santibañez-Aguilar et al. 2011, 2014; You et al. 2012). These approaches have been extended

J. M. Ponce-Ortega et al., *Sustainable Water Systems*, Water Resources Development and Management, https://doi.org/10.1007/978-3-031-72498-5_11

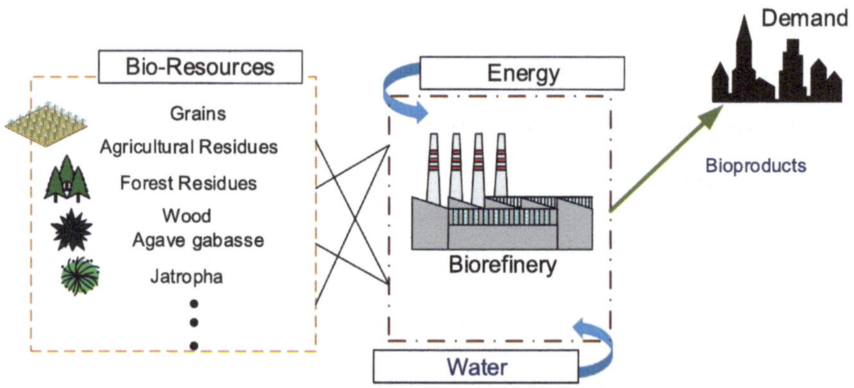

Fig. 11.1 A generic biorefinery. Printed with permission of "*Optimal location of biorefineries considering sustainable integration with the environment*", *Renewable Energy*, Copyright (2017), Elsevier

to include other sustainability criteria (e.g., Ba et al. 2015; Corsano et al. 2011; Giarola and Bezzo 2015; Santibañez-Aguilar et al. 2015a, b). Additionally, Guillén-Gosálbez et al. (2010); Kim et al. (2011), Ruiz-Femenia et al. (2013), and Tong et al. (2012) unified approaches to include uncertainty in optimizing biofuel supply chains. El-Halwagi et al. (2013) extended safety objectives in the design of biorefining supply chains. The socio-economic factors were integrated into the evaluation of biorefinery supply chains by Lehtonen and Okkonen (2016). Cucek et al. (2014) reported a multi-objective optimization approach for biorefineries. Noureldin and El-Halwagi (2015) reported an approach for combining biorefineries with hydrocarbon processing facilities through an industrial symbiosis network. A review of the challenges in optimizing biorefineries is given by Yue et al. (2014).

Nonetheless, the effectiveness and flexibility of the research efforts, do not directly tackle the important issue of macroscopic water management linked with the evolving biorefining supply chains. Considerable amounts of water are used and discharged in biomass cultivation, pretreatment, and conversion. Furthermore, withdrawal from and discharge into the surrounding watersheds can significantly affect the environmental quality of the water system. Furthermore, the design of the biorefining supply chain is impacted by the availability and quality of water resources and the environmental constraints at different points of the surrounding watershed. To address this issue, this chapter presents an integrated optimization approach for the incorporation of watershed models and constraints into the allocation and design of the building blocks of biorefining supply chains (López-Díaz et al. 2017). The interaction between the biorefining supply chain and the watershed is modeled using material flow analysis (MFA). The MFA has been successful in analyzing environmental problems and their interactions with industrial systems (e.g., El-Baz et al. 2005; Burgara-Montero et al. 2013; Lira-Barragán et al. 2013; Lopez-Villareal et al. 2014), where social, economic, and environmental connections with the surroundings are considered.

11.2 Problem Statement

The problem addressed in this chapter is defined as follows: Given a region where it is desired to locate a distributed biorefining supply chain (Figs. 11.2 and 11.3). Given are also potential feedstocks, cultivation areas, conversion technologies, facility locations, and product (biofuels, chemicals) demands. Also, a surrounding watershed is available to provide water for users and to receive wastewater discharges. Availability and environmental constraints must be satisfied at different points of the watershed. Afterward, the main objective is to synthesize a supply chain for meeting the market demands while determining the quantities and location of feedstocks, the type, size, and location of the conversion facilities, and the withdrawal from and discharge to the watershed.

Furthermore, to model the watershed and its interactions with the biorefining supply chain, the MFA approach is implemented. This approach considers water extractions and discharges linked with domestic, agricultural, and industrial uses and natural phenomena such as filtration, evaporation, and precipitation in the watershed. However, new water withdrawals and wastewater discharges that will arise from the biorefining supply chain are to be determined in terms of quantities, qualities, and location. To correctly simulate the watershed, the MFA approach requires splitting

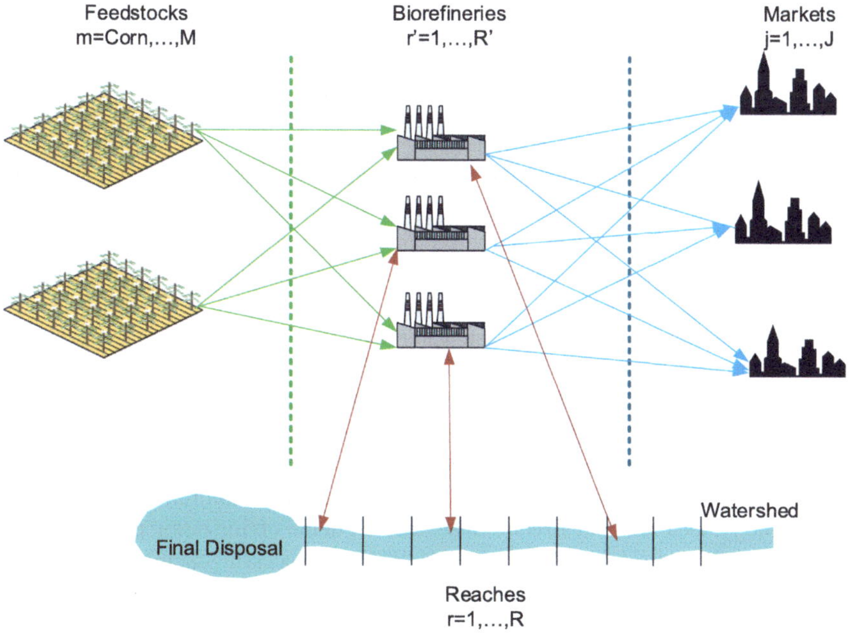

Fig. 11.2 General configuration of the proposed methodology. Printed with permission of "*Optimal location of biorefineries considering sustainable integration with the environment*", *Renewable Energy*, Copyright (2017), Elsevier

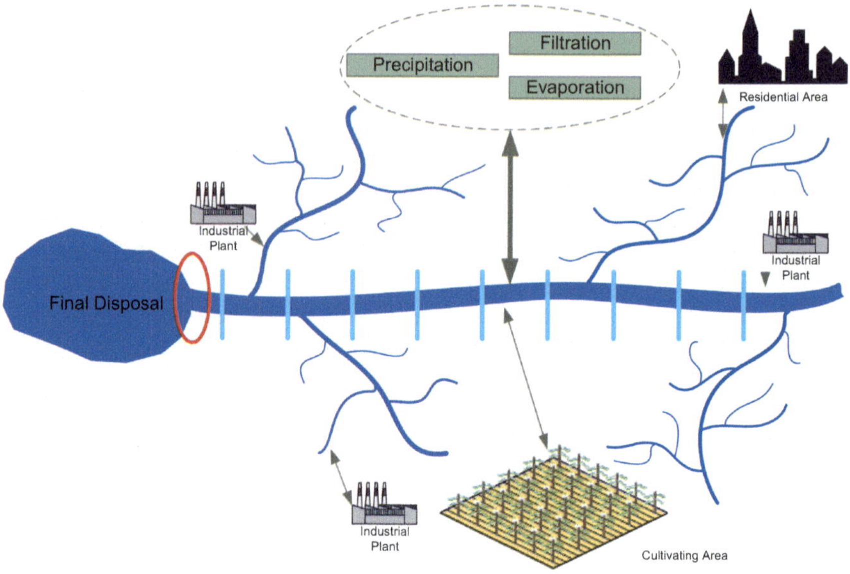

Fig. 11.3 A watershed system interacts with several extractions and discharges (users and natural phenomena). Printed with permission of "*Optimal location of biorefineries considering sustainable integration with the environment*", *Renewable Energy*, Copyright (2017), Elsevier

the watershed into sections called reaches. Within each reach (r), the flow and main characteristic are considered constant. The reaches represent zones where no big intermediate effluents are discharged and/or extracted. Afterward, it is required to identify geographically all the interactions existing in the watershed, particularly the largest water withdrawals and discharges to properly split a watershed into reaches and generate reaches with constant values for volume and flowrate. Also, numerous tributaries (index t) feed the reaches. So, this technique uses mass balances, property tracking, and physical, chemical, and biochemical phenomena for each reach. Due to there are numerous water uses and wastewater discharges in the biorefinery system (in the cultivation and harvesting sites, pretreatment, and processing facilities) that are location-dependent, it is imperative to optimize these uses and discharges in conjunction with the surrounding watershed. Therefore, to consider such interactions, the superstructure shown in Fig. 11.4 is implemented. It should be noted that there is a set of potential locations for the processing facilities to install the biorefineries. The type of technology, location, and size are to be decided in the optimization process. A set of maximum-capacity constraints is required on each processing facility based on the available infrastructure within each site. The maximum demand for each product is also known. Additionally, the implemented optimization formulation aims to maximize the net profit for the supply chain while accounting for the sales of biofuels and other products minus the supply chain costs (i.e., the cost for cultivation, harvesting, transportation, processing, and distribution). A key issue of the implemented optimization formulation is to consider the interaction of the biorefinery supply chain with the surrounding watershed.

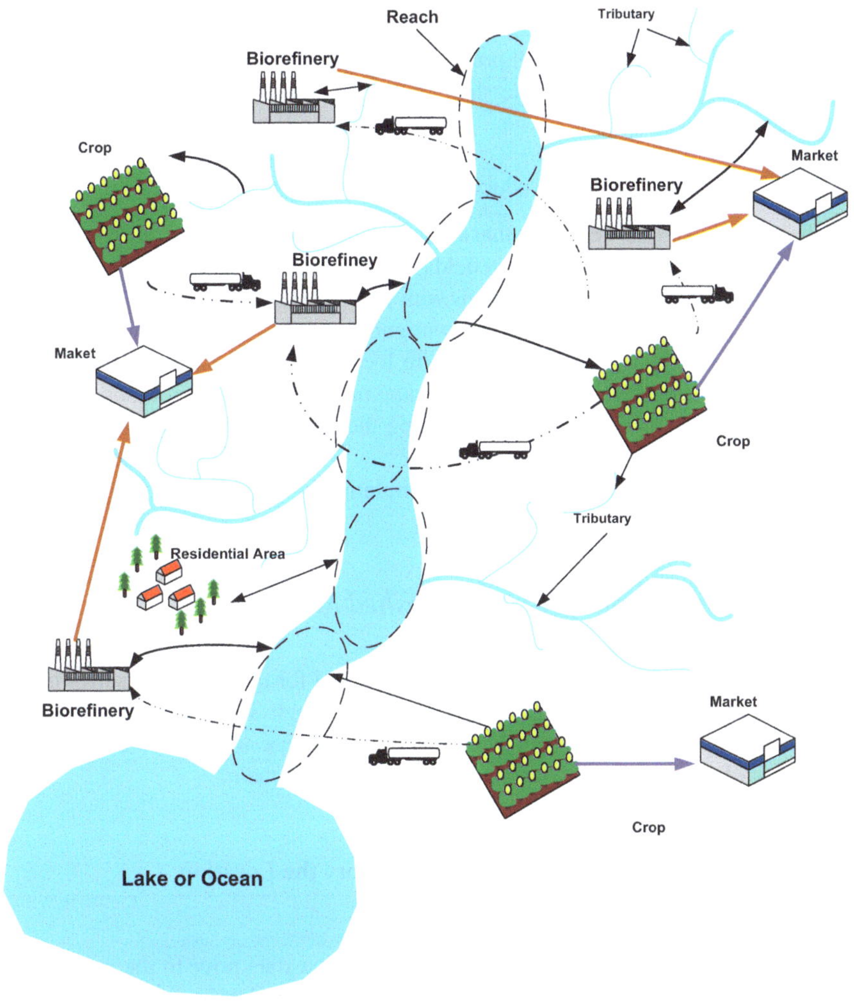

Fig. 11.4 Superstructure for the optimal location of biorefineries interacting with the surrounding watershed. Printed with permission of "*Optimal location of biorefineries considering sustainable integration with the environment*", *Renewable Energy*, Copyright (2017), Elsevier

11.3 Mathematical Model Formulation

The implemented optimization approach is based on the superstructure presented in Fig. 11.4. Firstly, the used indexes are defined as follows:

r: a reach and it is used to represent a cultivation site.
r′: a possible location to install processing facilities (i.e., biorefineries).
t: a tributary that feeds the watershed.
m: feedstock.
k: product.
j: markets to distribute the products.

The presented model considers that all the reaches have associated cultivation sites (where there are available cultivation fields and water); thus, each reach may produce feedstocks. Likewise, to avoid problems with water availability, the processing facilities (biorefineries) can be installed in contiguous areas of each reach; thus, all the reaches (r) are considered possible areas to locate a biorefinery (r′). Therefore, r′ is used to avoid confusion with r in the mathematical model formulation. Furthermore, this scheme is effective in determining the impact in terms of water for the implementation of the proposed project.

Nonetheless, the model optimization formulation is explained as follows.

11.3.1 Relationships for the Watershed Model

A material flow analysis formulation is implemented for modeling the surrounding watershed that relates the biorefinery system through water use and wastewater discharge. The following mathematical relationships are used to simulate the interaction of the watershed with the biorefinery supply chain.

11.3.1.1 Watershed at Initial Conditions Before the Location of the Biorefinery Supply Chain

The current conditions for the watershed represent the state prior to the installation of the biorefinery. Thus, $Q_r^{Initial-Out}$ represents the flowrate leaving reach *r*, which is equal to the sum of the flowrate entering in the reach *r* ($Q_r^{Initial-In}$), plus precipitation (P_r), direct industrial discharges (D_r), residential discharges (H_r), the sum of the effluents entering the reach from tributaries ($FT_{r,t}$) along the watershed, the sum of the effluents discharged from the cultivating land in the feedstock production ($w_{m,r}^{discharge-feedstocks\_Initial}$), the sum of extracted effluents to use in the cultivating land ($w_{m,r}^{used-feedstocks\_Initial}$), minus the losses (evaporation, filtration) (L_r) and the uses (U_r) in the section of the river *r*:

$$Q_r^{Initial-Out} = Q_r^{Initial-In} + P_r + D_r + H_r + \sum_{t=1}^{N_{(t)}} FT_{r,t} + \sum_m w_{m,r}^{discharge-feedstocks\_Initial} - \sum_m w_{m,r}^{used-feedstocks\_Initial} - L_r - U_r, \quad \forall r \tag{11.1}$$

$FT_{r,t}$ is the set of tributaries that inlet into the watershed in the reach r, where a tributary is a branch of the principal watershed.

The used water to produce the feedstock at the initial conditions ($w_{m,r}^{used-feedstocks\_Initial}$) is determined as the product of the conversion factor $\beta_{m,r}^{feedstocks}$ that signifies the consumed water to produce a unit mass of the feedstock per hectare, times the conversion factor $\alpha_{m,r}^{feddstock}$ that identifies the mass production of feedstock per hectare and the actual cultivated area for each crop:

$$w_{m,r}^{used-feedstocks\_Initial} = \beta_{m,r}^{feedstocks} \alpha_{m,r}^{feddstock} \left[a_{m,r}^{exist}\right], \quad \forall m, \ \forall r \tag{11.2}$$

The discharged water in the cultivated land for the feedstocks at the initial conditions is the product of the conversion factor $\gamma_{m,r}^{discharge}$ that is identified as the ratio between the discharged and used water to produce the feedstock times the used water for cultivating the feedstock.

$$w_{m,r}^{discharge-feedstock\_Initial} = \gamma_{m,r}^{discharge} w_{m,r}^{used-feedstock\_Initial}, \quad \forall m, \ \forall r \tag{11.3}$$

11.3.1.2 Watershed at Final Conditions After Siting the Biorefinery

The water flowrate that exits any reach at the final conditions afterwards sitting the biorefinery (Q_r^{Out}) is equal to the inlet flowrate (Q_r^{In}) plus the precipitation (P_r), industrial discharges in the reach (D_r), residential discharges (H_r), the sum of effluents that enter from tributaries ($FT_{r,t}$), discharged wastewater from biorefineries ($w_{k,r}^{discharge-biorefinery}$) and all the wastewater discharged from fields of feedstocks ($w_{m,r}^{discharge-feedstocks}$), minus the sum of used water for cultivating feedstocks ($w_{m,r}^{used-feedstocks}$) and the used water for processing in biorefineries ($w_{k,r}^{used-biorefinery}$), minus the extraction for natural processes such as evaporation and filtration (L_r) and the one used for other human activities as agriculture (U_r):

$$Q_r^{Out} = Q_r^{In} + P_r + D_r + H_r + \sum_{t=1}^{N_{(t)}} FT_{r,t} + \sum_m w_{m,r}^{discharge-feedstocks} + \sum_k w_{k,r'}^{discharge-biorefinery} - \sum_m w_{m,r}^{used-feedstocks} - \sum_k w_{k,r'}^{used-biorefinery} - L_r - U_r, \quad \forall r \tag{11.4}$$

11.3.1.3 Total Water Balance for Each Tributary

The water exiting from the tributary ($FT_{r,t}$) can be determined based on the untreated effluents discharged to the tributary ($S_{r,t}^{untreated}$), plus the treated discharges ($S_{r,t}^{treated}$), industrial discharges ($I_{r,t}$), precipitation ($P_{r,t}$), direct discharges ($D_{r,t}$), minus the losses for precipitation and evaporation ($L_{r,t}$) and the used water ($U_{r,t}$):

$$FT_{r,t} = S_{r,t}^{untreated} + S_{r,t}^{treated} + I_{r,t} + P_{r,t} + D_{r,t} - L_{r,t} - U_{r,t}, \quad \forall r, \quad \forall t \tag{11.5}$$

11.3.1.4 Agricultural Discharges and Uses

The uses and discharges for the agricultural activity are determined as follows:

$$D_{r,t} = \alpha_{r,t}^{agriculture} A_{r,t}^{agriculture}, \quad \forall r, \quad \forall t \tag{11.6}$$

$$U_{r,t} = \beta_{r,t}^{agriculture} A_{r,t}^{agriculture}, \quad \forall r, \quad \forall t \tag{11.7}$$

Here, $\alpha_{r,t}^{agriculture}$ corresponds to the discharged water per agricultural area, $\beta_{r,t}^{agriculture}$ represents a factor for the required water per agricultural area, and $A_{r,t}^{agriculture}$ corresponds to the area used for agricultural activities.

11.3.1.5 Discharged Wastewater from Fields Used for Cultivating Feedstocks

A portion of the used water for cultivating feedstock m in field r ($w_{m,r}^{used-feedstock}$) is discharged as wastewater ($w_{m,r}^{discharge-feedstock}$) depending on the conversion factor ($\gamma_{m,r}^{discharge}$):

$$w_{m,r}^{discharge-feedstock} = \gamma_{m,r}^{discharge} w_{m,r}^{used-feedstock}, \quad \forall m, \quad \forall r \tag{11.8}$$

The conversion factor $\gamma_{m,r}^{discharge}$ is function of the type of feedstock and the type and location of the field, which is calculated experimentally.

11.3.1.6 Used Water for Cultivating Feedstocks

The used water for feedstock m in a field r ($w_{m,r}^{used-feedstocks}$) depends on a conversion factor ($\beta_{m,r}^{feedstocks}$) multiplied by the produced feedstock ($f_{m,r}^{Prod-feedstocks}$):

$$w_{m,r}^{used-feedstocks} = \beta_{m,r}^{feedstocks} f_{m,r}^{Prod-feedstocks}, \quad \forall m, \quad \forall r \tag{11.9}$$

The conversion factor $\beta_{m,r}^{feedstocks}$ is a function of the considered feedstock as well as the location of the field (because there are involved the land type, weather conditions, and type of feedstock), and this is calculated experimentally.

11.3.1.7 Wastewater Discharge from New Biorefineries

The wastewater discharge of new biorefineries ($w_{k,r}^{discharge-biorefinery}$) is calculated by a conversion factor ($\alpha_{k,m}^{ww-biorefinery}$) and the yielded products ($P_{k,m,r}$):

$$w_{k,r}^{discharge-biorefinery} = \sum_{m} \alpha_{k,m}^{ww-biorefinery} P_{k,m,r}^{Bio}, \quad \forall k, \quad \forall r \tag{11.10}$$

the conversion factor for the use of water ($\alpha_{k,m}^{ww-biorefinery}$) depends on the type of product and the used feedstock.

11.3.1.8 Water Used for Biorefineries

The used water for the biorefineries ($w_{k,r}^{used-biorefinery}$) is calculated using a conversion factor ($\alpha_{k,m}^{used-biorefinery}$) and the production in this plant ($P_{k,m,r}^{Bio}$):

$$w_{k,r}^{used-biorefinery} = \sum_{m} \alpha_{k,m}^{used-biorefinery} P_{k,m,r}^{Bio}, \quad \forall k, \quad \forall r \tag{11.11}$$

the conversion factor $\alpha_{k,m}^{used-biorefinery}$ is a function of the type of product and the used feedstock.

11.3.1.9 Used Land for Cultivating Feedstocks

The maximum accessible area for cultivating feedstock m in location r ($a_{m,r}^{\max}$) must be larger than the existing cultivated area ($a_{m,r}^{exist}$) plus the new cultivated area ($a_{m,r}^{new}$):

$$a_{m,r}^{exist} + a_{m,r}^{new} \leq a_{m,r}^{\max}, \quad \forall m, \quad \forall r \tag{11.12}$$

11.3.1.10 Water Required to Provide the Installed Biorefineries

The availability of water for installing biorefineries must be calculated as the required water to meet the water demand for the new biorefinery and it is determined as follows:

$$Water^{required} = Q_{\text{final}}^{\text{Initial}} - Q_{final} \quad (11.13)$$

11.3.2 Modelling of the Biorefinery System

The following set of mathematical relationships represents the performance of the biorefinery system, which involves the production of feedstocks, distribution of biomass, processing in biorefineries, and distribution of products to markets, which is described as follows.

11.3.2.1 Production in the Fields

The produced feedstock m in a field r ($f_{m,r}^{Prod-feedstock}$) is the product of a conversion factor $\alpha_{m,r}^{feddstock}$ times the cultivated area (existing and new):

$$f_{m,r}^{Prod-feedstock} = \alpha_{m,r}^{fedstock}\left[a_{m,r}^{exist} + a_{m,r}^{new}\right], \quad \forall m, \quad \forall r \quad (11.14)$$

where $\alpha_{m,r}^{fedstock}$ depends on the type of feedstock and the quality of the field, and this can be calculated before optimization. Furthermore, the fields may produce additional agricultural products and the feedstock for the biorefineries are sub-products. For example, the main product can be agave for tequila production and the feedstock is the lignocellulosic residue, or the main product may be corn and the feedstock corresponds to the corresponding straw. This way, the main product g for a field r associated with the feedstock m ($f_{g^{(m)},r}^{Prod-grains}$) is calculated by multiplying the corresponding conversion factor ($\alpha_{g^{(m)},r}^{grains}$) times the cultivated area:

$$f_{g(m),r}^{Prod-grains} = \alpha_{g(m),r}^{grains}\left[a_{m,r}^{exist} + a_{m,r}^{new}\right], \quad \forall m, \quad \forall r \quad (11.15)$$

11.3.2.2 Distribution of Feedstocks to Biorefineries

The produced feedstock m in any field r ($f_{m,r}^{Prod-feedstock}$) may be distributed to any biorefinery in place r' ($f_{m,r,r'}$):

$$f_{m,r}^{Prod-feedstock} = \sum_{r'} f_{m,r,r'}, \quad \forall m, \quad \forall r \quad (11.16)$$

11.3.2.3 Mass Balance for Feedstocks

The flowrate for the feedstock m produced in the cultivation site related to the reach $r(F_{m,r})$ may be split and sent to any biorefinery located in other $r'(f_{m,r,r'}\, f_{m,r,r'})$:

$$F_{m,r} = \sum_{r'} f_{m,r,r'}, \quad \forall m, \quad \forall r \tag{11.17}$$

Here r and r' correspond to the same places, and the alias is used for modeling this issue.

11.3.2.4 Activation of Binary Variables for Biorefineries

To calculate the existence of a biorefinery, the binary variable $y_{m,r}^{Biorefinery}$ is used. If this binary variable is one, the biorefinery is required, else if the binary variable is zero the biorefinery is not needed. This is modeled as follows:

$$F_{m,r} \le \mathrm{F}_{\mathrm{m,r}}^{\mathrm{max}}\, y_{m,r}^{Biorefinery}, \quad \forall m, \quad \forall r \tag{11.18}$$

where the parameter $\mathrm{F}_{\mathrm{m,r}}^{\mathrm{max}}$ is an upper limit on the capacity of the plant.

11.3.2.5 Product Mass Balance for Biorefineries

The yielded products k in the biorefinery located in place r $(P_{k,r})$ depend on the used technology, which has accompanying a conversion factor $(\alpha_{k,m}^{bioref})$, and the used feedstock m $(F_{m,r})$:

$$P_{k,m,r}^{Bio} = \alpha_{k,m}^{Bioref} F_{m,r}, \quad \forall k, \quad \forall m, \quad \forall r \tag{11.19}$$

$$P_{k,r} = \sum_{m} P_{k,m,r}^{Bio}, \quad \forall k, \quad \forall r \tag{11.20}$$

11.3.2.6 Distribution of Products

The total product k in site r $(P_{k,r})$ is sent to markets j $(P_{k,r,j}^{prod-market})$:

$$P_{k,r} = \sum_{j} P_{k,r,j}^{prod-market}, \quad \forall k, \quad \forall r \tag{11.21}$$

11.3.2.7 Products Obtained in Markets

The product k in the market j ($\sum_r P_{k,r,j}^{prod-market}$) must be lower than the maximum demand ($\mathrm{P}_{\mathrm{k,j}}^{prod-market-Max}$) in such location:

$$\sum_r P_{k,r,j}^{prod-market} \leq \mathrm{P}_{\mathrm{k,j}}^{prod-market-Max}, \quad \forall k, \ \forall j \tag{11.22}$$

11.3.2.8 Distribution of Grains to Markets

Certain biomass involves components that can be sold separately. For instance, when maize is used as feedstock, the edible portion (corn grains) is sold as a sub-product while the stover (leaves, husks, cob, and stalks) are fed to the biomass conversion facility. To avoid the usage of food to produce fuels and chemicals, the model considers the produced grains as sub-products to be sold in markets. Thus, the required grains, g, in the market r ($f_{g,r}^{prod-grains}$) are sent from all cultivation sites r' ($f_{g,r',r}^{grain-market}$):

$$f_{g,r}^{Prod-grains} = \sum_{r'} f_{g,r',r}^{grain-market}, \quad \forall g, \ \forall r \tag{11.23}$$

11.3.2.9 Grains

The total grains sent to each market ($\sum_{r'} f_{g,r',r}^{grain-market}$) must be lower than the maximum demand in such location ($\mathrm{f}_{\mathrm{g,r}}^{grain-market-Max}$):

$$\sum_{r'} f_{g,r',r}^{grain-market} \leq \mathrm{f}_{\mathrm{g,r}}^{grain-market-Max}, \quad \forall g, \ \forall r \tag{11.24}$$

11.3.3 Involved Costs

This section presents the mathematical relationships needed to calculate the involved costs in the supply chain.

11.3.3.1 Gains for Products

The gains (sales) for products ($G^{products}$) are a function of the unit price for the products ($UC_k^{product}$) times the distributed product to each market ($P_{k,r}$):

$$G^{products} = \sum_k \sum_{r'} UC_k^{product} P_{k,r'} \tag{11.25}$$

11.3.3.2 Gains Obtained from Grain Sales

The obtained gains from grain (as sub-product) sales (G^{grains}) are get by multiplying the unit price for each grain g (UC_g^{grains}) times the produced grains g in each site r ($f_{g,r}^{prod-grains}$):

$$G^{grains} = \sum_m \sum_r UC_m^{grains} f_{m,r}^{Prod-grains} \tag{11.26}$$

11.3.3.3 Capital Costs for Biorefineries

The capital cost for biorefineries ($CapCost^{Biorefineries}$) has involved a fixed part ($FC_{m,r}^{Biorefinery}$) and a variable part ($VC_{m,r}^{Biorefinery} \left(F_{m,r}\right)^{\sigma_{m,r}^{Biorefinery}}$), and these costs are determined as follows:

$$CapCost^{Biorefineries} = k_F \left[\sum_m \sum_{r'} FC_{m,r'}^{Biorefinery} y_{m,r'}^{Biorefinery} + \sum_m \sum_{r'} VC_{m,r'}^{Biorefinery} \left(F_{m,r'}\right)^{\sigma_{m,r'}^{Biorefinery}} \right] \tag{11.27}$$

$y_{m.r}^{Biorefinery}$ is a binary variable that is activated when the biorefinery exists in such a location and k_F is an annualization factor for the capital cost.

11.3.3.4 Operating Costs for Biorefineries

The operating costs of the biorefineries ($OpCost^{Biorefineries}$) should be calculated as a function of the multiplication of the total unit cost of all biorefineries ($UC^{Biorefinery}$) times the processed feedstocks in each location r ($F_{m,r}$).

$$OpCost^{Biorefineries} = \sum_m \sum_{r'} UC^{Biorefinery} F_{m,r'} \tag{11.28}$$

11.3.3.5 Operating Costs for Producing Feedstocks

The operating cost for producing the required feedstocks in the biorefineries ($OpCost^{Feedstocks}$) is calculated by contemplating the sum of the unit costs associated with each feedstock m in each harvesting location r ($UC_{m,r}^{feedstocks}$) times the flowrate of the produced feedstocks ($f_{m,r}^{prod-feedstocks}$):

$$OpCost^{Feedstocks} = \sum_m \sum_r UC_{m,r}^{feedstocks} f_{m,r}^{Prod-feedstock} \tag{11.29}$$

11.3.3.6 Water Cost for Refineries

The cost related with the needed water in the biorefineries ($OpCost^{water-biorefineries}$) depends on the unit water cost ($UC_{k,r}^{water-used-biorefineries}$) and the total flowrate of used water ($w_{k,r}^{used-biorefineries}$) for producing product k in site r:

$$OpCost^{water-biorefineries} = \sum_k \sum_{r'} UC_{k,r'}^{water-used-biorefineries} w_{k,r'}^{used-biorefineries} \tag{11.30}$$

$UC_{k,r}^{water-used-biorefineries}$ is the unit cost that is a function of the type of product and location.

11.3.3.7 Costs Associated with the Used Water for Cultivating Feedstocks

The cost associated with the used water for producing feedstocks ($OpCost^{water-feedstocks}$) varies with the unit cost assigned to the feedstock m in site r ($UC_{m,r}^{water-feedstocks}$) and the needed flowrate to cultivate the feedstock m in location r ($w_{m,r}^{used-feedstocks}$):

$$OpCost^{water-feedstocks} = \sum_m \sum_r UC_{m,r}^{water-feedstocks} w_{m,r}^{used-feedstocks} \tag{11.31}$$

The unit cost $UC_{m,r}^{water-feedstocks}$ is a function of the type of feedstock and the location because of the weather conditions and type of land.

11.3.3.8 Transportation Cost for Feedstocks

The transportation costs for feedstocks ($TCost^{feedstocks}$) involve the unit transportation cost ($UTC_{m,r',r}^{feedstocks}$) multiplied by the amount of feedstock ($f_{m,r,r'}$) to be distributed in each biorefinery:

$$TCost^{feedstocks} = \sum_{m}\sum_{r}\sum_{r'} UTC_{m,r,r'}^{feedstocks} f_{m,r,r'} \tag{11.32}$$

11.3.3.9 Transportation Costs for Grains

The transportation costs for grains ($TCost^{grains}$) depend on the unit transportation cost ($UTC_{g,r,r'}^{grains}$) and the amount of transported grains from a site r to a site r' ($P_{k,r,r'}^{prod-market}$):

$$TCost^{grains} = \sum_{m}\sum_{r}\sum_{j} UTC_{m,r,j}^{grains} f_{m,r,j}^{grain-market} \tag{11.33}$$

11.3.3.10 Transportation Costs for Products

The transportation costs of the products ($TCost^{products}$) depend on the unit transportation costs for product k from site r to site r' ($UTC_{k,r,r'}^{products}$) and the distributed amount of products ($P_{k,r,r'}^{prod-market}$):

$$TCost^{products} = \sum_{k}\sum_{r'}\sum_{j} UTC_{k,r',j}^{products} P_{k,r',j}^{prod-market} \tag{11.34}$$

11.3.4 *Objective Function for the Total Annual Profit*

The total annual profit (*Profit*) is a function of the gains for the sales of products ($G^{products}$) and grains (G^{grains}) minus the operating costs accompanying biorefineries ($OpCost^{Biorefineries}$), produced feedstocks ($OpCost^{Feedstocks}$), water associated with feedstocks ($OpCost^{water-feedstocks}$), water associated with biorefineries ($OpCost^{water-biorefineries}$), transportation costs for feedstocks ($TCost^{feedstocks}$), transportation costs for grains ($TCost^{grains}$), transportation costs for products ($TCost^{products}$) and the capital costs associated with biorefineries ($CapCost^{Biorefineries}$):

$$\begin{aligned}\text{Max}\, Profit = {} & G^{products} + G^{grains} - CapCost^{Biorefineries} - OpCost^{Biorefineries} - OpCost^{Feedstocks} \\ & - OpCost^{water-biorefineries} - OpCost^{water-feedstocks} - TCost^{feedstocks} \\ & - TCost^{grains} - TCost^{products} \end{aligned} \tag{11.35}$$

The implemented optimization formulation is a mixed integer non-linear programming (MINLP) problem (where just nonlinear terms are included in the mathematical formulation corresponding to the exponent to consider the economies of scale for the capital cost of the biorefinery).

11.4 Results and Discussion

A case study is considered to evaluate the presented optimization approach. The location of a biorefinery in the central-west part of Mexico was considered. Because of the soil and weather conditions in this region, the potential feedstocks are lignocellulosic residues of crops such as corn, wheat, and sorghum for ethanol production, and Jatropha and palm oil for biodiesel production. To meet the energy demands for biofuels, it is necessary to consider the opportunity to cultivate additional areas around the watershed as Fig. 11.5 presents. The required data were obtained from SAGARPA (2011) and the needed data for the cultivation parameters were taken from published reports (SENER-IDB-GTZ 2006; Contran et al. 2013). In the considered region, the Balsas watershed is used as the main water source. Data on this watershed are available in CONAGUA (2013). For implementing the MFA approach, the watershed is segregated into 23 reaches (Fig. 11.6). The involved potential markets and the demands for biofuels are presented in Table 11.1. In this case, considering the market prices, the economic parameters that were considered to estimate costs and gains are presented in Table 11.2. The sale price of bioethanol is US \$ 0.9/L and for biodiesel is US\$1.13/L. Whereas, the coefficient of scale for variable cost to install a biorefinery ($\sigma_{m,r'}^{Biorefinery}$) has been fixed as 1 (this means that the proposed MINLP model is converted into a MILP model). The optimization model for this case study was coded in the software GAMS (Brooke et al. 2015) and solved with the solver CPLEX in a computer with an Intel Core i7-3612QM CPU processor at 2.10 GHz and 6.00 GB of RAM.

The formulated algebraic optimization problem consists of 19,342 constraints, 23 binary variables, and 6314 continuous variables, and it was solved in 0.141 s of CPU time. The optimal solution that maximizes the total net profit is presented in Fig. 11.6, where seven biorefineries along the watershed (which are called with letters A-G) are required to meet the biofuel demands in the markets with a total net annual profit of US\$1.0954 $\times$ 10^8/year, which considers the installation cost of the biorefineries with financing of 10 years (thus, the annualization factor for the capital cost was 0.1). It should be noted in Fig. 11.6 that biorefinery A should be installed

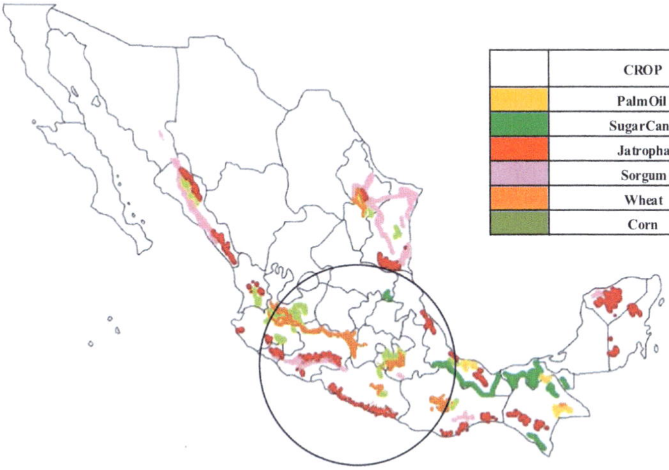

	CROP	POTENTIAL SURFACE (ha)
	PalmOil	252,651
	SugarCane	460,716
	Jatropha	2,009,122
	Sorgum	2,172,663
	Wheat	10,372,806
	Corn	11,571,690

Fig. 11.5 Potential areas to cultivate crops to produce biofuels in Mexico. Printed with permission of "*Optimal location of biorefineries considering sustainable integration with the environment*", *Renewable Energy*, Copyright (2017), Elsevier

Fig. 11.6 Distribution network of produced biofuel from biorefineries to markets. Printed with permission of "*Optimal location of biorefineries considering sustainable integration with the environment*", *Renewable Energy*, Copyright (2017), Elsevier

Table 11.1 Estimated demand for biofuels for each market of the case study

Estimated Demand (L/y)		
Market	Bioethanol	Biodiesel
1. Querétaro	3,236,729	3,999,526
2. Guadalajara	6,015,597	7,433,287
3. DF	35,602,846	43,993,334
4. Toluca	1,968,325	2,432,198
5. Cuernavaca	1,362,208	1,683,238
6. Morelia	2,403,467	2,969,890
7. Lázaro Cárdenas	719,285	888,798
8. Uruapan	1,268,484	1,567,427
9. León	4,983,835	6,158,370
10. Puebla	23,249,853	28,729,123

Table 11.2 Economic data for the case study

Crop	US$/Ton
Sale prices of grains	
Corn	224.4
Wheat	236.78
Sorghum	194.6
sugar cane	34.3
Jatropha	266.7
Palm oil	94.2
Feedstock prices	
Corn	11.22
Wheat	11.84
Sorghum	9.73
Sugar cane	3.623
Jatropha	13.33
Palm oil	4.71

in reach 1; whereas biorefineries B, C, D, E, F, and G should be in reaches 5, 6, 11, 14, 15, and 16, respectively. Furthermore, this solution considers that the total water required is 1.4125×10^8 m^3/year to properly implement this project (it should be noted that this value is calculated considering the flowrate leaving reach 23 at initial and final conditions). Likewise, it is important to mention that this value corresponds to a 0.6% concerning the total flow in Balsas system at initial conditions. Whereas, for this case are needed 47,862 has of additional area for wheat and jatropha crops; if this value is compared with existing total area (131,460 has), it means that there is required a 36.40% of new land to be used for agricultural purposes with respect to the current installed total capacity of agricultural area.

Moreover, the produced biofuels are distributed from the biorefineries and transported to the markets depending on the market demand for each reach and the economic feasibility as Fig. 11.7 presents for bioethanol and Fig. 11.8 for biodiesel. It should be noted in Fig. 11.7 that all the bioethanol demanded by all markets is satisfied; nevertheless, the biodiesel demand is only partially satisfied (which can be seen in Fig. 11.8). Similarly, it should be noted that Mexico City (i.e., market 3) has the highest demand for biofuels among the other markets, in addition, the market of Lazaro Cardenas is not receiving biodiesel. In most cases, the demands of the markets are satisfied through the nearest biorefinery. Fig. 11.9 shows the total amount per feedstock produced in each reach, where corn, wheat, and sorghum are harvested to produce bioethanol, while jatropha is used to produce biodiesel. This way, in the Balsas system there was no existing area for cultivating jatropha; then, to produce biodiesel, it is needed to cultivate new areas of jatropha. Fig. 11.10 presents the new land required to cultivate jatropha and wheat. Sugar cane and palm oil did not appear in the results because the regions to cultivate these crops are far from the Balsas River, which implies higher costs to produce biofuels due to the transportation of these feedstocks to the biorefineries.

Furthermore, the largest required cultivation area corresponds to reach 11 in the watershed, which is also the location for biorefinery D, which is the plant that produces bioethanol to meet the demand in the largest markets corresponding to Mexico City (market 3) and Cuernavaca (market 5). The main cultivation area is in the reaches 15 and 16, and the produced biodiesel is sent to the markets of Mexico City and Toluca. A high demand for biofuels requires a large amount of freshwater that is extracted from the watershed and involves the discharge of the generated wastewater The total wastewater discharges due to the biorefinery industrial activity is 1.82×10^5 m^3/year.

Two additional scenarios are proposed to determine the robustness of the model. Case A is called the optimistic one, where the biofuel and feedstock prices are greater, and the restriction for the demanded feedstock to produce biofuels is greater than the original problem. Case B (called the pessimist case) is the opposite of case A, where the biofuel and feedstock prices and the restriction of the maximum demand of feedstock by biorefineries are lower than in the original case. Solution for Case A yields a profit of 1.49×10^8 US$/year (40% greater than the original case study) with a total extracted water from the watershed of 1.46×10^8 m^3/year (which is 5.7% bigger respect to the original case), and it requires 49,447 has of additional area to cultivate wheat and jatropha (this represents an increase of 3.3% compared with the new area for the original case. Similarly, for case B, the profit is 5.97×10^7 US$/year (59.4% lower than the original case) and the total water required is 8.918×10^7 m^3/year (37% lower than the original case study) and finally the additional area

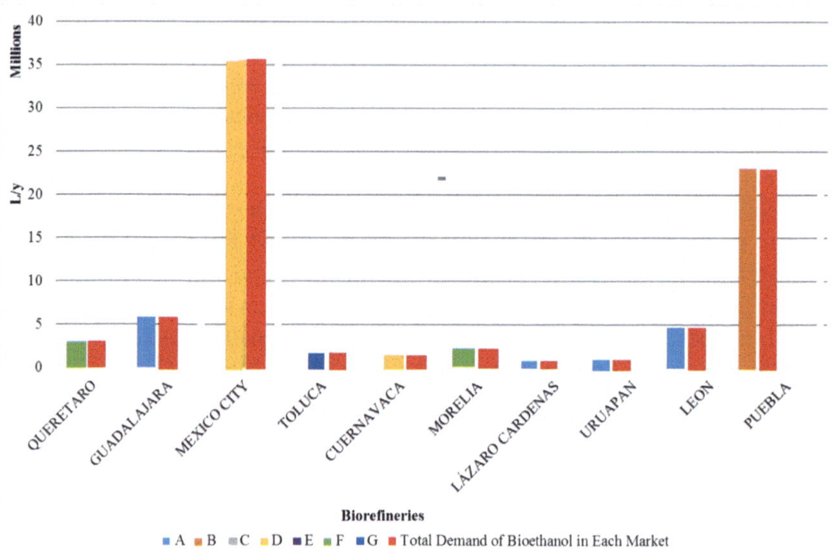

Fig. 11.7 Distribution of bioethanol in the markets. Printed with permission of "*Optimal location of biorefineries considering sustainable integration with the environment*", *Renewable Energy*, Copyright (2017), Elsevier

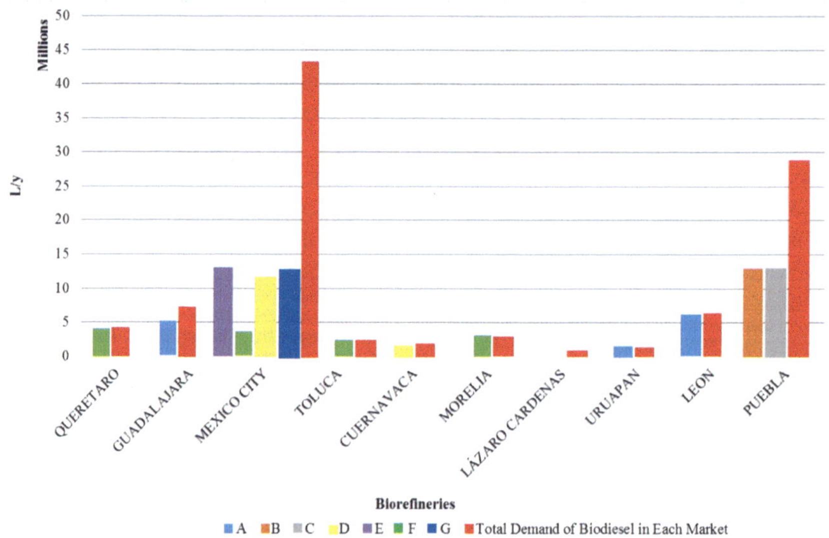

Fig. 11.8 Distribution of biodiesel in the markets. Printed with permission of "*Optimal location of biorefineries considering sustainable integration with the environment*", *Renewable Energy*, Copyright (2017), Elsevier

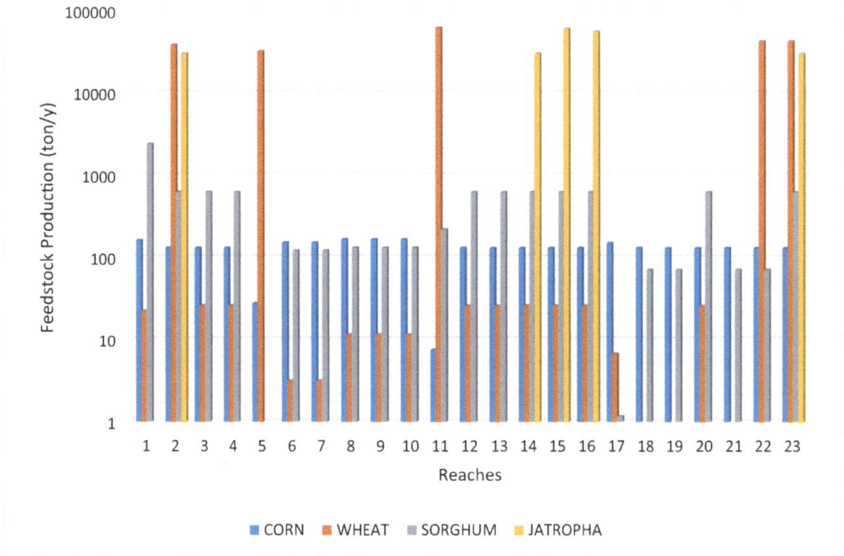

Fig. 11.9 Distribution of cultivated feedstock in the fields along the watershed. Printed with permission of "*Optimal location of biorefineries considering sustainable integration with the environment*", *Renewable Energy*, Copyright (2017), Elsevier

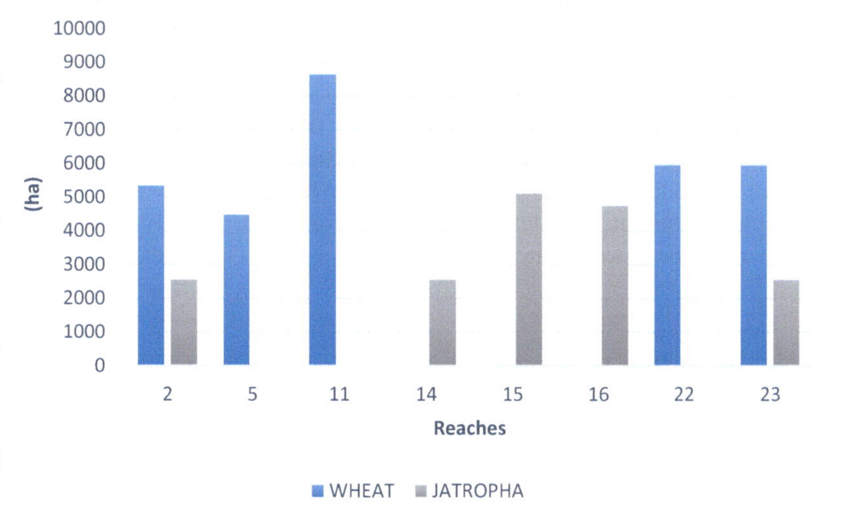

Fig. 11.10 Additional harvesting areas for feedstocks. Printed with permission of "*Optimal location of biorefineries considering sustainable integration with the environment*", *Renewable Energy*, Copyright (2017), Elsevier

is 23,860 has, (which represents a decrease of 50% concerning the original case to cultivate sorghum and jatropha). The locations and number of biorefineries do not change for cases A and B. The production of feedstocks is limited by the cost of water usage (environmental restriction to reduce the water consumption) along the watershed to generate lower effects and the available feedstock production according to the cultivating sites in each reach. However, case A presents a better condition to produce biofuels with the limited water demand in the model. On the other hand (case B), when the price of biofuel is lower to have a maximum profit the biorefinery system needs, a greater production of biofuels but it is not possible when the environmental conditions do not permit the affectation in the watershed.

Therefore, comparing these results there can be stated the following remarks:

- Case A (optimistic) considers a higher demand for products and better prices for biofuels, this creates a greater production of feedstocks and biofuels; subsequently, this yields a better performance for the economic goal (profit), but at the same time the environmental issues (water requirements and additional agricultural areas) are sacrificed.
- Case B (pessimist) corresponds to the opposite case, where lower demands and lower prices for biofuels generate a reduction in the profit (the economic objective is lower than the original case); but the water and new land requirements are reduced (this means that the environmental concerns are improved).

Even though, previous analyses represent an attractive comparison, the most important contribution of the presented approach is the quantification in terms of water and new land requirements to meet a determined demand for biofuels. Where most previous works on this topic have stated that the main disadvantages of biofuel production are the considerable amount of water and new land required; however, the present approach properly quantifies these aspects. Besides, it is important to highlight that there is a large uncertainty for several parameters involved in this model, such as the precipitation discharges, losses owing to filtration, evaporation, etc., and there are uncertainties in the industrial and residential discharges and even in the raw material costs to be processed by the biorefinery as well as the bioethanol price.

11.5 Concluding Remarks

An optimization approach for distributed biorefinery system design that considers the surrounding watershed has been provided in this chapter. The ideal placement for growing sites and biorefining facilities is incorporated into the presented mathematical model. The distribution of raw materials and finished products as well as the best choice of feedstocks are also included in the mathematical model. The best practices for water use and wastewater discharge are identified by directly addressing how the biorefinery supply chain interacts with the local watershed. The objective function takes into consideration the watershed's environmental limits to maximize net annual profit.

A case study from central-west Mexico has been taken into consideration. The findings demonstrate that it is possible to meet the watershed limits and produce the necessary biofuels while still reaping economic benefits. The ideal placement for the farming sites and biorefining facilities has also been established by the optimization formulation. Furthermore, the presented optimization formulation is a general approach that may be used for other case studies, provided that the information is gathered in accordance with the intended system.

Furthermore, it is evident from the results of the case study that it is feasible to generate substantial profits while meeting the need for biofuels and displacing fossil fuels. Environmental concerns were examined in addition to economic ones. A greater economic performance can then be achieved under other circumstances, but doing so also deteriorates the environment and vice versa. As a result, environmental and economic functions are incompatible. Ultimately, the most significant addition to the presented approach is the quantification of the new land and water needed to meet the demands for biofuels—a step that has not been taken by the majority of earlier studies.

Nomenclatures

Parameters

$a_{m,r}^{exist}$	Existing cultivated area of feedstock, ha
$a_{m,r}^{\max}$	Maximum area for cultivating feedstock, ha
$A_{r,t}^{agriculture}$	Cultivated area associated with the tributary t in the reach r, ha
D_r	Direct water discharges to the reach r, m^3/s
$D_{r,t}$	Agricultural water discharges in the tributary t of the reach r, m^3/s
$\mathrm{f}_{\mathrm{m,j}}^{\text{grain-market-Max}}$	Maximum grain flowrate of m sent to market j, ton/year
$\mathrm{F}_{\mathrm{m,r'}}^{\max}$	Maximum feedstock m that can be produced in the field r', ton/year
$FC_{m,r'}^{Biorefinery}$	Fixed cost to install biorefineries, US$
$FT_{r,t}$	Water discharge from the tributary t to the reach r, m^3/s
H_r	Total wastewater discharge (ie., industrial, sanitary) to the reach r, m^3/s
$I_{r,t}$	Industrial wastewater discharged to the tributary t of the reach r, m^3/s
L_r	Total water losses (filtration and evaporation) from the reach r, m^3/s
$L_{r,t}$	Total water losses (filtration and evaporation) in the tributary t to the reach r, m^3/s

$P_{k,j}^{prod\text{-}market\text{-}Max}$	Maximum product flowrate k sent to market j, l/year
P_r	Precipitation discharged to the reach r, m^3/s
$P_{r,t}$	Precipitation discharged in the tributary t of the reach r, m^3/s
$Q_r^{Initial}$	Initial flowrate leaving the reach r, m^3/s
$Q_{r-1}^{Initial}$	Initial flowrate leaving the reach r-1, m^3/s
Q_{r-1}	Flowrate leaving the reach r-1, m^3/s
$S_{r,t}^{untreated}$	Residual wastewater without treatment discharged to the tributary t of the reach r, m^3/s
$S_{r,t}^{treated}$	Residual treated wastewater discharged to the tributary t of the reach r, m^3/s
$TCost^{feedstocks}$	Transportation cost for feedstocks, US$/km
$TCost^{grains}$	Transportation cost for grains, US$/km
$TCost^{products}$	Transportation cost for products, US$/km
U_r	Water used from reach r, m^3/s
$U_{r,t}$	Water used from tributary t that discharges to reach r, m^3/s
$UC^{Biorefinery}$	Total unit processing cost for biorefinery, US$/ton/year
$UC_{m,r}^{feedstocks}$	Total unit cost to produce feedstock m in field r, US$/ton/year
UC_m^{grains}	Unit price for grain m, US$/ton
$UC_k^{product}$	Unit price for product k, US$/ton
$UC_{m,r}^{water-feedstocks}$	Unit cost for used water to cultivate feedstock m in field r, US/m^3
$UC_{k,r'}^{water-used-biorefineries}$	Unit cost for used water to operate the biorefinery k in location r', US$/$m^3$
$UTC_{m,r,r'}^{feedstocks}$ $UTC_{m,r,r'}^{feedstocks}$	Unit transportation cost for feedstock m from reach r to r', US$/km/ton/year
$UTC_{m,r,j}^{grains}$	Unit transportation cost for grain m produced in field r sent to market j, US$/km/ton/year
$UTC_{k,r',j}^{products}$	Unit transportation cost for product k produced in biorefinery r' send to market l, US$/km/ton/year
$VC_{m,r'}^{Biorefinery}$	Variable cost to install a biorefinery system, US$/ton/year
$w_{m,r}^{discharge-feedstocks\_Initial}$	Total initial discharged water flowrate to produce feedstock m in the reach r, m^3/s
$w_{m,r}^{used-feedstocks\_Initial}$	Total initial used water flowrate to produce feedstock m in the reach r, m^3/s

Greek symbols

$\alpha_{r,t}^{agriculture}$	Required agricultural flowrate related to the tributary t that discharges to the reach r, m^3/ha s
$\alpha_{k,m}^{Bioref}$	Conversion factor to produce fuel k from feedstock m
$\alpha_{m,r}^{feddstock}$	Conversion factor to produce feedstock m in the reach r
$\alpha_{m,r}^{grains}$	Conversion factor for cultivating grains m in the reach r
$\alpha_{k,m}^{ww-biorefinery}$	Conversion factor for the wastewater of biorefineries
$\alpha_{k,m}^{used-biorefinery}$	Conversion factor for the used water for biorefinery in r'
$\beta_{r,t}^{agriculture}$	Discharged agricultural flowrate related to the tributary t that discharges to the reach r, m^3/ha s
$\beta_{m,r}^{feedstocks}$	Conversion factor that represents the total used water to produce feedstock m in field r
$\gamma_{m,r}^{discharge}$	Conversion factor to represents the relation with the discharged water and used water to produce feedstock m in field r
$\sigma_{m,r'}^{Biorefinery}$	Coefficient of scale for variable cost to install a biorefinery

Variables

$a_{m,r}^{new}$	New area for cultivating feedstock, ha
$CapCost^{Biorefineries}$	Capital cost for the installation of biorefinery system, US$
$f_{m,r,r'}$	Produced feedstock m sent to biorefinery in location r' from field in r, ton/year
$f_{m,r}^{Prod-grains}$	Produced grain flowrate m in the field r, ton/year
$f_{m,r,j}^{grain-market}$	Grain flowrate m produced on the field in r sent to market j, ton/year
$f_{m,r}^{Prod-feedstocks}$	Feedstock m produced in the field r, ton/year
$F_{m,r}$	Feedstock m flowrate produced in the field r, ton/year
G^{grains}	Gains due to sell of grains, US$/year
$G^{products}$	Gains due to sell of products, US$/year
$OpCost^{Biorefineries}$	Operating cost for biorefineries, US$/year
$OpCost^{Feedstocks}$	Operating cost for feedstocks, US$/year
$OpCost^{water-biorefineries}$	Operating cost for consumed water in the biorefineries US$/$m^3$/year
$OpCost^{water-feedstocks}$	Operating cost for consumed water in the field to produce feedstock US$/$m^3$/year

$P_{k,r'}$	Yielded product flowrate *k* produced for the biorefinery located in the reach *r'*, ton/year
$P^{Bio}_{k,m,r'}$	Flowrate of biofuels *k* produced from feedstock *m* in the biorefinery in *r'*, l/year
$P^{prod-market}_{k,r',j}$	Total produced fuels *k* flowrate for biorefinery in *r'* sent to market *j*, l/year
Profit	Total annual profit, US$/year
Q^{Out}_{r}	Flowrate leaving the reach *r*, m^3/s
Q^{In}_{r}	Flowrate entering the reach *r*, m^3/s
$Water^{required}$	Total required water flowrate to satisfy the water demand for the biorefinery system, m^3/s
$w^{discharge-biorefinery}_{k,r'}$	Discharged water flowrate in biorefineries to produce biofuels *k* in the location *r'*, m^3/s
$w^{used-biorefinery}_{k,r'}$	Used water flowrate in the biorefineries to produce biofuels *k* in the location *r'*, m^3/s
$w^{discharge-feedstock}_{m,r}$	Discharged water flowrate to produce feedstock *m* in the field *r*, m^3/s
$w^{used-feedstock}_{m,r}$	Discharged water flowrate to produce feedstock *m* in field *r*, m^3/s

Binary Variables

$y^{Biorefinery}_{m,r'}$	Binary variable to biorefinery installation

Subscripts

i	Process sources
j	Market
k	Products (biofuels)
m	Feedstock
r	Reach
r'	Possible location of biorefinery
t	Tributary

References

Ba BH, Prins C, Prodhon C (2015) Models for optimization and performance evaluation of biomass supply chains: an operations research perspective. Renew Energy 87:977–989

Bowling IM, Ponce-Ortega JM, El-Halwagi MM (2011) Facility location and supply chain optimization for a biorefinery. Ind Eng Chem Res 50:6276–6286
Brooke A, Kendrick D, Meeruas A, Raman R (2015). GAMS-Language guide. GAMS Development Corp., Washington, DC, USA
Burgara-Montero O, El-Baz AA, Ponce-Ortega JM, El-Halwagi MM (2013) Optimal design of a distributed treatment system for increasing dissolved oxygen in watersheds through self-rotating discs. ACS Sustain Chem Eng 1(10):1267–1279
CONAGUA (2013) Water Atlas in Mexico. Mexican National Water Commission, Mexico City, Mexico
Contran N, Chessa L, Lubino M, Bellavite D, Roggero PP, Enne G (2013) State-of-the-art of the Jatropha curcas productive chain: from sowing to biodiesel and by-products. Ind Crops Prod 42:202–215
Corsano G, Vecchietti AR, Montagna JM (2011) Optimal design for sustainable bioethanol supply chain considering detailed plant performance model. Comput Chem Eng 35:1384–1398
Cucek L, Martín M, Grossmann IE, Kravanja Z (2014) Multi-period synthesis of optimally integrated biomass and bioenergy supply network. Comput Chem Eng 66:57–70
De Meyer A, Cattrysse D, Van Orshoven J (2015) Considering biomass growth and regeneration in the optimization of biomass supply chains. Renew Energy 87:990–1002
El-Baz AA, Ewida KT, Shouman MA, El-Halwagi MM (2005) Material flow analysis and integration of watersheds and drain systems II: integration and solution strategies with application to ammonium management in Bahr El-Baqar drain system. Clean Technol Environ Policy 7(1):78–86
El-Halwagi AM, Rosas V, Ponce-Ortega JM, Jiménez-Gutiérrez A, Mannan MS, El-Halwagi MM (2013) Multi-objective optimization of biorefineries with economic and safety objectives. AIChE J 59(7):2427–2434
Giarola S, Bezzo F (2015) Bioethanol supply chain and design and optimization: some achievements and future challenges for the development of sustainable biorefineries. Comput Aided Chem Eng 36:1570–7946. https://doi.org/10.1016/B978-0-444-63472-6.00022-7
Guillén-Gosálbez G, Grossmann I (2010) A global optimization strategy for the environmentally conscious design of chemical supply chains under uncertainty in the damage assessment model. Comp Chem Eng 34:42–58
Kelloway A, Daoutidis P (2014) Process synthesis of biorefineries: optimization of biomass conversion to fuels and chemicals. Ind Eng Chem Res 53:5261–5273
Kent S (2009) Biofuels in the U.S.—challenges and opportunities. Renew Energy 34:14–22
Kim JK, Realff MJ, Lee JH (2011) Optimal design and global sensitivity analysis of biomass supply chain networks for biofuels under uncertainty. Comput Chem Eng 35:1738–1751
Kokossis AC, Tsakalova M, Pyrgakis K (2015) Design of integrated biorefineries. Comput Chem Eng 81:40–56
Lehtonen O, Okkonen L (2016) Socio-economic impacts of a local bioenergy-based development strategy—the case of Pielinen Karelia, Finland. Renew Energy 85:610–619
Lira-Barragán LF, Ponce-Ortega JM, Nápoles-Rivera F, Serna-González M, El-Halwagi MM (2013) Incorporating property-based water networks and surrounding watersheds in site selection of industrial facilities. Ind Eng Chem Res 52(1):91–107
López-Díaz DC, Lira-Barragán LF, Rubio-Castro E, Ponce-Ortega JM, El-Halwagi MM (2017) Renew Energy 100:65–77
López-Villarreal F, Lira-Barragán LF, Rico-Ramírez V, Ponce-Ortega JM, El-Halwagi MM (2014) An MFA optimization approach for pollution treading considering the sustainability of the surrounding watershed. Comput Chem Eng 63:140–151
Marvin WA, Schmidt LD, Daoutidis P (2013) Biorefinery location and technology selection through supply chain organization. Ind Eng Chem Res 52:3192–3208

Murrillo-Alvarado PE, Santibañez-Aguilar JE, Ponce-Ortega JM, Castro-Montoya AJ, Serna-Gonzales M, El-Halwagi MM (2014) Optimization of the supply chain associated to the production of bioethanol from residues of agave from the tequila process in Mexico. Ind Eng Chem Res 53:5524–5538

Noureldin MMB, El-Halwagi MM (2015) Synthesis of C-H-O symbiosis networks. AIChE J 64(4):1242–1262

Okwo A, Thomas VM (2014) Biomass feedstock contracts: role of land quality and yield variability in near term feasibility. Energy Econ 42:67–80

Osmani A, Zhang J (2014) Economic and environmental optimization of a large scale sustainable dual feedstock lignocellulosic-based bioethanol supply chain in a stochastic environment. Appl Energy 114:572–587

Ruiz-Femenia R, Guillén-Gosálbez G, Jiménez L, Caballero JA (2013) Multi-objective optimization of environmentally conscious chemical supply chains under demand uncertainty. Chem Eng Sci 95:1–11

SAGARPA-SIAP (2011) Mexican system of information about agriculture and fishing. Advance of planting and harvesting for Mexico. Mexico City, Mexico. http://www.siap.gob.mx/index.php?option¼com_wrapper&;view¼wrapper&Itemid¼347. Accessed 01 Feb 2012

Sandun F, Sushil A, Chauda C, Naveen M (2006) Biorefineries: current status, challenges, and future direction. Energy Fuels 20(4):1727–1737

Santibañez-Aguilar JE, González-Campos JB, Ponce-Ortega JM, Serna-González M, El-Halwagi MM (2011) Optimal planning of a biomass conversion system considering economic and environmental aspects. Ind Eng Chem Res 50(14):8558–8570

Santibañez-Aguilar JE, González-Campos JB, Ponce-Ortega JM, Serna-Gonzalez M, El-Halwagi MM (2014) Optimal planning and site selection for distributed multiproduct biorefineries involving economic, environmental and social objectives. J Clean Prod 65:270–294

Santibañez-Aguilar JE, González-Campos JB, Ponce-Ortega JM, Serna-Gonzalez M, El-Halwagi MM. (2015). Supply chains and optimization for biorefineries. In: You F (ed) Sustainability of products, processes and supply chains-theory and applications, Chapter 19. Computer aided chemical engineering, vol 36. Elsevier, Amsterdam, Netherlands, pp 475–497

Santibañez-Aguilar JE, Rivera-Toledo M, Flores-Tlacuahuac A, Ponce-Ortega JM (2015b) A mixed-integer dynamic optimization approach for the optimal planning of distributed biorefineries. Comput Chem Eng 80:37–62

SENER-IDB-GTZ (2006) Potential use of bioethanol in Mexico. Ministry of Energy of Mexico, Mexico City

Stuart PR, El-Halwagi MM (2012) Integrated biorefineries: design, analysis, and optimization. Green chemistry and chemical engineering series. CRC Press, Boca Raton, USA

Tong K, Gong J, Yue D, You F (2012) Stochastic programming approach to optimal design and operations of integrated hydrocarbon biofuel and petroleum supply chains. Am Chem Soc Sustain Chem Eng 2:49–61

Yeh K, Whittaker G, Realff MJ, Lee JH (2015) Optimal harvest management adaptation for a new biorefinery investment in a timberlands supply chain using a modified cyclic scheduling model. Comput Aided Chem Eng 36:1570–7946

You F, Tao L, Graziano DJ, Snyder SW (2012) Optimal design of sustainable cellulosic biofuel supply chains: multiobjective optimization coupled with life cycle assessment and input-output analysis. AIChE J 58:1157–1180

Yue D, You F, Snyder SW (2014) Biomass-to-bioenergy and biofuels supply chain optimization: overview, key issues and challenges. Comput Chem Eng 66:36–56

Zhang FL, Johnson DM, Johnson MA (2012) Development of a simulation of biomass supply chain for biofuel production. Renew Energy 44:380–391

Chapter 12
Optimizing the Placement of New Industrial Facilities Dynamically with Watershed Sustainability Considerations

12.1 Introduction

The most significant natural water storage areas are watersheds, and these are where water streams are moved until they are disposed of (lake, sea, or ocean). Numerous human activities, including residential, commercial, and agricultural regions, are located close to watersheds and have a substantial impact on the quantity and quality of available water. Consequently, the primary cause of river contamination is these activities. Because of this, it is crucial to research the effects of new industrial discharges before choosing a site for new industrial facilities.

In this regard, economic factors are typically the only ones considered when deciding where to locate a new industrial plant; the environmental effects that the new industrial effluents will produce over the watershed are not taken into consideration (El-Baz et al. 2005a). Following the installation of the new facility, the authorities compelled the industry to comply with environmental standards; as a result, the wastewater streams must have treatment networks established. This additional expense could result in unfeasible operations. Accordingly, several approaches have been reported to deal with the issue of recycling and reusing inside industrial facilities to conserve resources (Dunn and El-Halwagi 2003; El-Halwagi 1997, 2006; El-Halwagi and Spriggs 1998). Additionally, approaches for resource conservation can be categorized as mathematical programming models and pinch techniques (Pillai and Bandyopadhyay 2007). However, the pinch technologies employ design-based heuristic criteria that are based on targeting and sequencing techniques (Wang and Smith 1994; Dhole et al. 1996; Polley and Polley 2000; Hallale 2002; Manan et al. 2004; El-Halwagi et al. 2003; Feng et al. 2007). Foo (2009) presented a review of the pinch strategies for the water network synthesis. Three further sets of approaches for conserving resources, based on algebraic techniques, have been published by Sorin and Bedard (1999), Agrawal and Shenoy (2006), and Gomes et al. (2007). Lastly, it has been reported that mathematical programming formulations can automate the creation of recycle and reuse networks within the plant (Savelski and Bagajewicz

J. M. Ponce-Ortega et al., *Sustainable Water Systems*, Water Resources Development and Management, https://doi.org/10.1007/978-3-031-72498-5_12

2000, 2001, 2003; Alva-Argaez et al. 1998; 1999; Benko et al. 2000; Teles et al. 2008; Kuo and Smith 1997; Doyle and Smith 1997; Galan and Grossmann 1998; Hernandez-Suarez et al. 2004; Gunaratnam et al. 2005; Karuppiah and Grossmann 2006; Putra and Amminudin 2008; Ponce-Ortega et al. 2009a; Nápoles-Rivera et al. 2010). However, the influence of wastewater streams when they enter the environment (i.e., watersheds) has not been taken into consideration, and all previous formulations have only considered the processes that take place inside industrial facilities, neglecting the processes that occur outside the plant.

However, during their course, watersheds are also damaged by other toxic discharges from other sources, such as agricultural areas, treated and untreated wastewater, precipitation, city zones, etc. Although it is important to note that these discharges significantly alter the stream conditions in the rivers, the natural degradation of pollutants occurs in the rivers because of the flora and fauna's chemical, biochemical, and physical changes (Brunner and Rechberg 2004). Then, Baccini and Brunner (1991) reported a material flow analysis (MFA) model for assessing ecosystems with human activities where mass, energy, and information are being exchanged with the surroundings about the pollutant behavior modeling in the watershed system. Furthermore, an MFA model was presented by Lampert and Brunner (1999) to monitor nutrient fluxes in the Danube River basin. An MFA model to monitor nitrogenous chemicals in the Bahr El-Baqar drainage system was presented by El-Baz et al. (2005a); an MFA model with mass integration was presented by El-Baz et al. (2005b). To ensure sustainable processes, Lovelady et al. (2009) presented a reverse problem formulation based on an MFA model that integrates industrial process discharges with macroscopic watershed systems. The main disadvantage of this model is that it uses a set of heuristic rules and does not simultaneously optimize the process. The key idea is to fix the maximum concentration of pollutants discharged into the environment to prevent hazardous accumulation at the final disposal.

Furthermore, to determine the best location for new industrial plants, Lira-Barragán et al. (2010, 2011, 2013) proposed innovative approaches that consider the impact of newly polluted discharges on the watersheds as well as the compositions for key pollutants or properties of interest. However, the previous methodologies have not considered the dynamic related to this type of process, which results in inadequate solutions because of all the interactions mentioned above between the watersheds and their surroundings, which are not constant and change over time. As a result, the conditions within the watersheds are also modified, and environmental restrictions should not be satisfied during critical periods. In this regard, reviews by Mendez et al. (2006), Floudas and Lin (2004), Kallrath (2002), and Barbosa-Povoa (2007) emphasized the significance of scheduling optimization methods in the process industry, while other authors have effectively used these techniques to address a range of manufacturing issues (e.g., Mouret et al. 2009, 2011; Sharda and Vazquez 2009; Terrazas-Moreno et al. 2012; Gutiérrez-Limón et al. 2012; Erdirik-Dogan and Grossmann 2008; Barbosa-Povoa and Macchietto 1994; Lin and Floudas 2001; Castro et al. 2005). Nevertheless, there is currently a lack of research on dynamically optimizing the best location for new industrial units in watersheds.

To address the shortcomings, this chapter presents a dynamic model formulation that combines the water treatment networks for new plants with the watershed system, where the pollutants from the plant are released while considering the possibility of installing the new plant in various locations. To determine the behavior of the watershed throughout the year, the model considers variations in the processes associated with the watershed. The best location for the new plant will determine the quantity and quality of wastewater that is released into the watershed. This is because the location will have a significant impact on the watershed, considering the interactions with other streams as well as the chemical and biological reactions that take place there. Additionally, the new plant's location will have a significant impact on the amount of treatment that is needed. The model that has been suggested is a disjunctive programming formulation that has been reformulated as a MINLP problem, considering the process and environmental regulations, the sustainability of the watershed system, and the minimization of the total annual cost for the recycle and reuse network and the installation of the new plant. The model can consider various contaminants present in waste streams and the choice of multiple solutions for their treatment (Lira-Barragán et al. 2015).

12.2 Problem Statement

The problem addressed in this chapter can be stated as follows. Given are:

- A watershed system with a mouth in final disposal (i.e., lake, sea, or ocean). The problem involves all the fluctuations and dynamic fluctuations in the subsystem that interact with the watershed such as industrial, residential, and agricultural discharges and extractions, and the seasonal variations for the precipitation, filtration, and evaporation. The watershed is divided into reaches ($R = \{r|r = 1, 2, \ldots, N_r\}$) and tributaries ($T = \{t|t = 1, 2, \ldots, N_t\}$) to simulate its behavior.
- A set of alternatives to site a new industrial facility ($P = \{p|p = 1, 2, \ldots, N_p\}$). Each authorized location p has associated an installation cost ($Cland_p$). This term involves the cost associated to the transportation of the raw materials, final products, services, and land cost.
- A set of pollutants ($C = \{c|c = 1, 2, \ldots, N_c\}$) discharged from the new industrial plant to the watershed.
- A set of interceptors to treat the first compound (INT^1), the second pollutant (INT^2), and the N component considered by the system, which is treated in the last set of interceptors (INT^N). Also, the operation of the treatment system is implemented under dynamic conditions accounting for the variations in the watershed. Each interceptor has accompanying a correction factor to simulate its capacity to remove the components.

- Lastly, a set of environmental constraints that meet the continuity in the typical activities carried out around the watershed and keep the system under sustainable conditions after the new industrial plant is operating.

The next challenge is to design and configure an interception network under dynamic operation that satisfies environmental regulations and environmental constraints for pollutants through the watershed system while also avoiding the accumulation of pollutants at the final disposal. This must be done in tandem with determining the best location for new industrial facilities. The objective function is to minimize the total annual cost (TAC), which includes the costs associated with treating the wastewater stream and installing the new plant, both of which are heavily influenced by the option that is chosen. However, the primary non-linearities in the mathematical formulation are in the component balances for the reaches, namely in the composition for each reach multiplied by the flowrate. The features of each reach in this context are known before the installation of new industrial facilities; the flowrate and the compositions of each reach become optimization variables because of the consideration of installing the new industrial plant and, subsequently, the potential discharge of this new wastewater to the watershed system (the waste stream's composition is a variable in the model). Furthermore, one of the most crucial variables to be determined by the optimization process is the pollutant concentration, which is affected by the treatment scheme chosen. This is because the new discharge must comply with environmental regulations imposed on industrial discharges and must not interfere with human activities occurring within the watershed (for example, if a water stream is extracted for agricultural use, the new polluted discharge must not cause the water to become inadequately clean for agricultural use). It also must prevent the accumulation of pollutants at the final disposal site, maintaining the system under sustainable conditions once the new industrial plant is operational. Figure 12.1 shows an illustration of the aforementioned. The watershed subdivision required by the MFA technique to model the system is also depicted in this figure. The river's main stream is divided into multiple reaches, which are represented by ovals. Each reach, which is a zone where no significant intermediate effluents are extracted or discharged, can receive multiple tributaries, which can be either a channel or a branch of the river. Typically, watersheds interchange water streams with various users along their path, including residential, commercial, and agricultural areas. These systems also experience natural phenomena including vaporization, filtration, and precipitation. The characteristics of the river are altered greatly by all the previous processes. Therefore, to accurately simulate the behavior of the river, the model that is selected for this task must incorporate the previously mentioned processes as well as a kinetic term that considers the chemical and biochemical interactions that occur between the pollutants and the aquatic flora and fauna, which can break down or produce hazardous materials.

Finally, a significant factor taken into consideration by this work is the variation concerning the time of all the flowrates and compositions incorporated in the mathematical formulation through the dynamic optimization, which produces annual profiles for the flowrates and composition of pollutants. Given that the conditions

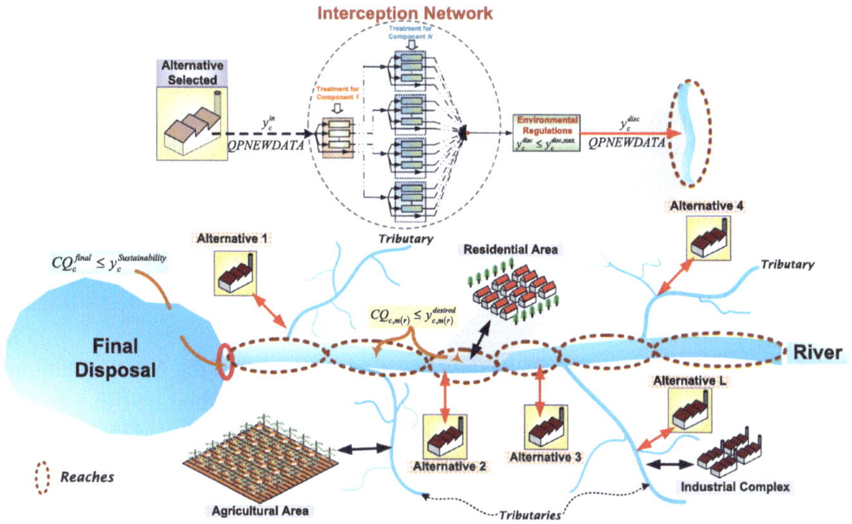

Fig. 12.1 Options to site the new industrial plant in a watershed system and the proposed treatment scheme. Printed with permission of "*Dynamic optimization for optimal location of new industrial facilities considering the sustainability of the watershed*", Sustainability of Products, Processes and Supply Chains, Copyright (2015), Elsevier

of the watershed system can vary greatly throughout the year, it can be very helpful to discuss the seasonal variations of its processes. In particular, the flowrate and component concentration are highly dependent on the presence or absence of the rainy season.

12.3 Model Formulation

Initially, the used sets in the mathematical model are defined for a better understanding: the subindex that represents the pollutant or component is c, thus N_c corresponds to the total number of pollutants and C is the set that contains all the components; r represents a subindex used to represent a section of the river, while N_r corresponds to the total number of sections of the river used for a specific problem, R represents the set that contains all r's; j corresponds to the subindex for the effluents discharged to the reach r, N_j represents the total number of effluents and J their set; also, the superindex int^N is the interceptors for the treatment of the pollutant N, while $Nint^N$ is the total number of interceptors for the treatment and INT^N is the set used to the interceptors; finally, the subindex representing the locations for the new plant is p, whereas N_p is the total number of possible locations, and P represents the set associated to the possible locations.

The mathematical model is based on the material flow analysis technique in the dynamic state to predict the watershed behavior during the time. Furthermore, the mathematical formulation contains the treatment for the wastewater stream considering multiple-pollutants. Lastly, the approach considers all the water quality restrictions for reaches where some water streams are extracted for human activities, and the objective function is established. The model generated is presented as follows.

12.3.1 Material Flow Analysis

The dynamic behavior of the watershed is simulated through the MFA technique to determine the variation of the pollutants along the watershed and during the time, considering all the existing interactions with the environment. This way, the watershed is segregated into reaches and tributaries to obtain proper modeling. The proposed approach only considers the possible location for the new industrial facilities in reaches (generating dynamic balances); thus, all the parameters involved in the tributary balances are known and consequently, these balances can be determined prior to the optimization process. Therefore, the required material and component balances are described as follows.

12.3.1.1 Overall Balance for Each Reach

This balance states that the volume variation in the reach r concerning the time (dv_r/dt) is equal to the inlet flowrate (Q_{r-1}), plus precipitation (P_r), direct industrial discharges (Id_r), agricultural discharges (D_r), residential discharges (Hd_r), the sum of all effluents entering to the reach ($FT_{r,t}$) and the potential discharge for the new plant when the alternative p to install a new industrial plant is near to the reach r ($QPNEW_{r(p)}$), minus the exit flowrate (Q_r), and the extractions for filtration and evaporation (L_r) and the used water (U_r) in the r section of the river. This dynamic balance can be stated as follows:

$$\frac{dV_r}{dt} = Q_{r-1}(t) + P_r(t) + Id_r(t) + D_r(t) + Hd_r(t) + \sum_{t=1}^{N_{t(r)}} FT_{r,t}(t) + QPNEW_{r(p)}(t) - Q_r(t) - L_r(t) - U_r(t), \quad \forall r \in R \tag{12.1}$$

$N_{t(r)}$ is to the total number of tributaries that are discharged to reach r.

To manipulate the differential equations, this chapter transforms the differential equations (required for the dynamic optimization) into a couple of algebraic equations using the orthogonal collocation method, and the finite element method. Therefore, the previous differential equation is discretized as follows:

$$V_{r,i,j} = V^o_{r,i} + h_i \sum_{k \in K} A_{k,j} \dot{V}_{r,i,k}, \quad \forall r \in R, \quad i \in I, \quad j \in J \tag{12.2}$$

$$\dot{V}_{r,i,j} = Q_{r-1,i,j} + P_{r,i,j} + Id_{r,i,j} + D_{r,i,j} + Hd_{r,i,j} + \sum_{t=1}^{N_{t(r)}} FT_{r,t,i,j} + QPNEW_{r(p),i,j}$$
$$- Q_{r,i,j} - L_{r,i,j} - U_{r,i,j}, \quad \forall r \in R, \quad i \in I, \quad j \in J \tag{12.3}$$

In previous equations, the time is discretized through the finite elements i and the internal collocation points j; furthermore h_i represents the finite element length, $A_{k,j}$ corresponds to the collocation matrix, $V^o_{r,i}$ is the volume of the reach r at the beginning of each element i, and $\dot{V}_{r,i,j}$ represents the first order derivative for the volume.

12.3.1.2 Component Balances for Each Reach

This balance is described as follows: the mass accumulation for each hazardous compound c in each reach r is equal to the mass at the inlet, plus the mass contained in the precipitation, industrial discharges, agricultural discharges, residential discharges, the sum of tributaries and the possible discharge from the new plant, minus the mass leaving the reach, loses, uses and the mass decomposed by the chemical reactions that are carried out in that section of the river. The reactive term involves the chemical and biochemical reactions that take place in the river by the flora and fauna because of the interaction between the components and the system, which can produce or eliminate the pollutants. Consequently, the component balance for each compound c in each section of the river r is given as follows:

$$\frac{d(V_r CQ_{c,r})}{dt} = Q_{r-1}(t)CQ_{c,r-1}(t) + P_r(t)CP_{c,r} + Id_r(t)CId_{c,r}(t) + D_r(t)CD_{c,r}(t)$$
$$+ Hd_r(t)CH_{c,r}(t) + \sum_{t=1}^{N_{t(r)}} FT_{r,t,i,j}(t)CT_{c,r,t,i,j}(t) - L_r(t)CL_{c,r}$$
$$- U_r(t)CU_{c,r}(t) - \int_{V=0}^{V_r} r_{c,r}(t)dV_r(t) - Q_r(t)CQ_{c,r}(t) + QPNEW_{r(p)}(t)CPNEW_{c,r(p)}(t),$$
$$\forall c \in C, \quad r \in R \tag{12.4}$$

Likewise to the overall balance for each reach, the concentration at each one of the discretized points ($CQ_{c,r,i,j}$) is approximated using a similar monomial basis representation:

$$CQ_{c,r,i,j} = CQ^o_{c,r,i} + h_i \sum_{k \in K} A_{k,j} \dot{CQ}_{r,i,k}, \quad \forall r \in R, \quad i \in I, \quad j \in J \tag{12.5}$$

$$\begin{aligned}
&V_{r,i,j}\,\dot{CQ}_{c,r,i,j}+\dot{V}_{r,i,j}\,CQ_{c,r,i,j}\\
&=Q_{r-1,i,j}CQ_{c,r-1,i,j}+P_{r,i,j}CP_{c,r,i,j}+Id_{r,i,j}CId_{c,r,i,j}+D_{r,i,j}CD_{c,r,i,j}\\
&+Hd_{r,i,j}CH_{c,r,i,j}+\sum_{t=1}^{N_{t(r)}}FT_{r,t,i,j}CT_{c,r,t,i,j}-L_{r,i,j}CL_{c,r,i,j}-U_{r,i,j}CU_{c,r,i,j}-\int_{V=0}^{V_r}r_{c,r,i,j}dV_{r,i,j}\\
&-Q_{r,i,j}CQ_{c,r,i,j}+QPNEW_{r(p),i,j}CPNEW_{c,r(p),i,j},\quad \forall c\in C,\ r\in R,\ i\in I,\ j\in J
\end{aligned} \tag{12.6}$$

$CQ^{o}_{c,r,i}$ is the concentration of the pollutant c for each reach r at the beginning of each element i and $\dot{CQ}_{r,i,k}$ is the first-order derivative for the pollutant concentrations.

12.3.1.3 Overall Balance for Each Tributary

The total flowrate discharged from the tributary t to the reach r ($FT_{r,t}$) is equal to the sum of the following types of discharges: discharges without ($S^{untreated}_{r,t}$)/with ($S^{treated}_{r,t}$) treatment, industrial discharges ($It_{r,t}$), pluvial discharges ($P_{r,t}$), direct discharges ($D_{r,t}$), minus the losses ($L_{r,t}$) and use or extraction ($U_{r,t}$) of water. Hence, the overall balance for the tributary t that discharges to the reach r is:

$$\begin{aligned}
FT_{r,t}&=S^{untreated}_{r,t}(t)+S^{treated}_{r,t}(t)+It_{r,t}(t)+P_{r,t}(t)+D_{r,t}(t)\\
&-L_{r,t}(t)-U_{r,t}(t),\quad \forall r\in R,\ t\in T
\end{aligned} \tag{12.7}$$

12.3.1.4 Component Balances for Each Tributary

The mass for the component c discharged from the tributary t to the reach r is given as follows:

$$\begin{aligned}
FT_{r,t}(t)CT_{c,r,t}(t)&=S^{untreated}_{r,t}(t)CS^{untreated}_{c,r,t}(t)+S^{treated}_{r,t}(t)CS^{treated}_{c,r,t}(t)+It_{r,t}(t)CIt_{c,r,t}(t)\\
&+P_{r,t}(t)CP_{c,r,t}(t)+D_{r,t}(t)CD_{c,r,t}(t)-L_{r,t}(t)CL_{c,r,t}(t)\\
&-U_{r,t}(t)CU_{c,r,t}(t)-\int_{V=0}^{V_{r,t}}r_{c,r,t}(t)dV_{r,t}(t),\quad \forall c\in C,\ r\in R,\ t\in T
\end{aligned} \tag{12.8}$$

The overall and component balances for tributaries only contain flowrates and concentrations known and do not include optimization variables; consequently, the flowrates and pollutant concentrations discharged by tributaries can be provided before the implementation of the approach. Furthermore, notice that most of the terms involved in the above balances have temporal variations, which leads to dynamic equations.

12.3.1.5 Reactive Terms

The reactive terms for the degradation of the pollutants in reaches and tributaries are given by the following kinetic reaction:

$$\int_{V=0}^{V_r} r_{c,r} dV_r = k_c \left[CQ_{c,r}(t) \right]^{\sigma_c} V_r(t), \quad \forall c \in C, \quad r \in R \tag{12.9}$$

$$\int_{V=0}^{V_{r,t}} r_{c,r,t} dV_{r,t} = k_c \left[CT_{c,r,t}(t) \right]^{\sigma_c} V_{r,t}(t), \quad \forall c \in C, \quad r \in R, \quad t \in T \tag{12.10}$$

k_c is the kinetic constant for the reactions carried out in the watershed, which is calculated experimentally, σ_c symbolizes the order reaction for the compound *c*, $CQ_{c,r,}$ and V_r are the concentration and Volume of each component of the reach *r*, respectively; while $CT_{c,r,t}$ and $V_{r,t}$ are the concentration and volume for tributaries.

12.3.1.6 Agricultural Discharges and Uses

Typically, water is extracted from watersheds to be supplied in some agricultural areas to achieve an adequate production of the field. The needed water for this purpose is a function of the area employed for the crops (which also can change during the year or even in some seasons when there is a growing area and for other seasons cannot exist the crop), as well as the water that returns to the watershed through filtration can be determined as a function of the cultivated area as follows:

$$D_{r,t}(t) = \alpha_{r,t} A_{r,t}(t), \quad \forall r \in R, \quad t \in T$$

$$U_{r,t}(t) = \beta_{r,t} A_{r,t}(t), \quad \forall r \in R, \quad t \in T$$

$\alpha_{r,t}$ is the water required per area for agricultural purposes and its units are m^3/ha s, $\beta_{r,t}$ represents the discharged flowrate per area and its units are m^3/ha s. $A_{r,t}$ represents the area used for agricultural purposes near the reach *r*.

12.3.2 Location of the New Plant

The presented approach involves a set of alternatives to place the new industrial plant *p*, and the optimization process must calculate the optimal location for the industrial facilities considering that the selected site depends on the installation cost and the impact of the new wastewater stream in the watershed. It has included a disjunctive model, which establishes that if the new industrial plant is in Site 1, then the wastewater and its composition associated with this site must be greater than zero; otherwise, if the new plant is not located in the position 1, the wastewater and its compositions discharged should be zero; the same logic is used for all options. This is simulated through the following disjunction:

$$\begin{bmatrix} Y_1 \\ CPNEW_{c,1}(t) \geq 0, QPNEW_1(t) \geq 0 \\ CPNEW_{c,2}(t) = 0, QPNEW_2(t) = 0 \\ \vdots \\ CPNEW_{c,p}(t) = 0, QPNEW_p(t) = 0 \end{bmatrix} \vee \begin{bmatrix} Y_2 \\ CPNEW_{c,1}(t) = 0, QPNEW_1(t) = 0 \\ CPNEW_{c,2}(t) \geq 0, QPNEW_2(t) \geq 0 \\ \vdots \\ CPNEW_{c,p}(t) = 0, QPNEW_p(t) = 0 \end{bmatrix} \vee \ldots$$

$$\vee \begin{bmatrix} Y_p \\ CPNEW_{c,1}(t) = 0, QPNEW_1(t) = 0 \\ CPNEW_{c,2}(t) = 0, QPNEW_2(t) = 0 \\ \vdots \\ CPNEW_{c,p}(t) \geq 0, QPNEW_p(t) \geq 0 \end{bmatrix}, \quad \forall c \in C \tag{12.11}$$

Y_p is the Boolean variable associated with the location p of the new plant, $QPNEW_p$ and $CPNEW_{c,p}$ are the flowrate and the concentration for the compound c of the new wastewater stream generated by the new industrial plant in location p, respectively. Furthermore, the disjunctive model states that if the optimal location for the new plant is on site 1, then the Boolean variable Y_1 is set as true and $CPNEW_{c,1}$ and $QPNEW_1$ must be greater than zero, while all the other Boolean variables must be false and the $CPNEW_{c,p}$ and $QPNEW_p$ associated to their locations must be set as zero. Similar situations occur for the case when the selected location is 2, 3 or so on for the p possible locations. Although the installation costs are not considered in the previous disjunction, this value affects the chosen alternative to locate the new industrial facilities, and the value for the flowrate of the wastewater and their concentrations of the new plant are powerfully influenced by the location through the material flow analysis model, satisfying the environmental regulations given to maintain under control the pollutants that include the environmental regulations, constraints to maintain the water quality for some specific reaches (restricting the pollutant concentrations for some reaches of interest) and the sustainability constraints (which implies that the compound concentrations at the final disposal are restricted). The main tool to satisfy all the environmental restrictions is the interception network. Lastly, the previous disjunction is reformulated using the convex hull technique (Raman and Grossmann, 1994; Ponce-Ortega et al., 2009b) as follows.

The Boolean variables (Y_p) are transformed into a set of binary variables (y_p), when the Boolean variables are true, the associated binary variables are one; or else, when the Boolean variables are false, the associated binary variables are zero. Thus, the following relationship allows selecting only one location to place the new industrial plant:

$$\sum_{p \in P} y_p = 1 \tag{12.12}$$

Furthermore, it is required upper limits for the variables $QPNEW_p$ and $CPNEW_{c,p}$.

$$QPNEW_p(t) \leq QPNEW^{Up} y_p, \quad \forall p \in P \tag{12.13}$$

$$CPNEW_{c,p}(t) \leq CPNEW^{Up} y_p, \quad \forall c \in C, \;\; p \in P \tag{12.14}$$

$QPNEW^{Up}$ and $CPNEW^{Up}$ represent the upper limits for $QPNEW_p$ and $CPNEW_{c,p}$, respectively. Therefore, considering Eq. 12.5 (where only one site is activated), previous relationships ensure that $QPNEW_p$ and $CPNEW_{c,p}$ take values greater than zero only for the selected position, also, $QPNEW_p$ and $CPNEW_{c,p}$ must be equal to zero when the alternative to place the new plant is not selected.

12.3.2.1 Superstructure Modeling for In-Plant Treatment System

The variables $QPNEW_p$ and $CPNEW_{c,p}$ are related to the variables used in the treatment system for the wastewater stream through the following relationships:

$$\sum_{p \in P} QPNEW_p(t) = QPNEWDATA(t) \tag{12.15}$$

$$\sum_{p \in P} CPNEW_{c,p}(t) = y_c^{disc}(t), \quad \forall c \in C \tag{12.16}$$

So, previous equations state that the flowrate ($QPNEW_p$) and pollutants composition ($CPNEW_{c,p}$) of the wastewater stream that receives the watershed in the selected position are the same than the flowrate ($QPNEWDATA$) and compositions (y_c^{disc}) discharged by the new industrial plant at the exit of the treatment system. Hence, previous equations connect the material flow analysis model with the treatment system model.

The presented superstructure for the in-plant treatment is depicted in Fig. 12.2, which is based on the works reported by Gabriel and El-Halwagi (2005) and Ponce-Ortega et al. (2010). The size for this formulation increases exponentially as a function of the number of treated compounds (N_c); however, the model obtained for this superstructure has a linear behavior, which is the main advantage of this formulation because it avoids the mixing of different streams, and it is useful for the optimization process. Here, for each pollutant to be treated there is a set of available treatment units to remove the compound c, and a bypass has been included to consider that a fraction of the total wastewater stream does not require to be treated (which is simulated as fictitious interceptor); consequently, for each component must be considered a column of interceptors, generating N columns of treatment units. There is a set of

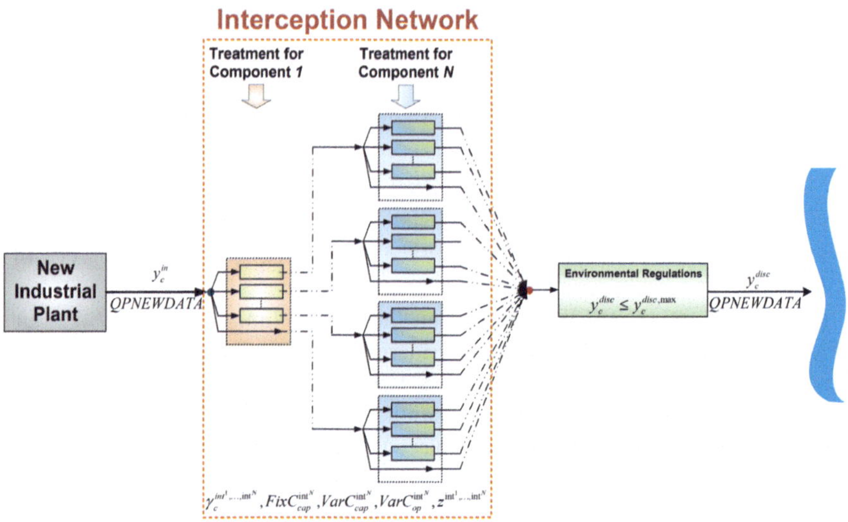

Fig. 12.2 Interceptors network for wastewater treatment. Printed with permission of "*Dynamic optimization for optimal location of new industrial facilities considering the sustainability of the watershed*", Sustainability of Products, Processes and Supply Chains, Copyright (2015), Elsevier

constraints for the wastewater stream discharged to the environment to meet the environmental regulations. Furthermore, the optimization process must indicate the variation of the required treatment concerning the time and the treatment requirements depend powerfully on the location of the new industrial plant to meet the environmental constraints imposed over the system.

Here, the waste stream leaving the new industrial facility is segregated to treat the first component in the set of interceptors (i.e., available technologies for the treatment of the specific pollutant). At the exit of the first treatment, the output streams are split again and sent to the following interceptors to treat the second component. The same sequence is used to treat all required components at the interception network, where the mixing of different streams is avoided in all the steps of the treatment. Lastly, all streams that leave the treatment units are mixed to be sent to the wastewater stream and discharged to the environment. The final wastewater streams should satisfy the environmental constraints.

A correction factor ($\gamma_p^{int^1,\ldots,int^N}$) calculated before the optimization process is used to simulate the interceptor performance. This approach leads to linear models for the interceptors. The correction factor depends on numerous variables such as the value of the concentration at the inlet conditions of the interceptor, design and operating parameters and the flowrate. The available treatment technologies considered in the interception network can be simulated (i.e., using process simulators such as ASPEN PLUS) to get their correction factors. The values for the concentrations of the components at the inlet conditions of the waste stream are also known before the optimization process (y_c^{in}). Then, the values for the concentrations of the components at the exit conditions of the interception network can be calculated before optimization as follows:

$$y_c^{out,\text{int}^1,\ldots,\text{int}^N}(t) = \gamma_c^{\text{int}^1,\ldots,\text{int}^N} y_c^{in}(t), \quad \forall c \in C, \ \text{int}^1 \in INT^1, \ldots, \text{int}^N \in INT^N \tag{12.17}$$

$y_c^{out,\text{int}^1,\ldots,\text{int}^N}$ represents the concentrations for the components at the exit of the interception network, according to the path of interceptors followed. On the other hand, the following relationships simulate the presented superstructure for the wastewater treatment in this work:

12.3.2.2 Splitting of the Waste Stream

The waste stream leaving the industrial plant is split to treat the first component on the first set of available interceptors.

$$QPNEWDATA(t) = \sum_{\text{int}^1 \in INT^1} f_1^{\text{int}^1}(t) \tag{12.18}$$

To treat the second component, the flowrate at the exit of the first line of interceptor is split and these are fed to the second set of interceptors.

$$f_1^{\text{int}^1}(t) = \sum_{int^2 \in INT^2} f_2^{\text{int}^1,\text{int}^2}(t), \quad \forall int^1 \in INT^1 \tag{12.19}$$

It does not allow mixing of different process streams. This approach eliminates the nonlinearities involved in the component balances for the wastewater stream (which is shown below), although the problem size increases exponentially.

Following the same sequence for N components to be intercepted, the next relationship simulates the segregations.

$$f_{N-1}^{int^1,\text{int}^2,\ldots,\text{int}^{N-1}}(t) = \sum_{int^N \in INT^N} f_N^{int^1,\text{int}^2,\ldots,\text{int}^N}(t), \quad \forall int^1 \in INT^1, \ldots, int^{N-1} \in INT^{N-1} \tag{12.20}$$

If the waste stream does not require any treatment, it is sent to a bypass or fictitious unit (located at the end of each set of interceptors) with effectiveness and cost equal to zero.

12.3.2.3 Overall Balance for the Waste Stream

The presented superstructure considers that there is no volume variation owing to the treatment, thus the flowrate of the wastewater stream discharged to the environment is equal to the waste stream leaving the industrial plant (*QPNEWDATA*). Therefore, all the streams leaving from the interception network ($f^{int^1,\ldots,int^N}$) are mixed to obtain the final wastewater stream.

$$QPNEWDATA(t) = \sum_{int^1 \in INT^1} \cdots \sum_{int^N \in INT^N} f^{int^1,\ldots,int^N}(t) \tag{12.21}$$

12.3.2.4 Component Balances for the Wastewater Stream

The following equation is useful to calculate the concentrations of the pollutants in the wastewater streams discharged into the environment.

$$QPNEWDATA(t)\ y_c^{disc}(t) = \sum_{int^1 \in INT^1} \cdots \sum_{int^N \in INT^N} f^{int^1,\ldots,int^N}(t) y_c^{out,int^1,\ldots,int^N}(t), \quad \forall c \in C \tag{12.22}$$

y_c^{disc} is the concentration of the component c discharged to the environment.

12.3.2.5 Environmental Regulations

The environmental regulations include a set of maximum concentrations that can be discharged into the environment:

$$y_c^{disc}(t) \le y_c^{disc,\max}, \quad \forall c \in C \tag{12.23}$$

The flowrates and concentrations involved in the previous equations to simulate the treatment of the wastewater stream depend powerfully on the time, because of the treatment requirements change according to the dynamic processes that occur in the watershed system.

12.3.3 Supplementary Environmental Constraints

In addition to the environmental regulations, the problem considers a set of environmental constraints imposed over specific reaches and another set of environmental constraints at the final disposal (called sustainability constraints) for the watershed system.

12.3.3.1 Constraints for the Quality of Specific Reaches

Frequently, the water of a watershed is used for different purposes (i.e., to be used in residential, industrial, or agricultural zones), and it becomes necessary to impose a set of concentration constraints for the specific reaches to guarantee the quality of the extracted water. The presented model can consider these constraints through the following relationship.

$$CQ_{c,m(r)}(t) \leq y_{c,m(r)}^{desired}, \quad \forall c \in C, \quad m(r) \in M(R) \tag{12.24}$$

where $M(R)$ is a subset of reaches requiring a specific water quality, while $y_{c,m(r)}^{desired}$ is the upper limit for the component concentrations to ensure desired conditions in the water quality for specific reaches. Although the concentration in the reach has a significant variation concerning the time ($CQ_{c,m(r)}$) the desired concentration must be maintained during all the time.

12.3.3.2 Sustainability Constraints

Before implementing the proposed approach, it must be carried out an analysis to know the maximum concentration of each compound c ($y_c^{sustainable}$) that can receive the final disposal (lake, sea, or ocean) to be degraded, avoiding putting at risk the sustainability of the watershed and its surroundings. In some cases, the environmental regulations are not enough to achieve this purpose.

$$CQ_c^{final}(t) \leq y_c^{sustainable} \tag{12.25}$$

CQ_c^{final} is the concentration for the component c at the final disposal after installing the new industrial facility. Although the concentration in the final reach has significant variation concerning the time (CQ_c^{final}) the sustainable concentration must be maintained all the time.

12.3.4 *Objective Function*

Lastly, the objective function consists in minimizing the total annual cost (TAC), which is constituted by the installation costs for the new plant and the treatment cost (including operational cost and variable capital cost).

$$Min\ TAC = \int_0^t \left[\sum_{int^1 \in INT^1} VarC_{op}^{int^1} f^{int^1}(t) + \cdots \sum_{int^N \in INT^N} VarC_{op}^{int^N} f^{int^1,\ldots,int^N}(t) \right] dt$$

$$+\int_0^t \left[\sum_{int^1 \in INT^1} VarC_{cap}^{int^1} f^{int^1}(t) + \cdots \sum_{int^N \in INT^N} VarC_{cap}^{int^N} f^{int^1,\ldots,int^N}(t) \right] dt$$
$$+ \sum_{p \in P} Cland_p y_p \tag{12.26}$$

H_Y is the hours per year that the new industrial plant operates and k_f is the factor used to annualize the inversion. Furthermore, a set of available technologies must be selected to treat the waste stream yielding capital ($VarC_{cap}^{int^N}$) and operational ($VarC_{op}^{int^N}$) costs for each one of the interceptors required. $Cland_p$ is the installation cost, which considers the land cost and the transportation costs for raw materials, products, and services, and lastly y_p is the binary variable associated with each alternative p to locate the new industrial facilities. The objective function includes integral terms; then a proper procedure to approximate numerically the integrals is required.

12.4 Results and Discussion

The following example is used to show the applicability of the proposed approach.

12.4.1 Case Study

The Balsas system represents one of the most important watersheds in México, it has a hydrologic surface of 117 305 km^2 covering 8 states and 420 municipalities. This watershed is affected by several cities, and industrial and agricultural zones that discharge their effluents over it; to adequately simulate the Balsas system, it was needed to take information from CONAGUA (2010). Figure 12.3 presents the schematic representation of the addressed problem. The residential, industrial, and agricultural zones shown in this figure are only representatives; however, there are an enormous number of interactions among these systems and the Balsas watershed. The pollutants nitrogen, phosphorous, and arsenic were considered as well as 11 possible positions to locate the new industrial facilities and the Balsas system has been divided into 23 reaches. Most of the information required by the model must be introduced with fluctuations concerning the time such as the precipitation, evaporation, filtration, and all the industrial, residential, and agricultural interactions with the system. Then, the precipitation data employed in this example has a seasonal variation following Fig. 12.4. The period June–September represents the rainy season and subsequently there is a substantial variation in the precipitation discharges during the year, which impacts in the watershed water quality.

Figures 12.5 and 12.6 present the flowrate and pollutant concentration for the new industrial plant before the inlet to the treatment network. Thus, there are three

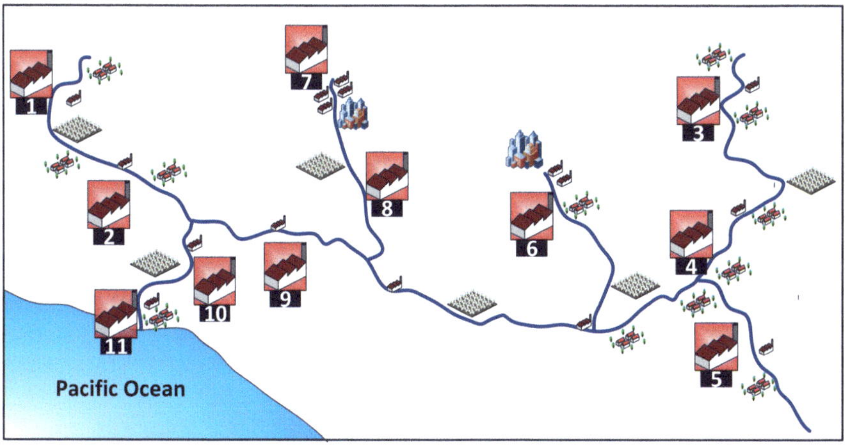

Fig. 12.3 Schematic representation of the Balsas system, including the alternatives to install the new industrial plant. Printed with permission of "*Dynamic optimization for optimal location of new industrial facilities considering the sustainability of the watershed*", Sustainability of Products, Processes and Supply Chains, Copyright (2015), Elsevier

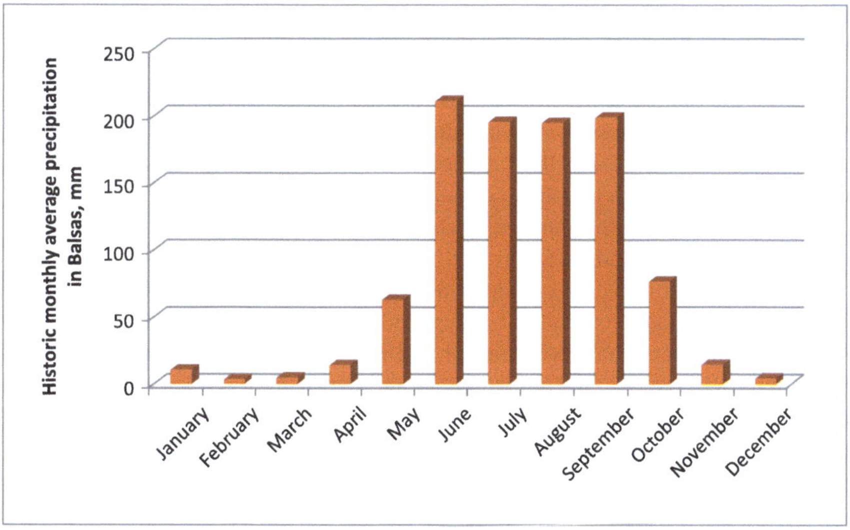

Fig. 12.4 Historic monthly average precipitation in the Balsas watershed. Printed with permission of "*Dynamic optimization for optimal location of new industrial facilities considering the sustainability of the watershed*", Sustainability of Products, Processes and Supply Chains, Copyright (2015), Elsevier

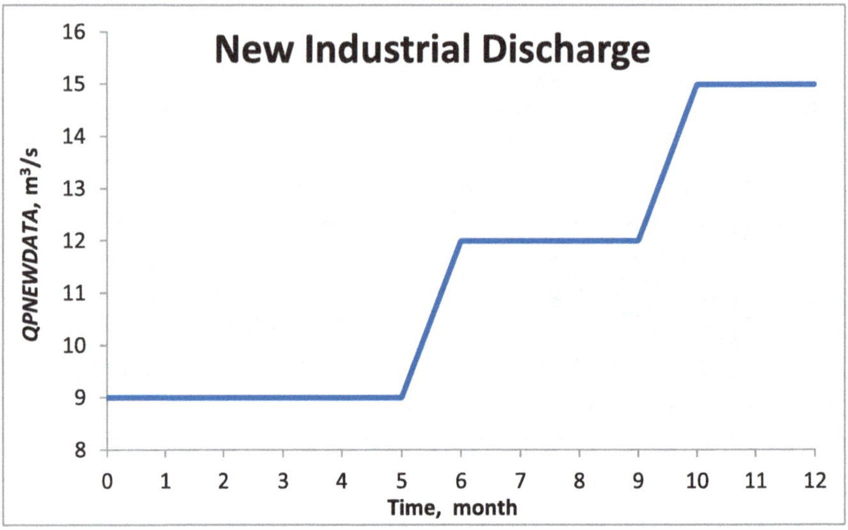

Fig. 12.5 Flowrate behavior for the new industrial plant. Printed with permission of "*Dynamic optimization for optimal location of new industrial facilities considering the sustainability of the watershed*", Sustainability of Products, Processes and Supply Chains, Copyright (2015), Elsevier

substantial alterations in the flowrate and pollutants concentration during the year. Arsenic has concentrations lower than nitrogen and phosphorous, for this reason and with the purpose to visualize the variation in the concentrations in the year its scale has been separated from the other components (Fig. 12.6).

Table 12.1 presents the annualized installation cost for each alternative considered for installing the new industrial project, whereas Table 12.2 presents the costs and efficiency factors for each technology available to treat the pollutants considered in this example. Each compound has been considered two interceptors, where the first option is always the most efficient and therefore is also the most expensive. Furthermore, the environmental regulations and sustainability constraints are shown in Table 12.3, which must be always satisfied. The environmental regulations restrict the maximum pollutant concentration that can be discharged by the new industrial plant; however, it will be necessary to consider the impact of this wastewater stream once enters the environment, in other words, the fulfillment of the environmental regulations does not guarantee to have sustainable systems.

Lastly, to implement this case study is also required the following parameters:

- The operation time for the new plant is 8000 h/year.
- The agricultural discharges and uses are: $\alpha_{r,t} = 0.000148$ m^3/ha s and $\beta_{r,t} = 0.000296$ m^3/ha s, respectively.
- The reactive term is represented through first-order kinetic reactions employing the following kinetic constants $k_{nitrogen} = 3.23764 \times 10^{-6}$ s^{-1}, $k_{phosphorous} = 1.492 \times 10^{-6}$ s^{-1} and $k_{arsenic} = 4.5205 \times 10^{-7}$ s^{-1}.

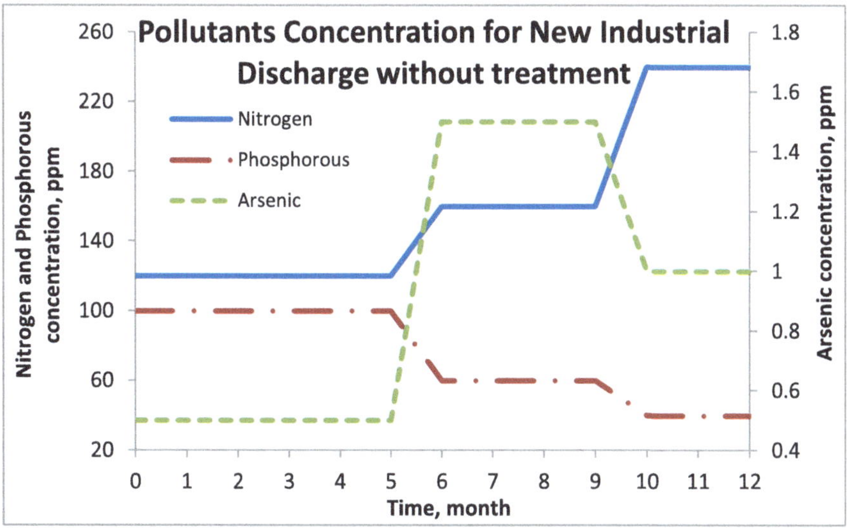

Fig. 12.6 Pollutant concentrations before the interception network for the new industrial plant. Printed with permission of "*Dynamic optimization for optimal location of new industrial facilities considering the sustainability of the watershed*", Sustainability of Products, Processes and Supply Chains, Copyright (2015), Elsevier

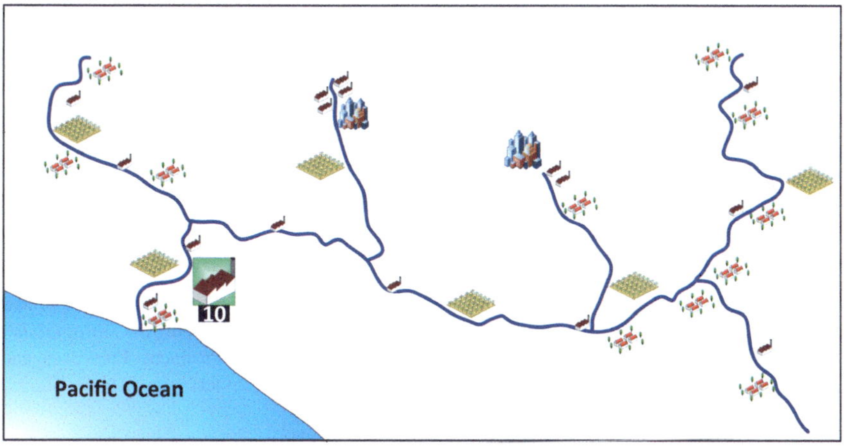

Fig. 12.7 Optimal solution for the considered example. Printed with permission of "*Dynamic optimization for optimal location of new industrial facilities considering the sustainability of the watershed*", Sustainability of Products, Processes and Supply Chains, Copyright (2015), Elsevier

Table 12.1 Installation costs for each site

Site	Annualized installation cost, \$/year
1	7.7×10^6
2	7.4×10^6
3	8.5×10^6
4	8.2×10^6
5	8×10^6
6	7.3×10^6
7	7.2×10^6
8	7×10^6
9	8.1×10^6
10	7.6×10^6
11	8×10^6

Printed with permission of "*Dynamic optimization for optimal location of new industrial facilities considering the sustainability of the watershed*", Sustainability of Products, Processes and Supply Chains, Copyright (2015), Elsevier

Table 12.2 Characteristics of the available interceptors

Interceptor	Efficiency factor	Cost, \$/month m^3
Nitrogen		
Rec^1	0.98	14,515.20
Rec^2	0.78	10,627.20
Phosphorous		
Dec^1	0.97	20,736
Dec^2	0.82	15,811.20
Arsenic		
Aer^1	0.99	16,848
Aer^2	0.85	5,702.40

Printed with permission of "*Dynamic optimization for optimal location of new industrial facilities considering the sustainability of the watershed*", Sustainability of Products, Processes and Supply Chains, Copyright (2015), Elsevier

Table 12.3 Environmental constraints for the considered example

Pollutant	Environmental regulation $y_c^{disc,\max}$, ppm	Sustainability constraint $y_c^{sustainable}$, ppm
Nitrogen	40	22
Phosphorous	20	11.1
Arsenic	0.2	0.083

Printed with permission of "*Dynamic optimization for optimal location of new industrial facilities considering the sustainability of the watershed*", Sustainability of Products, Processes and Supply Chains, Copyright (2015), Elsevier

All the above information is implemented in the software GAMS using the solver DICOPT; furthermore, it uses the orthogonal collocation over finite elements to discretize the differential equation into a set of algebraic equations, and the Lagrange polynomials were used to find the internal points and the collocation matrix. Once the optimization process is carried out, it is found that the optimal alternative to locate the new industrial facilities is site 10 (Fig. 12.7) with a *TAC* of \$ 11.928 MM/year, where \$ 7.6 MM/year corresponds to installation cost and the treatment cost is \$ 4.328 MM/year.

Figure 12.8 presents the optimal interception network to satisfy the environmental regulations and the sustainability constraints. Almost all the interceptors considered in the proposed superstructure are required in the optimal treatment configuration. Although this figure does not include the variability in flowrates and concentrations involved on it, these have been determined by the optimization process.

The variation in nitrogen concentration discharged to the final disposal (CQ^{final}) and for the new industrial discharge (y^{disc}) is presented in Fig. 12.9, where the environmental regulation and the sustainability constraint ($y^{sustainable}$) for this compound with the purpose of identifying the following analysis:

- During the rainy season (June–September) the watershed conditions allow the discharging of the waste stream from the new industrial plant satisfying only

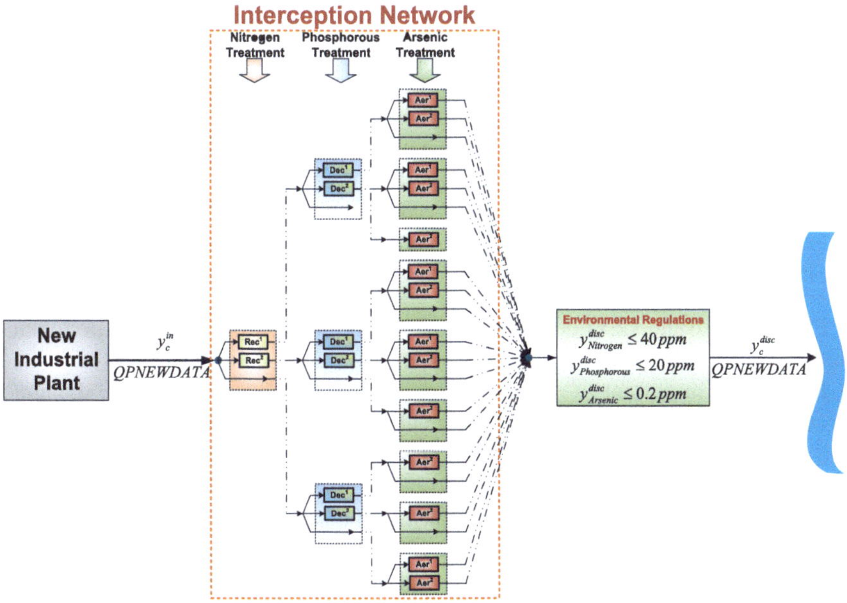

Fig. 12.8 Network for wastewater treatment obtained in the solution. Printed with permission of "*Dynamic optimization for optimal location of new industrial facilities considering the sustainability of the watershed*", Sustainability of Products, Processes and Supply Chains, Copyright (2015), Elsevier

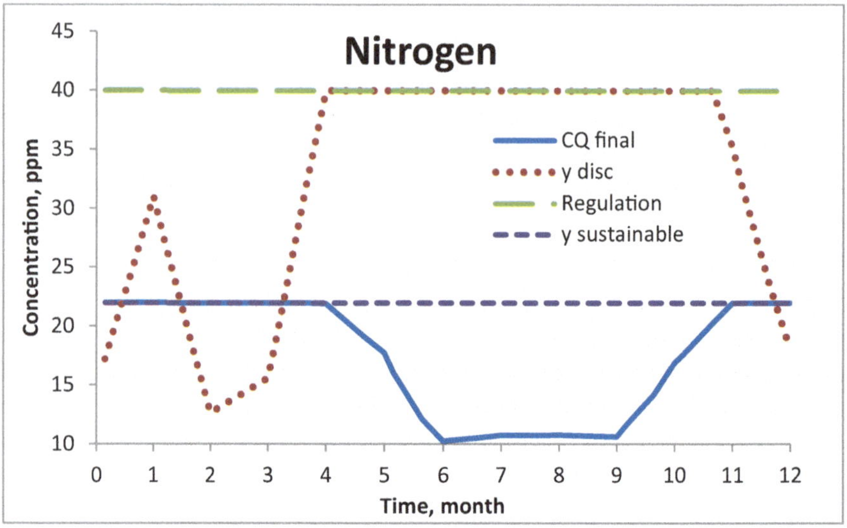

Fig. 12.9 Nitrogen concentration profiles. Printed with permission of "*Dynamic optimization for optimal location of new industrial facilities considering the sustainability of the watershed*", Sustainability of Products, Processes and Supply Chains, Copyright (2015), Elsevier

the environmental regulation; in other words, in this period the precipitation discharges dilute the pollutants discharged to the Balsas system and there are not problems related to the fulfillment of the sustainability constraint.

- When there are no significant precipitation discharges, the wastewater stream must be treated beyond the environmental regulation to satisfy the sustainability constraint; otherwise, if during this period the new waste stream only satisfies the environmental regulation, the system's capacity to degrade the pollutants will be exceeded. Subsequently, there will be pollutants accumulation at the final disposal putting in risk the watershed and surrounding areas.

Figures 12.10 and 12.11 determine that the same analysis occurs for the pollutants phosphorous and arsenic, respectively.

Lastly, it is important to remark on the importance of considering the dynamic processes in this type of system, which evidenced the problems that can be generated in critical conditions of the system. Typical models where the dynamic is not considered cannot detect these types of problems and their solutions are not satisfactory. Furthermore, this example has demonstrated that the environmental regulations for industrial wastewater streams are not enough to ensure the sustainability of the watersheds, and it is required to model the watershed to guarantee the sustainability of the system.

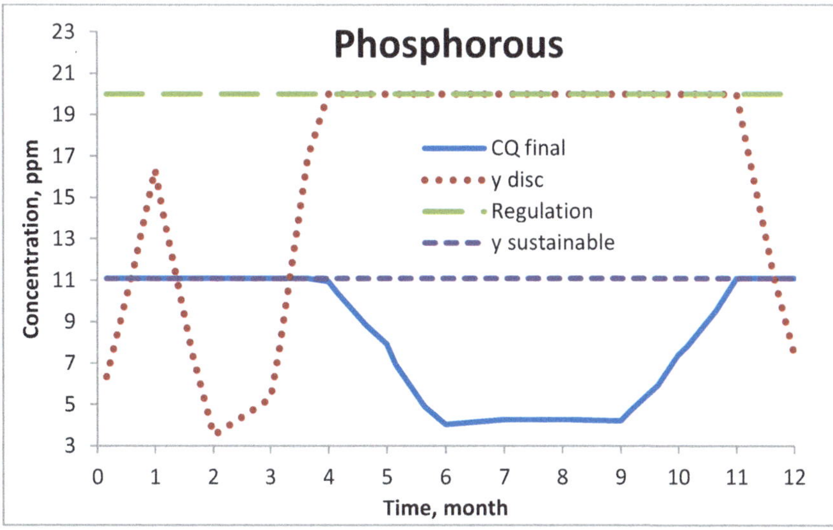

Fig. 12.10 Phosphorous concentration profiles. Printed with permission of "*Dynamic optimization for optimal location of new industrial facilities considering the sustainability of the watershed*", Sustainability of Products, Processes and Supply Chains, Copyright (2015), Elsevier

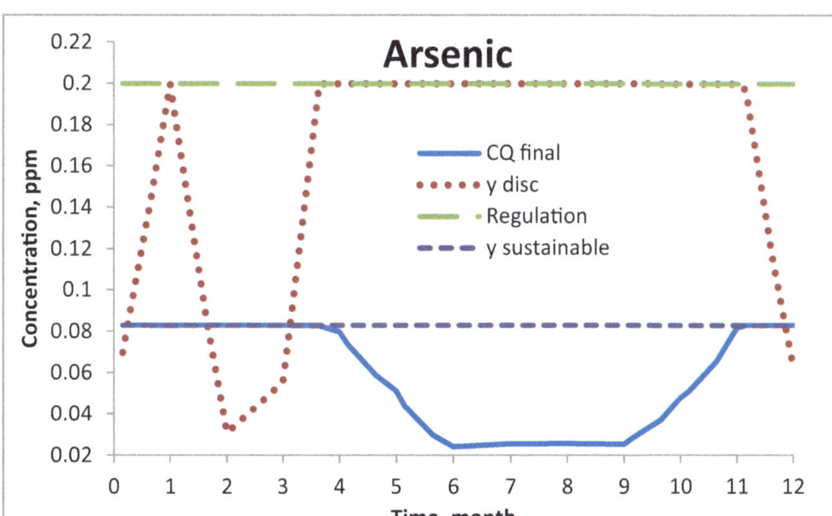

Fig. 12.11 Arsenic concentration profiles. Printed with permission of "*Dynamic optimization for optimal location of new industrial facilities considering the sustainability of the watershed*", Sustainability of Products, Processes and Supply Chains, Copyright (2015), Elsevier

12.5 Concluding Remarks

To ensure the sustainability of the watershed and its surroundings, a new dynamic mixed integer non-linear programming model has been presented in this chapter. It is intended to be used to determine the best location for new industrial facilities while adhering to environmental regulations, sustainability constraints imposed at the final disposal, and restrictions for specific river sites with water extractions. The objective function is to minimize the annual total cost, which is made up of the expenditures associated with installing the new industrial plant and treating wastewater. The material flow analysis technique is used to forecast how the watershed will behave when it is exposed to new polluted discharges. It also considers all the interactions that currently exist within the watershed, including precipitation, treated and untreated discharges, industrial, residential, and agricultural discharges, as well as natural phenomena like evaporation, filtration, and chemical and biological reactions that occur in the rivers. The dynamic model makes it possible to pinpoint the intervals of time during which the circumstances in the watershed are crucial and the necessary treatment needs to be modified to meet all environmental requirements. To accomplish this goal, the MFA technique is paired with a disjunctive model for the ideal location of a new industrial facility, and the mass integration within the industrial plant is taken into consideration together with the surrounding watershed. The case study's solution demonstrates how the need for treatment to produce a sustainable process is greatly impacted by the location of a new industrial plant. Incorporating dynamic processes to visualize the watershed's actual behavior throughout the year, including seasonal fluctuations and identifying critical periods associated with overall sustainability, is another important aspect of obtaining holistic solutions. If this is not done in this form, it may result in inadequate solutions because critical periods may not be taken into consideration. Ultimately, the findings demonstrate that, occasionally, the environmental standards for industrial wastewater streams are insufficient to guarantee the watersheds' sustainability; thus, more precise approaches—such as the methods suggested here—are needed to accomplish this goal.

Nomenclatures

Indexes

c	Component
i	Finite element
int	Interceptor
In	Inlet
j	Internal collocation point
max	Maximum
min	Minimum

N_c	Number of pollutants to be treated
$m(r)$	Specific reaches that require a composition constraint
p	Site to locate the new plant
r	Reach
t	Tributary
UP	Upper value

Sets

C	Set for the components, $\{c\|c = 1, \ldots, N_c\}$
I	Set for the finite elements, $\{i\|i = 1, \ldots, N_i\}$
J	Set for the internal collocation points, $\left\{j\|j = 1, \ldots, N_j\right\}$
INT	Set for the interceptor, $\{\text{int}\|\text{int} = 1, \ldots, N_{\text{int}}\}$
P	Set for the sites to locate the new plant, $\left\{p\|p = 1, \ldots, N_p\right\}$
R	Set for the reaches, $\{r\|r = 1, \ldots, N_r\}$
R	Set for the tributaries

Parameters

$A_{r,t}$	Area cover by effluent t in reach r, acre or ha
$A_{k,j}$	Collocation matrix
$CD_{c,r}$	Pollutant concentration of agricultural discharges to the reach r, ppm
$CD_{c,r,t}$	Pollutant concentration of agricultural discharges to the tributary t to the reach r, ppm
$CH_{c,r}$	Pollutant concentration of industrial discharge to the reach r, ppm
$CId_{c,r}$	Pollutant concentration of industrial discharge to the reach r, ppm
$CIt_{c,r,t}$	Pollutant concentration of industrial discharge from the tributary t to the reach r, ppm
CQ_c^{final}	Pollutant concentration discharged to the final disposal, ppm
$CP_{c,r}$	Pollutant concentration of precipitation discharged to the reach r, ppm
$CP_{c,r,t}$	Pollutant concentration of precipitation discharged to the tributary t to the reach r, ppm
$CT_{c,r,t}$	Pollutant concentration of the tributary t to the reach r, ppm
$Cland_p$	Annualized installation cost for the new plant in site p, \$/year
$CL_{c,r}$	Pollutant concentration of total losses (filtration and evaporation) from the reach r, ppm
$CL_{c,r,t}$	Pollutant concentration of total losses (filtration and evaporation) from tributary t to the reach r, ppm

$CS_{c,r,t}^{untreated}$	Pollutant concentration of residual wastewater discharged without treatment to the reach r for tributary t, ppm
$CS_{c,r,t}^{treated}$	Pollutant concentration of residual treated wastewater discharged to the reach r for tributary t, ppm
$CU_{c,r}$	Pollutant concentration of water used from reach r, ppm
$CU_{c,r,t}$	Pollutant concentration of water used from tributary t discharge to reach r, ppm
D_r	Agricultural discharges to the reach r, m^3/s
$D_{r,t}$	Agricultural discharges to the tributary t to the reach r, m^3/s
$FT_{r,t}$	Discharge for the tributary t to the reach r, m^3/s
Hd_r	Residential discharge to the reach r, m^3/s
h_i	Finite element length
Id_r	Industrial discharge from the reach r, m^3/s
$It_{r,t}$	Industrial discharge from the tributary t to the reach r, m^3/s
k_c	Kinetic constant for the degradation for pollutant c
L_r	Total losses (filtration and evaporation) from the reach r, m^3/s
$L_{r,t}$	Total losses (filtration and evaporation) from tributary t of the reach r, m^3/s
$M(R)$	Subset for the specific reaches that require composition constraints
N_c	Total number of components
N_i	Total number of finite elements
N_{int}	Total number of interceptors
N_j	Total number of internal collocation points
N_p	Total number of sites allowable to locate the new plant
N_r	Total number of reaches
N_t	Total number of tributaries
P_r	Precipitation discharged to the reach r, m^3/s
$P_{r,t}$	Precipitation discharged for the tributary t to the reach r, m^3/s
$QPNEWDATA$	Flowrate discharged for the new plant
$S_{r,t}^{untreated}$	Residual wastewater discharged without treatment to the reach r for tributary t, m^3/s
$S_{r,t}^{treated}$	Residual treated wastewater discharged to the reach r for tributary t, m^3/s
$V_{r,t}$	Volume for tributary t from reach r, m^3
$VarC_{cap}^{\text{int}^N}$	Capital variable cost for interceptor int^N, $\$/m^3$
$VarC_{op}^{\text{int}^N}$	Operational variable cost for interceptor int^N, $\$/m^3$
$y_{c,m(r)}^{desired}$	Desired concentrations in some reaches, ppm
$y_c^{disc,\max}$	Environmental regulation for the pollutant concentration to be discharged by the new plant, ppm
y_c^{in}	Pollutant concentration at the inlet to interception network, ppm
$y_c^{sustainable}$	Sustainability constraints at the final disposal, ppm

Greek symbols

$\alpha_{r,t}$	Agricultural flowrate per area, m^3/ha*s
$\beta_{r,t}$	Agricultural use of water from tributary t, m^3/ha s
γ_c	Efficiency factor to remove the pollutant c
σ_c	Order reaction for the compound c

Variables

$CQ_{c,r}$	Pollutants concentration of flowrate exit to the reach r, ppm
$CQ_{c,r,i,j}$	Discretized pollutant concentration of flowrate exit from the reach r, ppm
$CQ^{o}_{c,r,i}$	Concentration of the pollutant c for each reach r at the beginning of each element i, ppm
$\dot{CQ}_{r,i,k}$	First order derivative for the pollutant concentration, ppm
$CQ_{c,r-1}$	Pollutant concentration of flowrate inlet to the reach r, ppm
$CPNEW_{c,p}$	Concentration for the contaminants discharged for the new plant installed in site p, ppm
dv_r/dt	Volume variation in the reach r with respect to the time, m^3/s
f_N	Segregated flowrate for the treatment network, m^3/s
Q_r	Outflow from the reach r, m^3/s
Q_{r-1}	Inflow to the reach r, m^3/s
$QPNEW_p$	Flowrate discharged for the new plant for the location p
$r_{c,r}$	Reaction carried out in the reach r, g/s
$r_{c,r,t}$	Reaction carried out in the tributary t that discharges to reach r, g/s
TAC	Total annual cost, \$/year
U_r	Water used from reach r, m^3/s
$U_{r,t}$	Water used from tributary t discharged to reach r, m^3/s
$V^{o}_{r,i}$	Volume of the reach r at the beginning of each element i, m^3
V_r	Volume for reach r, m^3
$V_{r,i,j}$	Discretized volume for reach r, m^3
$\dot{V}_{r,i,j}$	First order derivative for the volume, m^3/s
Y_p	Boolean variable for the location of the new plant
y_p	Binary variable for the location of the new plant
y_c^{disc}	Pollutant concentration for the final discharge of the new plant to the river, ppm

References

Agrawal V, Shenoy UV (2006) Unified conceptual approach to targeting and design of water and hydrogen networks. AIChE J 52(3):1071–1082

Alva-Argaez A, Kokossis AC, Smith R (1998) Wastewater minimization of industrial systems using an integrated approach. Comput Chem Eng 22(Suppl):S741–S744

Alva-Argaez A, Vallianatos A, Kokossis A (1999) A multi-contaminant transshipment model for mass exchange networks and wastewater minimization problems. Comput Chem Eng 23(10):1439–1453

Baccini P, Brunner P (1991) Metabolism of the anthroposphere. Springer-Verlag, Berlin

Barbosa-Povoa A (2007) A critical review on the design and retrofit of batch plants. Comput Chem Eng 31–7:833–855

Barbosa-Povoa A, Macchietto S (1994) Detailed design of multipurpose batch plants. Comput Chem Eng 18(11–12):1013–1042

Benko N, Rev E, Fonyo Z (2000) The use of nonlinear programming to optimal water allocation. Chem Eng Commun 178:67–101

Brunner PH, Rechberg H (2004) Practical handbook of material flow analysis. CRC

Castro P, Barbosa-Povoa A, Novais A (2005) Simultaneous design and scheduling of multipurpose plants using resource task network based continuous-time formulations. Ind Eng Chem Res 44–2:343–357

CONAGUA (2010) Mexican National Water Commission. Water statistics in the watershed of Balsas River 2010

Dhole VR, Ramchandani N, Tainsh RA, Wasilewski M (1996) Make your process water pay for itself. Chem Eng 103(1):100–103

Doyle SJ, Smith R (1997) Targeting water reuse with multiple contaminants. Process Saf Environ Prot 75(B3):181–189

Dunn RF, El-Halwagi MM (2003) Process integration technology review: background and application in chemical process industry. J Chem Technol Biotechnol 78(9):1011–1021

El-Baz AA, Ewida KT, Shouman MA, El-Halwagi MM (2005a) Material flow analysis and integration of watersheds and drainage systems: I. Simulation and application to ammonium management in Bahr El-Baqar drainage system. Clean Technol Environ Policy 7(1):51–61

El-Baz AA, Ewida KT, Shouman MA, El-Halwagi MM (2005b) Material flow analysis and integration of watersheds and drain systems: II. Integration and solution strategies with application to ammonium management in Bahr El-Baqar drain system. Clean Technol Environ Policy 7(1):78–86

El-Halwagi MM (1997) Pollution prevention through process integration: systematic design tools. Academic Press, San Diego

El-Halwagi MM (2006) Process integration. Academic Press, New York

El-Halwagi MM, Spriggs HD (1998) Solve design puzzles with mass integration. Chem Eng Prog 94(8):25–44

El-Halwagi MM, Gabriel F, Harell D (2003) Rigorous graphical targeting for resource conservation via material recycle/reuse networks. Ind Eng Chem Res 42(19):4319–4328

Erdirik-Dogan M, Grossmann IE (2008) Simultaneous planning and scheduling of single-stage multiproduct continuous plants with parallel lines. Comput Chem Eng 32(11):2664–2683

Feng X, Bai J, Zheng XS (2007) On the use of graphical method to determine the targets of single-contaminant regeneration recycling water systems. Chem Eng Sci 62(8):2127–2138

Floudas CA, Lin X (2004) Continuous-time versus discrete-time approaches for scheduling of chemical processes: a review. Comput Chem Eng 28(11):2109–2129

Foo CY (2009) State-of-the-art review of pinch analysis techniques for water network synthesis. Ind Eng Chem Res 48:5125–5159

Gabriel FB, El-Halwagi MM (2005) Simultaneous synthesis of waste interception and material reuse networks: problem reformulation for global optimization. Environ Prog 24(2):171–180

Galan B, Grossmann IE (1998) Optimal design of distributed wastewater treatment networks. Ind Eng Chem Res 37(10):4036–4048
Gomes JFS, Queiroz EM, Pessoa FLP (2007) Design procedure for water/wastewater minimization: single contaminant. J Clean Prod 15(5):474–485
Gunaratnam M, Alva-Argaez A, Kokossis A, Kim JK, Smith R (2005) Automated design of total water systems. Ind Eng Chem Res 44(3):588–599
Gutiérrez-Limón MA, Flores-Tlacuahuac A, Grossmann IE (2012) A multiobjetive optimization approach for simultaneous single line scheduling and control of CSTRs. Ind Eng Chem Res 51(17):5881–5890
Hallale N (2002) A new graphical targeting method for water minimization. Adv Environ Res 6(3):377–390
Hernandez-Suarez R, Castellanos-Fernandez J, Zamora JM (2004) Superstructure decomposition and parametric optimization approach for the synthesis of distributed wastewater treatment networks. Ind Eng Chem Res 43(9):2175–2191
Kallrath J (2002) Planning and scheduling in the process industry. Or Spectrum 24(3):219–250
Karuppiah R, Grossmann IE (2006) Global optimization for the synthesis of integrated water systems in chemical processes. Comput Chem Eng 30(4):650–673
Kuo WCJ, Smith R (1997) Effluent treatment system design. Chem Eng Sci 52(23):4273–4290
Lampert C, Brunner PH (1999) Material accounting as a policy tool for nutrient management in the Danube Basin. Water Sci Technol 40(10):43–49
Lin X, Floudas C (2001) Design, synthesis and scheduling of multipurpose batch plants via an effective continuous-time formulation. Comput Chem Eng 25(4–6):665–674
Lira-Barragán LF, Ponce-Ortega JM, Serna-González M, El-Halwagi MM (2010) An MINLP model for the optimal location of a new industrial plant with simultaneous consideration of economic and environmental criteria. Ind Eng Chem Res 50(2):953–964
Lira-Barragán LF, Ponce-Ortega JM, Serna-González M, El-Halwagi MM (2011) Synthesis of water networks considering the sustainability of the surrounding watershed. Comput Chem Eng 35(12):2837–2852
Lira-Barragán LF, Ponce-Ortega JM, Nápoles-Rivera F, Serna-González M, El-Halwagi MM (2013) Incorporating property-based water networks and surrounding watersheds in site selection of industrial facilities. Ind Eng Chem Res 52(1):91–107
Lira-Barragán LF, Flores-Tlacuahuac A, Rivera-Toledo M, Ponce-Ortega JM, Serna-González M (2015) Dynamic optimization for the optimal location of new industrial facilities considering the sustainability of the watershed. In: Sustainability of products, processes and supply chains—theory and applications. Elsevier, Amsterdam, Netherlands
Lovelady EM, El-Baz AA, El-Monayeri D, El-Halwagi MM (2009) Reverse problem formulation for integrated process discharges with wastewater and drainage systems: managing phosphorus in lake Manzala. J Ind Ecol 13(6):914–927
Manan ZA, Tan YL, Foo DCY (2004) Targeting the minimum water rate using water cascade analysis technique. AIChE J 50(12):3169–3183
Mendez CA, Cerda J, Grossmann IE, Harjunkoski I, Fahl M (2006) State-of-the-art review of optimization methods for short-term scheduling of batch processes. Comput Chem Eng 30(6–7):913–946
Mouret S, Grossmann IE, Pestiaux P (2009) A novel priority-slot based continuous-time formulation for crude-oil scheduling problems. Ind Eng Chem Res 48(18):8515–8528
Mouret S, Grossmann IE, Pestiaux P (2011) A new Lagrangian decomposition approach applied to the integration of refinery planning and crude-oil scheduling. Comput Chem Eng 35(12):2750–2766
Nápoles-Rivera F, Ponce-Ortega JM, El-Halwagi MM, Jiménez-Gutiérrez A (2010) Global optimization of mass and property integration networks with in-plant property interceptors. Chem Eng Sci 65(15):4363–4377
Pillai HK, Bandyopadhyay S (2007) A rigorous targeting algorithm for resource allocation networks. Chem Eng Sci 62:6212–6221

Polley GT, Polley HL (2000) Design better water networks. Chem Eng Prog 96(2):47–52
Ponce-Ortega JM, Hourtua AC, El-Halwagi MM, Jiménez-Gutiérrez A (2009a) A property-based optimization of direct recycle networks and wastewater treatment processes. AIChE J 55(9):2329–2344
Ponce-Ortega JM, Serna-González M, Jiménez-Gutiérrez A (2009b) A disjunctive programming model for simultaneous synthesis and detailed design of cooling networks. Ind Eng Chem Res 48(6):2991–3003
Ponce-Ortega JM, El-Halwagi MM, Jiménez-Gutiérrez A (2010) Global optimization of property-based recycle and reuse networks including environmental constraints. Comput Chem Eng 34(3):318–330
Putra ZA, Amminudin KA (2008) Two-step optimization approach for design a total water system. Ind Eng Chem Res 47(16):6045–6057
Raman R, Grossmann IE (1994) Modeling and computational techniques for logic based integer programming. Comput Chem Eng 18(7):563–578
Savelski MJ, Bagajewicz MJ (2000) On the optimality conditions of water utilization systems in process plants with single contaminants. Chem Eng Sci 55(21):5035–5048
Savelski MJ, Bagajewicz MJ (2001) Algorithmic procedure to design water utilization systems featuring a single contaminant in process plants. Chem Eng Sci 56(5):1897–1911
Savelski M, Bagajewicz M (2003) On the necessary conditions of optimality of water utilization systems in process plants with multiple contaminants. Chem Eng Sci 58(23–24):5349–5362
Sharda B, Vazquez A (2009) Evaluating capacity and expansion opportunities at tank farm: a decision support system using discrete event simulation. In Proceedings of the 2009 winter simulation conference. Institute of Electrical and Electronic Engineers, New York, pp 2218–2224
Sorin M, Bedard S (1999) The global pinch point in water reuse networks. Process Saf Environ Prot 77(B5):305–308
Teles J, Castro PM, Novals AQ (2008) LP-based solution strategies for the optimal design of industrial water networks with multiple contaminants. Chem Eng Sci 63(2):376–394
Terrazas-Moreno S, Grossmann IE, Wassick JM (2012) A mixed-integer linear programming model for optimization the scheduling and assignment of tank farm operation. Ind Eng Chem Res 51(18):6441–6454
Wang YP, Smith R (1994) Wastewater minimization. Chem Eng Sci 49(7):981–1006

Chapter 13
Optimizing the Recycling of Flowback Water in Hydraulic Fracturing with Consideration of Seasonal and Environmental Limitations

13.1 Introduction

A lot of shale gas reserves have been discovered around the world. Hydraulic fracturing technologies help the production of shale gas trapped in tight formations. According to the Energy Information Administration (*EIA*), by the year 2035 shale gas is expected to provide about half of the total natural gas supply in the U.S (EIA 2012). The *EIA* predicts that the U.S. has enough natural and shale gas to meet domestic electricity demands for 575 years. Without a doubt, this data emphasizes the significance and vast potential of shale gas in the future. Siirola (2014) evaluated the impacts of shale gas on the chemical industry and the natural gas market. Weijermars (2014) predicted the gas production up to 2025 in the six major shale plays of U.S. under uncertain conditions to determine different scenarios of production. Cafaro and Grossmann (2014) reported a mathematical programming model for planning the shale gas supply and determining the number of wells, size of gas processing plants, section and length of pipelines for gathering raw gas and delivering processed gas and by-products, the power of gas compressors, and the amount of freshwater required from reservoirs for drilling and hydraulic fracturing to maximize the net present value. He and You (2014) incorporated shale gas processing with ethylene production to increase effectiveness. Yang et al. (2014a, b) implemented a design to incorporate an oil shale refinery with reforming retorting gas; this approach allows to production of the needed hydrogen in the hydrogenation part and provides economic benefits. Wang and Xu (2014) reported a new approach for the simultaneous operation of shale gas, NGL recovery, and LNG re-gasification involving the uncertainty of the shale gas feed rate. Knudsen et al. (2014) reported an optimization approach for well scheduling based on a Lagrangian relaxation. Pirzadeh et al. (2014) studied hydraulic fracturing additives that can generate H_2S under hydrothermal conditions or delay production.

A significant obstacle facing the shale gas sector is related to water-related problems, which encompass the following:

J. M. Ponce-Ortega et al., *Sustainable Water Systems*, Water Resources Development and Management, https://doi.org/10.1007/978-3-031-72498-5_13

- The supply of water requirements for the hydraulic fracturing process demands enormous amounts of water to be successfully implemented. Furthermore, drilling needs significant amounts of water.
- The treatment for the flowback water is collected after the completion phase. Afterward, the outlet stream should be reused or disposed of. If the reuse option is selected, a storage system is required.
- Lastly, there are several environmental worries associated with wastewater disposal and the risks of contamination of water bodies and reservoirs (including surface water and groundwater).

To address the environmental issues of shale gas production, some approaches have been implemented. In this context, Vengosh et al. (2014) reported a serious review of the risks related to the water resources in shale gas operations, explicitly for the pollution of shallow aquifers and surface water, the accumulation of toxic and radioactive elements near disposals, and the over-extraction of freshwater for hydraulic fracturing. Rahm et al. (2013) evaluated some wastewater management practices in the Marcellus shale play. Lavoie et al. (2014) assessed possible risks in the Utica play for aquifers associated with shale gas plays. Warner et al. (2013) evaluated the water quality and isotopic compositions of wastewaters, surface waters, and sediments linked with a treatment facility located in western Pennsylvania. Gordalla et al. (2013) measured the impact of hydraulic fracturing on groundwater in Germany accounting for that this is the most important source of drinking water in European countries. Finally, Manda et al. (2014) implemented an approach to determine that the change from single-well pads to multi-well pads has drastically affected wastewater volumes and yielded environmental violations.

Recycling and reusing flowback water is implemented to reduce wastewater streams and freshwater consumption. Figure 13.1 presents a general undesired configuration for water required in hydraulic fracturing operations, where typical practices use freshwater in hydraulic fracturing and dispose of the flowback water without reuse. Water scarcity is a huge problem for several regions around the world, and because of the enormous amount of freshwater required in shale gas plays, there is a need to consider flowback water reuse options as shown in Fig. 13.2.

It is important to implement the segregation and mixing of flowback water to be treated and then reused in the hydraulic fracturing process. The water reuse network can help in reducing the total wastewater discharged to the environment and at the same time reduce the consumption of freshwater sources. Also, it is possible to decrease the total annual cost of water usage and discharge. Numerous research

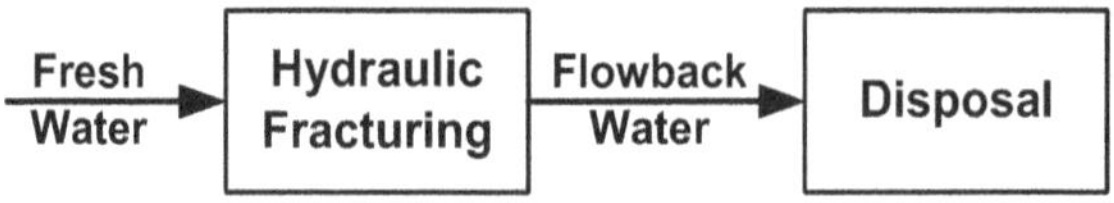

Fig. 13.1 Undesired configuration for water management in shale gas. Printed with permission of "*Optimal reuse of flowback wastewater in hydraulic fracturing including seasonal and environmental constraints*", *AIChE Journal*, Copyright (2015), Wiley

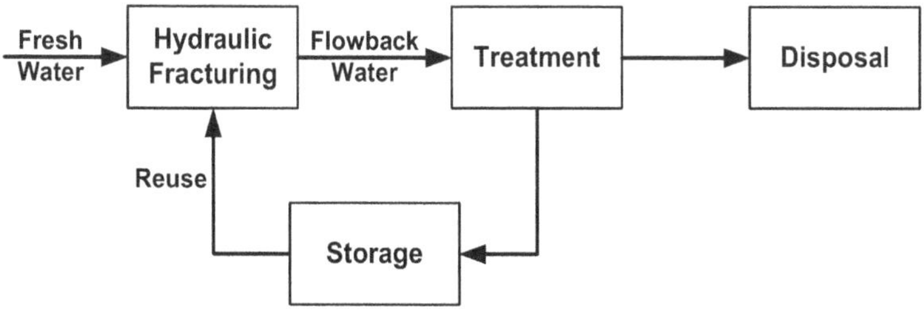

Fig. 13.2 Water management in shale gas including flowback water reuse. Printed with permission of "*Optimal reuse of flowback wastewater in hydraulic fracturing including seasonal and environmental constraints*", *AIChE Journal*, Copyright (2015), Wiley

efforts have been implemented for synthesizing water networks in industrial facilities. Similarly, it can be useful to consider recycling flowback in shale gas operations. These industrial water-management techniques have been classified as graphical, algebraic, and mathematical approaches. Among the graphical and algebraic approaches are the pinch point techniques (Wang and Smith 1997; Dhole et al. 1996; El-Halwagi and Spriggs 1998; Hallale 2002; Manan et al. 2004; El-Halwagi et al. 2003; Feng et al. 2007; Foo 2009). The mathematical techniques consider formal programming approaches (Pillai and Bandyopadhyay 2007; Savelski and Bagajewicz 2003; Teles et al. 2008; Gabriel and El-Halwagi 2005; Gunaratnam et al 2005; Karuppiah and Grossmann 2006; Putra and Amminudin 2008; Ponce-Ortega et al. 2009; 2010; Nápoles-Rivera et al. 2010; Lira-Barragán et al. 2011a, 2011b, 2013). Even though these approaches have been implemented for traditional industrial processes, valuable insights have been obtained that can be extended to the area of flowback water management.

Contemplating that shale gas production is a new industry, few publications have incorporated flowback water reuse strategies. Best and Lowry (2014) evaluated the possible effects of high-volume water extractions from groundwater in the Marcellus shale play. Jiang et al. (2014) implemented a life cycle analysis for water consumption and wastewater impact in a Marcellus shale gas well from its construction to the end of life. Clark et al. (2013) evaluated the consumed freshwater during shale gas production, involving the different stages of production and flowback water reuse. Theodori et al. (2014) estimated the performance of hydraulic fracturing and water reuse.

Yang et al. (2014a, b) implemented an optimization technique to evaluate the optimal fracturing schedule and minimize the associated costs involving transportation, treatment, storage, and disposal while considering reusing flowback water. Nevertheless, some of the most relevant aspects were not accounted for in the previous model. These aspects include the capital costs for treatment and storage systems and disposal. Besides, wastewater streams were disposed of without treatment. The transportation costs for some internal trajectories (wells-treatment, treatment-storage, and storage-wells) and the maximum capacities for the storage units and disposal were not involved. Therefore, the number of treatment technologies, storage units, and disposals required in the solution were not accounted for.

To overcome the previous limitations, this chapter reported an optimization formulation to find the optimal flowback water management system that is aimed at minimizing the total annual cost, which is integrated into the costs of freshwater, treatment, storage, disposals (including operating and capital costs), and transportation. The presented mathematical approach involves freshwater consumption, flowback water treatment, storage, reuse, and disposal. Furthermore, the model considers the seasonal variation in the freshwater availability and the impact on the completion phase for the given wells. Lastly, the approach involves a fixed scheduling for the completion operations (Lira-Barragán et al. 2016).

13.2 Problem Statement

The addressed problem in this chapter can be stated as follows.

Given is a scheduling on the horizon for the completion of a set of shale wells ($N = \{n \mid n = 1,2, \ldots, N_n\}$), which is implemented by one or several hydraulic fracturing crews that are able to operate concurrently. The water required during this phase, the operation time for completion and the flowback water are identified. Also, there is a freshwater source with seasonal restrictions, and unit and unit transportation costs (involving the distance between the locations of freshwater sources and the geographical sites for the wells). Furthermore, a set of treatment technologies is given ($I = \{i \mid i = 1,2, \ldots, N_i\}$), which are available to treat the flowback water and yield adequate conditions for its reuse or to discharge to the environment for its final disposal. Each treatment unit has associated a unit operating cost, fixed and variable (accounting for the exponent associated with the economies of scale) costs for the capital cost function and the maximum capacity of the flowrate that can be treated in each unit. Furthermore, a set of storage/pits ($S = \{s \mid s = 1,2, \ldots, N_s\}$) should be installed to store the treated flowback water. The exponent, the fixed and variable unit costs for the capital cost function, and the maximum volume for each storage unit are known. Additionally, there is a set of final disposals ($D = \{d \mid d = 1,2, \ldots, N_d\}$), where the treated flowback water can be disposed of. Likewise, each final disposal incorporates the known parameters required in the capital cost functions and a maximum volume allowed. Lastly, it is needed the unit transportation cost for the other trajectories (wells-treatments, treatments-disposals, treatments-storage, and storage-wells).

Afterward, the problem consists of determining the optimal configuration for the water network that allows operating the completion step for all the wells under the projected constraints in the freshwater availability and simultaneously minimizing the total annual cost. Hence, the optimization approach must decide the number of treatment technologies, storage/pits, and disposals required in the solution, and their optimal size as well as the internal flowrates with their variations over time. The objective function consists of the minimization of the total annual cost (*TAC*), which

Fig. 13.3 Shale gas well lifecycle. Printed with permission of "*Optimal reuse of flowback wastewater in hydraulic fracturing including seasonal and environmental constraints*", *AIChE Journal*, Copyright (2015), Wiley

is constituted by the total operating cost (*TOC*) and the total capital cost (*TCC*). *TOC* includes the freshwater costs, operating costs for treatment units, and transportation costs; while the *TCC* involves the capital costs linked to the treatment units, storage units, and disposals.

Furthermore, it has been projected that the water management cost in the shale gas industry represents 5–15% of overall shale gas drilling and completion (Slutz et al. 2012); therefore, Fig. 13.3 presents the typical shale gas lifecycle. The first step in the lifecycle for a shale gas well (once the authorization has been obtained and the location is determined) corresponds to site preparation, which needs a projected time of three weeks. Subsequently, there are implemented the drilling, completion, flowback, and production processes (Fig. 13.3 represents the needed time for each stage). It is noteworthy that hydraulic fracturing is carried out during completion; furthermore, water acquisition is required by drilling and completion processes; nevertheless, most of the total water needed is employed in the completion phase. After the completion process, and for 2–3 weeks, flowback water returns to the surface. Lastly, the well can produce natural gas for several years or even decades.

This way, implementing a review of the previous works on this topic, the water required for the completion phase in each well is nearby $8–26 \times 10^3$ m^3; also, the flowback water that can be collected at the end of this process is 10–40% concerning the total water injected and the rest of the fluid remains in the formation. Then, the flowback water reuse signifies a valuable resource to decrease freshwater consumption and the associated costs.

Additionally, to implement an approach that involves the abovementioned issues, first, it is required to know the number of wells for a given field; also considering a set of hydraulic fracturing crews available to operate in this field and the seasonal constraints for the freshwater availability, there is determined the scheduling for the completion phases of the wells during a fixed horizon. After this, the implemented mathematical optimization approach can be applied. This way, the superstructure to carry out the flowback water management in this chapter is presented in Fig. 13.4. Freshwater is the primary resource to complete the water demands in wells; however, flowback water can be reused in hydraulic fracturing operations. It should be noted that all the obtained flowback water is sent to the interception network, which is constituted of a set of treatment technologies. At the exit of the treatment network, the fluid can be disposed of or stored (this option is useful for the periods when the availability of freshwater is severely reduced).

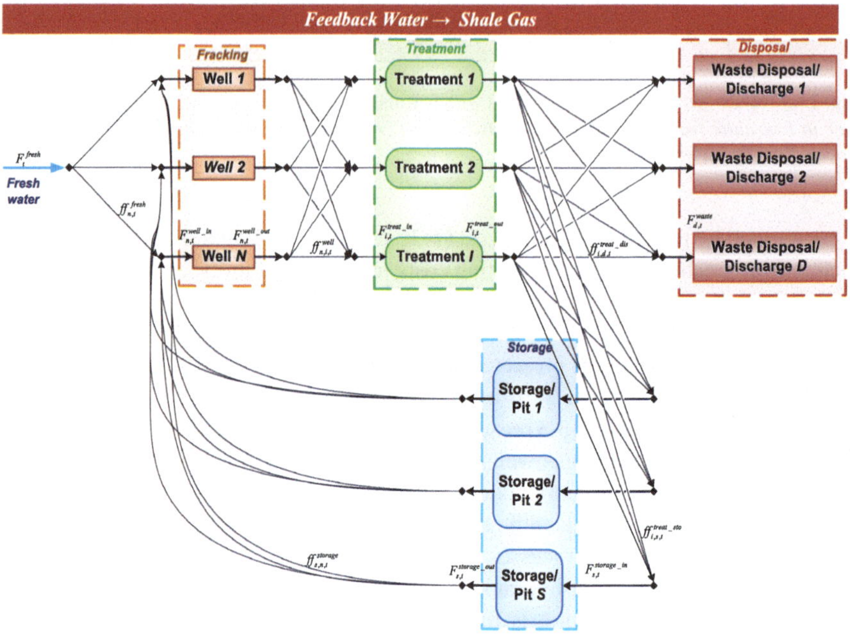

Fig. 13.4 Schematic representation of the proposed superstructure for water management. Printed with permission of "*Optimal reuse of flowback wastewater in hydraulic fracturing including seasonal and environmental constraints*", *AIChE Journal*, Copyright (2015), Wiley

13.3 Implemented Mathematical Model Formulation

To obtain the desired targets, the proposed optimization approach needs a fixed scheduling for the exploitation of the well pads; this way, the approach can contemplate the projected scheduling for one or several hydraulic fracturing crews, which if the conditions are satisfactory can operate concurrently in different wells. Afterward, it is possible to identify and determine the periods when will be required water acquisition to operate and carry out the hydraulic fracturing (i.e., the completion phase) for each well. Accordingly, for a given time horizon, it can plan the periods when water will be needed; precisely during the hydraulic fracturing phase for each well. Furthermore, previous works have approximated experimentally the water needed to complete each well; hence, the flowrate entering the well n in the time t ($F_{n,t}^{well\_in}$) corresponds to a given parameter even if several hydraulic fracturing crews are operating concurrently.

Furthermore, at the end of the hydraulic fracturing process, there will be flowback water coming from the wells. This way, some reports have stated experimental data about the amount of flowback water produced from each well; consequently, the flowback water for the well n over the period t is represented by $F_{n,t}^{well\_out}$, which is a known parameter for the proposed mathematical optimization approach. It is noteworthy to mention that the flowback water should be treated and stored to be reused or sent directly to the final disposals (Fig. 13.4).

Consequently, the implemented optimization approach involves the mathematical relationships needed to simulate the aspects previously mentioned and the limitations of the water available in the freshwater source, and the associated costs to the water management involved in the shale gas industry. The required relationships are the following.

13.3.1 Availability Constraints for Freshwater

The primary problem and weakness of the shale gas industry is the availability of the freshwater needed to complete each well, especially in regions with severe problems of water scarcity. Also, one of the most important contributions of the model of this chapter is that it considers the available freshwater (in some cases it must be authorized and regulated by the government) to be used during the hydraulic fracturing phase for each well

$$F_t^{fresh} \leq \left(\frac{AvailV_t^{fresh\_\max}}{H^{time}} \right), \quad \forall t \tag{13.1}$$

F_t^{fresh} is the freshwater consumption over the period t, $AvailV_t^{fresh\_\max}$ represents the water availability over the period t and H^{time} corresponds to a time conversion factor.

13.3.2 Segregation of Freshwater

Freshwater is split and sent to the wells n during the completion process:

$$F_t^{fresh} = \sum_n ff_{n,t}^{fresh}, \quad \forall t \tag{13.2}$$

13.3.3 Water Source to Wells

The water needs during the hydraulic fracturing phase in each well n ($F_{n,t}^{well\_in}$) are supplied by freshwater source ($ff_{n,t}^{fresh}$) and the water arriving from the storage system ($ff_{s,n,t}^{storage}$):

$$F_{n,t}^{well\_in} = ff_{n,t}^{fresh} + \sum_{s} ff_{s,n,t}^{storage}, \quad \forall n, \ \forall t \tag{13.3}$$

13.3.4 Flowback Water

After finishing the completion phase in each well, flowback water is produced ($F_{n,t}^{well\_out}$), which is split and sent to a set of treatment units ($ff_{n,i,t}^{well}$)

$$F_{n,t}^{well\_out} = \sum_{i} ff_{n,i,t}^{well}, \quad \forall n, \ \forall t \tag{13.4}$$

All flowback water is treated, avoiding the storage or disposal of untreated hydraulic fracturing fluid.

13.3.5 Water Inlet to Treatment Units

The implemented mathematical approach involves an interception treatment network composed of various treatment units usually employed to treat the hydraulic fracturing fluid. This way, the literature shows technical reports with relevant information on the treatment units used in existing shale gas plays (Slutz et al. 2012). Consequently, the technologies tested efficiently for the treatment of flowback water can be considered even if they involve different costs and capacities. The only requirement to consider any treatment technology in this approach is that the outlet stream grants good quality and adequate conditions to be disposed of or reused without generating environmental issues or technical drawbacks in the process. Lastly, the model can select the technologies required (if it considers different units) and decide the number of treatment technologies and size for each one.

Therefore, after the flowback is split, the balance for the mixer at the inlet of each treatment unit indicates that the inlet flowrate for unit i over period t ($F_{i,t}^{treat\_in}$) is obtained from the sum of the flowback water separated at the exit of the wells at the same time and sent to the unit ($ff_{n,i,t}^{well}$)

$$F_{i,t}^{treat\_in} = \sum_{n} ff_{n,i,t}^{well}, \quad \forall i, \ \forall t \tag{13.5}$$

13.3.6 Water Outlet from Treatment Units

Considering the implemented superstructure, there is a splitter at the exit of each treatment unit, where the outlet stream ($F_{i,t}^{treat\_out}$) should be sent to the storage units ($ff_{i,s,t}^{treat\_sto}$) or to the final disposal ($ff_{i,d,t}^{treat\_dis}$):

$$F_{i,t}^{treat\_out} = \sum_{s} ff_{i,s,t}^{treat\_sto} + \sum_{d} ff_{i,d,t}^{treat\_dis}, \quad \forall i, \ \forall t \tag{13.6}$$

13.3.7 Water Balance for Waste Disposal and Discharge

The wastewater streams generated ($F_{d,t}^{waste}$) are obtained from the interception network, where the water quality satisfies the environmental constraints for the pollutants:

$$F_{d,t}^{waste} = \sum_{i} ff_{i,d,t}^{treat\_dis}, \quad \forall d, \ \forall t \tag{13.7}$$

13.3.8 Water Inlet to Storage Units

Similarly, the storage system ($F_{s,t}^{storage\_in}$) is fed by wastewater streams treated in the interception network to satisfying the quality needs for its reuse:

$$F_{s,t}^{storage\_in} = \sum_{i} ff_{i,s,t}^{treat\_sto}, \quad \forall s, \ \forall t \tag{13.8}$$

13.3.9 Water Outlet from Storage Units

After the treated wastewater is stored in impoundments or pits, the flowrate exit from the storage unit s is recycled to the well n over the period t ($ff_{s,n,t}^{storage}$):

$$F_{s,t}^{storage\_out} = \sum_{n} ff_{s,n,t}^{storage}, \quad \forall s, \ \forall t \tag{13.9}$$

13.3.10 Balances in Treatment Units

Many treatment units consider water losses during their operation; this feature is simulated as follows:

$$F_{i,t}^{treat\_out} = \alpha_i^{treat} F_{i,t}^{treat\_in}, \quad \forall i, \ \forall t \tag{13.10}$$

α_i^{treat} is the efficiency factor in terms of volumetric flowrate for the unit i.

13.3.11 Balances in Storage Units

The usual well pad development includes the completion phase where the hydraulic fracturing process is implemented. This step requires enormous amounts of water to be injected into the wells as hydraulic fracturing fluid, which frequently takes 1–3 months (Yang et al. 2014a, b). After this stage has ended, an important portion of the total water injected is produced in the wells as flowback water. The flowback water can be reused in the hydraulic fracturing processes to decrease the freshwater consumption, but this stream must be treated and stored. Afterward, considering scheduling for the completion phases of a set of wells implemented by various hydraulic fracturing crews, the position of the storage involves saving the treated flowback water to be used during the periods where hydraulic fracturing process is implemented; this water also corresponds to a backup resource for periods with low freshwater availability.

This way, the accumulation balance for storage pits at the states that the volume in the pit s at the period t ($V_{s,t}^{storage}$) is equal to the one at the end of the previous period ($V_{s,t-1}^{storage}$), plus the difference of the inlet and outlet flowrates multiplied by the time conversion factor:

$$V_{s,t}^{storage} = V_s^{storage\_initial} + H^{time}\left(F_{s,t}^{storage\_in} - F_{s,t}^{storage\_out}\right), \quad \forall s, \ \forall t = 1 \tag{13.11}$$

$$V_{s,t}^{storage} = V_{s,t-1}^{storage} + H^{time}\left(F_{s,t}^{storage\_in} - F_{s,t}^{storage\_out}\right), \quad \forall s, \ \forall t > 1 \tag{13.12}$$

At the first period, there is considered a known initial pit volume ($V_s^{storage\_initial}$). Also, it is needed to satisfy the continuity of the cycles. This way, the initial volume is equal to the last volume in the pit s:

$$V_s^{storage\_initial} = V_{s,t}^{storage}, \quad \forall t = t^{final} \tag{13.13}$$

13.3.12 Capacity and Existence of Treatment Units

The implemented optimization model determines the number of treatment units, treatment technologies and final disposals needed in the optimal solution and the size for each one. Afterward, first for the treatment units the following couple of relationships determine if the unit i is needed or not (the existence of the treatment units is modeled through the binary variable y_i^{treat}); furthermore, the operational capacity for the treatment is represented through $F_i^{treat\_cap}$:

$$F_i^{treat\_cap} \geq F_{i,t}^{treat\_in}, \quad \forall i, \ \forall t \tag{13.14}$$

$$F_i^{treat\_cap} \leq F_i^{treat\_\max} y_i^{treat}, \quad \forall i \tag{13.15}$$

$F_i^{treat\_\max}$ is an upper limit for the operational capacity associated to the unit i (this information can be provided by sellers of treatment technologies). The optimal capacity ($F_i^{treat\_cap}$) must be used in the cost function associated to the treatment costs.

13.3.13 Capacity and Existence of Storage Units

Likewise, for the storage/pits units, the following relationships are required:

$$V_s^{storage\_cap} \geq V_{s,t}^{storage}, \quad \forall s, \ \forall t \tag{13.16}$$

$$V_s^{storage\_cap} \leq V_s^{storage\_\max} y_s^{storage}, \quad \forall s \tag{13.17}$$

$V_s^{storage\_cap}$ is the optimal capacity for the storage unit; whereas $V_s^{storage\_\max}$ is the maximum capacity for the pit s, and $y_s^{storage}$ corresponds to a binary variables used to model the existence of the storage/pits units.

13.3.14 Capacity and Existence of Disposals

Finally, the final disposals accumulate all the flowrate discharged on them during all the given time horizon. This way, the operational capacity for the disposal d ($V_d^{waste\_cap}$) is determined as follows:

$$V_d^{waste\_cap} = H^{time} \sum_t F_{d,t}^{waste}, \quad \forall d \tag{13.18}$$

$$V_d^{waste\_cap} \leq V_d^{waste\_\max} y_d^{waste}, \quad \forall d \tag{13.19}$$

$V_d^{waste\_\max}$ represents the maximum capacity for each disposal and the existence of each disposal is simulated through the binary variable y_d^{waste}. Therefore, it can determine the number of treatment technologies, storages/pits and disposals in the optimal solution and their capacities.

13.3.15 Total Water Used

Even though the water constraints can be overcome, it is important to determine the total water required to complete all the proposed wells. Consequently, the next relationship calculates the total water requirements (*TWR*):

$$TWR = H^{time} \sum_t F_t^{fresh} \tag{13.20}$$

Considering the existing water concerns, it is imperative to know the total water needs of the shale gas industry to exploit a certain number of wells since the environmental point of view the minimization of the used freshwater has a greater relevance than the economic aspects. This way, the best economic solution frequently corresponds to the minimum freshwater consumption; nevertheless, for some cases where the storage costs are menacingly greater than the costs associated to the freshwater and final disposal, it would be possible that the environmental and economic objectives contradict each other, and as consequence, Pareto curves can be useful to determine solutions that compensate the contradicting objectives.

13.3.16 Operating Costs

The operating costs considered in this project correspond to the freshwater cost, operating costs for treatment units and transportation costs (including all the trajectories), which are described as follows.

13.3.16.1 Freshwater Cost

The freshwater cost ($Cost^{fresh}$) is determined by multiplying the sum of the water needed during all periods with the unit cost for the freshwater (UC^{fresh}) and the time conversion factor:

$$Cost^{fresh} = H^{time} UC^{fresh} \sum_t F_t^{fresh} \tag{13.21}$$

13.3.16.2 Operating Costs for Treatment Units

The operating costs for the treatment ($Cost^{op\_treat}$) is obtained as follows:

$$Cost^{op\_treat} = H^{time} \sum_i \sum_t UOC_i^{treat} F_{i,t}^{treat\_in} \tag{13.22}$$

UOC_i^{treat} represents the unit operating cost for the treatment unit i.

13.3.16.3 Freshwater Transportation Cost

The freshwater transportation cost ($Cost^{trans\_fresh}$) can be determined multiplying the flowrates of freshwater sent to the well n during period t times the unit transportation cost (UTC_n^{fresh}):

$$Cost^{trans\_fresh} = H^{time} \sum_n \sum_t UTC_n^{fresh} ff_{n,t}^{fresh} \tag{13.23}$$

The unit costs are different for each well to consider the geographical distance between the freshwater source with each well. For some existing shale plays, the freshwater resource is available at important distances from the wells; consequently, in cases like Marcellus shale region the freshwater transportation is several times higher than freshwater cost (Yang et al. 2014a, b).

13.3.16.4 Flowback Water Transportation Cost

The flowback water produced in the wells is transported to the interception network and the cost associated to this operation is obtained through the next expression:

$$Cost^{trans\_fb} = H^{time} \sum_n \sum_i \sum_t UTC_{n,i}^{used} ff_{n,i,t}^{well} \tag{13.24}$$

$UTC_{n,i}^{used}$ represents the unit transportation cost for the flowback water in the path generated by the well n and the treatment i. Now, also the unit cost can change for each trajectory.

13.3.16.5 Transportation Cost for the Treated Water to Disposals

The flowrate exiting the treatment can be sent to the disposals and the transportation costs, which is modeled as follows:

$$Cost^{trans\_treat\_dis} = H^{time} \sum_{i} \sum_{d} \sum_{t} UTC_{i,d}^{treat\_dis} ff_{i,d,t}^{treat\_dis} \quad (13.25)$$

$UTC_{i,d}^{treat\_dis}$ represents the unit transportation cost for the treated water from the treatment i to the disposal d.

13.3.16.6 Transportation Cost for the Treated Water to Storage

Furthermore, the treated water can be sent to the storage system to obtain the following costs:

$$Cost^{trans\_treat\_sto} = H^{time} \sum_{i} \sum_{s} \sum_{t} UTC_{i,s}^{treat\_sto} ff_{i,s,t}^{treat\_sto} \quad (13.26)$$

$UTC_{i,s}^{treat\_sto}$ is the unit transportation cost for the treated water from the treatment i to the storage/pit s.

13.3.16.7 Transportation Cost for Reused Water to Well

The stored water in the storage/pits is transported to the wells for reuse and the associated cost is accounted in the following relationship:

$$Cost^{trans\_sto\_well} = H^{time} \sum_{s} \sum_{n} \sum_{t} UTC_{s,n}^{sto\_well} ff_{s,n,t}^{storage} \quad (13.27)$$

$UTC_{s,n}^{sto\_well}$ represents the unit transportation cost for the reused water from the pit s to the well n.

It should be noted that all the unit transportation costs can be different according to each path to consider geographical locations.

13.3.16.8 Total Operating Cost

Lastly, the sum of all operating costs previously described yields the total operating cost (TOC) for the project:

$$TOC = Cost^{fresh} + Cost^{op\_treat} + Cost^{trans\_fresh} + Cost^{trans\_fb}$$

$$+ Cost^{trans\_treat\_dis} + Cost^{trans\_treat\_sto} + Cost^{trans\_sto\_well} \tag{13.28}$$

13.3.17 Capital Costs

The capital of the treatment units, storage units and the construction of final disposals yields capital which are involved in the economic objective function.

13.3.17.1 Capital Cost for Treatment Units

The capital cost for treatment units considers a fixed charge (FC^{treat}) and a variable charge (VC^{treat}) that depends on the treated flowrate as follows:

$$CapCost^{treatment} = k_F \sum_i \left[FC^{treat} y_i^{treat} + VC^{treat} \left(F_i^{treat\_cap} \right)^{\beta^{treat}} \right] \tag{13.29}$$

k_F represents a factor used to annualize the inversion, while β^{treat} is the exponent associated to the economies of scales for the treatment units.

13.3.17.2 Capital Cost for Storage Units and Disposals

Correspondingly for the storage units and disposals:

$$CapCost^{storage} = k_F \sum_s \left[FC^{storage} y_s^{storage} + VC^{storage} \left(V_s^{cap} \right)^{\beta^{storage}} \right] \tag{13.30}$$

$$CapCost^{waste} = k_F \sum_d \left[FC^{waste} y_d^{waste} + VC^{waste} \left(V_d^{waste\_cap} \right)^{\beta^{waste}} \right] \tag{13.31}$$

β is the only nonlinear terms included in the proposed mathematical model.

13.3.17.3 Total Capital Cost

Furthermore, the total capital cost (TCC) involves the capital costs for treatment units, storage units and final disposals:

$$TCC = CapCost^{treat} + CapCost^{waste} + CapCost^{storage} \tag{13.32}$$

The fixed cost included in the capital cost function is a function of the generated costs that do not depend on the flowrate such as site preparation to install the treatment/pit/disposal/etc.

13.3.18 Total Annual Cost

Lastly, the total annual cost (*TAC*) is determined by the sum of *TOC* and *TCC:*

$$Min\ TAC = TOC + TCC \tag{13.33}$$

13.4 Results and Discussion

The applicability of the presented mathematical model is exposed through the following example, where the MINLP model consists of 10 binary variables, 9446 continuous variables, and 3665 constraints. Furthermore, the solver DICOPT in conjunction with the solvers CONOPT and CPLEX were implemented in the General Algebraic Modeling System (Brooke et al. 2015).

13.4.1 Case Study

United States has a large production of shale gas, currently USA is the number one natural gas producer in the world. This way, Marcellus (located mainly in Pennsylvania) and Barnett (located in Texas) plays are the largest shale gas reserves. Most of the information needed to implement the presented methodology in this case study has been taken from technical reports with updated data for the Marcellus and Barnett regions (Slutz et al. 2012; Hayes 2009; Shaffer et al. 2013).

The explanation of this example problem, where the water management is optimized in periods of a week with a time horizon of 52 weeks (a year) and the completion for each well takes 5 weeks, begins defining the amount of water needed to complete each well. This way, Slutz et al. (2012) considered a range of 80,000–120,000 bbl (12,700–19,000 m^3) per hydraulic fracturing, Shaffer et al. (2013) estimated a volume of 8000–27,000 m^3 for each well in the Marcellus shale region. Hayes (2009) indicated that there are needed 71,400 – 95,200 bbl (11,335–15,114 m^3) to successful complete each horizontal well. This way, an average value of 15,000 m^3 is used in this chapter for the water needs in the completion of each well. While this last technical report also has established that approximately 25% of the water injected during the hydraulic fracturing is collected over several days following to the hydraulic fracturing (mainly in the two weeks after the completion process).

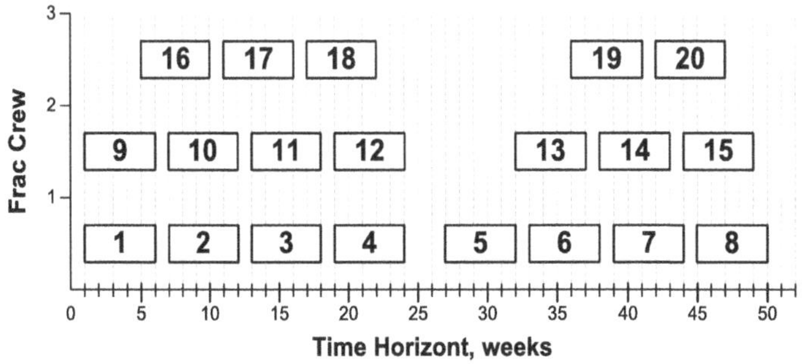

Fig. 13.5 Scheduling for the completion phase in each well. Printed with permission of "*Optimal reuse of flowback wastewater in hydraulic fracturing including seasonal and environmental constraints*", *AIChE Journal*, Copyright (2015), Wiley

Therefore, the flowrate needed in each well during the completion time is 428.57 m^3/d; whereas the flowback water obtained after this phase has a value of 342.86 m^3/d, 150 m^3/d and 42.85 m^3/d only for the first three weeks, respectively.

Furthermore, 20 wells should be completed by three hydraulic fracturing crews according to the scheduling presented in Fig. 13.5, where the hydraulic fracturing crew 1 completes the wells 1–8, hydraulic fracturing crew 2 works the wells 9–15 and finally crew 3 operates the wells 16–20. Also, at the beginning two hydraulic fracturing crews are working; after a few weeks the three crews can operate. However, approximately in the middle of the year the operation of any crew is not allowed owing to the water scarcity and progressively when the water availability permits the hydraulic fracturing crews restart the operations. The scheduling performance is powerfully dependent on the historical limitations of the available freshwater. Likewise, the transition time required to move the hydraulic fracturing crew from a well to the next one is a week.

Furthermore, the time conversion factor (H^{time}) is 7 d/w, $k_F = 0.1$, the volumetric efficiency factor (α_i^{treat}) for the treatment units is 0.9 and there is not initial volume in the storage/pits units ($V_s^{storage\_initial}$). Furthermore, the treatment units, storage units and final disposals considered in this example are usually employed in the existing shale gas plays. Consequently, the treatment unit accounted ensures a quality for the outlet stream enough to be reused or disposed under adequate environmentally conditions. Also, between the common options to store the flowback water are pits and hydraulic fracturing tanks; however, this last represents an unfeasible option for this specific case owing to the enormous volumes that represent the flowback water storage of 20 wells compared with the hydraulic fracturing tanks capacity. While the final disposal is implemented through typical Class II wells. Table 13.1 presents some data related to the treatment and storage units and disposals.

Furthermore, it requires a geographical subdivision for the wells, where the first subset of wells $n1$ corresponds to the wells 1–8 (which are operated by hydraulic fracturing crew 1); whereas the subsets $n2$ and $n3$ are the wells 9–15 (related to crew 2) and 16–20 (associated to crew 3), respectively. In that way, the unit transportation

Table 13.1 Data used in the example problem

Concept/Case	Treatment	Storage	Disposal
Fixed cost (FC^{case}), $	95,000	10,000	52,000
Variable cost (V^{case}), $/m^3	4.1	6.3	13.5
Maximum capacity (V^{case_max}), m^3/d or m^3	500	12,000	20,000
Exponent for the economies of scale (β^{case})	0.65	0.72	0.9

Printed with permission of "*Optimal reuse of flowback wastewater in hydraulic fracturing including seasonal and environmental constraints*", AIChE Journal, Copyright (2015), Wiley

costs from the fresh source to wells $n1$, $n2$ and $n3$ are 4, 7 and 12 $/m^3, respectively. The unit transportation costs for the wells $n1$, $n2$ and $n3$ to the interception network are 0.39, 0.72 and 0.94 $/m^3, respectively; while these costs for the segments from the treatment to storage/pits and disposals are 1.41 and 8.3 $/m^3, respectively. Lastly, the unit transportation costs from the storage to the wells $n1$, $n2$ and $n3$ are 0.98, 1.23 and 1.65 $/m^3, respectively.

The unit operating cost for the treatment technology is 12.50 $/m^3 and a value of 2 $/m^3 is considered for the unit cost of fresh water. Additionally, Fig. 13.6 presents the maximum available volume of freshwater. This aspect is an important point because the considerable differences in the water availability during the year (it should be noted that the maximum and minimum values). Thus, if a comparison among Figs. 13.5 and 13.6 is implemented, the critical conditions in the water availability during the period corresponding to the 24–27 weeks interrupt the completion operations for the wells. Furthermore, the total availability of freshwater (summing all the values during the year) is 276,890 m^3. If this value is compared with the total water requirements for the hypothetical case where the storage and reuse is not allowed and all the flowback water is disposed of as well as all the total water requirements must be supplied only by the fresh sources (20 × 15,000 m^3 = 300,000 m^3), there is a tremendous water deficit; otherwise, because the water scarcity, only freshwater resources are not enough to exploit all the proposed wells if the optimal water management (flowback water storage and its reuse) is not accounted. This analysis presents a key point for the importance of the proposed approach.

To enriching the discussion, three cases with little differences among them are considered. It should be noted that all the parameters previously reported were employed for **Cases A**, **B** and **C** with exception of those highlighted as follows:

13.4.2 Case A

This case corresponds to the solution of the base problem considering all the provided data, and the freshwater availability presented in Fig. 13.6.

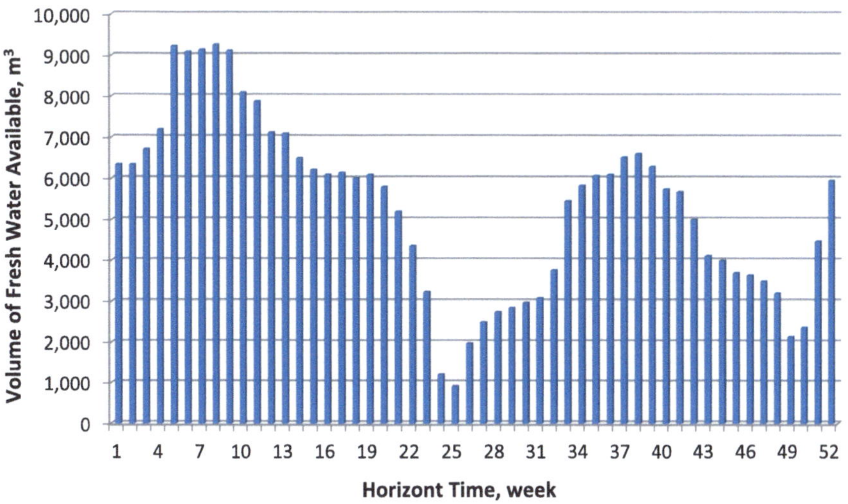

Fig. 13.6 Volume of fresh water available in the example problem. Printed with permission of "*Optimal reuse of flowback wastewater in hydraulic fracturing including seasonal and environmental constraints*", *AIChE Journal*, Copyright (2015), Wiley

13.4.3 *Case B*

In this scenario the constraints for the freshwater availability are removed and it considers that there are no problems of water availability.

13.4.4 *Case C*

This case eliminates the storage and reuse for the flowback water; thus, the total water needs are provided by fresh resources, consuming the maximum amount of fresh water (300,000 m^3). Next, to make feasible this case, it requires the elimination of the freshwater availability.

Next, implementing the presented optimization approach to the indicated cases, the results obtained are presented in Table 13.2, where the most important details for each solution such as the number of treatment and storage units and disposals required as well as the *TWR* and *TAC* (including the individual costs) are presented. It should be noted that there are significant differences between the *TOC*s concerning the *TCCs*; thus, the *TAC* is tremendously influenced by the operating costs. This way, the most important term that impacts the *TOC* is the transportation cost to take the freshwater from its source to the wells. This aspect corresponds to an operational problem for some existing shale gas plays and if the distance between the freshwater sources and wells is large, this cost is significantly increased.

Table 13.2 Results obtained for the example problem

Concept/case	Case A	Case B	Case C
Number of treatment units required	2	2	2
Number of storage/pits units required	2	2	0
Number of disposals required	1	1	4
Fresh water cost, $/y	474,179	474,179	600,000
Operating cost for treatment units, $/year	937,493	937,493	937,493
Transportation cost fresh source-wells, $/year	1,474,927	1,396,918	2,114,990
Transportation cost wells-treatments, $/year	48,225	48,225	48,225
Transportation cost treatments-disposals $/year	38,096	38,096	560,245
Transportation cost treatments-storages $/year	88,702	88,702	0
Transportation cost storages-wells, $/year	94,160	100,711	0
Total operating cost *TOC*, $/year	3,155,782	3,084,324	4,260,953
Capital cost for treatment units, $/year	19,037	19,037	19,037
Capital cost for disposals, $/ysear	5663	5663	26,742
Capital cost for storage/pits, $/year	2763	2644	0
Total capital cost *TCC*, $/year	27,463	27,344	45,779
Total annual cost TAC, $/year	3,183,245	3,111,668	4,306,732
Total water requirements TWR, m^3	237,089	237,089	300,000
Are the freshwater limitations considered?	Yes	No	No
Is the flowback water reuse considered?	Yes	Yes	No

Furthermore, the capacities for treatment units, storage/pits and disposals are presented in Table 13.3.

Figures 13.7, 13.8, and 13.9 present the total water volumes needed in all the internal trajectories for each week in the optimal solutions of **Cases A**, **B**, and **C**, respectively. For the three solutions in the first weeks only freshwater is used in the wells; while at the last part of the horizon, there is no required freshwater. Furthermore, important differences in the operation during the year can be esteemed. Figure 13.9 does not include water movements for treatment-storage and reuse because **Case C** does not include the flowback water management and all the values for these trajectories must be equal to zero.

It should be noticed the huge amount of the flowback water management through its storage and reuse (in economic terms), which is implemented through a direct comparison between the previously indicated cases.

The main difference between **Cases A** and **B** is the consideration of the freshwater availability; but the cost of **Case A** is 2.30% greater than the one of **Case B**. Consequently, it can be stated that the freshwater limitations yield an increase in the *TAC* by 2.30%. Whereas the match **A** versus **C** shows the contribution of the flowback

Table 13.3 Optimal capacities for treatment units, storage pits and disposal for each case considered

No	Case A	Case B	Case C
Treatment unit, m^3/d			
1	500	500	500
2	228.57	228.57	228.57
Storage/pit, m^3			
1	12,000	12,000	–
2	3375	1125	–
Disposal, m^3			
1	4590	4590	20,000
2	–	–	20,000
3	–	–	20,000
4	–	–	7500

Printed with permission of "*Optimal reuse of flowback wastewater in hydraulic fracturing including seasonal and environmental constraints*", AIChE Journal, Copyright (2015), Wiley

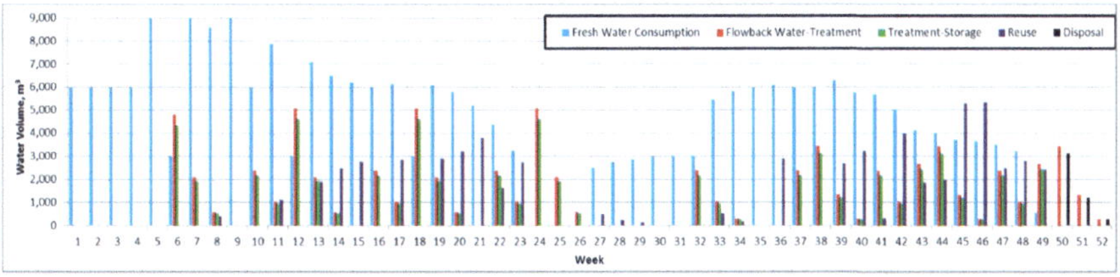

Fig. 13.7 Total water volumes in the solution for **Case A**. Printed with permission of "*Optimal reuse of flowback wastewater in hydraulic fracturing including seasonal and environmental constraints*", *AIChE Journal*, Copyright (2015), Wiley

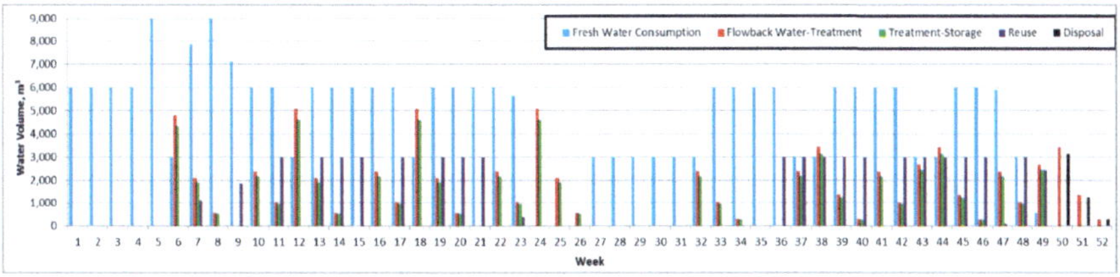

Fig. 13.8 Total water volumes in the solution for **Case B**. Printed with permission of "*Optimal reuse of flowback wastewater in hydraulic fracturing including seasonal and environmental constraints*", *AIChE Journal*, Copyright (2015), Wiley

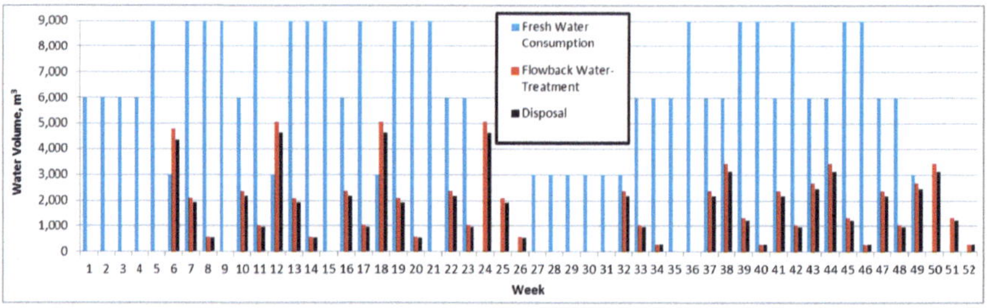

Fig. 13.9 Total water volumes in the solution for **Case C**. Printed with permission of "*Optimal reuse of flowback wastewater in hydraulic fracturing including seasonal and environmental constraints*", *AIChE Journal*, Copyright (2015), Wiley

water management evidencing considerable economic savings, specifically in **Case A** the *TAC* is 26% lower than the one of **Case C**; therefore, the flowback water reuse can achieve substantial cost reductions. Nonetheless, this effect is better appreciated when **Case B** is compared with **Case C** (owing to both do not include the freshwater limitations); thus, the cost decrease is approximately of 28%.

It should be noted that all the inner flowrates associated with the proposed superstructure have been determined as well as their fluctuations concerning the time.

13.5 Concluding Remarks

This chapter presented a mathematical programming model for optimal water management in shale gas production considering economic and environmental issues. The economic objective function consists of the minimization of the total annual cost which is composed of the operating and capital costs. The environmental objective involves minimizing the total freshwater requirements (through synthesizing recycle/reuse water networks and determining the optimal operation and scheduling) and the discharged waste streams while meeting the environmental constraints. The presented approach determines the number of treatment technologies, storage pits, and disposal required in the optimal configuration and their capacities. The optimal inner flowrates are determined. The model involves the limitations associated with the available freshwater over different seasons throughout the year. This aspect affects the scheduling for the completion of the wells. The presented superstructure allows efficient flowback water management that can save freshwater and reuse flowback water over periods with water scarcity problems.

The presented optimization approach was applied to scenarios associated with Marcellus and Barnett shale plays. The results feature the benefits of considering the presented approach for flowback water management as it allows a greater number of wells that can be completed with lower amounts of freshwater and minor costs.

It should be noted that the reuse strategy results in up to 27% of savings in the total costs.

Nomenclature

Parameters

$AvailV_t^{fresh\_\max}$	Maximum water availability, m^3/week
FC^{treat}	Fixed charge for the treatment unit in the capital cost function, \$
$FC^{storage}$	Fixed charge for the storage unit in the capital cost function, \$
FC^{waste}	Fixed charge for the disposal in the capital cost function, \$
$F_{n,t}^{well\_in}$	Inlet flowrate required in the wells, m^3/d
$F_{n,t}^{well\_out}$	Flowback water leaving from the wells, m^3/d
H^{time}	Time conversion factor, d/ week
k_F	Factor used to annualize the inversion, $year^{-1}$
UC^{fresh}	Unit cost for the fresh water, \$/m^3
UOC_i^{treat}	Unit operating cost for the treatment units, \$/m^3
UTC_n^{fresh}	Unit transport cost for the fresh water, \$/m^3
$UTC_{s,n}^{sto\_well}$	Unit transport cost for the reused water from the storage to wells, \$/m^3
$UTC_{i,d}^{treat\_dis}$	Unit transport cost for the treated water from the treatment to disposal, \$/m^3
$UTC_{i,s}^{treat\_sto}$	Unit transport cost for the treated water from the treatment to storage, \$/m^3
$UTC_{n,i}^{used}$	Unit transport cost for the flowback water from wells to treatment, \$/m^3
VC^{treat}	Variable charge for the treatment unit in the capital cost function, \$/m^3/d
$VC^{storage}$	Variable charge for the storage unit in the capital cost function, \$/m^3
VC^{waste}	Variable charge for the disposal in the capital cost function, \$/m^3

Greek Symbols

α_i^{treat}	Volumetric efficiency factor for the treatment unit
β^{treat}	Exponent associated to the economies of scale for the treatment units
$\beta^{storage}$	Exponent associated to the economies of scale for the storage/pits
β^{waste}	Exponent associated to the economies of scale for the disposals

Variables

$CapCost^{treatment}$	Capital cost for treatment, \$
$CapCost^{storage}$	Capital cost for storage, \$
$CapCost^{waste}$	Capital cost for disposal, \$
$Cost^{fresh}$	Freshwater cost, \$
$Cost^{op\_treat}$	Operating costs for the treatment, \$
$Cost^{trans\_fb}$	Transportation cost for the flowback water from wells to treatment, \$
$Cost^{trans\_fresh}$	Freshwater transportation cost, \$
$Cost^{trans\_sto\_well}$	Transportation cost for reused water from storage to wells, \$
$Cost^{trans\_treat\_dis}$	Transportation cost for treated water from treatment to disposals, \$
$Cost^{trans\_treat\_sto}$	Transportation cost for treated water from treatment to storage, \$
F_t^{fresh}	Freshwater flowrate, m^3/d
$F_{s,t}^{storage\_in}$	Flowrate entering to the storage units, m^3/d
$F_{s,t}^{storage\_out}$	Flowrate leaving to the storage units, m^3/d
$F_i^{treat\_cap}$	Capacity flowrate for the treatment units, m^3/d
$F_{i,t}^{treat\_in}$	Inlet flowrate to the treatment units, m^3/d
$F_i^{treat\_\max}$	Maximum flowrate for the treatment units, m^3/d
$F_{i,t}^{treat\_out}$	Flowrate leaving the treatment units, m^3/d
$F_{d,t}^{waste}$	Flowrate entering to the treatment units, m^3/d
$ff_{n,t}^{fresh}$	Segregated flowrate of fresh water to wells, m^3/d
$ff_{s,n,t}^{storage}$	Segregated flowrate from storage system to wells, m^3/d
$ff_{i,d,t}^{treat\_dis}$	Segregated flowrate from treatment system to disposals, m^3/d
$ff_{i,s,t}^{treat\_sto}$	Segregated flowrate from treatment system to storage system, m^3/d
$ff_{n,i,t}^{well}$	Segregated flowrate from wells to treatment system, m^3/d
TAC	Total annual cost, \$
TCC	Total capital cost, \$
TOC	Total operating cost, \$
TWR	Total water requirements, m^3
$V_{s,t}^{storage}$	Storage/pits volume, m^3
$V_s^{storage\_cap}$	Capacity for storage/pits, m^3
$V_s^{storage\_initial}$	Initial storage/pits volume, m^3
$V_s^{storage\_\max}$	Maximum volume for storage/pits, m^3
$V_d^{waste\_cap}$	Capacity for disposals, m^3
$V_d^{waste\_\max}$	Maximum volume for disposals, m^3
$y_s^{storage}$	Binary variable used to model the existence of the storage units
y_i^{treat}	Binary variable used to model the existence of the treatment units
y_d^{waste}	Binary variable used to model the existence of the disposals

Sets

I	$\{i \mid i$ Is a treatment unit$\}$
D	$\{d \mid d$ Is a disposal$\}$
N	$\{n \mid n$ Is a well$\}$
S	$\{s \mid s$ Is a storage/pit$\}$
T	$\{t \mid t$ Is a time period

Subscripts and Superscripts

i	Treatment unit
d	Disposal
n	Well
s	Storage/pit
t	Time period

References

Best LC, Lowry CS (2014) Quantifying the potential effects of high-volume water extractions on water resources during natural gas development: Marcellus Shale, NY. J Hydrol: Reg Stud 1:1–16

Brooke A, Kendrick D, Meeruas A, Raman R (2015) GAMS: a users guide. GAMS Development Corporation, Washington, DC, USA

Cafaro DC, Grossmann IE (2014) Strategic planning, design, and development of the shale gas supply chain network. AIChE J 60(6):2122–2142

Clark CE, Horner RM, Harto CB (2013) Life cycle water consumption for shale gas and conventional natural gas. Environ Sci Technol 47(20):11829–11836

Dhole VR, Ramchandani N, Tainsh RA, Wasilewski M (1996) Make your process water pay for itself. Chem Eng 103(1):100–103

EIA (2012) U.S. 2012 annual energy outlook with projects to 2035. US Department of Energy, Washington, DC

El-Halwagi MM, Gabriel F, Harell D (2003) Rigorous graphical targeting for resource conservation via material recycle/reuse networks. Ind Eng Chem Res 42(19):4319–4328

El-Halwagi MM, Spriggs HD (1998) Solve design puzzles with mass integration. Chem Eng Prog 94(8):25–44

Feng X, Bai J, Zheng XS (2007) On the use of graphical method to determine the targets of single-contaminant regeneration recycling water systems. Chem Eng Sci 62(8):2127–2138

Foo CY (2009) State-of-the-art review of pinch analysis techniques for water network synthesis. Ind Eng Chem Res 48:5125–5159

Gabriel FB, El-Halwagi MM (2005) Simultaneous synthesis of waste interception and material reuse networks: Problem reformulation for global optimization. Environ Prog 24(2):171–180

Gordalla BC, Ewers U, Frimmel FH (2013) Hydraulic fracturing: a toxicological threat for groundwater and drinking-water? Environ Earth Sci 70(8):3875–3893

Gunaratnam M, Alva-Argaez A, Kokossis A, Kim JK, Smith R (2005) Automated design of total water systems. Ind Eng Chem Res 44(3):588–599

Hallale N (2002) A new graphical targeting method for water minimization. Adv Environ Res 6(3):377–390

Hayes T (2009) Sampling and analysis of water streams associated with the development of Marcellus shale gas. Marcellus Shale Initiative Publications Database, 10. http://energyindepth.org/wp-content/uploads/marcellus/2012/11/MSCommission-Report.pdf

He C, You F (2014) Shale gas processing integrated with ethylene production: novel process designs, exergy analysis, and techno-economic analysis. Ind Eng Chem Res 53(28):11442–11459

Jiang M, Hendrickson CT, VanBriesen JM (2014) Life cycle water consumption and wastewater generation impacts of a Marcellus shale gas well. Environ Sci Technol 48(3):1911–1920

Karuppiah R, Grossmann IE (2006) Global optimization for the synthesis of integrated water systems in chemical processes. Comput Chem Eng 30(4):650–673

Knudsen BR, Grossmann IE, Foss B, Conn AR (2014) Lagrangian relaxation based decomposition for well scheduling in shale-gas systems. Comput Chem Eng 63:234–249

Lavoie D, Rivard C, Lefebvre R, Séjourné S, Thériault R, Duchesne MJ, Ahad JME, Wang B, Benoit N, Lamontagne C (2014) The Utica shale and gas play in southern Quebec: geological and hydrogeological syntheses and methodological approaches to groundwater risk evaluation. Int J Coal Geol 126:77–91

Lira-Barragán LF, Ponce-Ortega JM, Nápoles-Rivera F, Serna-González M, El-Halwagi MM (2013) Incorporating property-based water networks and surrounding watersheds in site selection of industrial facilities. Ind Eng Chem Res 52(1):91–107

Lira-Barragán LF, Ponce-Ortega JM, Serna-González M, El-Halwagi MM (2011a) An MINLP model for the optimal location of the new industrial plant with simultaneous consideration of economic and environmental criteria. Ind Eng Chem Res 50(2):953–964

Lira-Barragán LF, Ponce-Ortega JM, Serna-González M, El-Halwagi MM (2011b) Synthesis of water networks considering the sustainability of the surrounding watershed. Comput Chem Eng 35:2837–2852

Lira-Barragán LF, Ponce-Ortega JM, Serna-González M, El-Halwagi MM (2016) Optimal reuse of wastewater in hydraulic fracturing including seasonal and environmental constraints. AIChE J 62(5):1634–1645

Manan ZA, Tan YL, Foo DCY (2004) Targeting the minimum water rate using water cascade analysis technique. AIChE J 50(12):3169–3183

Manda AK, Heath JL, Klein WA, Griffin MT, Montz BE (2014) Evolution of multi-well pad development and influence of well pads on environmental violations and wastewater volumes in the Marcellus shale (USA). J Environ Manage 142:36–45

Nápoles-Rivera F, Ponce-Ortega JM, El-Halwagi MM, Jiménez-Gutiérrez A (2010) Global optimization of mass and property integration networks with in-plant property interceptors. Chem Eng Sci 65(15):4363–4377

Pillai HK, Bandyopadhyay S (2007) A rigorous targeting algorithm for resource allocation networks. Chem Eng Sci 62:6212–6221

Pirzadeh P, Lesage KL, Marriott RA (2014) Hydraulic fracturing additives and the delayed onset of hydrogen sulfide in shale gas. Energy Fuels 28(8):4993–5001

Ponce-Ortega JM, El-Halwagi MM, Jiménez-Gutiérrez A (2010) Global optimization of property-based recycle and reuse networks including environmental constraints. Comput Chem Eng 34(3):318–330

Ponce-Ortega JM, Hortua AC, El-Halwagi MM, Jiménez-Gutiérrez A (2009) A property-based optimization of direct recycle networks and wastewater treatment processes. AIChE J 55(9):2329–2344

Putra ZA, Amminudin KA (2008) Two-step optimization approach for design a total water system. Ind Eng Chem Res 47(16):6045–6057

Rahm BG, Bates JT, Bertoia LR, Galford AE, Yoxtheimer DA, Riha SJ (2013) Wastewater management and Marcellus shale gas development: trends, drivers, and planning implications. J Environ Manage 120:105–113

Savelski M, Bagajewicz M (2003) On the necessary conditions of optimality of water utilization systems in process plants with multiple contaminants. Chem Eng Sci 58(23–24):5349–5362

Shaffer DL, Arias-Chavez LH, Ben-Sasson M, Romero-Vargas Castrillón S, Yip NY, Elimelech M (2013) Desalination and reuse of high-salinity shale gas produced water: drivers, technologies, and future directions. Environ Sci Technol 47(17):9569–9583

Siirola JJ (2014) The impact of shale gas in the chemical industry. AIChE J 4(60):810–819

Slutz J, Anderson J, Broderick R, Horner P (2012) Key shale gas water management strategies: an economic assessment tool. In: SPE/APPEA International conference on health, safety, and environment in oil and gas exploration and production. Perth, Australia

Teles J, Castro PM, Novals AQ (2008) LP-based solution strategies for the optimal design of industrial water networks with multiple contaminants. Chem Eng Sci 63(2):376–394

Theodori GL, Luloff AE, Willits FK, Burnett DB (2014) Hydraulic fracturing and the management, disposal, and reuse of frac flowback waters: views from the public in the Marcellus Shale. Energy Res Soc Sci 2:66–74

Vengosh A, Jackson RB, Warner N, Darrah TH, Kondash A (2014) A critical review of the risks to water resources from unconventional shale gas development and hydraulic fracturing in the United States. Environ Sci Technol 48(15):8334–8348

Wang M, Xu Q (2014) Optimal design and operation for simultaneous shale gas NGL recovery and LNG re-gasification under uncertainties. Chem Eng Sci 112:130–142

Wang YP, Smith R (1997) Wastewater minimization. Chem Eng Sci 49(7):981–1006

Warner NR, Christie CA, Jackson RB, Vengosh A (2013) Impacts of shale gas wastewater disposal on water quality in western Pennsylvania. Environ Sci Technol 47(20):11849–11857

Weijermars R (2014) US shale gas production outlook based on well roll-out rate scenarios. Appl Energy 124:283–297

Yang S, Zhang J, Yang Q, Qian Y (2014a) Development of an integrated oil shale refinery with retorting gas steam reforming for hydrogen production. Energy Fuels 28(8):5557–5564

Yang L, Grossmann IE, Manno J (2014b) Optimization models for shale gas water management. AIChE J 60(10):3490–3501

Chapter 14
Strategic Water Management for Shale Gas Production Under Uncertainty

14.1 Introduction

Advances in drilling have produced substantial growth in the extraction and production of shale gas in North America. The Energy Information Administration (*EIA*) has projected that 46% of the total natural gas production in 2035 will be provided by shale gas in U.S (EIA 2012). An indispensable step to correctly extract shale gas is hydraulic fracturing, which is called fracturing. This procedure demands enormous amounts of freshwater to prepare the fracturing fluid that is injected into the wells to release the gas. Numerous challenges are open and still merit additional attention. These include water problems related to shale gas production, such as sustaining the water required by the completion phase, collection of flowback water, optimal equipment design and operation for treating and storing wastewater, the optimal disposition of wastewater and lastly optimizing transportation. Treating and storing flowback wastewater to be reused may help to reduce freshwater consumption and wastewater streams, which may help decline the associated costs. These worries are extremely analogous to those presented in the synthesis of water networks in the downstream processing industries. Numerous works for the synthesis of water networks in industrial processing plants have been proposed to recycle and reuse wastewater streams. These approaches can be categorized as graphical, algebraic, or mathematical approaches. The approaches based on the pinch point techniques correspond to the graphical ones (Wang and Smith 1994; Kazantzi and El-Halwagi 2005; Foo 2012; Deng et al. 2011; Bandyopadhyay et al. 2010). The works of Almutlaq et al. (2005) and Qin et al. (2004) represent graphical approaches. Mathematical approaches use proper mathematical programming techniques (Gabriel and El-Halwagi 2005; Yang and Grossmann 2012; Yu et al. 2014; Ponce-Ortega et al. 2009; 2010; Nápoles-Rivera et al. 2010; Lira-Barragán et al. 2011a, 2011b, 2013). Previous approaches have been implemented for traditional industrial process plants and cannot be smoothly applied to shale gas production.

J. M. Ponce-Ortega et al., *Sustainable Water Systems*, Water Resources Development and Management, https://doi.org/10.1007/978-3-031-72498-5_14

Uncertainty is an intrinsic condition in the design and planning of energy systems and water management networks. This way, uncertainty has been considered in the supply chain planning to control the associated risk (Gebreslassie et al. 2012; You et al. 2009). Besides, this critical aspect has been extensively investigated for the synthesis of water networks. Therefore, Nápoles-Rivera et al. (2015) presented an approach to sustainable use of water under parametric uncertainty in macroscopic water networks, Broad et al. (2015) included uncertainty in an optimization approach for water distribution systems, while Khor et al. (2014) combined the synthesis water networks involving uncertainty with risk management. Additionally, Ahn and Kang (2014) proposed long-term planning for sustainable water supply involving future uncertainties in water demands and source accessibility. Islam et al. (2014) investigated the uncertainties present in traditional water supply systems. Also, Yang et al. (2014a, b) reported an approach for water networks involving realistic and unit-specific shortcut models and the uncertainty in the mass load of pollutants.

It should be noted that previous approaches have not addressed the water network synthesis under uncertainty for optimal water management in hydraulic fracturing operations to shale gas production. Shale gas production is an innovative technology, thus there is scarcely published on water reuse strategies for shale gas processing. This way, Jiang et al. (2014) evaluated the life cycle water consumption and wastewater generation impacts of a Marcellus shale gas well from its construction to the end of life. Theodori et al. (2014) presented experimental information about hydraulic fracturing and the management, disposal, and reuse of fracturing wastewater. Clark et al. (2013) quantified the water consumed over the life cycle of conventional and shale gas production, considering several stages of production and flowback water reusing. Best and Lowry (2014) evaluated the hypothetical effects of high-volume water extractions by applying groundwater flow modeling for the Marcellus shale play. Nonetheless, these approaches are experimental studies and do not include proper optimization approaches. This way, Gao and You (2015a, b) reported an approach for the optimal design of a shale gas supply chain network involving the life cycle analysis through an MINLP model formulation incorporating a global optimization algorithm. However, this work did not account for the water network associated with hydraulic fracturing operations to determine the optimal treatment, storage, and disposal to exploit shale gas resources. Furthermore, Yang et al. (2014a, b) presented an optimization model for scheduling fracturing operations while minimizing the associated costs with transportation, treatment, storage, and disposal, and considering reusing of flowback wastewater; Gao and You (2015a, b) reported a mathematical programming model for the optimal design of water supply networks for shale gas production involving multiple transportation modes and water management options. However, the most significant aspect that was not considered in previous approaches is that these methodologies rely on deterministic models that neglect the different uncertainty sources affecting the calculations in water management in shale gas.

Furthermore, the most critical uncertainties in water management for shale gas are the water needed to complete each well and the percentage of this water collected as flowback water once the completion has finished. Table 14.1 indicates that there are substantial differences in the values of water requirements and flowback water. This information is expressed as intervals (these values cannot be perfectly known beforehand, so a most likely interval within which they may fall is provided instead of a single nominal value). This establishes a considerable difficulty when one needs to predict the amount of water that will be required and the quantity of water that should be collected as flowback. This information is required to calculate the total costs, or the total water needed to complete a given number of wells. It has an impact on the equipment capacities in terms of treatment technologies, storage units, and disposals necessary by the optimal water network to correctly manage the water (Fig. 14.1). Decision-makers must calculate the optimal equipment size to be acquired and installed before fracturing operations, thus the main aim of this chapter is to present an optimization approach for helping the associated decision-making problem.

This chapter presents an optimization model for the synthesis of water networks to determine the water requirements and the flowback water considering the most critical uncertainties involved in fracturing operations associated with shale gas production (Lira-Barragán et al. 2016). The considered uncertainties are modeled through stochastic distributions from which representative scenarios showing different profiles for costs and water requirements and equipment capacities for treating, storing, and disposing of wastewater are generated through sampling methods. This approach is a suitable tool for decision-makers to choose the final capacities. Therefore, the objective function consists of minimizing the total annual cost, which includes the capital costs (equipment acquisition) and operating costs (freshwater, treatment, and transportation costs). Furthermore, the model of this chapter considers that all the flowback water is treated before being stored (and reused) or disposed of, generating an environmentally friendly way to alleviate the contamination of groundwater bodies. Lastly, the presented approach involves a fixed scheduling for the

Table 14.1 Previous reports for the main uncertainties in shale gas processing

Report/Concept	Amount of water required to complete each well	Part of the water injected returns as flowback water (%)
Hayes (2009)	3–4 MM gallons ($11{,}356$–$15{,}142\ m^3$)	25
Slutz et al. (2012)	80,000–120,000 bbl ($12{,}700$–$19{,}000\ m^3$)	10–40
Vidic et al. (2013)	2–7 MM gallons ($7{,}570$–$26{,}500\ m^3$)	9–53
Zammerilli et al. (2014)	2–6 MM gallons ($7{,}570$–$22{,}712\ m^3$)	30–70
Silva (2012)	2–6.5 MM gallons ($7{,}570$–$24{,}605\ m^3$)	15–25

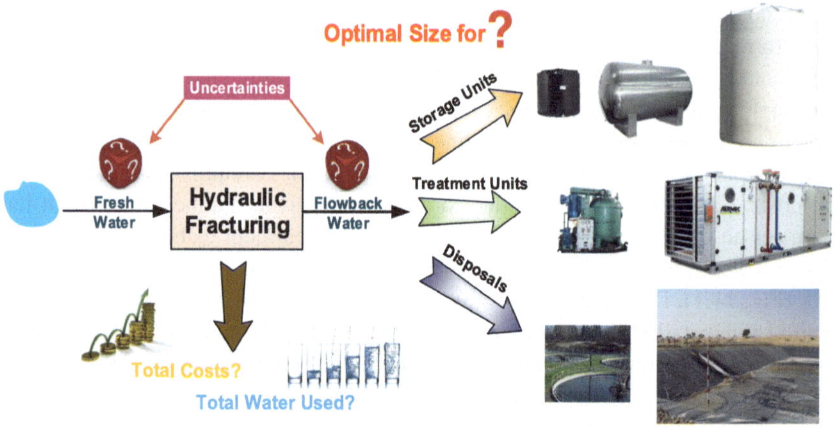

Fig. 14.1 Schematic representation of the uncertainties associated with hydraulic fracturing and their impact on the determination of equipment size. Printed with permission of "*Optimal water management under uncertainty for shale gas production*", *Industrial and Engineering Chemistry Research*, Copyright (2016), American Chemical Society

completion operations of each well to implement the approach, which can incorporate simultaneous operations by several fracturing crews.

14.2 Problem Statement

The problem addressed in this chapter is described as follows. First, given are:

- A set of shale gas wells to be completed using hydraulic fracturing operations, where water requirements and flowback water obtained operate under uncertain conditions. This process can be implemented at the same time by several fracturing crews. The operation time needed to fracture each well is established.
- A freshwater source exists to be the primary resource supplied to wells. The unit costs for freshwater consumption and transportation (contemplating the distance between the fresh water source site and the geographical location for the wells) are also given.
- A set of treatment technologies offered to treat all the flowback water collected to achieve appropriate conditions for its reuse or fulfill the environmental regulations in terms of pollutant concentrations for its final disposal. Each treatment unit has an associated unit operating cost and fixed and variable charges for the capital cost function. The maximum capacity or size for each unit is given.
- Given is a set of storage/pits to keep the treated streams coming from the interception network. The storage equipment is necessary to correctly reuse the flowback water in hydraulic fracturing. Each storage unit contains fixed and variable costs for the capital cost function, the maximum capacity existing for each unit, and

a volumetric efficiency (flowrate leaving the unit concerning to flowrate at the inlet).

- A set of disposals for the treated flowback water. Likewise, each disposal counts fixed and variable costs and a maximum volume.
- The unit transportation costs for the rest of the trajectories obtained in each step are given (wells-treatments, treatments-disposals, treatments-storage, and storage-wells) and fixed scheduling for the fracturing operation of all wells.
- The uncertain parameters (i.e., the water requirements to complete each well and the percentage of water collected as flowback water) are given through probability functions that are discretized via sampling methods. Consequently, a set of equally likely scenarios is generated to represent the uncertainties inherent to water management in shale gas operations.

Subsequently, the objective is to find a strategic planning that considers the uncertainties and accompanying risks in the water management in shale gas operations. The objective function consists of minimizing the total annual cost (*TAC*), which includes the total operating cost (*TOC*) and the total capital cost (*TCC*). *TOC* considers the freshwater costs, the operating costs for treatment units, and the transportation costs; while *TCC* involves the capital costs for the acquisition of treatment, storage, and disposal units.

14.3 Implemented Mathematical Model Formulation

To appropriately develop and implement the proposed mathematical model, the superstructure shown in Fig. 14.2 was considered. A given number of wells to be exploited N_n must be fixed and a fixed scheduling for the operations to complete each well (which can be carried out by one or several fracturing crews). The water required to complete each well ($F_{n,t,s}^{well\_in}$) and the flowback water ($F_{n,t,s}^{well\_out}$) are given parameters, and their involved uncertainties are described through scenarios. If these values change according to the scenario, subsequently the total water required, costs and all the internal flowrates considered in the proposed configuration depend on each scenario. Thus, the implemented model is a two-stage stochastic programming one, where in the first stage decisions are taken before the uncertainty is unveiled. The objective is to determine the optimal values of the first-stage and second-stage variables that optimize the expected economic performance of the overall system.

Figure 14.2 shows that the water required by the fracturing operations is provided mainly by freshwater resources and by the reused streams leaving the storage/pits. Moreover, the flowback water collected is treated in the interception network to avoid future contamination of groundwater; nonetheless, the treated streams have two options: the fluid can be stored (and later reused in the wells) or disposed of. In both cases, the water has been previously treated.

Furthermore, the sets used in the mathematical model are identified for a better understanding. The subscript n is a well, N_n represents the total number of wells,

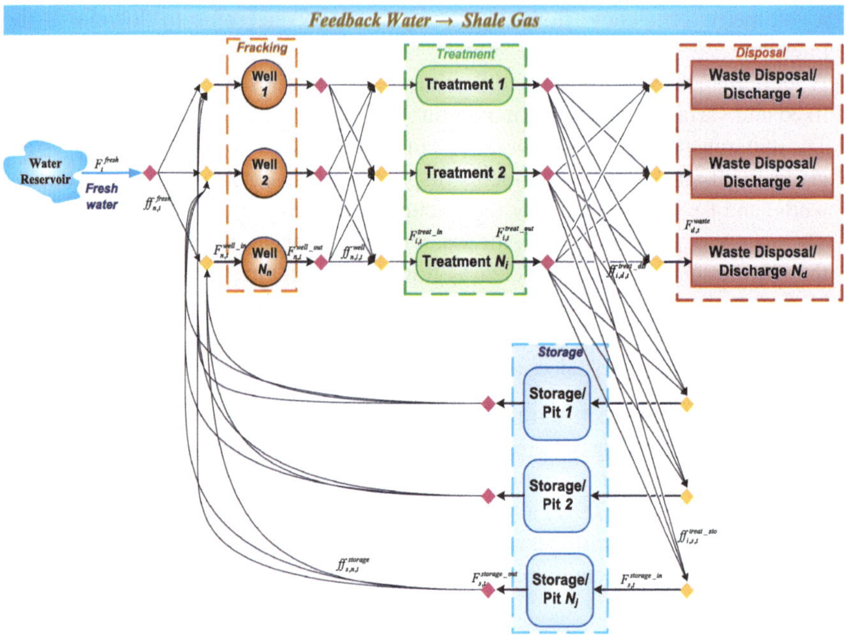

Fig. 14.2 Proposed superstructure for the water management in fracturing operations. Printed with permission of "*Optimal water management under uncertainty for shale gas production*", *Industrial and Engineering Chemistry Research*, Copyright (2016), American Chemical Society

and N is the set used to denote the wells; i represents the subscript for treatment units, N_i is the total number of treatment units and I represents the set for treatment technologies; j is the storage units, while the total number of storage units corresponds to N_j and its set is J; furthermore, d represents the disposals, the total number of disposals is N_d and D corresponds to the set for the disposals; whereas the subscript that represents the period is t, N_t and T are the total periods and the associated set for the periods, respectively. Lastly, s indicates the subscript for a scenario that considers the uncertainty associated with the water requirements for the completion of each well and the portion of flowback water obtained, N_s represents the total number of scenarios and the set representing the scenarios is S.

The mathematical formulation is composed of mass balances, capacity limitations, and design constraints, total water requirements, and the objective function and these are described as follows.

14.3.1 Water Balances Based on the Superstructure

This section shows the equations needed to model the mixers and splitters considered in the presented superstructure and the balances for the storage units, treatment technologies, and disposals. Furthermore, because of the water flowrate required by the wells and the flowback water changeover for each scenario, all the flowrates and equations belonging to the superstructure model change in each scenario.

14.3.1.1 Segregation of Freshwater

Freshwater resources arrive from a water reservoir accessible to extract the liquid needed to prepare the fracturing fluid and release the gas. Afterward, the water stream extracted from the water body in period t and scenario s ($F_{t,s}^{fresh}$) is separated and sent to each well n that needs water during the same period ($ff_{n,t,s}^{fresh}$):

$$F_{t,s}^{fresh} = \sum_{n} ff_{n,t,s}^{fresh}, \quad \forall t, \quad \forall s \tag{14.1}$$

14.3.1.2 Water Supply to Wells

Added to fresh resources ($ff_{n,t,s}^{fresh}$), water required by well n in period t and scenario s ($F_{n,t,s}^{well\_in}$) can be obtained by the reused wastewater coming from the storage j ($ff_{j,n,t,s}^{storage}$):

$$F_{n,t,s}^{well\_in} = ff_{n,t,s}^{fresh} + \sum_{j} ff_{j,n,t,s}^{storage}, \quad \forall n, \quad \forall t, \quad \forall s \tag{14.2}$$

14.3.1.3 Flowback Water

After the completion phase has finished in each well, a certain amount of water returns to the surface mainly in the first two weeks, which is called "*flowback water*"; but, previous reports have specified important variations for the percentage of flowback water concerning the total water injected even for adjacent wells. Therefore, the flowback water in well n during period t and for scenario s ($F_{n,t,s}^{well\_out}$) can be split to be sent to any treatment i ($ff_{n,i,t,s}^{well}$):

$$F_{n,t,s}^{well\_out} = \sum_{i} ff_{n,i,t,s}^{well}, \quad \forall n, \quad \forall t, \quad \forall s \tag{14.3}$$

This configuration proposes that all flowback water is treated and can be stored or disposed of. Furthermore, the water requirement and the flowback water in the wells are parameters under uncertainty, so their values vary from one scenario to another.

14.3.1.4 Water Inlet to Treatment Units

The model of this chapter just considers treatment technologies for the interception network with proven efficiency to treat the flowback water and generate outlet water streams under proper quality and adequate conditions to be disposed of or reused without generating environmental issues or technical drawbacks in the process, where each treatment unit can consider diverse costs and volumetric efficiency factors. Therefore, in existing shale gas plays some treatment technologies meet the restriction needed by the proposed units (Slutz et al. 2012). Furthermore, all the flowback water is treated, which represents an environmental improvement that evades the disposal of untreated flowback. Hence, the following relationship states that the stream entering the treatment unit i in tie period t and scenario s ($F_{i,t,s}^{treat\_in}$) is supplied by the outlet streams from the wells n ($ff_{n,i,t,s}^{well}$):

$$F_{i,t,s}^{treat\_in} = \sum_{n} ff_{n,i,t,s}^{well}, \quad \forall i, \ \forall t, \ \forall s \tag{14.4}$$

14.3.1.5 Water Outlet from Treatment Units

The stream treated with technology I during period t and scenario s ($F_{i,t,s}^{treat\_out}$) is segregated to be sent to any storage unit j ($ff_{i,j,t,s}^{treat\_sto}$) and to any disposal d ($ff_{i,d,t,s}^{treat\_dis}$):

$$F_{i,t,s}^{treat\_out} = \sum_{j} ff_{i,j,t,s}^{treat\_sto} + \sum_{d} ff_{i,d,t,s}^{treat\_dis}, \quad \forall i, \ \forall t, \ \forall s \tag{14.5}$$

14.3.1.6 Wastewater Balance for Disposal and Discharge

The split wastewater in treatment i-disposal d in the period t for the scenario s ($f_{i,d,t,s}^{treat\_dis}$) is collected to yield the waste streams ($F_{d,t,s}^{waste}$):

$$F_{d,t,s}^{waste} = \sum_{i} ff_{i,d,t,s}^{treat\_dis}, \quad \forall d, \ \forall t, \ \forall s \tag{14.6}$$

14.3.1.7 Water Inlet to Storage Units

The streams coming from treatment I and sent to storage/pits j in period t and scenario s ($f_{i,j,t,s}^{treat\_sto}$) are obtained to yield the inlet flowrate to each storage unit ($F_{j,t,s}^{storage\_in}$):

$$F_{j,t,s}^{storage\_in} = \sum_{i} ff_{i,j,t,s}^{treat\_sto}, \quad \forall j, \quad \forall t, \quad \forall s \tag{14.7}$$

14.3.1.8 Water Outlet from Storage Units

The water is stored in pits (because of their large capacities), which show a large residence time; nonetheless, water needs to be available for reuse according to the fracturing fluid condition for the wells. This way, the flowrate at the exit of storage j over period t and scenario s ($F_{j,t,s}^{storage\_out}$) is split to be transported to any well n needing water ($ff_{j,n,t,s}^{storage}$):

$$F_{j,t,s}^{storage\_out} = \sum_{n} ff_{j,n,t,s}^{storage}, \quad \forall j, \quad \forall t, \quad \forall s \tag{14.8}$$

14.3.1.9 Balances in Treatment Units

During the treatment there are usually significant water losses; this feature is simulated through an efficiency factor in terms of the volume associated with the technology i (α_i^{treat}):

$$F_{i,t,s}^{treat\_out} = \alpha_i^{treat} F_{i,t,s}^{treat\_in}, \quad \forall i, \quad \forall t, \quad \forall s \tag{14.9}$$

This factor can be calculated before the optimization based on experimental information.

14.3.1.10 Balances in Storage Units

Mass balance models the storage units, wherever it is considered an initial volume for each unit ($V_j^{storage\_initial}$) and a fixed density (which allows for defining the balances on a volumetric basis). Later, the volume for storage/pit j over period t in scenario s ($V_{j,t,s}^{storage}$) is calculated from the previous volume in time period t-1 ($V_{j,t-1,s}^{storage}$) plus the product of the time conversion factor (H^{time}) and the difference between the inlet ($F_{j,t,s}^{storage\_in}$) and outlet ($F_{j,t,s}^{storage\_out}$) flowrates, which is modeled as follows:

$$V_{j,t,s}^{storage} = V_j^{storage\_initial} + H^{time}\left(F_{j,t,s}^{storage\_in} - F_{j,t,s}^{storage\_out}\right), \quad \forall j, \quad \forall t = 1, \quad \forall s \tag{14.10}$$

$$V_{j,t,s}^{storage} = V_{j,t-1,s}^{storage} + H^{time}\left(F_{j,t,s}^{storage\_in} - F_{j,t,s}^{storage\_out}\right), \quad \forall j, \quad \forall t > 1, \quad \forall s \tag{14.11}$$

$$V_j^{storage\_initial} = V_{j,t,s}^{storage}, \quad \forall j, \ \forall t = t^{final}, \ \forall s \tag{14.12}$$

Equation (14.12) is needed to guarantee the continuity of the cycles.

14.3.2 Capacity and Design Constraints for the Involved Units

The mathematical formulation must decide whether a treatment technology, storage unit, and disposal unit exist or not. Likewise, the model optimizes the capacity for each unit required, which must be lower than the maximum capacity available in each case (i.e., the highest capacity of each unit). Hence, the optimization calculates the total number of treatment technologies, storage/pits, and disposals required. Next, the capacities of the treatment, storage, and disposal units are calculated as follows.

14.3.2.1 Treatment Units

For the treatment units, the following relationships determine their existence and capacity in the optimal solution.

$$F_i^{treat\_cap} \geq F_{i,t,s}^{treat\_in}, \quad \forall i, \ \forall t, \ \forall s \tag{14.13}$$

$$F_i^{treat\_cap} \leq F_i^{treat\_max} y_i^{treat}, \quad \forall i \tag{14.14}$$

$$TotCap^{treatment} = \sum_i F_i^{treat\_cap} \tag{14.15}$$

$F_i^{treat\_cap}$ represents the optimal capacity for technology *i*, $F_i^{treat\_max}$ corresponds to the upper limit for the capacity of unit *i*, while y_i^{treat} represents a binary variable that simulates the existence of treatment unit *i*. Finally, $TotCap^{treatment}$ represents the total capacity of treatment technologies that must be installed for the operation of the project. The maximum capacity of the treatment technologies can be obtained by the equipment manufacturers.

14.3.2.2 Storage Units

For the storage/pits the following relationships are required:

$$V_j^{storage\_cap} \geq V_{j,t,s}^{storage}, \quad \forall j, \ \forall t, \ \forall s \tag{14.16}$$

$$V_j^{storage\_cap} \leq V_j^{storage\_\max} y_j^{storage}, \quad \forall j \tag{14.17}$$

$$TotCap^{storage} = \sum_j V_j^{storage\_cap} \tag{14.18}$$

$V_j^{storage\_cap}$ is the capacity for the storage/pit unit j, $V_j^{storage\_\max}$ corresponds to the upper capacity of the unit j, whereas $y_j^{storage}$ represents a binary variable used to simulate the existence of storage unit j, and $TotCap^{storage}$ is the total capacity of storage/pit units needed to store the fluid.

14.3.2.3 Disposals

Lastly, the following relationships are used to calculate the existence of disposals, total number of disposals, the capacity for each one, and their total capacity. Equation (14.19) determines the optimal capacity for disposal d in scenario s ($V_{d,s}^{waste\_scen}$):

$$V_{d,s}^{waste\_scen} = H^{time} \sum_t F_{d,t,s}^{waste}, \quad \forall d, \quad \forall s \tag{14.19}$$

$$V_d^{waste\_cap} \geq V_{d,s}^{waste\_scen}, \quad \forall d, \quad \forall s \tag{14.20}$$

$$V_d^{waste\_cap} \leq V_d^{waste\_max} y_d^{waste}, \quad \forall d \tag{14.21}$$

$$TotCap^{disposal} = \sum_d V_d^{waste\_cap} \tag{14.22}$$

$V_d^{waste\_cap}$ represents the optimal capacity for disposal d, $V_d^{waste\_max}$ corresponds to the maximum capacity of disposal d, y_d^{waste} represents a binary variable used to simulate the existence of disposal d and $TotCap^{disposal}$ is the total capacity for disposals.

It should be noted that the existence, number, unit, and total capacity for treatment and storage units and disposals do not depend on the scenario, because these decisions are first-stage ones that should be made here and now before the uncertainty is unveiled. The model should then calculate the optimal values of these first-stage variables considering the uncertainty in the input parameters, whose values change from one scenario to another.

14.3.3 Total Water Requirements

It is important to quantify the total freshwater requirements (TWR_s) to exploit accurately the fracturing wells, which represents an environmental concern. Thus, freshwater requirements are determined by the following equation:

$$TWR_s = H^{time} \sum_t F_{t,s}^{fresh}, \quad \forall s \tag{14.23}$$

TWR_s is influenced by the water requirement of the well and by the amount of flowback water that should be obtained to be reused; next, considering that these two parameters are uncertain, the profile of water consumption is expected to fluctuate significantly between the scenarios. Lastly, the expected total freshwater requirement ($ETWR$) is calculated by the following equation:

$$ETWR = \sum_s prob_s TWR_s \tag{14.24}$$

14.3.4 Objective Function

The presented optimization model is a mixed-integer linear programming (MILP) problem, whose objective is to minimize the expected total annual cost (economic criteria). Furthermore, the model should minimize the total freshwater needed (environmental objective). Nevertheless, for the presented approach, both objectives tend to be equivalent (i.e., freshwater can be reduced by minimizing the cost). Hence, when these two objectives contradict each other, multi-objective optimization tools should be used to generate Pareto curves and visualize the tradeoffs among the contradicting objectives.

14.3.4.1 Expected Total Annual Cost (*ETAC*)

The main objective for the presented optimization approach is to minimize the expected total annual cost ($ETAC$). Because a different economic performance is obtained for each scenario s, the resulting cost distribution is calculated as follows:

$$Min \ \ ETAC = \sum_s prob_s TAC_s \tag{14.25}$$

$prob_s$ is the probability of scenario s and TAC_s corresponds to the total annual cost associated with the same scenario s, which is constituted by the total operating cost (TOC_s) and the total capital cost (TCC):

$$TAC_s = TOC_s + TCC, \quad \forall s \tag{14.26}$$

The operating costs change over the scenario, but the capital costs do not depend on the scenario. Furthermore, the operating costs include the freshwater cost, operating costs for treatment units, and the transportation costs for all the trajectories considered (freshwater-wells, flowback water-treatment, treatment-storage, storage-wells, and treatment-disposals). Therefore, the total operating costs are determined through the following equation:

$$TOC_s = H^{time}\left[\begin{array}{c} UC^{fresh}\sum_t F_{t,s}^{fresh} + \sum_i\sum_t UOC_i^{treat} F_{i,t,s}^{treat\_in} + \sum_n\sum_t UTC_n^{fresh} ff_{n,t,s}^{fresh} \\ +\sum_n\sum_i\sum_t UTC_{n,i}^{used} ff_{n,i,t,s}^{well} + \sum_i\sum_j\sum_t UTC_{i,j}^{treat\_sto} ff_{i,j,t,s}^{treat\_sto} \\ +\sum_j\sum_n\sum_t UTC_{j,n}^{sto\_well} ff_{j,n,t,s}^{storage} + \sum_i\sum_d\sum_t UTC_{i,d}^{treat\_dis} ff_{i,d,t,s}^{treat\_dis} \end{array}\right], \quad \forall s \tag{14.27}$$

UC^{fresh} is the unit cost for the freshwater, UOC_i^{treat} represents the unit operating cost for the treatment unit i; while UTC_n^{fresh}, $UTC_{n,i}^{used}$, $UTC_{i,j}^{treat\_sto}$, $UTC_{j,n}^{sto\_well}$, and $UTC_{i,d}^{treat\_dis}$ are the unit transportation costs by sending material between freshwater source-wells, flowback water-treatment, treatment-storage, storage-wells and treatment-disposals, respectively.

Additionally, the capital cost represents the equipment acquisition associated with the treatment technologies, storage units, and the creation of final disposals. Therefore, the total capital cost is determined as follows:

$$TCC = k_F\left\{\begin{array}{c} \sum_i\left[FC^{treat} y_i^{treat} + VC^{treat}\left(F_i^{treat\_cap}\right)\right]+ \\ \sum_j\left[FC^{storage} y_j^{storage} + VC^{storage}\left(V_j^{storage\_cap}\right)\right]+ \\ \sum_d\left[FC^{waste} y_d^{waste} + VC^{waste}\left(V_d^{waste\_cap}\right)\right] \end{array}\right\} \tag{14.28}$$

FC^{treat}, $FC^{storage}$, and FC^{waste} are fixed charges included in the capital cost functions for treatment units, storage units, and disposals, respectively, whereas the variable charges are given by VC^{treat}, $VC^{storage}$, and VC^{waste}.

14.4 Results and Discussion

The applicability of the presented optimization approach is illustrated in the following example problem. This example uses technical information collected by some of the most important existing shale plays in the United States (mainly Marcellus and Barnett) (Hayes 2009; Slutz et al. 2012; Vidic et al. 2013; Zammerilli 2014; Silva 2012). Furthermore, Table 14.1 shows the mean values of 15,000 m^3 and 25% for water required to complete each well, and the portion of the water injected that returns as flowback water during the following three weeks, respectively. Slutz et al. (2012) reported a range of 12,700–19,000 m^3 for water required to complete each well. Therefore, if the mean value (15,000 m^3) has a standard deviation of 10%, it means that approximately 96% of the data can be found between 12,000 and 18,000 m^3; whereas for a deviation of 20% the equivalent range is 9,000–21,000 m^3. Thus, if these values are compared with the intervals contained in Table 14.1, it is possible to see that the selected values for the standard deviations (i.e., 10 and 20%) are proper to simulate the common uncertainties under study. Nevertheless, the model is general enough to deal with any type of scenario regardless of how they are generated. 100 equally likely (i.e., all with the same probability of occurrence) scenarios were generated via Monte Carlo sampling and used to build the multi-scenario optimization model.

Furthermore, twenty wells were completed by three fracturing crews according to the scheduling presented in Fig. 14.3. The time horizon is a year, and each period consists of a week. The time required by the fracturing crew for the ends and starts of operations well-well is at least one week because of the transition time to move from one well to the next one. Figure 14.3 shows a period where water scarcity leads to low fracturing operations (approximately during weeks 23–35).

The unit operating cost in the treatment units is 12.50 $/$m^3$ and the freshwater cost is 2 $/$m^3$. Subsets $n1$, $n2$ and $n3$ are used to facilitate the simulation of the transportation costs: $n1$ represents wells 1–8, while $n2$ and $n3$ wells 9–15 and 16–20, respectively. Hence, the unit transportation costs are 4, 7 and 12 $/$m^3$ from the fresh

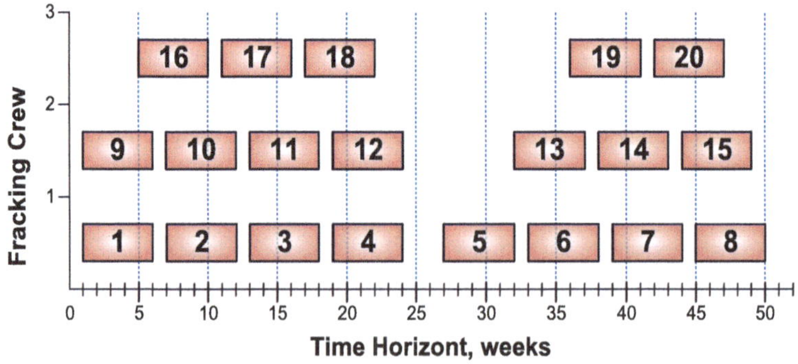

Fig. 14.3 Scheduling for the completion phase in each well. Printed with permission of "*Optimal water management under uncertainty for shale gas production*", *Industrial and Engineering Chemistry Research*, Copyright (2016), American Chemical Society

source to wells $n1$, $n2$ and $n3$, respectively. The transportation costs from the wells to the interception network are 0.39, 0.72, and 0.94 \$/m^3, respectively. Furthermore, the unit transportation costs for the path treatment-storage/pits and treatment-disposals are 1.41 and 8.3 \$/m^3, respectively. Lastly, the transportation costs from the storage to wells $n1$, $n2$, and $n3$ are 0.98, 1.23, and 1.65 \$/m^3, respectively.

Additionally, the interception network of this case study is composed of a set of identical treatment units, where the characteristics of the considered technologies have been previously registered (Slutz et al. 2012). The solution of this example shows the number of treatment units needed. It should be noted that each treatment unit should generate clean water streams available to be reused or disposed of under environmental conditions (this type of treatment technology is typically employed in shale gas plays, and it accomplishes the mentioned requirements). The disposals considered are typical Class II wells; nonetheless, it is important to remark that the proposed configuration just considers treated wastewater streams. The fixed and variable costs employed in the capital cost functions and the maximum capacities are shown in Table 14.2. Lastly, the conversion factor H^{time} and the annualization factor k_F are 7 days/week and 0.1 year^{-1}, respectively; while the volumetric efficiency for the treatment units α_i^{treat} and the initial volume for the storage/pits units $V_s^{storage\_initial}$ are 0.9 and 0, respectively.

The MILP problem associated with the case study has 17 binary variables, 1,629,342 continuous variables, and 465,923 constraints and each solution consumes around 1019 s using a computer with an i7 processor at 2.1 GHz with 8 MB of RAM. The presented approach was implemented in the General Algebraic Modeling System (GAMS (Brooke et al. 2015)) and solved using the solver CPLEX.

Figures 14.4 and 14.5 present the cumulative probability curves for the total cost and freshwater consumption, respectively, for standard deviations of 10 and 20% in the uncertain parameters. Therefore, the vertical axis of the figures provides the probability of performance less or equal to the target value given in the horizontal axis. For example, the 20% curve has a probability of 43% of being below \$3.2 MM/y, while the 10% curve has a probability of 77%. Lastly, if the objective is more ambitious (i.e., \$3 MM/y), the possibilities are 1% and 8% for the 10% and 20% curves, respectively. Similarly, for a standard deviation of 20%, the probability of satisfying the total water requirements with an amount of 0.23 hm^3 is 28%; nevertheless, if the standard deviation is 10%, the probability drops to 5%.

Table 14.2 Data used in the example problem

Concept/Case	Treatment	Storage	Disposal
Fixed cost (FC^{case}), \$	95,000	10,000	52,000
Variable cost (V^{case}), \$/m^3	4.1	6.3	13.5
Maximum capacity ($V^{case\_\max}$), m^3/d or m^3	500	12,000	20,000

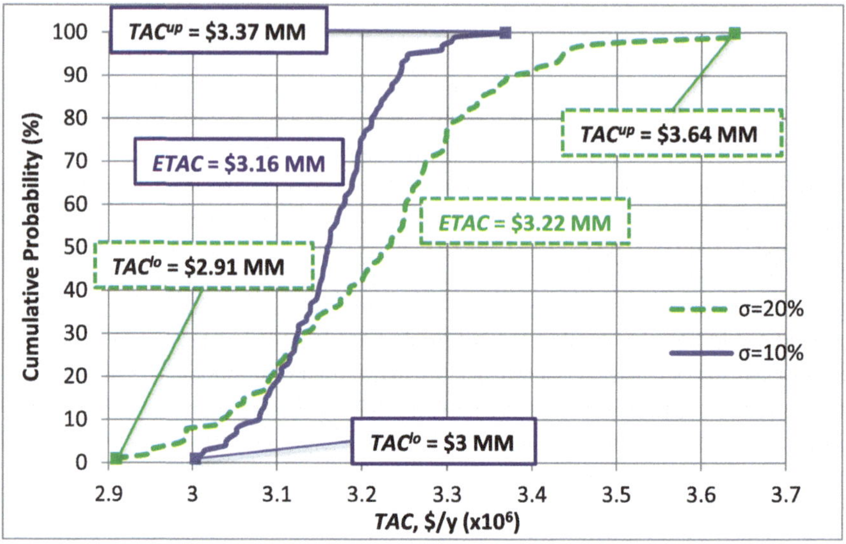

Fig. 14.4 Cumulative probability curves for the economic performance when the equipment capacity is fixed. Printed with permission of "*Optimal water management under uncertainty for shale gas production*", *Industrial and Engineering Chemistry Research*, Copyright (2016), American Chemical Society

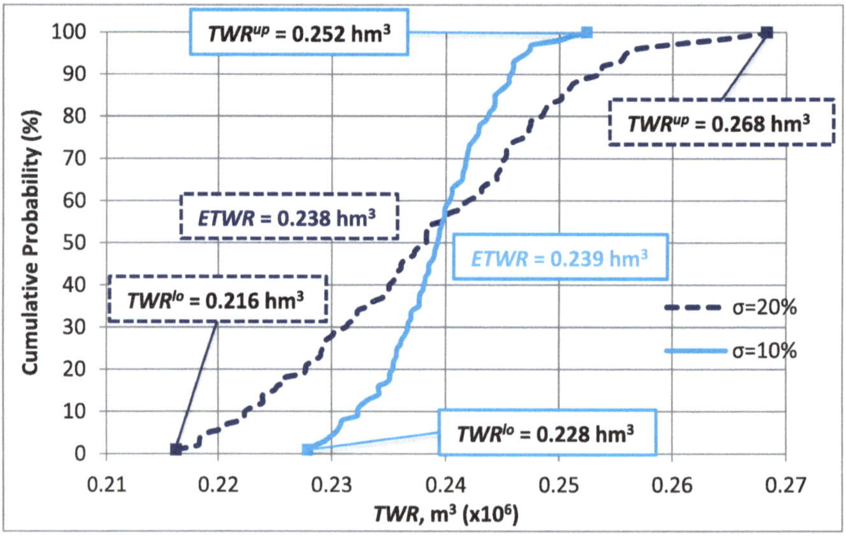

Fig. 14.5 Cumulative probability curves for the total water requirements. Printed with permission of "*Optimal water management under uncertainty for shale gas production*", *Industrial and Engineering Chemistry Research*, Copyright (2016), American Chemical Society

Table 14.3 Details for solutions when the equipment capacity is fixed

Concept/Case	σ = 10%	σ = 20%
Total capacity for treatments, m^3/d	894.37	1,108.77
Total capacity for storage/pits, dam^3	14.38	14.20
Total capacity for disposals, dam^3	5.90	7.37

These figures depict the expected performance of each solution and the maximum and minimum costs attained by them over all the scenarios. The expected total annual costs are \$3.16 MM/y and \$3.22 MM/y for standard deviations of 10% and 20%, respectively. Next, if a comparison of these values concerning the extreme cases, it can be established that the worst scenarios augment the costs 6.65% (\$3.37 MM/y) and 13.04% (\$3.64 MM/y), respectively; while the best scenarios can obtain discounts of 5.06% (\$3 MM/y) and 9.63% (\$2.91 MM/y), respectively. Furthermore, although 20% of the standard deviation curve is wider than the other one, the expected values are close (0.238 hm^3 for the 10% curve and 0.0239 hm^3 for the 20% curve). Lastly, as the level of uncertainty increases, solutions with more variability and worse expected performance are obtained.

Table 14.3 presents the equipment capacities for this solution, where it is noted that an increase in the uncertainty (from 10 to 20%), generates larger capacities for treatment and disposals by 24% and 25%, respectively, and an insignificant decrement in the total storage capacity.

Previous results were obtained considering that the equipment capacity (i.e., the capacity for treatment technologies, storage units, and disposals) is fixed. This way, the economic and water profiles change according to the variations in water requirements for each of the wells and to the amount of flowback water (i.e., the uncertainties considered). This approach provides equipment sizes acceptable for the worst scenarios; nevertheless, this configuration could generate oversized equipment capacities for most of the scenarios and consequently undesired solutions in the economic performance. Considering this aspect and to carry out a detailed analysis for the optimal equipment capacities, the value of full information for this approach is included. This way, the total capacities for treatment technologies, storage units, and for disposal are calculated in each scenario (i.e., the equipment capacities are optimized according to the scenario). This means that the design found in the optimal solution is different for each scenario. Then, the previous model formulation that consists of constraints (14.1)–(14.28) is restricted to obtain the second mathematical model, which needs to remove Eqs. (14.13)–(14.22), (14.24) and (14.26) to include the following relationships:

Treatment units. Constraints assigned to model the capacity and existence for the treatment units are modified to allow that the capacity ($F_{i,s}^{treat\_cap}$) and existence ($y_{i,s}^{treat}$) for the treatment unit i and the total capacity of treatment units ($TotCap_s^{treatment}$) can be calculated in each scenario s:

$$F_{i,s}^{treat\_cap} \geq F_{i,t,s}^{treat\_in}, \quad \forall i, \ \forall t, \ \forall s \tag{14.29}$$

$$F_{i,s}^{treat\_cap} \leq F_{i}^{treat\_max} y_{i,s}^{treat}, \quad \forall i, \ \forall s \tag{14.30}$$

$$TotCap_{s}^{treatment} = \sum_{i} F_{i,s}^{treat\_cap}, \quad \forall s \tag{14.31}$$

$$ETotCap^{treatment} = \sum_{s} prob_s TotCap_{s}^{treatment} \tag{14.32}$$

It is included the expected total capacity ($ETotCap^{storage}$) in analogy to the expected costs to determine a mean value for the total capacity of treatment technologies that should be installed.

Storage units. In the same way for the storage/pits:

$$V_{j,s}^{storage\_cap} \geq V_{j,t,s}^{storage}, \quad \forall j, \ \forall t, \ \forall s \tag{14.33}$$

$$V_{j,s}^{storage\_cap} \leq V_{j}^{storage\_max} y_{j,s}^{storage}, \quad \forall j, \ \forall s \tag{14.34}$$

$$TotCap_{s}^{storage} = \sum_{j} V_{j,s}^{storage\_cap}, \quad \forall s \tag{14.35}$$

$$ETotCap^{storage} = \sum_{s} prob_s TotCap_{s}^{storage} \tag{14.36}$$

Disposals. Lastly, for the disposals:

$$V_{d,s}^{waste\_cap} = H^{time} \sum_{t} F_{d,t,s}^{waste}, \quad \forall d, \ \forall s \tag{14.37}$$

$$V_{d,s}^{waste\_cap} \leq V_{d}^{waste\_max} y_{d,s}^{waste}, \quad \forall d, \ \forall s \tag{14.38}$$

$$TotCap_{s}^{disposal} = \sum_{d} V_{d,s}^{waste\_cap}, \quad \forall s \tag{14.39}$$

$$ETotCap^{disposal} = \sum_{s} prob_s TotCap_{s}^{disposal} \tag{14.40}$$

Objective function. The capital costs (TCC_s) included in the objective function vary for each scenario s:

$$TCC_s = k_F \left\{ \begin{array}{c} \sum_i \left[FC^{treat} y_{i,s}^{treat} + VC^{treat} \left(F_{i,s}^{treat\_cap} \right) \right] + \\ \sum_j \left[FC^{storage} y_{j,s}^{storage} + VC^{storage} \left(V_{j,s}^{storage\_cap} \right) \right] + \\ \sum_d \left[FC^{waste} y_{d,s}^{waste} + VC^{waste} \left(V_{d,s}^{waste\_cap} \right) \right] \end{array} \right\} \quad (14.41)$$

$$TAC_s = TOC_s + TCC_s, \quad \forall s \quad (14.42)$$

Equations (14.24) and (14.26) do not consider the variability scenario-scenario for the capital costs. Then, the second mathematical programming model is composed of Eqs. (14.1)–(14.12), (14.23), (14.25), and (14.27)–(14.42). The presented approach is a MILP model that consists of 1700 binary variables, 1633,104 continuous variables, 468,002 constraints, and each solution consumes around 281.54 s of CPU time. The results for the economic performance are presented in Fig. 14.6. This way, the expected costs present a small improvement concerning the previous case (for σ = 10%, \$3.12 MM/y vs \$3.16 MM/y, and for σ = 20%, \$3.15 MM/y vs \$3.22 MM/y) and even for the extreme scenarios, which is because of the equipment capacities are not fixed. The extreme and intermediate solutions (with a probability of 50%) have been highlighted through Points A, B, and C for σ = 10%, and A', B' and C' for σ = 20% to make a wider discussion among these cases.

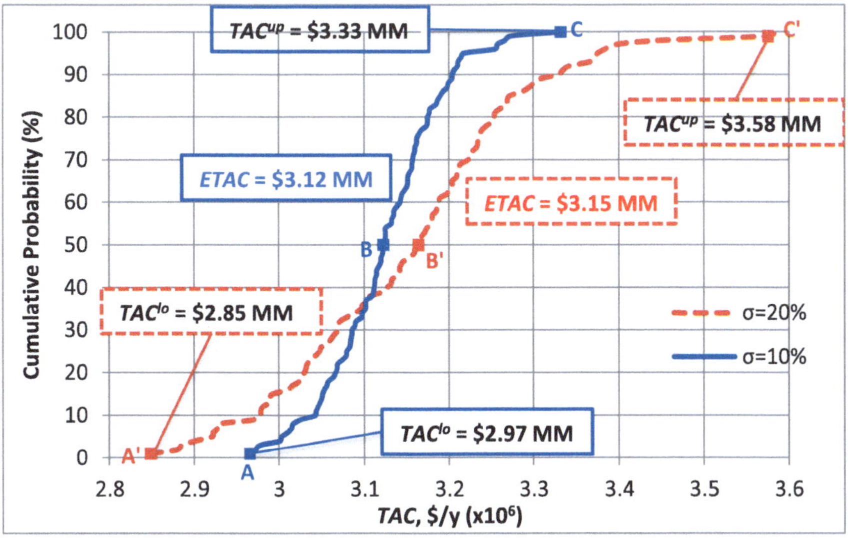

Fig. 14.6 Cumulative probability curves for the economic performance when the equipment capacity can change in each scenario. Printed with permission of "*Optimal water management under uncertainty for shale gas production*", *Industrial and Engineering Chemistry Research*, Copyright (2016), American Chemical Society

Table 14.4 presents the differences in ***TAC***, ***TWR*** and the optimal total capacities for treatment units, storage units, and disposals. An increment in the uncertainty (from $\sigma = 10$ to 20%) may affect minimally the cost by 1.28% (***TAC*** augments $0.04 MM/y) for the solutions with 50% of probabilities (B vs B') and at the same time, the total capacity needed for storage units is significantly diminished by 24.30% (13.95 vs 10.56 dam^3). Furthermore, these extreme and intermediate cases do not correspond to their equivalent extreme and intermediate solutions for the ***TWR*** curves (Fig. 14.5 and Table 14.4).

Figures 14.7, 14.8, and 14.9 show the amount of water moving in each route included by the superstructure for points A, B, and C, respectively (these points are the extreme and intermediate solutions for the curve with $\sigma = 10\%$). The corresponding solutions A', B' and C' (points belonging to solutions with a profile of uncertainty $\sigma = 20\%$) are presented in Figs. 14.10, 14.11, and 14.12, respectively. All the configurations are different among them, and the water needed in each step yields different values.

Subsequently, a similar analysis of cumulative probabilities is implemented for the optimal equipment capacities with the purpose that this study can be useful for decision-makers associated with the selection of total size for treatment technologies, storage units, and disposals that must be acquired and installed around the fracturing wells for proper operations. Figure 14.13 presents the cumulative probability curves for the total capacity of treatment units with the same two cases for the standard

Table 14.4 Details for the extreme and intermediate solutions

Concept/Case	A	B	C	A'	B'	C'
Total capacity for treatments, m^3/d	787.00	754.90	758.11	736.10	759.84	926.64
Total capacity for storage/pits, dam^3	13.31	13.95	13.89	11.85	10.56	15.76
Total capacity for disposals, dam^3	4.17	4.55	5.30	3.84	4.65	5.44
TWR, hm^3	0.230	0.240	0.251	0.218	0.237	0.266
TAC, $ MM/y	2.97	3.12	3.33	2.85	3.16	3.58

Printed with permission of "*Optimal water management under uncertainty for shale gas production*", Industrial and Engineering Chemistry Research, Copyright (2016), American Chemical Society

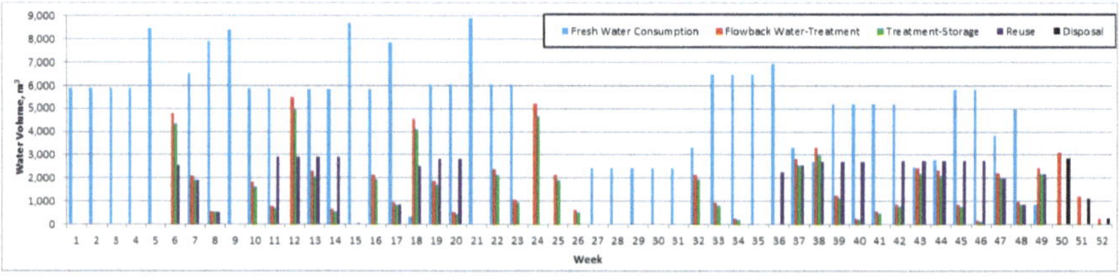

Fig. 14.7 Total water volumes in the solution for **Case A**. Printed with permission of "*Optimal water management under uncertainty for shale gas production*", *Industrial and Engineering Chemistry Research*, Copyright (2016), American Chemical Society

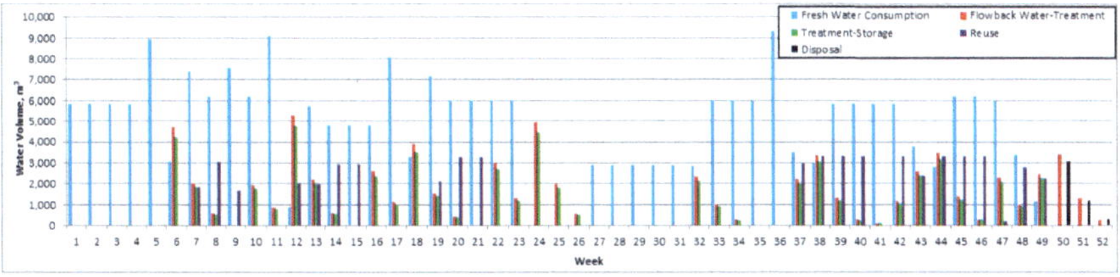

Fig. 14.8 Total water volumes in the solution for **Case B**. Printed with permission of "*Optimal water management under uncertainty for shale gas production*", *Industrial and Engineering Chemistry Research*, Copyright (2016), American Chemical Society

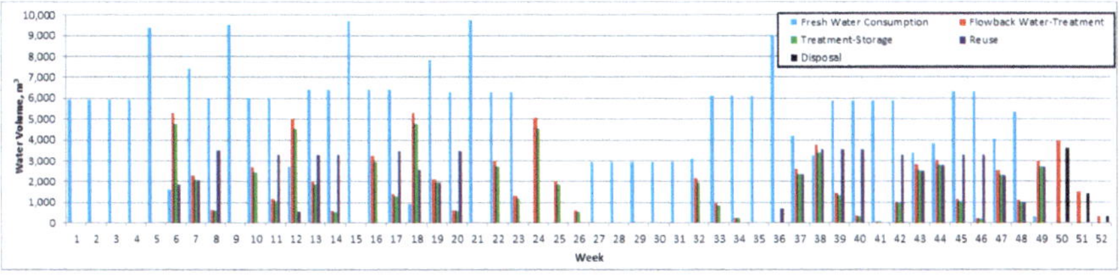

Fig. 14.9 Total water volumes in the solution for **Case C**. Printed with permission of "*Optimal water management under uncertainty for shale gas production*", *Industrial and Engineering Chemistry Research*, Copyright (2016), American Chemical Society

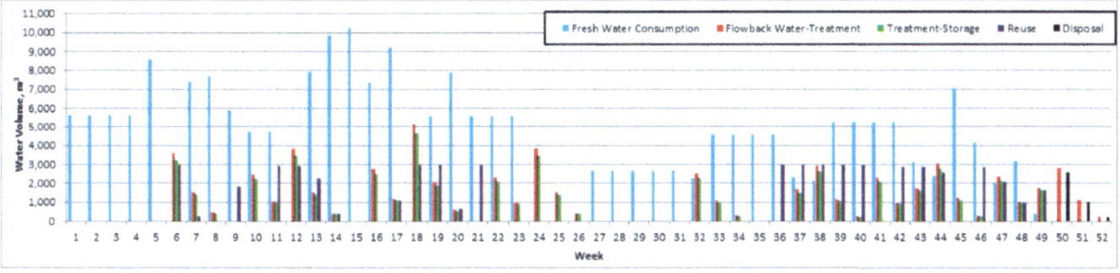

Fig. 14.10 Total water volumes in the solution for **Case A'**. Printed with permission of "*Optimal water management under uncertainty for shale gas production*", *Industrial and Engineering Chemistry Research*, Copyright (2016), American Chemical Society

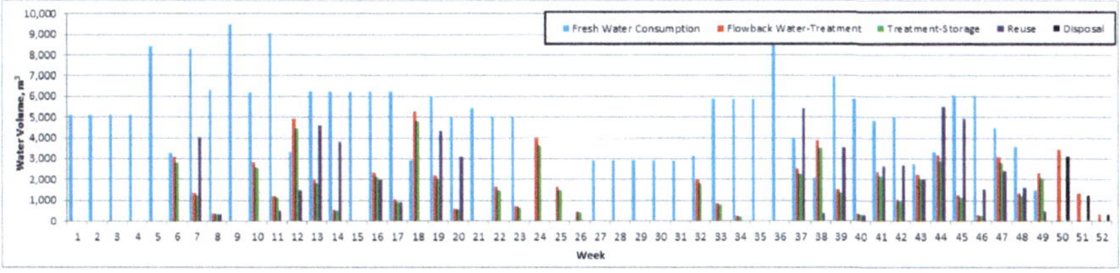

Fig. 14.11 Total water volumes in the solution for **Case B'**. Printed with permission of "*Optimal water management under uncertainty for shale gas production*", *Industrial and Engineering Chemistry Research*, Copyright (2016), American Chemical Society

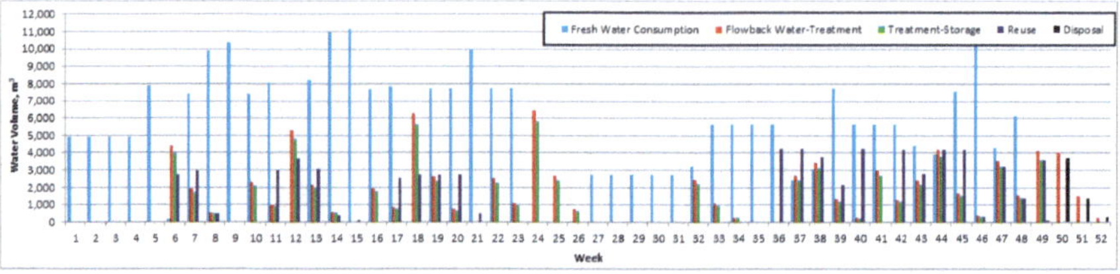

Figure. 14.12 Total water volumes in the solution for **Case C'**. Printed with permission of "*Optimal water management under uncertainty for shale gas production*", *Industrial and Engineering Chemistry Research*, Copyright (2016), American Chemical Society

deviations (Fig. 14.9). Therefore, for $\sigma = 10\%$ the minimum capacity needed is 696 m^3/d, the maximum case requires 894 m^3/d and the expected value is 792 m^3/d; but for $\sigma = 20\%$ the minimum, expected and maximum capacities are 608, 867, and 1,109 m^3/d, respectively. In addition, if the decision-maker chooses a total capacity for treatment units of 800 m^3/d, it means that there is a 57% possibility that this equipment is enough to treat the flowback water if the uncertainties related to the water requirements and flowback water follow a $\sigma = 10\%$; however, if this uncertainty follows a $\sigma = 20\%$, then the probability that 800 m^3/d of treatment technologies are adequate to treat the flowback water decreases to 27%. Furthermore, if a value of 900 m^3/d is selected, then it is completely guaranteed (probability of 100%) the flowback water treatment if $\sigma = 10\%$; nonetheless, if $\sigma = 20\%$, then the possibility diminishes to 60%.

Figure 14.14 presents the cumulative probability curves for the total capacity of storage units, where two details are identified: for the area of the 20% curve where it does not overlap with the other one, it conducts to lower total capacities for storage units and even for almost all overlapped zone the 20% curve has a higher probability to require lower total capacities for storage units than the 10% curve. This way, the 20% curve leads to total capacities for storage units lower than the 10% curve.

Additionally, if a total capacity for storage units of 12 dam^3 is selected, so the 10% curve has a 9% possibility of storing properly the flowback water; however, the 20% curve increases this value to reach 54% probability. A similar analysis can be carried out for the total capacity for disposals according to the cumulative probability curves shown in Fig. 14.15 For all the cumulative probabilities reported, the curve for $\sigma = 20\%$ is always wider than the $\sigma = 10\%$ curve. Furthermore, the extreme and intermediate solutions highlighted in ***TAC*** curves (points A, A', B, B', C, and C') do not represent their analogous solutions for any of the total capacity curves presented (comparing Table 14.4).

Lastly, the presented optimization approach incorporates useful tools for the decision-makers to calculate and select proper equipment capacities for the treatment technologies, storage units, and disposals for the water management in wells (this decision should be taken before the start of operations in the wells) according to the historic or a projected uncertainty in the water requirements to implement the fracturing process and the flowback water returned.

14.5 Concluding Remarks

To account for uncertainties, this chapter has presented mathematical programming models for the synthesis of water networks in shale gas production. The water needed to fracture each well and the amount of water that returns as flowback water are the factors that are taken into consideration while calculating uncertainty. By creating scenarios, these uncertainties were controlled. The water integration is based on a superstructure that enables effective flowback water management to lower freshwater consumption, and wastewater streams are released in a manner that prevents groundwater contamination. Water requirements and cost probability curves are produced by the process.

Using technical information from the Marcellus and Barnett shale fields, a case study was resolved. The evaluation of every aspect covered in the results considered that the uncertainties had standard deviations of 10% and 20% and were distributed normally. This allowed for the examination of how the economics, water requirements, and equipment capacity performances changed as the uncertainty increased. Ultimately, the presented methodology is all-encompassing, and its resolution process doesn't pose any noteworthy numerical or convergence issues.

Nomenclatures

Parameters

FC^{treat}	Fixed charge for the treatment unit in the capital cost function, \$
$FC^{storage}$	Fixed charge for the storage unit in the capital cost function, \$
FC^{waste}	Fixed charge for the disposal in the capital cost function, \$
$F_{n,t,s}^{well\_in}$	Flowrate entering in the wells *n* in the time *t* for the scenario *s*, m^3/d
$F_{n,t,s}^{well\_out}$	Flowback water leaving the well *n* in the time *t* for the scenario *s*, m^3/d
H^{time}	Time conversion factor, d/ week
k_F	Factor used to annualize the inversion, $year^{-1}$
$prob_s$	Probability of the scenario *s*
UC^{fresh}	Unit cost for the freshwater, \$/m^3
UOC_i^{treat}	Unit operating cost for the treatment *i*, \$/m^3
UTC_n^{fresh}	Unit transport cost for the fresh water to the well *n*, \$/m^3
$UTC_{j,n}^{sto\_well}$	Unit transport cost for the reused water from the storage *j* to well *n*, \$/m^3
$UTC_{i,d}^{treat\_dis}$	Unit transport cost for the treated water from the treatment *i* to disposal *d*, \$/m^3

$UTC_{i,j}^{treat\_sto}$	Unit transport cost for the treated water from the treatment i to storage j, \$/m^3
$UTC_{n,i}^{used}$	Unit transport cost for the flowback water from well n to treatment i, \$/m^3
VC^{treat}	Variable charge for the treatment unit in the capital cost function, \$/m^3/d
$VC^{storage}$	Variable charge for the storage unit in the capital cost function, \$/m^3
VC^{waste}	Variable charge for the disposal in the capital cost function, \$/m^3

Greek Symbols

α_i^{treat}	Volumetric efficiency factor for the treatment i

Variables

$ETAC$	Expected total annual cost, \$
$ETotCap^{treatment}$	Expected total capacity of treatment technologies that must be installed, m^3/d
$ETotCap^{storage}$	Expected total capacity of storage units that must be installed, m^3
$ETotCap^{disposal}$	Expected total capacity of disposals that must be installed, m^3
$ETWR$	Expected total water requirements, m^3
$F_{t,s}^{fresh}$	Freshwater flowrate consumed in the time t for the scenario s, m^3/d
$F_{j,t,s}^{storage\_in}$	Flowrate entering to the storage j in the time t for the scenario s, m^3/d
$F_{j,t,s}^{storage\_out}$	Flowrate leaving the storage j in the time t for the scenario s, m^3/d
$F_i^{treat\_cap}$	Capacity flowrate for the treatment i, m^3/d
$F_{i,s}^{treat\_cap}$	Capacity flowrate for the treatment i for the scenario s, m^3/d
$F_{i,t,s}^{treat\_in}$	Inlet flowrate to the treatment i in the time t for the scenario s, m^3/d
$F_i^{treat\_\max}$	Maximum flowrate for the treatment units, m^3/d
$F_{i,t,s}^{treat\_out}$	Flowrate leaving the treatment i in the time t for the scenario s, m^3/d
$F_{d,t,s}^{waste}$	Flowrate sent to disposal d in the time t for the scenario s, m^3/d
$ff_{n,t,s}^{fresh}$	Segregated flowrate of fresh water to well n in the time t for the scenario s, m^3/d

$ff_{j,n,t,s}^{storage}$	Segregated flowrate from storage j to well n in the time t for the scenario s, m^3/d
$ff_{i,d,t,s}^{treat\_dis}$	Segregated flowrate from treatment i to disposal d in the time t for scenario s, m^3/d
$ff_{i,j,t,s}^{treat\_sto}$	Segregated flowrate from treatment i to storage j in the time t for scenario s, m^3/d
$ff_{n,i,t,s}^{well}$	Segregated flowrate from well n to treatment i in the time t for scenario s, m^3/d
TAC_s	Total annual cost for scenario s, \$
TCC	Total capital cost, \$
TCC_s	Total capital cost for scenario s, \$
TOC_s	Total operating cost for scenario s, \$
$TotCap^{treatment}$	Total capacity of treatment technologies that must be installed, m^3/d
$TotCap_s^{treatment}$	Total capacity of treatment technologies that must be installed for scenario s, m^3/d
$TotCap^{storage}$	Total capacity of storage units that must be installed, m^3
$TotCap_s^{storage}$	Total capacity of storage units that must be installed for scenario s, m^3
$TotCap^{disposal}$	Total capacity of disposals that must be installed, m^3
$TotCap_s^{disposal}$	Total capacity of disposals that must be installed for scenario s, m^3
TWR_s	Total water requirements for the scenario s, m^3
$V_{j,t,s}^{storage}$	Storage/pit j volume in the time t for the scenario s, m^3
$V_j^{storage\_cap}$	Capacity for storage/pit j, m^3
$V_{j,s}^{storage\_cap}$	Capacity for storage/pit j for the scenario s, m^3
$V_j^{storage\_initial}$	Initial storage/pit j volume, m^3
$V_j^{storage\_\max}$	Maximum volume for storage/pit j, m^3
$V_d^{waste\_cap}$	Capacity for disposal d, m^3
$V_{d,s}^{waste\_cap}$	Capacity for disposal d for the scenario s, m^3
$V_d^{waste\_\max}$	Maximum volume for disposals, m^3
$V_{d,s}^{waste\_scen}$	Capacity for disposal d in for scenario s, m^3
$y_j^{storage}$	Binary variable used to model the existence of the storage j
$y_{j,s}^{storage}$	Binary variable used to model the existence of the storage j in scenario s
y_i^{treat}	Binary variable used to model the existence of the treatment i
$y_{i,s}^{treat}$	Binary variable used to model the existence of the treatment i in scenario s
y_d^{waste}	Binary variable used to model the existence of the disposal d
$y_{d,s}^{waste}$	Binary variable used to model the existence of the disposals d in scenario s

Sets

I	$\{i \mid i$ is a treatment unit$\}$
J	$\{j \mid j$ is a storage/pit$\}$
D	$\{d \mid d$ is a disposal$\}$
N	$\{n \mid n$ is a well$\}$
S	$\{s \mid s$ is a scenario$\}$
T	$\{t \mid t$ is a period

Subscripts and Superscripts

i	Treatment unit
j	Storage/pit
d	Disposal
n	Well
s	Scenario
t	Period

References

Ahn J, Kang D (2014) Optimal planning of water supply system for long-term sustainability. J Hydro-Environ Res 8(4):410–420

Almutlaq A, Kazantzi V, El-Halwagi MM (2005) An algebraic approach to targeting waste discharge and impure-fresh usage via material recycle/reuse networks. Clean Technol Environ Policy 7(4):294–305

Bandyopadhyay S, Sahu GC, Foo DCY, Tan RR (2010) Segregated targeting for multiple resource networks using decomposition algorithm. AIChE J 56(5):1235–1248

Best LC, Lowry CS (2014) Quantifying the potential effects of high-volume water extractions on water resources during natural gas development: Marcellus Shale, NY. J Hydrol: Reg Stud 1:1–16

Broad DR, Dandy GC, Maier HR (2015) A systematic approach to determining metamodel scope for risk-based optimization and its application to water distribution system design. Environ Model Softw 69:382–395

Brooke A, Kendrick D, Meeruas A, Raman R (2015) GAMS: a users guide. GAMS Dev Corporation, Washington, DC, USA

Clark CE, Horner RM, Harto CB (2013) Life cycle water consumption for shale gas and conventional natural gas. Environ Sci Technol 47(20):11829–11836

Deng C, Feng X, Ng DKS, Foo DCY (2011) Process-based graphical approach for simultaneous targeting and design of water network. AIChE J 57(11):3085–3104

EIA (2012) U.S. Energy Information Administration. U.S. 2012 annual energy outlook with projects to 2035. US Department of Energy, Washington, DC

Foo DCY (2012) Process integration for resource conservation. CRC Press, Boca Raton, Florida, USA

Gabriel F, El-Halwagi MM (2005) Simultaneous synthesis of waste interception and material reuse networks: problem reformulation for global optimization. Environ Prog 24(2):171–180

Gao J, You F (2015a) Optimal design and operations of supply chain networks for water management in shale gas production: MILFP model and algorithms for the water energy nexus. AIChE J 61(4):1184–1208

Gao J, You F (2015b) Shale gas supply chain design and operations toward better economic and life cycle environmental performance: MINLP model and global optimization algorithm. ACS Sustain Chem Eng 3(7):1282–1291

Gebreslassie BH, Yao Y, You F (2012) Design under uncertainty of hydrocarbon biorefinery supply chains: multiobjective stochastic programming models, decomposition algorithm, and a comparison between CVaR and downside risk. AIChE J 58(7):2155–2179

Hayes T (2009) Sampling and analysis of water streams associated with the development of Marcellus shale gas. Marcellus Shale Initiative Publications Database, 10. http://energyindepth.org/wp-content/uploads/marcellus/2012/11/MSCommission-Report.pdf

Islam SM, Sadiq R, Rodriguez MJ, Najjaran H, Hoorfar M (2014) Reliability assessment for water supply systems under uncertainties. J Water Resour Plan Manage 140(4):468–479

Jiang M, Hendrickson CT, VanBriesen JM (2014) Life cycle water consumption and wastewater generation impacts of a Marcellus shale gas well. Environ Sci Technol 48(3):1911–1920

Kazantzi V, El-Halwagi MM (2005) Targeting material reuse via property integration. Chem Eng Prog 101(8):28–37

Khor CS, Chachuat B, Shah N (2014) Fixed-flowrate total water network synthesis under uncertainty with risk management. J Clean Prod 77:79–93

Lira-Barragán LF, Ponce-Ortega JM, Nápoles-Rivera F, Serna-González M, El-Halwagi MM (2013) Incorporating property-based water networks and surrounding watersheds in site selection of industrial facilities. Ind Eng Chem Res 52(1):91–107

Lira-Barragán LF, Ponce-Ortega JM, Serna-González M, El-Halwagi MM (2011a) An MINLP model for the optimal location of the new industrial plant with simultaneous consideration of economic and environmental criteria. Ind Eng Chem Res 50(2):953–964

Lira-Barragán LF, Ponce-Ortega JM, Serna-González M, El-Halwagi MM (2011b) Synthesis of water networks considering the sustainability of the surrounding watershed. Comput Chem Eng 35:2837–2852

Lira-Barragán LF, Ponce-Ortega JM, Guillén-Gosálbez G, El-Halwagi MM (2016) Optimal water management under uncertainty for shale gas production. Ind Eng Chem Res 55:1322–1335

Nápoles-Rivera F, Ponce-Ortega JM, El-Halwagi MM, Jiménez-Gutiérrez A (2010) Global optimization of mass and property integration networks with in-plant property interceptors. Chem Eng Sci 65(15):4363–4377

Nápoles-Rivera F, Rojas-Torres MG, Ponce-Ortega JM, Serna-González M, El-Halwagi MM (2015) Optimal design of macroscopic water networks under parametric uncertainty. J Clean Prod 88:172–184

Ponce-Ortega JM, El-Halwagi MM, Jiménez-Gutiérrez A (2010) Global optimization of property-based recycle and reuse networks including environmental constraints. Comput Chem Eng 34(3):318–330

Ponce-Ortega JM, Hortua AC, El-Halwagi MM, Jiménez-Gutiérrez A (2009) A property-based optimization of direct recycle networks and wastewater treatment processes. AIChE J 55(9):2329–2344

Qin X, Gabriel F, Harell D, El-Halwagi MM (2004) Algebraic techniques for property integration via componentless design. Ind Eng Chem Res 43:3792–3798

Silva JM (2012) RPSEA produced water pretreatment for water recovery and salt production. Final report, 08122-36. http://www.rpsea.org/media/files/project/18621900/08122-36-FR-Pretreatment_Water_Mgt_Frac_Water_Reuse_Salt-01-26-12.pdf

Slutz J, Anderson J, Broderick R, Horner P (2012) Key shale gas water management strategies: an economic assessment tool. In: SPE/APPEA International conference on health, safety, and environment in oil and gas exploration and production. Perth, Australia

Theodori GL, Luloff AE, Willits FK, Burnett DB (2014) Hydraulic fracturing and the management, disposal, and reuse of frac flowback waters: views from the public in the Marcellus Shale. Energy Res Soc Sci 2:66–74

Vidic RD, Brantley SL, Vandenbossche JM, Yoxtheimer D, Abad JD (2013) Impact of shale gas development on regional water quality. Science 340(6134):1235009

Wang YP, Smith R (1994) Wastewater minimisation. Chem Eng Sci 49:981–1006

Yang L, Grossmann IE (2012) Water targeting models for simultaneous flowsheet optimization. Ind Eng Chem Res 52(9):3209–3224

Yang L, Grossmann IE, Manno J (2014a) Optimization models for shale gas water management. AIChE J 60(10):3490–3501

Yang L, Salcedo-Diaz R, Grossmann IE (2014b) Water network optimization with wastewater regeneration models. Ind Eng Chem Res 53(45):17680–17695

You F, Wassick JM, Grossmann IE (2009) Risk management for a global supply chain planning under uncertainty: models and algorithms. AIChE J 55(4):931–946

Yu JQ, Chen Y, Shao S, Zhang Y, Liu SL, Zhang SS (2014) A study on establishing an optimal water network in a dyeing and finishing industrial park. Clean Technol Environ Policy 16(1):45–57

Zammerilli A, Murray RC, Davis T, Littlefield J (2014) Environmental impacts of unconventional natural gas development and production. DOE/NETL-2014/1651, (800) 553-7681.gure

Chapter 15
Comprehensive Planning for Optimal Management of Different Types of Shale Gas Wastewater

15.1 Introduction

The energy portfolio (Al-Douri et al. 2017) has been greatly impacted by the recent findings of significant shale gas deposits, which should directly enhance electrical power planning (Serrano-Arevalo et al. 2020). For example, it's anticipated that in the US, the share of energy produced from shale or natural gas will rise from 20% in 2020 to roughly 37% in 2050 (EIA 2021). The main method for producing shale gas has been hydraulic fracturing technology (Xu et al. 2018). The complexity of producing, treating, and distributing unconventional gas means that most nations lack the infrastructure required for producing it on a big scale (Arredondo-Ramírez et al. 2016; Cao et al. 2020; Oke et al. 2020; Al-Aboosi and El-Halwagi 2019). In particular, the absence of adequate infrastructure, sustainable technology, and resources makes hydraulic fracturing's water usage and outflow a significant concern. As a result, to fulfill a certain gas quality standard, an approach to decide the best selection and technologies considering economic elements must be developed (Martinez-Gomez et al. 2017). Studies have concentrated on assessing the water needs to be related to hydraulic fracturing (Lira-Barragán et al. 2016c). Features of wastewater from shale gas fields influence the choices used for treatment to reduce the discharge's negative environmental effects (Sun et al. 2019; Estrada and Bhamidimarri 2016). Various methods (such as reverse osmosis and membrane distillation) have been explored for the treatment of wastewater from shale gas extraction (Al-Aboosi and El-Halwagi 2018; Oke et al. 2019; Shaffer et al. 2013). Additional research has focused on more specialized physical attributes like mineral content, clay mineral ionic exchange, water–rock reactions versus reservoir fluid mixing on chemical concentration, mechanisms governing the chemistry of flowback water, and water-shale interaction—the latter of which is the process that causes structural damage when water and shale gas interact (Wang et al. 2020; Phan et al. 2018; Zolfaghari et al. 2015). Furthermore, by mathematical programming formulations for optimal water management taking into

J. M. Ponce-Ortega et al., *Sustainable Water Systems*, Water Resources Development and Management, https://doi.org/10.1007/978-3-031-72498-5_15

account economic and environmental aspects (Lira-Barragán et al. 2016b), decision-making tools have been employed for gas field fracture planning in storage and treatment (Tavakkoli et al. 2020). Water management planning has made use of several multi-objective optimization techniques, such as mixed-integer linear programming models that combine supply chain design and planning with water management (Laguna-Martínez et al. 2020). Wood (2020) implemented machine learning algorithms in this context, Yang et al. (2020) conducted a hierarchical optimization model to optimize resource allocation, Zhang et al. (2017) presented a stochastic fracturing wastewater management model, and Hernández-Pérez et al. (2020) presented hybrid models with deterministic and metaheuristic techniques.

The exploitation of unconventional deposits holds significant importance in terms of safety and public health due to the increased interaction between the people living in the vicinity of shale-gas production and extraction activities (Lira-Barragán et al. 2016a). Numerous issues are brought up by drilling operations being close to residential areas, such as noise pollution, dangers of dust and emissions, diminished quality of life, public safety, and effects on property values and community aesthetics (Fry 2013). As a result, it is critical to strike a balance between policies, the environment, and the economy (Fernández et al. 2020; Hasaneen and El-Halwagi 2017; Herrera-Pantoja and Hiscock 2015).

The type and characteristics of shale gas wastewater represent a main role in devising water-management strategies. Upon the injection of water (F_t^{fres}) in hydraulic fracturing, the initial wastewater returning with gas is referred to as "flowback" (which contains leached pollutants and chemicals used in preparing the fracturing fluid). Following, wastewater is referred as "transition water" (which includes some contaminants in moderate concentrations), and finally "produced water" is generated (this is polluted by total dissolved solids (TDS), thorium, metals, volatile organic compounds (VOC) such as kerosene, benzene, and toluene (Manzanares-Rivera 2014) and radioactive substances such as uranium, radio 226 and 228 (Haluszczak et al. 2013)). Due to the significant impact of wastewater characteristics on the development of an optimal long-term management strategy, this chapter presents an approach for the simultaneous minimization of the use of freshwater and the minimum cost involved in hydraulic fracturing for shale gas extraction, considering the different types of wastewater streams that depend on the time they return to the surface (flowback, transition and produce water). The proposed approach avoids the mixing of different types of wastewater to improve their reuse. The presented approach chooses the set of different treatment technologies, reuses, and capacities required for the correct operation of the project. Fixed scheduling is selected for the completion of a set of wells to fix the periods where the water is needed and where the water is returning to the surface. The presented multi-objective, multi-period optimization formulation addresses the environmental target via the minimization of total freshwater requirements (Serrano-Arevålo et al. 2022). The returning wastewater is stored, reused, or recirculated. The solution of the optimization formulation also calculates the number of treatment technologies and storage wells needed in the optimal configuration and their capabilities. The model involves the limitations associated with freshwater available in different seasons during the year.

15.2 Problem Statement

The addressed problem in this chapter can be stated as follows:
Given:

- A set of wells (n = 1, 2, … N) to be completed (involving hydraulic fracturing operations) according to a fixed scheduling involving several fracking crews (so, it is conceivable to complete concurrently several wells).
- A freshwater source to satisfy the water requirements in shale gas operations. It is needed the unit cost and the unit transportation cost.
- Several treatment technologies to clean all the wastewater streams returning to the surface; therefore, in accordance with the time (after the completion phase finishes) the wastewater is categorized as flowback, transition, and produced water. This way, a set of treatment technologies is given for the flowback (f = 1, 2, … F), for transition water (m = 1, 2, … M), and for produced water (p = 1, 2, … P). Furthermore, it is needed to determine the efficiencies to eliminate the pollutants, the unit operating costs, and the unit fix and variable costs.
- Furthermore, a set of buckets for the flowback (c = 1, 2, … C), transitional water (d = 1, 2, … D), and produced water (g = 1, 2, … G) should be installed to store reused water, each tank has a fixed and variable cost associated with it, and a maximum volume for each storage of the fixed unit.
- There is a set of possible uses for the treated water from flowback (b = 1, 2, … B), transitional (a = 1, 2, … A), and produced water (e = 1, 2, … E), where treated wastewater should be used in a different activity where water quality is acceptable.
- Furthermore, a set of uses for each type of wastewater stream (flowback, transition, and produced) with its quality conditions. Also, it must be known the sale price.
- lastly, the unit transportation cost is needed for all the trajectories (wells-treatments, treatments-reuses, well-storage, and storage-wells).

Then, the problem consists in determining the optimal configuration for water management to implement the completion phase in shale gas fields involving economic (minimizing the total cost) and environmental (minimizing the freshwater consumption) objectives. After solving the proposed approach, it should be determined the number and capacities of treatment technologies (for each type of wastewater returning to the surface), storage units, and reuses needed in the optimal solution and the internal flows. The total cost (*TC*) is constituted by the total operating cost (*TOC*), the total capital cost (*TCC*), and the profits generated by the sale of wastewater streams to be reused (*TVC*). The *TOC* involves freshwater costs, operating costs of treatment units, as well as transportation costs of the trajectories; whereas the *TCC* involves the costs associated with the acquisition of treatment and storage units, and the *TVC* is the total sale of wastewater.

Furthermore, a novel characteristic presented in this chapter is related to the split of the type of wastewater streams captured and sent to the treatment technologies; while it is not allowed mixing flowback, transitional or produced water evading that clean or poor streams are mixed with rich streams, which can increase the treatment costs. This way, the water required for the completion phase in each well varies between 8000 and 26,000 m^3, and at the end of this process, a significant portion of the demanded water returns to the surface, which can be collected and reused achieving an important reduction in freshwater consumption. Lastly, the well can produce natural gas for several years ranging from 20 to 40 (Haluszczak et al. 2013). Nevertheless, according to the period t (to come back to the surface), wastewater streams are classified as flowback water, transition, and produced. Flowback water returns the next 3 weeks after the hydraulic fracture is completed; while the transition water returns the following 12 weeks after all the flowback water is captured, and lastly the produced water returns the next 17 weeks. Furthermore, the flowback water represents 20.59% of the total injected water, while the transition is 12.13%, and produced water is 6.58% concerning the total water requirements (Bai et al. 2015). The reuse of wastewater streams corresponds to an appreciated resource for reducing freshwater consumption and the total costs associated with the project.

15.3 Model Formulation

The presented optimization approach consists of a set of mathematical relationships to simulate the addressed project. Two key aspects that represent significant contributions concerning previous attempts are the consideration of different types of wastewater streams returning to the surface (flowback, transition, and produced water), and the different reuses avoiding the final disposal of residual water streams (generating economic and environmental benefits). This way, some options for water reusing are incorporated, such as irrigation of crops, seaweed cultivation, and cooling water in cooling towers; nonetheless, additional reuses should be included in this approach without major changes in the mathematical optimization model. A key goal to be resolved by the optimization process is to determine the appropriate selection of treatment technologies for outgoing wastewater streams coming from wells (after the completion phase). Therefore, the model should fix the optimal size for treatment technologies and for storage pits, where a trade-off solution approach has been incorporated to analyze the dependence among minimizing the cost of the water network and freshwater for fracturing. First, it presents a disjunctive optimization model to select the set of diverse wastewater treatment technologies and reservoirs that are needed for meeting the extraction of unconventional gas considering as a variable the schedule of fracture of wells that are operated at the same time in a planning horizon (Fig. 15.1).

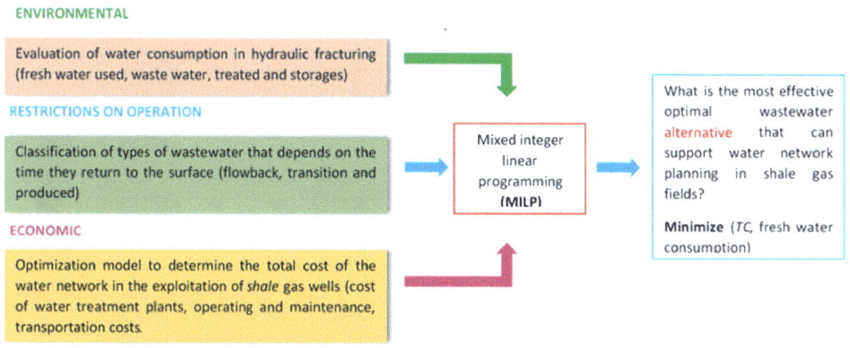

Fig. 15.1 Proposed methodology to evaluate the compensation between total cost and water consumption. Printed with permission of "*Strategic planning for the optimal management of different types of shale gas wastewater*", *ACS Sustainable Chemistry and Engineering*, Copyright (2022), American Chemical Society

The presented optimization model is based on the superstructure shown in Fig. 15.2, where there are mixers, splitters, treatment equipment, etc. Furthermore, it is needed an important freshwater source for satisfying water demands in wells. Wastewater can be stored for recirculation in hydraulic fracturing operations (this is useful for periods when the available freshwater is severely reduced) and the remaining water is sent to a set of treatment technologies to get proper quality conditions in the fluid to be reused achieving a holistic approach that manages and reuse wastewater streams in other areas.

Thus, the proposed optimization model involves the mathematical relationships needed to model the aforementioned aspects and constraints for water available in the fresh source, and the costs associated with water management involved in the shale gas industry. The required relationships are stated as follows.

15.3.1 Constraints for Freshwater Availability

Freshwater availability is important in several parts of the world, and it represents a major challenge in most of the regions with great shale gas reserves.

$$F_t^{fresh} \leq \left(\frac{\mathrm{AvailV}_t^{\mathrm{fresh\text{-}max}}}{\mathrm{H}^{\mathrm{time}}} \right), \quad \forall t \tag{15.1}$$

F_t^{fresh} represents the consumption of freshwater during the time period t (m^3/day), $AvailV_t^{fresh\_max}$ is the availability of freshwater during the time period t (m^3/week), and H^{time} is a conversion factor (day/week).

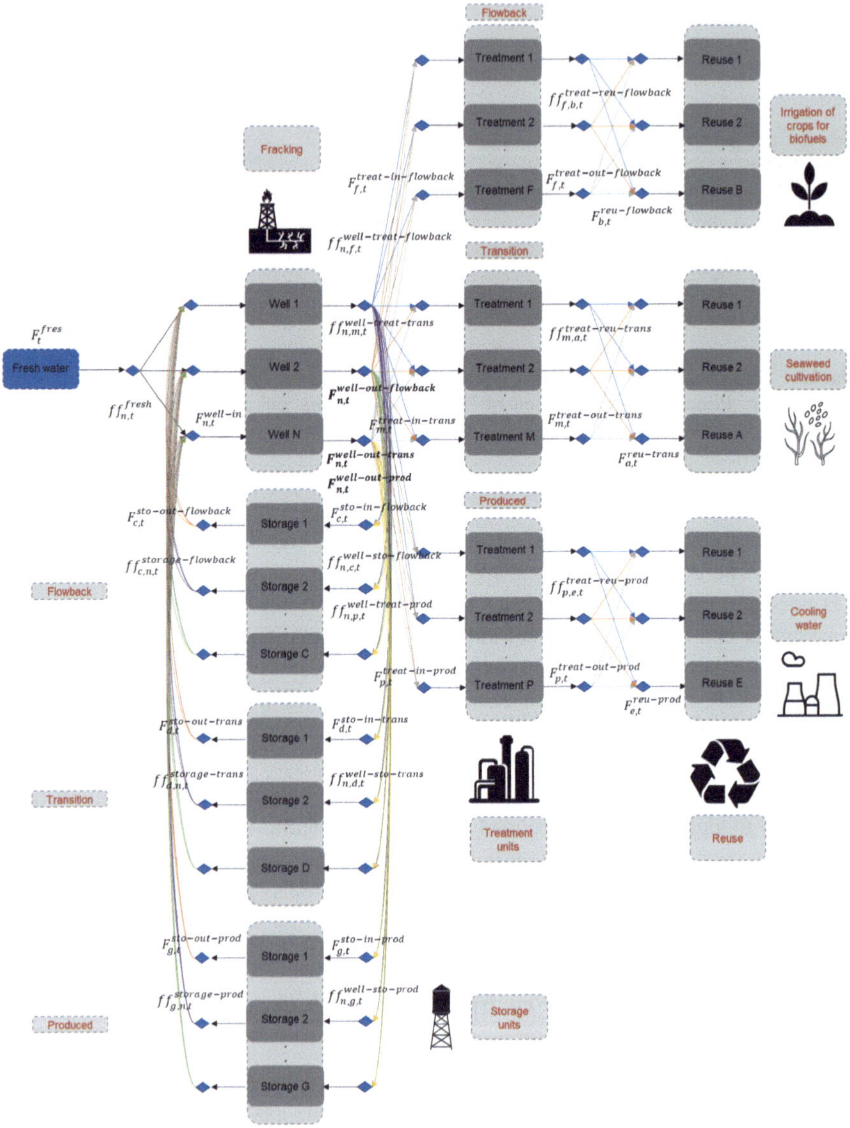

Fig. 15.2 Schematic representation of the proposed superstructure for flowback, transition, and produced water management during hydraulic fracturing. Printed with permission of "*Strategic planning for the optimal management of different types of shale gas wastewater*", *ACS Sustainable Chemistry and Engineering*, Copyright (2022), American Chemical Society

15.3.2 Freshwater Segregation

Freshwater is split and sent to each n well throughout the hydraulic fracturing process:

$$F_t^{fresh} = \sum_n ff_{n,t}^{fresh}, \quad \forall t \tag{15.2}$$

15.3.3 Water Supply to Wells

The water needs throughout the hydraulic fracturing phase in each n $F_{n,t}^{well-in}$ are obtained from the freshwater source $ff_{n,t}^{fresh}$ plus the water from the flowback $ff_{c,n,t}^{storage-flowback}$ water storage system, transition $ff_{d,n,t}^{storage-trans}$ and produced $ff_{g,n,t}^{storage-prod}$:

$$\mathrm{F}_{\mathrm{n,t}}^{\mathrm{well\text{-}in}} = ff_{n,t}^{fresh} + \sum_c ff_{c,n,t}^{storage-flowback} + \sum_d ff_{d,n,t}^{storage-trans} + \sum_g ff_{g,n,t}^{storage-prod}, \quad \forall n, \ \forall t \tag{15.3}$$

15.3.4 Wastewater Streams After Hydraulic Fracturing

The residual water that returns to the surface subsequently hydraulic fracturing is classified according to the time. To get the amounts of outlet water for the flowback, transition and produced water ($\mathrm{F}_{\mathrm{n,t}}^{\mathrm{well\text{-}out\text{-}flowback}}$, $\mathrm{F}_{\mathrm{n,t}}^{\mathrm{well\text{-}out\text{-}trans}}$, and $\mathrm{F}_{\mathrm{n,t}}^{\mathrm{well\text{-}out\text{-}}prod}$) are computed by the factors $\beta_{n,t}^{flowback}$, $\beta_{n,t}^{trans}$, and $\beta_{n,t}^{prod}$ (which represent the portion of the total injected water returning as flowback, transition and produced water, respectively) multiplied by the total water requirements in each well $\mathrm{F}_{n,t}^{\mathrm{well\text{-}in}}$, which is modeled as follows:

$$\mathrm{F}_{\mathrm{n,t}}^{\mathrm{well\text{-}out\text{-}flowback}} = \beta_{n,t}^{flowback} \mathrm{F}_{\mathrm{n,t}}^{\mathrm{well\text{-}in}}, \quad \forall n, \ \forall t \tag{15.4}$$

$$\mathrm{F}_{\mathrm{n,t}}^{\mathrm{well\text{-}out\text{-}}trans} = \beta_{n,t}^{trans} \mathrm{F}_{\mathrm{n,t}}^{\mathrm{well\text{-}in}}, \quad \forall n, \ \forall t \tag{15.5}$$

$$\mathrm{F}_{\mathrm{n,t}}^{\mathrm{well\text{-}out\text{-}prod}} = \beta_{n,t}^{prod} \mathrm{F}_{\mathrm{n,t}}^{\mathrm{well\text{-}in}}, \quad \forall n, \ \forall t \tag{15.6}$$

The aforementioned amounts of water returning to the surface ($\mathrm{F}_{\mathrm{n,t}}^{\mathrm{well\text{ - }out\text{ - }flowback}}$, $\mathrm{F}_{\mathrm{n,t}}^{\mathrm{well\text{ - }out\text{ - }}\mathit{trans}}$, and $\mathrm{F}_{\mathrm{n,t}}^{\mathrm{well\text{ - }out\text{ - }}\mathit{prod}}$) are known parameters for the proposed mathematical optimization model.

15.3.5 Segregation of Wastewater Streams

Then, each type of water stream leaving the well is sent to the treatment or the storage without mixing. This way, the balance for the water at the outlet of each well is stated as follows.

$$\mathrm{F}_{\mathrm{n,t}}^{\mathrm{well\text{ - }out\text{ - }flowback}} = \sum_{f} ff_{n,f,t}^{well-treat-flowback} + \sum_{c} ff_{n,c,t}^{well-sto-flowback}, \quad \forall n, \ \forall t \tag{15.7}$$

$$\mathrm{F}_{\mathrm{n,t}}^{\mathrm{well\text{ - }out\text{ - }trans}} = \sum_{m} ff_{n,m,t}^{well-treat-trans} + \sum_{d} ff_{n,d,t}^{well-sto-trans}, \quad \forall n, \ \forall t \tag{15.8}$$

$$\mathrm{F}_{\mathrm{n,t}}^{\mathrm{well\text{ - }out\text{ - }prod}} = \sum_{p} ff_{n,p,t}^{well-treat-prod} + \sum_{g} ff_{n,g,t}^{well-sto-prod}, \quad \forall n, \ \forall t \tag{15.9}$$

where the subscripts *f, m, p* are the treatment units for flowback, transition, and produced water, respectively. Similarly for the storage units *c, d, g*.

15.3.6 Water Inlet to Treatment Units

The presented optimization model incorporates an interception network constituted by several treatment units ussually used to treat the fracturing fluid. This way, established technologies for treating wastewater can be considered even if they involve different costs and capacities. The main requirement to consider any treatment in this approach is that the output stream has a good quality and appropriate conditions to be reused without generating environmental problems or technical inconveniences in the process. Lastly, the model can choice the required technologies (if it considers different units) and determine the number of treatment technologies and size for each one. Thus, the inlet flowrate to any treatment unit *f, m, p* during the period *t* $\left(F_{f,t}^{treat-in-flowback}, F_{m,t}^{treat-in-trans}, F_{p,t}^{treat-in-prod}\right)$ is obtained by the sum of the wastewater coming from the wells and requiring treatment $\left(ff_{n,f,t}^{well-treat-flowback}, ff_{n,m,t}^{well-treat-trans}, ff_{n,p,t}^{well-treat-prod}\right)$. This is modeled in Eqs. (15.10), (15.11), and (15.12).

$$F_{f,t}^{treat-in-flowback} = \sum_{n} ff_{n,f,t}^{well-treat-flowback}, \quad \forall f, \quad \forall t \tag{15.10}$$

$$F_{m,t}^{treat-in-trans} = \sum_{n} ff_{n,m,t}^{well-treat-trans}, \quad \forall m, \quad \forall t \tag{15.11}$$

$$F_{p,t}^{treat-in-prod} = \sum_{n} ff_{n,p,t}^{well-treat-prod}, \quad \forall p, \quad \forall t \tag{15.12}$$

15.3.7 Water Streams Leaving from Treatment Units

The considered treatment units are located onsidering the type of wastewater stream able to treat. The water coming out of these units is sent to diverse reuses, which depend on the water quality. Therefore, the treated water leaving the flowback, transition, and produced technologies is sent to a set of available reuses.

$$F_{f,t}^{treat-out-flowback} = \sum_{b} ff_{f,b,t}^{treat-reu-flowback}, \quad \forall f, \quad \forall t \tag{15.13}$$

$$F_{m,t}^{treat-out-trans} = \sum_{a} ff_{m,a,t}^{treat-reu-trans}, \quad \forall m, \quad \forall t \tag{15.14}$$

$$F_{p,t}^{treat-out-prod} = \sum_{e} ff_{p,e,t}^{treat-reu-prod}, \quad \forall p, \quad \forall t \tag{15.15}$$

15.3.8 Water Balance for Water Reusing

At the inlet of the different reuses is considered a mixer to mix the treated streams coming from different technologies; this way, the following mathematical equations are needed in the model formulation.

$$F_{b,t}^{reu-flowback} = \sum_{f} ff_{f,b,t}^{treat-reu-flowback}, \quad \forall b, \quad \forall t \tag{15.16}$$

$$F_{a,t}^{reu-trans} = \sum_{m} ff_{m,a,t}^{treat-reu-trans}, \quad \forall a, \quad \forall t \tag{15.17}$$

$$F_{e,t}^{reu-prod} = \sum_{p} ff_{p,e,t}^{treat-reu-prod}, \quad \forall e, \quad \forall t \tag{15.18}$$

where treatment guarantees that the water quality satisfies the environmental regulations for the reuses.

15.3.9 Water Streams Entering Storage Units

The presented superstructure allows wastewater streams leaving the wells can be sent to treatment units and to storage pits (with the purpose of being directly recycled). Therefore, the balances for the mixers at the inlet of the storages *c*, *d*, and *g* are written as follows:

$$F_{c,t}^{sto-in-flowback} = \sum_{n} ff_{n,c,t}^{well-sto-flowback}, \quad \forall c, \quad \forall t \tag{15.19}$$

$$F_{d,t}^{sto-in-trans} = \sum_{n} ff_{n,d,t}^{well-sto-trans}, \quad \forall d, \quad \forall t \tag{15.20}$$

$$F_{g,t}^{sto-in-prod} = \sum_{n} ff_{n,g,t}^{well-sto-prod}, \quad \forall g, \quad \forall t \tag{15.21}$$

$F_{c,t}^{sto-in-flowback}$, $F_{d,t}^{sto-in-trans}$, and $F_{d,t}^{sto-in-prod}$ are the total flowrate entering the pits *c*, *d*, and *g* at period *t*.

15.3.10 Water Outlet to Storage Units

After the wastewater is stored, it can be recirculated to the hydraulic fracture of any well. Therefore, the wastewater stream leaving the pits *c*, *d*, and *g* is recirculated ($ff_{c,n,t}^{storage-flowback}$, $ff_{d,n,t}^{storage-trans}$ and $ff_{g,n,t}^{storage-prod}$) to the well *n*.

$$F_{c,t}^{sto-out-flowback} = \sum_{n} ff_{c,n,t}^{storage-flowback}, \quad \forall c, \quad \forall t \tag{15.22}$$

$$F_{d,t}^{sto-out-trans} = \sum_{n} ff_{d,n,t}^{storage-trans}, \quad \forall d, \quad \forall t \tag{15.23}$$

$$F_{g,t}^{sto-out-prod} = \sum_{n} ff_{g,n,t}^{storage-prod}, \quad \forall g, \quad \forall t \tag{15.24}$$

15.3.11 Balance in Treatment Units

Several treatment units have related water losses during their operation, this issue is simulated through the following equations.

$$F_{f,t}^{treat-out-flowback} = \alpha_{\mathrm{f}}^{\mathrm{treat\ -\ flowback}} F_{f,t}^{treat-in-flowback}, \quad \forall f, \ \forall t \tag{15.25}$$

$$F_{m,t}^{treat-out-trans} = \alpha_{\mathrm{m}}^{treat-trans} F_{m,t}^{treat-in-trans}, \quad \forall m, \ \forall t \tag{15.26}$$

$$F_{p,t}^{treat-out-prod} = \alpha_{\mathrm{p}}^{\mathrm{treat\ -\ prod}} F_{p,t}^{treat-in-prod}, \quad \forall p, \ \forall t \tag{15.27}$$

$\alpha_f^{treat-flowback}$, $\alpha_m^{treat-trans}$, and $\alpha_p^{treat-prod}$ are the efficiency factor in terms of the volumetric flow for the unit f, m, and p.

15.3.12 Balance in Storage Units

The existing shale gas fields and even zones with significant shale gas reserves but without exploiting this resource are associated with water availability. Therefore, the importance of storage pits is to save wastewater to be used during periods where the freshwater availability is low. The following volume accumulation balance in the storage units c, d, g, at time t states that the volume for the pits c, d, g at time t $\left(V_{c,t}^{sto-flowback}, V_{d,t}^{sto-trans}, V_{g,t}^{sto-prod}\right)$ is equal to the volume in the previous period $\left(V_{c,t-1}^{sto-flowback}, V_{d,t-1}^{sto-trans}, V_{g,t-1}^{sto-prod}\right)$, plus the difference between the inlet and outlet flowrates multiplied by the time conversion factor H^{time}. This is presented in Eqs. (15.28)–(15.33).

$$V_{c,t}^{sto-flowback} = \mathrm{V}_{\mathrm{c}}^{\mathrm{sto\ -\ initial\ -\ flowback}} + \mathrm{H}^{\mathrm{time}}\left(F_{c,t}^{sto-in-flowback} - F_{c,t}^{sto-out-flowback}\right), \quad \forall c, \ \forall t = 1 \tag{15.28}$$

$$V_{c,t}^{sto-flowback} = V_{c,t-1}^{sto-flowback} + \mathrm{H}^{\mathrm{time}}\left(F_{c,t}^{sto-in-flowback} - F_{c,t}^{sto-out-flowback}\right), \quad \forall c, \ \forall t > 1 \tag{15.29}$$

$$V_{d,t}^{sto-trans} = \mathrm{V}_{\mathrm{d}}^{\mathrm{sto\ -\ initial\ -\ trans}} + \mathrm{H}^{\mathrm{time}}\left(F_{d,t}^{sto-in-trans} - F_{d,t}^{sto-out-trans}\right), \quad \forall d, \ \forall t = 1 \tag{15.30}$$

$$V_{d,t}^{sto-trans} = V_{d,t-1}^{sto-trans} + \mathrm{H}^{\mathrm{time}}\left(F_{d,t}^{sto-in-trans} - F_{d,t}^{sto-out-trans}\right), \quad \forall d, \ \forall t > 1 \tag{15.31}$$

$$V_{g,t}^{sto-prod} = \mathrm{V}_{g}^{\mathrm{sto\ -\ initial\ -\ prod}} + \mathrm{H}^{\mathrm{time}}\left(F_{g,t}^{sto-in-prod} - F_{g,t}^{sto-out-prod}\right), \quad \forall g, \ \forall t = 1 \tag{15.32}$$

$$V_{g,t}^{sto-prod} = V_{g,t-1}^{sto-prod} + \mathrm{H}^{\mathrm{time}}\left(F_{g,t}^{sto-in-prod} - F_{g,t}^{sto-out-prod}\right), \quad \forall g, \ \forall t > 1 \tag{15.33}$$

For the first time period, a known initial well volume is considered $\left(V_c^{sto-initial-flowback}, V_d^{sto-initial-trans}, V_g^{sto-initial-prod}\right)$. Furthermore, it is needed to guarantee the continuity of the cycles; this is the reason to equalize the initial volume to the last volume in the buffer *c, d, g*:

$$\mathrm{V}_{\mathrm{d}}^{\mathrm{sto\text{ - }initial\text{ - }trans}} = V_{d,t}^{sto-trans}, \quad \forall t = t^{final} \tag{15.34}$$

$$\mathrm{V}_{\mathrm{c}}^{\mathrm{sto\text{ - }initial\text{ - }flowback}} = V_{c,t}^{sto-flowback}, \quad \forall t = t^{final} \tag{15.35}$$

$$\mathrm{V}_{\mathrm{g}}^{\mathrm{sto\text{ - }initial\text{ - }prod}} = V_{g,t}^{sto-prod}, \quad \forall t = t^{final} \tag{15.36}$$

15.3.13 Capacity and Existence of Treatment and Storage Units

Key objectives involved by the optimization model are determining the number of storage units, treatment technologies, and their corresponding capacities. Therefore, the following relationships select whether the unit *f, m, p* is required or not (the existence of the treatment units is simulated through the binary variables $y_f^{treat-flowback}, y_m^{treat-trans}, y_p^{treat-prod}$); also, the operational capacity for each treatment is represented as follows:

$$F_f^{treat-cap-flowback} \geq F_{f,t}^{treat-in-flowback}, \quad \forall f, \ \forall t \tag{15.37}$$

$$F_f^{treat-cap-flowback} \leq \mathrm{F}_{\mathrm{f}}^{\mathrm{treat\text{ - }max\text{ - }flowback}} y_f^{treat-flowback}, \quad \forall f \tag{15.38}$$

$$F_m^{treat-cap-trans} \geq F_{m,t}^{treat-in-trans}, \quad \forall m, \ \forall t \tag{15.39}$$

$$F_m^{treat-cap-trans} \leq \mathrm{F}_{\mathrm{m}}^{\mathrm{treat\text{ - }max\text{ - }trans}} y_m^{treat-trans}, \quad \forall m \tag{15.40}$$

$$F_p^{treat-cap-prod} \geq F_{p,t}^{treat-in-prod}, \quad \forall p, \ \forall t \tag{15.41}$$

$$F_p^{treat-cap-prod} \leq \mathrm{F}_{\mathrm{p}}^{\mathrm{treat\text{ - }max\text{ - }prod}} y_p^{treat-prod}, \quad \forall p \tag{15.42}$$

max is an upper limit for operational capacity related to the treatment unit (this information can be provided by the seller of treatment technologies). The optimal capacities $F_f^{treat-cap-flowback}$, $F_m^{treat-cap-trans}$, $F_p^{treat-cap-prod}$ should be used by the cost function linked to treatment costs.

15.3.14 Capacity and Existence of Storage Units

Likewise, the following relationships are required for storage units:

$$V_c^{sto-cap-flowback} \geq V_{c,t}^{sto-flowback}, \quad \forall c, \quad \forall t \tag{15.43}$$

$$V_c^{sto-cap-flowback} \leq \mathrm{V}_{\mathrm{c}}^{\mathrm{sto\text{ - }max\text{ - }flowback}} y_c^{sto-flowback}, \quad \forall c \tag{15.44}$$

$$V_d^{sto-cap-trans} \geq V_{d,t}^{sto-trans}, \quad \forall d, \quad \forall t \tag{15.45}$$

$$V_d^{sto-cap-trans} \leq \mathrm{V}_{\mathrm{d}}^{\mathrm{sto\text{ - }max\text{ - }trans}} y_d^{sto-trans}, \quad \forall d \tag{15.46}$$

$$V_g^{sto-cap-prod} \geq V_{g,t}^{sto-prod}, \quad \forall g, \quad \forall t \tag{15.47}$$

$$V_g^{sto-cap-prod} \leq \mathrm{V}_{\mathrm{g}}^{\mathrm{sto\text{ - }max\text{ - }prod}} y_g^{sto-prod}, \quad \forall g \tag{15.48}$$

$V_c^{sto-cap-flowback}$, $V_d^{sto-cap-trans}$, $V_g^{sto-cap-prod}$ are the optimal capacity for storage units; and $y_c^{sto-flowback}$, $y_d^{sto-trans}$, $y_g^{sto-prod}$ are binary variables used to simulate the existence of the units.

15.3.15 Total Used Water

Even though water limitations can be satisfied, it is imperative to quantify the total water required to complete all planned wells. Consequently, the following equation calculates the total water needs (*TWR*):

$$TWR = \mathrm{H}^{\mathrm{time}} \sum_t F_t^{fresh} \tag{15.49}$$

The previous equation embodies the environmental objective considered by this project.

15.3.16 Operating Costs

Operating costs correspond to the cost of freshwater, the operating costs of treatment units and transportation costs (including all trajectories), which are defined in detail as follows.

15.3.16.1 Cost of Freshwater

The cost of freshwater $\left(\mathrm{Cos}t^{fresh}\right)$ is given by multiplying the sum of the required water during all periods with the freshwater unit cost $\left(UC^{fresh}\right)$ and considering the time conversion factor:

$$\mathrm{Cos}t^{fresh} = \mathrm{H}^{\mathrm{time}} UC^{\mathrm{fresh}} \sum_{t} F_{t}^{fresh} \tag{15.50}$$

15.3.16.2 Operating Costs for Treatment Units

The operating costs for the treatment of units are defined as follows:

$$\mathrm{Cos}t^{op-treat-flowback} = \mathrm{H}^{\mathrm{time}} \sum_{f} \sum_{t} \mathrm{UOC}_{\mathrm{f}}^{\mathrm{treat\text{ - }flowback}} F_{f,t}^{treat-in-flowback} \tag{15.51}$$

$$\mathrm{Cos}t^{op-treat-trans} = \mathrm{H}^{\mathrm{time}} \sum_{m} \sum_{t} \mathrm{UOC}_{\mathrm{m}}^{\mathrm{treat\text{ - }trans}} F_{m,t}^{treat-in-trans} \tag{15.52}$$

$$\mathrm{Cos}t^{op-treat-prod} = \mathrm{H}^{\mathrm{time}} \sum_{p} \sum_{t} \mathrm{UOC}_{\mathrm{p}}^{\mathrm{treat\text{ - }prod}} F_{p,t}^{treat-in-prod} \tag{15.53}$$

UOC^{treat} represents the unit operating cost of the treatment unit associated with the technologies treating flowback, transitional, and produced water.

15.3.16.3 Cost of Transporting Freshwater

The cost for the freshwater transportation $\left(Cost^{trans-fresh}\right)$ can be determined by multiplying the sum of the freshwater flowrates sent to well n in the period t by the unit transport cost $\left(UTC_{n}^{fresh}\right)$:

$$Cost^{trans-fresh} = \mathrm{H}^{\mathrm{time}} \sum_{n} \sum_{t} \mathrm{UTC}_{\mathrm{n}}^{\mathrm{fresh}} ff_{n,t}^{fresh} \tag{15.54}$$

Unit costs are different for each well when considering the geographic distance between the source of freshwater with each well. The freshwater resource is available at significant distances from wells.

15.3.16.4 Cost of Transporting Exit Water from Wells in Hydraulic Fracturing to Treatment

The exit water leaving the wells is transported to the interception network and the associated cost is considered through the next equations:

$$Cost^{trans-flowback} = \mathrm{H}^{\mathrm{time}} \sum_{n} \sum_{f} \sum_{t} \mathrm{UTC}_{\mathrm{n,f}}^{\mathrm{used\text{ - }flowback}} ff_{n,f,t}^{well-treat-flowback} \quad (15.55)$$

$$Cost^{trans-trans} = \mathrm{H}^{\mathrm{time}} \sum_{n} \sum_{m} \sum_{t} \mathrm{UTC}_{\mathrm{n,m}}^{\mathrm{used\text{ - }trans}} ff_{n,m,t}^{well-treat-trans} \quad (15.56)$$

$$Cost^{trans-prod} = \mathrm{H}^{\mathrm{time}} \sum_{n} \sum_{p} \sum_{t} \mathrm{UTC}_{\mathrm{n,p}}^{\mathrm{used\text{ - }prod}} ff_{n,p,t}^{well-treat-prod} \quad (15.57)$$

$UTC_{n,f}^{used-flowback}$, $UTC_{n,m}^{used-trans}$, $UTC_{n,p}^{used-prod}$ represent the unit transportation costs for the wastewater in the route generated by well n and treatment f, m, p. The unit cost is different for each route.

15.3.16.5 Cost of Transporting Treated Water to Reusing

The flow rate that leaves the treatment is directed to the reuses and the transportation costs generated are given as follows:

$$Cost^{trans-treat-reu-flowback} = \mathrm{H}^{\mathrm{time}} \sum_{f} \sum_{b} \sum_{t} \mathrm{UTC}_{\mathrm{f,b}}^{\mathrm{treat\text{ - }reu\text{ - }flowback}} ff_{f,b,t}^{treat-reu-flowback} \quad (15.58)$$

$$Cost^{trans-treat-reu-trans} = \mathrm{H}^{\mathrm{time}} \sum_{m} \sum_{a} \sum_{t} \mathrm{UTC}_{\mathrm{m,a}}^{\mathrm{treat\text{ - }reu\text{ - }trans}} ff_{m,a,t}^{treat-reu-trans} \quad (15.59)$$

$$Cost^{trans-treat-reu-prod} = \mathrm{H}^{\mathrm{time}} \sum_{p} \sum_{e} \sum_{t} \mathrm{UTC}_{\mathrm{p,e}}^{\mathrm{treat\text{ - }reu\text{ - }prod}} ff_{p,e,t}^{treat-reu-prod} \quad (15.60)$$

$UTC_{f,b}^{treat-reu-flowback}$, $UTC_{m,a}^{treat-reu-trans}$, $UTC_{p,e}^{treat-reu-prod}$ are the unit transportation costs for water from treatment f to reuse b, treatment m to reuse a, and treatment p to reuse e, respectively.

15.3.16.6 Cost of Transporting Well Outlet Water in Hydraulic Fracturing to Storage

Equations (15.61)–(15.63) are used to model the transportation costs of the different types of flowback, transition, and produced water coming from the wells to the storage pits.

$$Cost^{trans-well-sto-flowback} = \mathrm{H}^{\mathrm{time}} \sum_{n}\sum_{c}\sum_{t} \mathrm{UTC}_{\mathrm{n,c}}^{\mathrm{well\text{ - }sto\text{ - }flowback}} ff_{n,c,t}^{well-sto-flowback} \quad (15.61)$$

$$Cost^{trans-well-sto-trans} = \mathrm{H}^{\mathrm{time}} \sum_{n}\sum_{d}\sum_{t} \mathrm{UTC}_{\mathrm{n,d}}^{\mathrm{well\text{ - }sto\text{ - }trans}} ff_{n,d,t}^{well-sto-trans} \quad (15.62)$$

$$Cost^{trans-well-sto-prod} = \mathrm{H}^{\mathrm{time}} \sum_{n}\sum_{g}\sum_{t} \mathrm{UTC}_{\mathrm{n,g}}^{\mathrm{well\text{ - }sto\text{ - }prod}} ff_{n,g,t}^{well-sto-prod} \quad (15.63)$$

$UTC_{n,c}^{well-sto-flowback}$, $UTC_{n,d}^{well-sto-trans}$, $UTC_{n,g}^{well-sto-prod}$ are the unit transportation costs for water from well n to storage c, d, g.

15.3.16.7 Cost of Transport from Storage to Well

The next relationships are used to determine the transportation costs from the storage pits to wells:

$$\mathrm{Cos}t^{trans-sto-well-flowback} = \mathrm{H}^{\mathrm{time}} \sum_{c}\sum_{n}\sum_{t} \mathrm{UTC}_{\mathrm{c,n}}^{\mathrm{sto\text{ - }well\text{ - }flowback}} ff_{c,n,t}^{storage-flowback} \quad (15.64)$$

$$\mathrm{Cos}t^{trans-sto-well-trans} = \mathrm{H}^{\mathrm{time}} \sum_{d}\sum_{n}\sum_{t} \mathrm{UTC}_{\mathrm{d,n}}^{\mathrm{sto\text{ - }well\text{ - }trans}} ff_{d,n,t}^{storage-trans} \quad (15.65)$$

$$\mathrm{Cos}t^{trans-sto-well-prod} = \mathrm{H}^{\mathrm{time}} \sum_{g}\sum_{n}\sum_{t} \mathrm{UTC}_{\mathrm{g,n}}^{\mathrm{sto\text{ - }well\text{ - }prod}} ff_{g,n,t}^{storage-prod} \quad (15.66)$$

$UTC_{c,n}^{sto-well-flowback}$, $UTC_{d,n}^{sto-well-trans}$, $UTC_{g,n}^{sto-well-prod}$ are the unit transport costs for the wastewater that goes from the storage pits c, d, g to well n, this varies considering each trajectory and taking into account the geographical location.

15.3.17 Total Operating Cost

The sum of all the operating costs presented before produces the *TOC* for the project:

$$TOC = Cost^{fresh} + Cost^{op-treat-flowback} + Cost^{op-treat-trans}$$
$$+ Cost^{op-treat-prod} + Cost^{trans-fresh} + Cost^{trans-flowback} + Cost^{trans-trans}$$
$$+ Cost^{trans-prod} + Cost^{trans-treat-reu-flowback} + Cost^{trans-treat-reu-trans}$$
$$+ Cost^{trans-treat-reu-prod} + Cost^{trans-well-sto-flowback} + Cost^{trans-well-sto-trans}$$
$$+ Cost^{trans-well-sto-prod} + Cost^{trans-sto-well-flowback}$$
$$+ Cost^{trans-sto-well-trans} + Cost^{trans-sto-well-prod} \quad (15.67)$$

15.3.18 Capital Costs for Treatment Units

The acquisition of treatment units and storage units produces capital costs, which are involved in the economic objective function.

15.3.18.1 Capital Cost for Treatment Units

The capital costs for the treatment technologies related to flowback, transition, and produced water consider two terms; the first one corresponds to a fixed charge $\left(FC^{treat-flowback}, FC^{treat-trans}, FC^{treat-prod}\right)$ and to be activated depends only on the existence of the unit (this is multiplied by the binary variable $y_f^{treat-flowback}, y_m^{treat-trans}, y_p^{treat-prod}$), and the other must multiply the capacity of the unit with the unit cost $VC^{treat-flowback}, VC^{treat-trans}, VC^{treat-prod}$, as follows:

$$CapCost^{treat-flowback} = \mathrm{k}_{\mathrm{Ffb}} \sum_f \begin{bmatrix} \mathrm{FC}^{\text{treat - flowback}} y_f^{treat-flowback} + \\ \mathrm{VC}^{\text{treat - flowback}} \left(F_f^{treat-cap-flowback}\right) \end{bmatrix} \quad (15.68)$$

$$CapCost^{treat-trans} = \mathrm{k}_{\mathrm{Ftrans}} \sum_m \begin{bmatrix} \mathrm{FC}^{\text{treat - trans}} y_m^{treat-trans} + \\ \mathrm{VC}^{\text{treat - trans}} \left(F_m^{treat-cap-trans}\right) \end{bmatrix} \quad (15.69)$$

$$CapCost^{treat-prod} = \mathrm{k}_{Fprod} \sum_p \begin{bmatrix} \mathrm{FC}^{treat-prod} y_p^{treat-prod} + \\ \mathrm{VC}^{treat-prod} \left(F_p^{treat-cap-prod}\right) \end{bmatrix} \quad (15.70)$$

$k_{Ffb}, k_{Ftrans}, k_{Fprod}$ are factors used to annualize the investment. In the previous relationships, the fixed and variable charges, FC and VC, are calculated before the optimization process.

15.3.18.2 Capital Cost for Storage Units

Likewise, for storage units:

$$CapCost^{storage-flowback} = \mathrm{k_{Ffb}} \sum_{c} \begin{bmatrix} \mathrm{FC}^{\text{storage - flowback}} y_c^{storage-flowback} + \\ \mathrm{VC}^{\text{storage - flowback}} \left(V_c^{sto-cap-flowback}\right) \end{bmatrix} \quad (15.71)$$

$$CapCost^{storage-trans} = \mathrm{k_{Ftrans}} \sum_{d} \begin{bmatrix} \mathrm{FC}^{\text{storage - trans}} y_d^{storage-trans} + \\ \mathrm{VC}^{\text{storage - trans}} \left(V_d^{sto-cap-trans}\right) \end{bmatrix} \quad (15.72)$$

$$CapCost^{storage-prod} = \mathrm{k_{Fprod}} \sum_{g} \begin{bmatrix} \mathrm{FC}^{\text{storage - prod}} y_g^{storage-prod} + \\ \mathrm{VC}^{\text{storage - prod}} \left(V_g^{sto-cap-prod}\right) \end{bmatrix} \quad (15.73)$$

15.3.19 Total Capital Cost

The total capital cost (TCC) involves the capital costs for treatment and storage units:

$$\begin{aligned} TCC = {} & CapCost^{treat-flowback} + CapCost^{treat-trans} + CapCost^{treat-prod} \\ & + CapCost^{storage-flowback} + CapCost^{storage-trans} + CapCost^{storage-prod} \end{aligned} \quad (15.74)$$

15.3.20 Sales Associated with Reused Water

Treated wastewater reuse involves economic (owing to these streams being sold) and environmental (avoiding the final disposal of these streams) paybacks. Therefore, the following equation is needed to include the economic incomed through the sale of treated wastewater streams.

$$TVC = \mathrm{H}^{\text{time}} \begin{bmatrix} \sum_{b} \sum_{t} \mathrm{UVC}_{\mathrm{b}}^{\text{reu - flowback}} F_{b,t}^{reu-flowback} + \\ \sum_{a} \sum_{t} \mathrm{UVC}_{a}^{\text{reu - }trans} F_{a,t}^{reu-trans} + \\ \sum_{e} \sum_{t} \mathrm{UVC}_{e}^{\text{reu - }prod} F_{e,t}^{reu-prod} \end{bmatrix} \quad (15.75)$$

The unit sale costs for the different reuses are determined by $UVC_b^{reu-flowback}$, $UVC_a^{reu-trans}$, $UVC_e^{reu-prod}$.

15.3.21 Total Cost

The total cost (*TC*) is determined by the sum of the total operating cost, total capital cost, and the difference in the total sale of reused water.

$$TC = TOC + TCC - TVC \tag{15.76}$$

This mathematical formulation corresponds to a multi-objective problem, and it is desired to minimize both objectives simultaneously.

$$O.F. = \min\{TC, TWR\} \tag{15.77}$$

Consequently, the proposed mathematical model corresponds to a multi-objective mix-integer linear programming (MILP) problem, and a trade-off solution method is implemented to solve this and determine trade-offs between the solutions (minimum *TC* and minimum *TWR*).

15.4 Results and Discussion

To show the implementation of the presented approach, a case study is shown. This way, the optimization model was coded in the GAMS software® and it was resolved on a processor AMD Radeon™ of 2.3 0 GHz and 8 GB of RAM in 0.25 s of CPU time using the CPLEX solver in the General Algebraic Modeling System. The problem involves 26,236 continuous variables, 9 binary variables, and 10,930 constraints. The results are discussed In the next section.

15.4.1 Case Study

This example is a hypothetical case study; nevertheless, real data is used to obtain results near reality. To implement the proposed approach, the needed parameters have been taken from technical reports with up-to-date data for the Marcellus and Barnett regions (Shaffer et al. 2013; Slutz et al. 2012; Hayes 2009).

This way, the United States has grown significantly in shale gas production in recent years, even though currently the USA is the number one natural gas producer in the world. This way, Marcellus (located mainly in Pennsylvania) and Barnett (located in Texas) have the largest reserves of shale gas. Furthermore, other data were taken from the Mexican energy agency (SENER 2018).

In this case, 20 wells must be completed with three fracturing crews (able to carry out the hydraulic fracturing simultaneously) according to a given schedule, the drilling takes approximately 5 weeks per well, and the water management is optimized in periods of one week with a time horizon of 90 weeks to complete the process that includes storages, treatment units and 32 weeks to complete the exit of wastewater after the fracture. After the drilling is completed, the wastewater begins to return, sorting into: flowback, transition, and produced (according to the period in which the water begins to return). The flowback water returns the first 3 weeks after the drilling is finished, the transition water the next 12 weeks (after finishing the flowback water) and the produced water returns the next 17 weeks (after finishing returning the transition water). Consequently, it is possible to establish the following periods to capture and collect each type of wastewater streams; if the week where the fracture is completed corresponds to the time zero, then, flowback is returning in periods 1–3, the transition water is collected in the weeks 4–15 and produced water returns during the weeks 16–32. Furthermore, the percentages of the wastewater streams returning to the surface concerning the total injected water are the following: for the flowback is 20.59%, transition represents 12.13%, and produced corresponds to 6.58% (Bai et al. 2015). This way, all internal flows associated with the presented superstructure have been determined, as well as their fluctuations concerning time.

Furthermore, the water needs for each well rounded between 7.6 and 15 million liters (7600 m^3 and 15,000 m^3) per well because the wells are usually drilled in multiple fracturing stages (Manzanares-Rivera 2014) and this range can be wider such as for the Marcellus shale region where the water needs have been projected in 8000–27,000 m^3 (Shaffer et al.2013). This way, in this chapter, an average value of 15,000 m^3 is considered for the water needed to complete each well. It should be noted that the fracture requires 5 weeks, next the needed water flowrate is 428.57 m^3/day. Likewise, it can be projected the wastewater flowrates, which correspond to 273.51 m^3/day, 140 m^3/day 27.7 m^3/day for the flowback water in weeks 1, 2, and 3, respectively. The transitional water is 26.4 m^3/day for weeks 4–6, 23.5 m^3/day for weeks 7–9, 20.9 m^3/day for weeks 10–12 and 15.8 m^3/day for weeks 13–15. Lastly, the produced water is 13.2 m^3/day for weeks 16–20, 9.1 m^3/day for weeks 21–25, 4.9 m^3/day for weeks 26–30 and finally 2.5 m^3/day for weeks 31–32. A time conversion factor H^{time} (7 days/week) is used for the consistency of units, while the factors to annualize the investment of flowback, transition, and produced treatment units (k_{Ffb}, $k_{Ftrans,}$ and k_{Fprod}), have a value of 0.1. In this chapter, there are two types of treatment technologies: membrane distillation and osmosis, which can guarantee water quality for the outlet streams in conditions to be reused (meeting the environmental constraints); but, the volumetric efficiency depends on the type of wastewater stream (because of the treatment of polluted streams incites lower efficiency concerning cleaner streams). For storage units, there is no initial volume in storage units ($V_c^{sto-initial-flowback} = V_d^{sto-initial-trans} = V_g^{sto-initial-prod} = 0$), maximum volumes for these units are different for each classification (flowback, transition, and produced water), fixed, variable, and operational costs for equipment, cost of unit transport for all the trajectories (wells-treatments, treatments-reuses, storage wells, and storage wells), the foregoing is included in the parameters used in the case studies.

The considered reuses are: irrigation of plants for biofuels, seaweed cultivation, and cooling water in power plants; while the requirements are different for each one. Principally, in this case, it is proposed to reuse the treated flowback water in irrigation for biofuels (because it requires high-quality water, almost free of pollutants); the treated transitional water can be reused by algae cultivation (since the quality requirements are lower) and finally, the treated produced water can be sent to power plants for the reason that the pollutant conditions are laxer than the previous ones and in power plants, the water is not in direct contact with living matter as in the case of irrigation or cultivation. Subsequently, the unit prices are 1, 0.8 and 0.6 $/m^3 for flowback, transition, and produced water, respectively. The sale prices are lower than the sale price of freshwater (2 $/m^3) because of the quality of the water.

Additionally, to visualize the importance of the scheduling to carry out the hydraulic fracturing, different scenarios have been considered (where the only aspect changing among them is the scheduling) as follows:

15.4.2 Scenario A

The fixed scheduling for this case is presented in Table 15.1, where, fracking crew 1 is aimed to complete 8 wells, while crew 2 can frack to 7 wells and crew 3 can deal with 6 wells. Also, two hydraulic fracturing crews work at the beginning, and in week 5 crew 3 is incorporated to operate. However, approximately mid-year (weeks 24–26) no crew is allowed to operate because of water scarcity; gradually when water availability permits to operate the hydraulic fracturing crews restart operations. It should be noted that the proposed scheduling is severely affected by the freshwater limitations. Likewise, the needed time to move the crew fracture equipment from one well to the next is one week.

15.4.3 Scenario B

Table 15.2 shows the given scheduling to be considered by this scenario. Crew 1 involves wells 1–7, crew 2 considers wells 8–14, and crew 3 involves wells 15–20. This way, the three crews start to work simultaneously since week 1. But, approximately at mid-year (weeks 24–26) no crew is allowed to operate due to water scarcity.

Table 15.1 Scheduling for the completion phase at each well in scenario A

Well	FH (weeks)	Departure (week)	Flowback (1–3)	Transition (4–15)	Produced (16–32)
Crew 1					
1	1–5	6	6–8	9–20	21–37
2	7–11	12	12–14	15–26	27–43
3	13–17	18	18–20	21–32	33–49
4	19–23	24	24–26	27–38	39–55
5	27–31	32	32–34	35–46	47–63
6	33–37	38	38–40	41–52	53–69
7	39–43	44	44–46	47–58	59–75
8	45–49	50	50–52	53–64	65–81
Crew 2					
9	1–5	6	6–8	9–20	21–37
10	7–11	12	12–14	15–26	27–43
11	13–17	18	18–20	21–32	33–49
12	19–23	24	24–26	27–38	39–55
13	32–36	37	37–39	40–51	52–68
14	38–42	43	43–45	46–57	58–74
15	44–48	49	49–51	52–63	64–80
Crew 3					
16	5–9	10	10–12	13–24	25–41
17	11–15	16	16–18	19–30	31–47
18	17–21	22	22–24	25–36	37–53
19	36–40	41	41–43	44–55	56–72
20	42–46	47	47–49	50–61	62–78

15.4.4 Scenario C

The schedule for this case is shown in Table 15.3. The most important difference in this scenario is related to the operation of crew 3, which starts operations at mid-year. For this reason, crew 3 only operates 4 wells. Crews 1 and 2 operate simultaneously from week 1 to week 49, and both consider water scarcity.

Table 15.2 Programming the completion phase at each well in scenario B

Well	FH (weeks)	Departure (week)	Flowback (1–3)	Transition (4–15)	Produced (16–32)
Crew 1					
1	1–5	6	6–8	9–20	21–37
2	7–11	12	12–14	15–26	27–43
3	13–17	18	18–20	21–32	33–49
4	19–23	24	24–26	27–38	39–55
5	27–31	32	32–34	35–46	47–63
6	33–37	38	38–40	41–52	53–69
7	39–43	44	44–46	47–58	59–75
Crew 2					
8	1–5	6	6–8	9–20	21–37
9	7–11	12	12–14	15–26	27–43
10	13–17	18	18–20	21–32	33–49
11	19–23	24	24–26	27–38	39–55
12	27–31	32	32–34	35–46	47–63
13	33–37	38	38–40	41–52	53–69
14	39–43	44	44–46	47–58	59–75
Crew 3					
15	1–5	6	6–8	9–20	21–37
16	7–11	12	12–14	15–26	27–43
17	13–17	18	18–20	21–32	33–49
18	19–23	24	24–26	27–38	39–55
19	27–31	32	32–34	35–46	47–63
20	33–37	38	38–40	41–52	53–69

15.4.5 Scenario D

This case considers the operation presented in Table 15.4, and similarly to Scenario B, the three fracking crews initiate the operation at week 1; but, in this case, crew 3 only works the first semester.

After the presented mathematical model is implemented considering all the previous information, there can be obtained the following results. Initially, Table 15.5 presents the results obtained in terms of *TWR* and *TC* (including its detailed composition) for the four scenarios (A, B C, and D). In Scenario, D the consumption of freshwater is the lowest; consequently, scenario D involves the highest amount of water stored for recycling (to hydraulic fracturing) and as a consequence, the lowest water steams in treating and reused units. This scenario has associated the lowest

Table 15.3 Programming the completion phase at each well in scenario C

Well	FH (weeks)	Departure (week)	Flowback (1–3)	Transition (4–15)	Produced (16–32)
Crew 1					
1	1–5	6	6–8	9–20	21–37
2	7–11	12	12–14	15–26	27–43
3	13–17	18	18–20	21–32	33–49
4	19–23	24	24–26	27–38	39–55
5	27–31	32	32–34	35–46	47–63
6	33–37	38	38–40	41–52	53–69
7	39–43	44	44–46	47–58	59–75
8	45–49	50	50–52	53–64	65–81
Crew 2					
9	1–5	6	6–8	9–20	21–37
10	7–11	12	12–14	15–26	27–43
11	13–17	18	18–20	21–32	33–49
12	19–23	24	24–26	27–38	39–55
13	27–31	32	32–34	35–46	47–63
14	33–37	38	38–40	41–52	53–69
15	39–43	44	44–46	47–58	59–75
16	45–49	50	50–52	53–64	65–81
Crew 3					
17	27–31	32	32–34	35–46	47–63
18	33–37	38	38–40	41–52	53–69
19	39–43	44	44–46	47–58	59–75
20	45–49	50	50–52	53–64	65–81

value for *TC* (this can be explained considering that it needs the lowest amount of *TWR* thanks to the recirculation and although the treated water can be reused, its sale price does not reward the costs associated with the capital and operation for treatment and storage). Otherwise, the solution for Scenario C involves the greatest value for *TWR* and at the same time the biggest *TC* and highest amount of reused water (avoiding as much as possible the recirculation, which represents a significant disadvantage). In this sense, Scenario C also is the scenario with the greatest added value for the treated water owing to the sale of treated wastewater streams. Scenarios A and B represent intermediate solutions between optimistic and pessimistic scenarios. The outlet water leaving the wells is the same for all scenarios because in all the scenarios ultimately there are fractured 20 wells with the same water requirements and water amounts returning to the surface. Consequently, scenario D uses the lowest amount

Table 15.4 Programming the completion phase at each well in scenario D

Well	FH (weeks)	Departure (week)	Flowback (1–3)	Transition (4–15)	Produced (16–32)
Crew 1					
1	1–5	6	6–8	9–20	21–37
2	7–11	12	12–14	15–26	27–43
3	13–17	18	18–20	21–32	33–49
4	19–23	24	24–26	27–38	39–55
5	27–31	32	32–34	35–46	47–63
6	33–37	38	38–40	41–52	53–69
7	39–43	44	44–46	47–58	59–75
8	45–49	50	50–52	53–64	65–81
Crew 2					
9	1–5	6	6–8	9–20	21–37
10	7–11	12	12–14	15–26	27–43
11	13–17	18	18–20	21–32	33–49
12	19–23	24	24–26	27–38	39–55
13	27–31	32	32–34	35–46	47–63
14	33–37	38	38–40	41–52	53–69
15	39–43	44	44–46	47–58	59–75
16	45–49	50	50–52	53–64	65–81
Crew 3					
17	1–5	6	6–8	9–20	21–37
18	7–11	12	12–14	15–26	27–43
19	13–17	18	18–20	21–32	33–49
20	19–23	24	24–26	27–38	39–55

of freshwater, and 32.43% of freshwater is saved. This way, scenario C uses only 212,983.61 m^3 of freshwater, thereby saving 29.00%. It includes the results in terms of the wastewater streams that leave the wells and reused, where scenario D corresponds to the solution with the lowest amount of reused water (12.26%). On the other hand, scenario C reuses 18.30% of the total water that comes out of the wells; since it uses a greater quantity of freshwater and in turn does not recirculate a large quantity, it has the option of treating and immediately reusing more amount of wastewater (consider that the original amount required without recirculation in all sceneries is 300,000.00 m^3 at the end of the exploitation of all the wells and that it must be reduced by using the recirculation system).

Figure 15.3 shows a comparison of the amount of water required by considering recirculation systems or when these recirculation options are not used in scenario

Table 15.5 Resulting configuration for each scenario with a minimum *TC*

Concept	Scenario A	Scenario B	Scenario C	Scenario D
Fresh water used without recirculation (m^3)	300,000.00	300,000.00	300,000.00	300,000.00
Fresh water used with recirculation (*TWR*) (m^3)	204,686.65	206,707.41	212,983.61	202,700.96
Percentage of freshwater saved (%)	31.77	31.09	29.00	32.43
TOC ($)	2.69E + 06	2.67E + 06	2.82E + 06	2.66E + 06
TCC ($)	41,833.91	49,235.91	48,177.73	49,216.53
TVC ($)	11,938.61	13,424.50	17,731.93	11,840.83
TC ($)	2.720E + 06	2.708E + 06	2.850E + 06	2.697E + 06
Flowback reused water (m^3)	3847.23	5455.46	8288.18	5455.46
Transition reused water (m^3)	6583.25	6065.98	7375.54	4917.02
Produced reused water (m^3)	4707.98	5193.55	5905.54	4075.68
Total reused water (m^3)	15,138.45	16,714.99	21,569.25	14,448.16
Total output water from the *n* wells (m^3)	117,881.40	117,881.40	117,881.40	117,881.40
Percentage of reused water concerning the outlet (%)	12.84	14.18	18.30	12.26

A. It should be noted that from week 49 to 90 the requirements of freshwater are zero because these weeks were not completed any wells according to the proposed schedule, where the last well to be fractured is well number 8 for crew 1; likewise, the week with the greatest amount of water needs is the week 5 with 8999.97 m^3 (in this week, the 3 crews operate simultaneously). Noted that during weeks 24–26, freshwater is not needed because no crew operates due to water shortage. The amount required with recirculation in scenario A is 204,686.65 m^3 with a saving of fresh unused water of 95,312.35 m^3. Analogous to Fig. 15.3, Figs. 15.4, 15.5, and 15.6 present the comparisons between the consideration or not for the recirculating systems for scenarios B, C, and D, respectively. This way, Fig. 15.4 presents the freshwater needs with recirculation, which is 206,707.41 m^3, achieving a saving of 93,291.59 m^3. Furthermore, the last week in which water is required is in 43 because this is the week where the last well is completed. Although Fig. 15.5 presents the amount of water saved with the recirculation system in pits is 87,015.39 m^3 for scenario C. Lastly, Fig. 15.6 presents that the amount of water required with the recirculation system is 202,700.96 m^3; it should be noted that the freshwater needs from week 50 to the last are zero. The water needs for weeks 27–49 are 3000 m^3 and before the lean period, the requirements vary between 6000 and 9000 m^3.

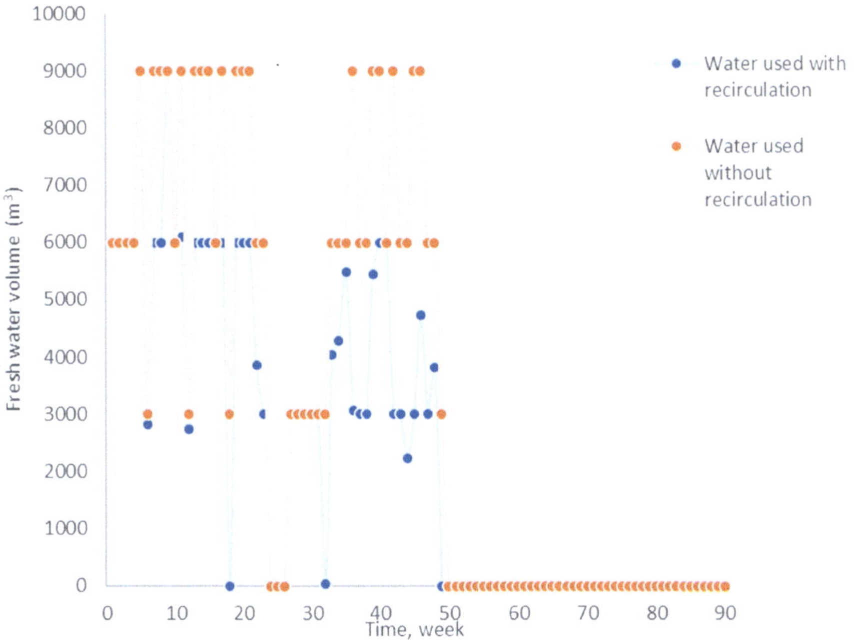

Fig. 15.3 Consumption of freshwater in scenario A. Printed with permission of "*Strategic planning for the optimal management of different types of shale gas wastewater*", *ACS Sustainable Chemistry and Engineering*, Copyright (2022), American Chemical Society

Table 15.6 presents the amount of wastewater entering each storage, treatment, and reuse system after it returns to the surface during the total time horizon. According to the total water requirements and the estimated portions of water returning to the surface as flowback, transition, and produced water a total of 117,881.40 m^3 of wastewater is recovered for each scenario. Table 15.6 establishes the destination of each wastewater stream in each option and for each scenario. This information can be useful to calculate the environmental impact (it specifies the amounts of water being reused or recycled for each type of wastewater). Likewise, scenario A presents a higher percentage of reused water (during weeks 50 to 64), where 6583.25 m^3 comes from transition water, and the total reused water for the three types is 15,138.45 m^3, which signifies 12.84% of the total water returned at the end of the hydraulic fracturing of the 20 wells. Scenario B involves to reuse 16,714.99 m^3 of the total wastewater, which is 14.18%, where 6,065.98 m^3 corresponds to transitional water during weeks 44 to 58. Scenario C reuses the largest amount of flowback water (8288.18 m^3) in weeks 50 to 52, with 18.3% of the total water recovered at the outlet of the set of wells. But, the scenario that reuses the least amount of water is D with 12.26%, which is 14,448.16 m^3 of the total outlet water streams (117,881.40 m^3), where the highest supplier of this reused water is the flowback stream (during weeks 50 to 52) and represents 5455.46 m^3 of wastewater.

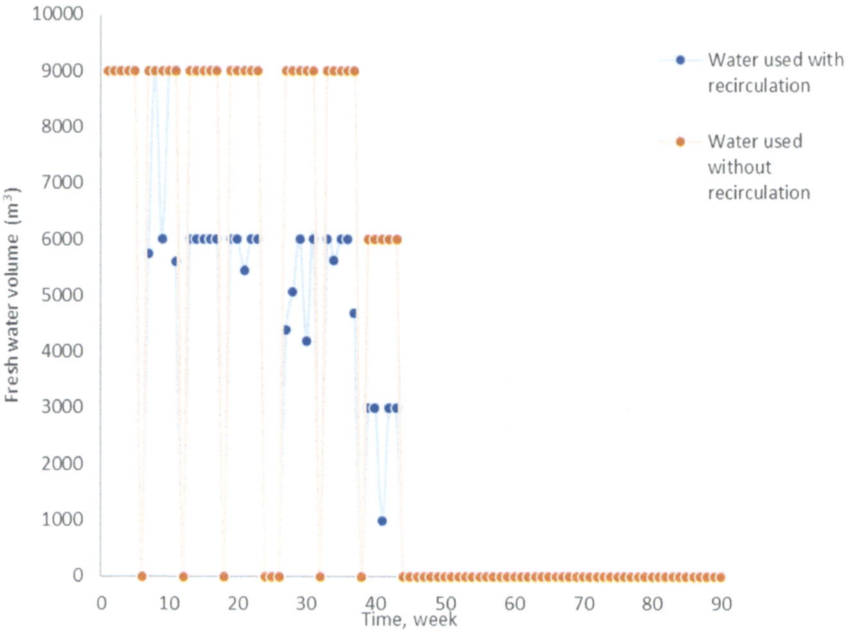

Fig. 15.4 Consumption of freshwater in scenario B. Printed with permission of "*Strategic planning for the optimal management of different types of shale gas wastewater*", *ACS Sustainable Chemistry and Engineering*, Copyright (2022), American Chemical Society

Furthermore, to compare economic and environmental multi-objective solutions (*TC* vs *TWR*), a trade-off solution approach was implemented to determine the following diagram (Fig. 15.7), and the information to make this figure can be found in Table 15.5. This diagram is built as follows: the point with the minimum *TC* and even with the minimum *TWR* is identified by solution D (this is the best scenario). Then, the opposite case is represented by point C, where the *TWR* and *TC* values significantly deteriorate, and it is the worst case. Although solutions A and B represent intermediate solutions. Figure 15.7 presents the behavior of the curve from D to C, where the economic objective decreases if the environmental objective decreases, this situation implies that both objectives aim in the same direction.

Another important aspect involved in the project is the number of treatment and storage units required to assurance the appropriate operation of the hydraulic fracturing process in the aforementioned case studies. Table 15.7 presents the needed treatment units that must be installed for each scenario, where treatment unit 1 corresponds to membrane distillation and 2 is osmosis, and the number of treatment and storage units for the different types of water.

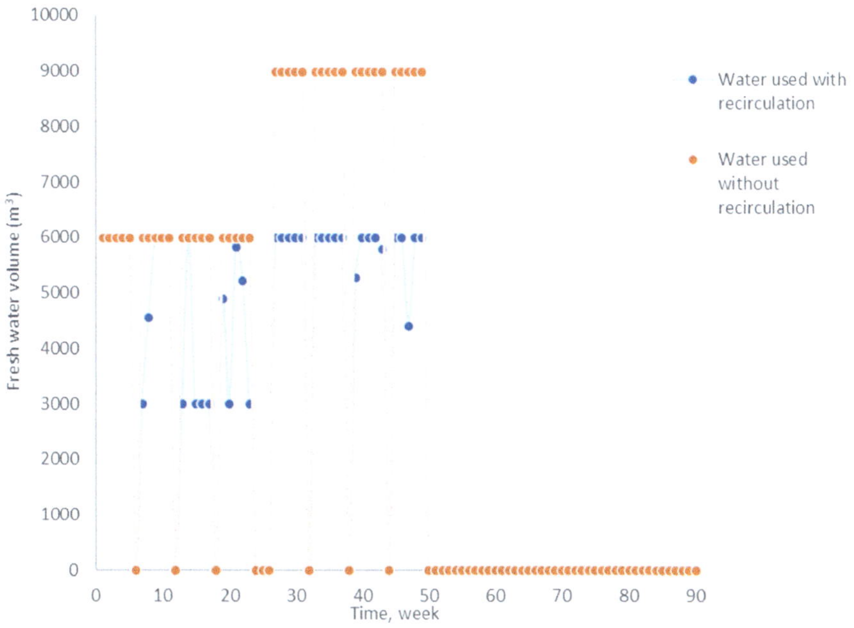

Fig. 15.5 Consumption of freshwater in scenario C. Printed with permission of "*Strategic planning for the optimal management of different types of shale gas wastewater*", *ACS Sustainable Chemistry and Engineering*, Copyright (2022), American Chemical Society

15.5 Concluding Remarks

To optimize water network planning for hydraulic fracturing operations in the shale gas sector and to consider both environmental and economic issues, this chapter has presented a mathematical programming approach. The presented model calculates the freshwater requirements for the project and mandates set scheduling for the teams executing hydraulic fracturing. The goal of the economic objective function is to minimize the overall cost, which is made up of capital and operating expenses. The goal of the environmental objective is to reduce the overall amount of freshwater needed for hydraulic fracturing, with the returned water being stored, recycled, or circulated. The number of treatment technologies and storage wells needed in the ideal arrangement, as well as their capacities, are determined by the optimization process. The model considers the restrictions related to the amount of freshwater that is available throughout the year. This factor influences the well completion schedule as well. When there are issues with water scarcity, the presented superstructure enables effective wastewater management that can conserve fluid and reuse it for extended periods. The presented method was used while taking into account various timetables and circumstances related to the Marcellus and Barnett fields.

The findings demonstrate the advantages of taking into account the presented methodology for wastewater management since it makes it possible to drill more

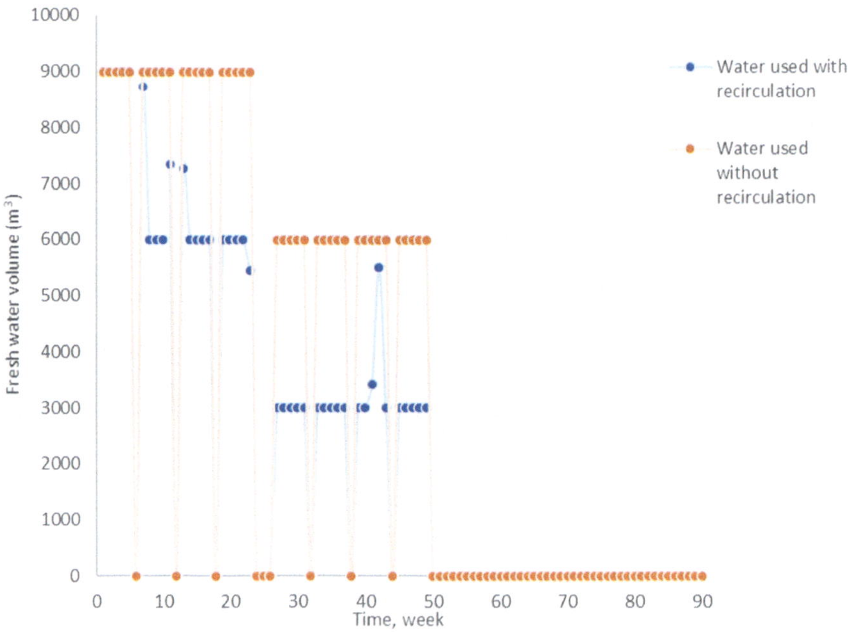

Fig. 15.6 Consumption of freshwater in scenario D. Printed with permission of "*Strategic planning for the optimal management of different types of shale gas wastewater*", *ACS Sustainable Chemistry and Engineering*, Copyright (2022), American Chemical Society

wells for less freshwater at a cheaper cost. When comparing the financial outcomes of the various scenarios, it is more practical for the three teams to work concurrently, finishing the wells right away (scenario D), rather than waiting for the drought to pass (weeks 24–26). Scenario D has the least amount of freshwater use in terms of the environment since it recycles as much water as possible. On the other hand, scenario C is the most water-reusing solution, but it also has the highest freshwater requirements overall. A and B are examples of intermediate solutions.

The presented method significantly lowers both the total cost (TC) and the total water requirements (TWR); in particular, scenario D reduces the freshwater requirements by 32.43% when 12.26% of the water produced in the wells is recycled. Furthermore, compared to the worst-case scenario (C), a total cost savings of $160,000 can be achieved. The trade-off graphic demonstrates that achieving both environmental and economic benefits is feasible. The findings back up the strategy for utilizing shale gas with minimal impact on the environment and cost savings.

Table 15.6 Wastewater at minimum *TC*

Classification	Storage	Treatment	Reuse	Total	% Reused
Scenario A					
Flowback (m^3)	57,397.63	4371.85	3847.23	61,769.48	6.23
Transition (m^3)	27,228.60	9143.40	6583.25	36,372.00	18.10
Produced (m^3)	10,686.20	9053.80	4707.98	19,740.00	23.85
Scenario B					
Flowback (m^3)	55,592.46	6176.94	5455.46	61,769.40	8.83
Transition (m^3)	27,946.80	8425.20	6065.98	36,372.00	16.68
Produced (m^3)	9752.40	9987.60	5193.55	19,740.00	26.31
Scenario C					
Flowback (m^3)	52,503.99	9265.41	8288.18	61,769.40	13.42
Transition (m^3)	26,128.20	10,243.80	7375.54	36,372.00	20.28
Produced (m^3)	8383.20	11,356.80	5905.54	19,740.00	29.92
Scenario D					
Flowback (m^3)	55,592.46	6176.94	5455.46	61,769.40	8.83
Transition (m^3)	29,542.80	6829.20	4917.02	36,372.00	13.52
Produced (m^3)	11,881.80	7858.20	4075.68	19,740.00	20.65

Printed with permission of "*Strategic planning for the optimal management of different types of shale gas wastewater*", *ACS Sustainable Chemistry and Engineering*, Copyright (2022), American Chemical Society

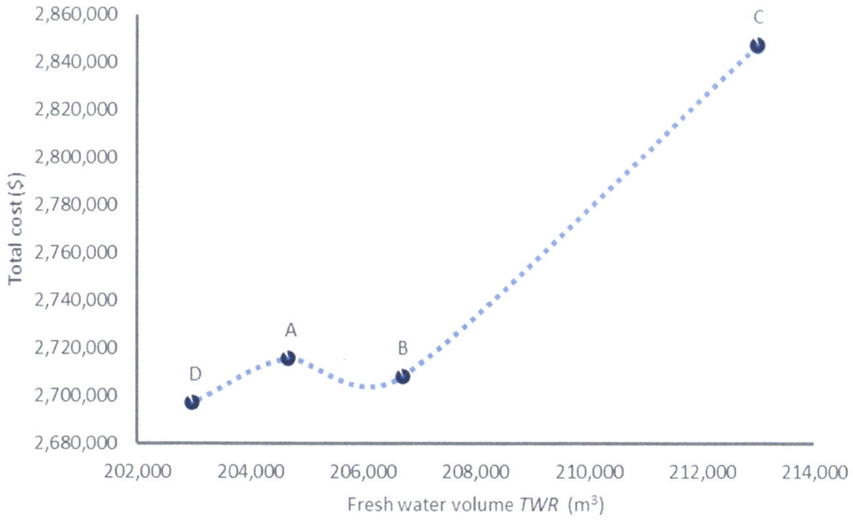

Fig. 15.7 Trade-off curve *TC* vs *TWR* for scenarios A-D. Printed with permission of "*Strategic planning for the optimal management of different types of shale gas wastewater*", *ACS Sustainable Chemistry and EngineeriSSng*, Copyright (2022), American Chemical Society

Table 15.7 Number of existing storage and treatment units

Concept	Scenario A	Scenario B	Scenario C	Scenario D
Existence of units				
Flowback treatment unit				
1	No	Yes	Yes	Yes
2	Yes	Yes	Yes	Yes
Transition treatment unit				
1	No	No	No	No
2	Yes	Yes	Yes	Yes
Produced treatment unit				
1	No	No	No	No
2	Yes	Yes	Yes	Yes
Number of units				
Flowback treatment units	1	2	2	2
Transition treatment units	1	1	1	1
Produced treatment units	1	1	1	1
Flowback storage units	1	1	1	1
Transition storage units	1	1	1	1
Produced storage units	1	1	1	1

Nomenclatures

Sets

A	{a | a is a reuse transition}
B	{b | b is a reuse flowback}
C	{c | c is a storage flowback}
D	{d | d is a storage transition}
E	{e | e is a reuse produced}
F	{f | f is a treatment unit flowback}
G	{g | g is a storage produced}
M	{m | m is a treatment unit transition}
N	{n | n is a well}
P	{p | p is a treatment unit produced}
T	{t | t is a time period}

Subscripts and Superscripts

a	Transition reuse
b	Flowback reuse
c	Flowback storage
d	Transition storage
e	Produced reuse
f	Flowback treatment unit
g	Produced storage
m	Transition treatment unit
n	Well
p	Produced treatment unit
t	Period

Binary Variables

$y_f^{treat-flowback}$	Binary variable used to model the existence of flowback treatment units
$y_m^{treat-trans}$	Binary variable used to model the existence of transition treatment units
$y_p^{treat-prod}$	Binary variable used to model the existence of produced treatment units
$y_c^{sto-flowback}$	Binary variable used to model the existence of flowback storage units
$y_d^{sto-trans}$	Binary variable used to model the existence of transition storage units
$y_g^{sto-prod}$	Binary variable used to model the existence of produced storage units

Variables

$CapCost^{treat-flowback}$	Capital cost for flowback treatment, \$
$CapCost^{treat-trans}$	Capital cost for transition treatment, \$
$CapCost^{treat-prod}$	Capital cost for produced treatment, \$
$CapCost^{storage-flowback}$	Capital cost for flowback storage, \$
$CapCost^{storage-trans}$	Capital cost for transition storage, \$
$CapCost^{storage-prod}$	Capital cost for produced storage, \$
$Cost^{fresh}$	Freshwater cost, \$
$Cost^{op-treat-flowback}$	Flowback treatment operating costs, \$
$Cost^{op-treat-trans}$	Transition treatment operating costs, \$

$Cost^{op-treat-prod}$	Produced treatment operating costs, \$
$Cost^{trans-fresh}$	Freshwater transportation cost, \$
$Cost^{trans-flowback}$	Cost of transporting water from well to flowback treatment, \$
$Cost^{trans-trans}$	Cost of transporting water from well to transition treatment, \$
$Cost^{trans-prod}$	Cost of transporting water from well to produced treatment, \$
$Cost^{trans-treat-reu-flowback}$	Cost of transporting water from treatment to flowback reuse, \$
$Cost^{trans-treat-reu-trans}$	Cost of transporting water from treatment to transition reuse, \$
$Cost^{trans-treat-reu-prod}$	Cost of transporting water from treatment to produced reuse, \$
$Cost^{trans-well-sto-flowback}$	Cost of transporting water from well to flowback storage, \$
$Cost^{trans-well-sto-trans}$	Cost of transporting water from well to transition storage, \$
$Cost^{trans-well-sto-prod}$	Cost of transporting water from well to produced storage, \$
$Cost^{trans-sto-well-flowback}$	Cost of transporting water from flowback storage to well, \$
$Cost^{trans-sto-well-trans}$	Cost of transporting water from transition storage to well, \$
$Cost^{trans-sto-well-prod}$	Cost of transporting water from produced storage to well, \$
$ff_{n,f,t}^{well-treat-flowback}$	Segregated flow from well to flowback treatment, m^3/day
$ff_{n,t}^{fresh}$	Segregated freshwater flow to wells, m^3/day
$ff_{c,n,t}^{storage-flowback}$	Segregated flow from the flowback storage system to wells, m^3/day
$ff_{d,n,t}^{storage-trans}$	Segregated flow from the transition storage system to wells, m^3/day
$ff_{g,n,t}^{storage-prod}$	Segregated flow from the produced storage system to wells, m^3/day
$ff_{f,b,t}^{treat-reu-flowback}$	Segregated flow from treatment to reuse flowback, m^3/day
$ff_{m,a,t}^{treat-reu-trans}$	Segregated flow from treatment to reuse transition, m^3/day
$ff_{p,e,t}^{treat-reu-prod}$	Segregated flow from treatment to reuse produced, m^3/day
$ff_{n,c,t}^{well-sto-flowback}$	Segregated flow from well to flowback storage system, m^3/day

$ff_{n,m,t}^{well-treat-trans}$	Segregated flow from well to transition treatment, m^3/day
$ff_{n,d,t}^{well-sto-trans}$	Segregated flow from well to storage system transition, m^3/day
$ff_{n,p,t}^{well-treat-prod}$	Segregated flow from well to treatment produced, m^3/day
$ff_{n,g,t}^{well-sto-prod}$	Segregated flow from well to storage system produced, m^3/day
$F_{f,t}^{treat-in-flowback}$	Flow entering the flowback treatment units, m^3/day
$F_{m,t}^{treat-in-trans}$	Flow entering the transition treatment units, m^3/day
$F_{p,t}^{treat-in-prod}$	Flow entering the treatment units produced, m^3/day
$F_{f,t}^{treat-out-flowback}$	Flow leaving the flowback treatment units, m^3/day
$F_{m,t}^{treat-out-trans}$	Flow leaving the transition treatment units, m^3/day
$F_{p,t}^{treat-out-prod}$	Flow leaving the treatment units produced, m^3/day
$F_{b,t}^{reu-flowback}$	Flow entering the water reuse flowback, m^3/day
$F_{a,t}^{reu-trans}$	Flow entering the transition water reuse, m^3/day
$F_{e,t}^{reu-prod}$	Flow entering the transition water produced, m^3/day
$F_{c,t}^{sto-in-flowback}$	Flow entering the flowback storage units, m^3/day
$F_{d,t}^{sto-in-trans}$	Flow entering the transition storage units, m^3/day
$F_{g,t}^{sto-in-prod}$	Flow entering the produced storage units, m^3/day
$F_{c,t}^{sto-out-flowback}$	Flow leaving the flowback storage units, m^3/day
$F_{d,t}^{sto-out-trans}$	Flow leaving the transition storage units, m^3/day
$F_{g,t}^{sto-out-prod}$	Flow leaving the produced storage units, m^3/day
F_t^{fresh}	Freshwater flow, m^3/day
$F_f^{treat-cap-flowback}$	Capacity flow of the flowback water treatment units, m^3/day
$F_m^{treat-cap-trans}$	Capacity flow of the transition water treatment units, m^3/day
$F_p^{treat-cap-prod}$	Capacity flow of the produced water treatment units, m^3/day
TWR	Total water requirements, m^3
TC	Total cost, $
TCC	Total capital cost, $
TOC	Total operating cost, $
TVC	Total sale of reused water, $
$V_{c,t}^{sto-flowback}$	Flowback water storage volume, m^3
$V_{d,t}^{sto-trans}$	Transition water storage volume, m^3
$V_{g,t}^{sto-prod}$	Produced water storage volume, m^3
$V_c^{sto-cap-flowback}$	Flowback water storage capacity, m^3
$V_d^{sto-cap-trans}$	Transition water storage capacity, m^3
$V_g^{sto-cap-prod}$	Produced water storage capacity, m^3

$\beta_{n,t}^{flowback}$ Flowback wastewater recovery factor
$\beta_{n,t}^{trans}$ Transition wastewater recovery factor
$\beta_{n,t}^{prod}$ Produced wastewater recovery factor

Parameters

$\mathrm{AvailV}_t^{fresh-max}$ Maximum water availability, m^3/week
$\mathrm{FC}^{storage-prod}$ Fixed charge for the produced storage unit in cost of capital function, \$
$\mathrm{FC}^{storage-trans}$ Fixed charge for the transition storage unit in cost of capital function, \$
$\mathrm{FC}^{treat-flowback}$ Fixed charge for the flowback treatment unit in cost of capital function, \$
$\mathrm{FC}^{treat-trans}$ Fixed charge for the transition treatment unit in cost of capital function, \$
$\mathrm{FC}^{treat-prod}$ Fixed charge for the produced treatment unit in cost of capital function, \$
$\mathrm{FC}^{storage-flowback}$ Fixed charge for the flowback storage unit in cost of capital function, \$
$\mathrm{F}_{n,t}^{well-in}$ Required inlet flow in wells, m^3/day
$\mathrm{F}_{n,t}^{well-out-flowback}$ Flowback water coming out of wells, m^3/day
$\mathrm{F}_{n,t}^{well-out-trans}$ Transition water coming out of wells, m^3/day
$\mathrm{F}_{n,t}^{well-out-prod}$ Produced water coming out of wells, m^3/day
$\mathrm{F}_f^{treat-max-flowback}$ Maximum flow of the flowback treatment units, m^3/day
$\mathrm{F}_m^{treat-max-trans}$ Maximum flow of the transition treatment units, m^3/day
$\mathrm{F}_p^{treat-max-prod}$ Maximum flow of the produced treatment units, m^3/day
H^{time} Time conversion factor, day/week
k_{Ffb} Factor used to annualize the investment of flowback water, $year^{-1}$
k_{Ftrans} Factor used to annualize the investment of transition water, $year^{-1}$
k_{Fprod} Factor used to annualize the investment of produced water, $year^{-1}$
UC^{fresh} Unit cost of freshwater, \$/m^3
$\mathrm{UOC}_f^{treat-flowback}$ Unit operating cost of flowback treatment units, \$/m^3
$\mathrm{UOC}_m^{treat-trans}$ Unit operating cost of transition treatment units, \$/m^3
$\mathrm{UOC}_p^{treat-prod}$ Unit operating cost of produced treatment units, \$/m^3
UTC_n^{fresh} Unit cost of transporting freshwater, \$/m^3
$\mathrm{UTC}_{n,f}^{used-flowback}$ Unit cost of transporting flowback water from wells to treatment, \$/m^3

$UTC_{n,m}^{used-trans}$	Unit cost of transporting transition water from wells to treatment, \$/m^3
$UTC_{n,p}^{used-prod}$	Unit cost of transporting produced water from wells to treatment, \$/m^3
$UTC_{f,b}^{treat-reu-flowback}$	Unit cost of transporting flowback water from treatment to reuse, \$/m^3
$UTC_{m,a}^{treat-reu-trans}$	Unit cost of transporting transition water from treatment to reuse, \$/m^3
$UTC_{p,e}^{treat-reu-prod}$	Unit cost of transporting produced water from treatment to reuse, \$/m^3
$UTC_{n,c}^{well-sto-flowback}$	Unit cost of transporting flowback water from wells to storage, \$/m^3
$UTC_{n,d}^{well-sto-trans}$	Unit cost of transporting transition water from wells to storage, \$/m^3
$UTC_{n,g}^{well-sto-prod}$	Unit cost of transporting produced water from wells to storage, \$/m^3
$UTC_{c,n}^{sto-well-flowback}$	Unit cost of transporting flowback water from storage to well, \$/m^3
$UTC_{d,n}^{sto-well-trans}$	Unit cost of transporting transition water from storage to well, \$/m^3
$UTC_{g,n}^{sto-well-prod}$	Unit cost of transporting produced water from storage to well, \$/m^3
$UVC_{b}^{reu-flowback}$	Unit cost of sale of flowback reuse water, \$/m^3
$UVC_{a}^{reu-trans}$	Unit cost of sale of transition reuse water, \$/m^3
$UVC_{e}^{reu-prod}$	Unit cost of sale of produced reuse water, \$/m^3
$V_{c}^{sto-max-flowback}$	Maximum volume for flowback storage, m^3
$V_{d}^{sto-max-trans}$	Maximum volume for transition storage, m^3
$V_{g}^{sto-max-prod}$	Maximum volume for produced storage, m^3
$V_{c}^{sto-initial-flowback}$	Initial volume of storage water flowback, m^3
$V_{d}^{sto-initial-trans}$	Initial volume of storage water transition, m^3
$V_{g}^{sto-initial-prod}$	Initial volume of storage water produced, m^3
$VC^{storage-prod}$	Unit variable charge for the produced storage unit in cost of capital function, \$/m^3
$VC^{treat-flowback}$	Unit variable charge for the flowback treatment unit in cost of capital function, \$/m^3
$VC^{treat-trans}$	Unit variable charge for the transition treatment unit in cost of capital function, \$/m^3
$VC^{treat-prod}$	Unit variable charge for the produced treatment unit in cost of capital function, \$/m^3
$VC^{storage-flowback}$	Unit variable charge for the flowback storage unit in cost of capital function, \$/m^3
$VC^{storage-trans}$	Unit variable charge for the transition storage unit in cost of capital function, \$/m^3

$\alpha_f^{treat-flowback}$ Volumetric efficiency factor of the flowback treatment unit

$\alpha_m^{treat-trans}$ Volumetric efficiency factor of the transition treatment unit

$\alpha_p^{treat-prod}$ Volumetric efficiency factor of the produced treatment unit

References

Al-Aboosi FY, El-Halwagi MM (2018) An integrated approach to water-energy nexus in shale gas production. Proceses 6(5):52. https://doi.org/10.3390/pr6050052

Al-Aboosi FY, El-Halwagi MM (2019) A stochastic optimization approach to the design of shale gas/oil wastewater treatment systems with multiple energy sources under uncertainty. Sustainability 11(18):4865–4906. https://doi.org/10.3390/su11184865

Al-Douri A, Sengupta D, El-Halwagi MM (2017) Shale gas monetization—a review of downstream processing to chemicals and fuels. J Nat Gas Sci Eng 45:436–455. https://doi.org/10.1016/j.jngse.2017.05.016

Arredondo-Ramírez K, Ponce-Ortega JM, El-Halwagi MM (2016) Optimal planning and infrastructure development for shale gas production. Energy Convers Manage 119:91–100. https://doi.org/10.1016/j.enconman.2016.04.038

Bai B, Carlson K, Prior A, Douglas C (2015) Sources of variability in flowback and produced water volumes from shale oil and gas wells. J Unconv Oil Gas Resourvoirs 12:1–5. https://doi.org/10.1016/j.juogr.2015.07.001

Cao K, Siddhamshetty P, Ahn Y, El-Halwagi MM, Kwon JS (2020) Evaluating the spatiotemporal variability of water recovery ratios of shale gas wells and their effects on shale gas development. J Clean Prod 276:123171. https://doi.org/10.1016/j.jclepro.2020.123171

EIA (2021) Energy Information Administration (EIA) annual energy outlook 2021. US Department of Energy, Washington, DC. https://www.eia.gov/pressroom/presentations/AEO2021_Release_Presentation.pdf. Accessed 20 Jun 2021

Estrada JM, Bhamidimarri R (2016) A review of the issues and treatment options for wastewater from shale gas extraction by hydraulic fracturing. Fuel 182:292–303. https://doi.org/10.1016/j.fuel.2016.05.051

Fernández JE, Alcon F, Diaz-Espejo A, Hernandez-Santana V, Cuevas MV (2020) Water use indicators and economic analysis for on-farm irrigation decision: a case study of a super high density olive tree orchard. Agric Water Manag 237:106074. https://doi.org/10.1016/j.agwat.2020.106074

Fry M (2013) Urban gas drilling and distance ordinances in the Texas Barnett Shale. Energy Policy 62:79–89. https://doi.org/10.1016/j.enpol.2013.07.107

Haluszczak LO, Rose AW, Kump LR (2013) Geochemical evaluation of flowback brine from Marcellus gas wells in Pennsylvania USA. Appl Geochem 28:55–61. https://doi.org/10.1016/j.apgeochem.2012.10.002

Hasaneen R, El-Halwagi MM (2017) Integrated process and microeconomic analyses to enable effective environmental policy for shale gas in the United States. Clean Technol Environ Policy 19(6):1775–1789. https://doi.org/10.1007/s10098-017-1366-5

Hayes T (2009) Sampling and analysis of water streams associated with the development of marcellus shale gas. Marcellus Shale Initiative Publications Database, 10. https://www.water-research.net/naturalgasPA/pdffiles/MSCommission-Report.pdf. Accessed 15 May 2021

Hernández-Pérez LG, Lira-Barragán LF, Ponce-Ortega JM (2020) Hybrid multiobjective optimization using deterministic and metaheuristic techniques for flowback water reusing in hydraulic fracturing processes. Ind Eng Chem Res 59:15298–15308. https://doi.org/10.1021/acs.iecr.0c02867

Herrera-Pantoja M, Hiscock KM (2015) Projected impacts of climate change on water availability indicators in a semi-arid region of central Mexico. Environ Sci Policy 54:81–89. https://doi.org/10.1016/j.envsci.2015.06.020

Laguna-Martinez MG, Garibay-Rodriguez J, Rico-Ramirez V, Castrejon-Gonzalez EO, Ponce-Ortega JM (2020) Water impact of an optimal natural gas production and distribution system: an MILP model and the case-study of Mexico. Chem Eng Res Des 153:887–906. https://doi.org/10.1016/j.cherd.2019.11.028

Lira-Barragán LF, Martinez-Gomez J, Ponce-Ortega JM, Serna-González M, El-Halwagi MM (2016a) Optimal reuse of flowback wastewater in shale gas fracking operations considering economic and safety aspects. Comput Aided Chem Eng 38:943–948. https://doi.org/10.1016/B978-0-444-63428-3.50162-4

Lira-Barragán LF, Ponce-Ortega JM, Guillén-Gosálbez G, El-Halwagi MM (2016b) Optimal water management under uncertainty for shale gas production. Ind Eng Chem Res 55(5):1322–1335. https://doi.org/10.1021/acs.iecr.5b02748

Lira-Barragán LF, Ponce-Ortega JM, Serna-González M, El-Halwagi MM (2016c) Optimal reuse of flowback wastewater in hydraulic fracturing including seasonal and environmental constraints. AIChE J 62(5):1634–1645. https://doi.org/10.1002/aic.15167

Serrano-Areválo TI, Lira-Barragán LF, Halwagi MM, Ponce-Ortega JM (2022) Strategic planning for optimal management of different types of shale gas wastewater. ACS Suatain Chem Eng 10:1451–1470

Manzanares-Rivera JL (2014) Uso de agua en la extracción de gas de lutitas en el noreste de México: Retos de regulación ambiental. Estudios sociales (Hermosillo, Son.) 22(44):172–197. http://www.scielo.org.mx/scielo.php?script=sci_arttext&pid=S0188-45572014000200007

Martinez-Gomez J, Napoles-Rivera F, Ponce-Ortega JM, El-Halwagi MM (2017) Optimization of the production of syngas from shale gas with economic and safety considerations. Appl Therm Eng 110:678–685. https://doi.org/10.1016/j.applthermaleng.2016.08.201

Oke D, Mukherjee R, Sengupta D, Majozi T, El-Halwagi MM (2020) On the optimization of water-energy nexus in shale gas network under price uncertainties. Energy 203:117770. https://doi.org/10.1016/j.energy.2020.117770

Oke D, Mukherjee R, Sengupta D, Majozi T, El-Halwagi MM (2019) Optimization of water-energy nexus in shale gas exploration: from production to transmission. Energy 183:651–669. https://doi.org/10.1016/j.energy.2019.06.104

Phan TT, Paukert Vankeuren AN, Hakala JA (2018) Role of water-rock interaction in the geochemical evolution of Marcellus Shale produced waters. Int J Coal Geol 191:95–111. https://doi.org/10.1016/j.coal.2018.02.014

SENER (2018) Balance general de energía. https://www.gob.mx/cms/uploads/attachment/file/528054/Balance_Nacional_de_Energ_a_2018.pdf. Accessed 20 Jun 2021

Serrano-Arévalo TI, Juárez-García M, Ponce-Ortega JM (2020) Optimal planning for satisfying future electricity demands involving simultaneously economic, emissions, and water concerns. Process Integr Optim Sustain 4:379–389. https://doi.org/10.1007/s41660-020-00125-8

Shaffer DL, Arias-Chavez LH, Ben-Sasson M, Romero-Vargas Castrillón S, Yip NY, Elimelech M (2013) Desalination and reuse of high salinity shale gas produced water: drivers, technologies, and future directions. Environ Sci Technol 47(17):9569–9583. https://doi.org/10.1021/es401966e

Slutz JA, Anderson JA, Broderick R, Horner PH (2012) Key shale gas water management strategies: an economic assessment tool. In: SPE/APPEA International conference on health, safety, and environment in oil and gas exploration and production. Perth, Australia. https://doi.org/10.2118/157532-MS

Sun Y, Wang D, Tsang DCW, Wang L, Ok YS, Feng Y (2019) A critical review of risks, characteristics, and treatment strategies for potentially toxic elements in wastewater from shale gas extraction. Environ Int 125:452–469. https://doi.org/10.1016/j.envint.2019.02.019

Tavakkoli S, Likare O, Vidic R, Khanna V (2020) Shale gas produced water management using membrane distillation: an optimization-based approach. Resour Conserv Recycl 158:104803. https://doi.org/10.1016/j.resconrec.2020.104803

Wang Y, Liu X, Liang L, Xiong J (2020) Experimental study on the damage of organic-rich shale during water-shale interaction. J Nat Gas Sci Eng 74:103103. https://doi.org/10.1016/j.jngse.2019.103103

Wood DA (2020) Brittleness index predictions from Lower Barnett Shale well-log data applying an optimized data matching algorithm at various sampling densities. Geosci Front 101087. https://doi.org/10.1016/j.gsf.2020.09.016

Xu J, Zhai C, Qin L, Liu S (2018) Pulse hydraulic fracturing technology and its application in coalbed methane extraction. Int J Oil, Gas Coal Technol 19(1):115–133. https://doi.org/10.1504/IJOGCT.2018.093962

Yang X, Liu S, Zhang L, Su J, Ye T (2020) Design and analysis of a renewable energy power system for shale oil exploitation using hierarchical optimization. Energy 206:118078. https://doi.org/10.1016/j.energy.2020.118078

Zhang X, Sun AY, Duncan IJ, Vesselinov VV (2017) Two-stage fracturing wastewater management in shale gas development. Ind Eng Chem Res 56(6):1570–1579. https://doi.org/10.1021/acs.iecr.6b03971

Zolfaghari A, Tang Y, Holyk J, Binazadeh M, Dehghanpour H, Bearinger D (2015) Advances in flowback chemical analysis of gas shales. Paper presented at the SPE annual technical conference and exhibition, Houston, Texas. https://doi.org/10.2118/175154-MS

Chapter 16
Optimizing the Water-Energy Relationship in Hydraulic Fracturing by Utilizing CO_2 and Water as Fracturing Agents

16.1 Introduction

The ongoing rise in energy consumption is currently one of the primary issues in social and economic development (Karavalakis et al. 2012). Global energy consumption is expected to rise by 28% between 2015 and 2040, according to the US. Energy Information Administration research (EIA 2012). Although there is annual progress in the development of renewable energies, soon the usage of non-renewable fuels (natural gas, carbon, and oil) will continue to be dominant and indispensable (Durand 2019). Since natural gas has a higher energy content (55.7 kJ/g if it is entirely based on methane as its main component) than coal (39.3 kJ/g), it is of particular interest among non-renewable resources (fossil resources). As a result, natural gas has a lower carbon footprint than coal, 30% less than oil and 50% less than coal. In addition, there are a lot of natural gas reserves in the world (BP 2019). As a result, natural gas consumption rose dramatically (by 24%) during the previous ten years, ending in 2017, and it is anticipated to overtake coal and oil as the primary energy source for the next 20 years (Zou et al. 2016). As a result, a significant amount of research and development has been done recently about the use of natural gas in a variety of industries, including transportation, industry, electricity, and domestic (Petrov et al. 2018). Nations all over the world have put in place methods and regulations for shale gas production because it has been a significant energy source for the past 10 years (Melikoglu 2014). Numerous researchers have published their methods for producing shale gas. There have been proposals to create networks for reusing water. These methods might be categorized as mathematical, algebraic, or graphical approaches. Almutlaq et al. (2005) and Qin et al. (2004) have shown a few instances of algebraic techniques. Numerous academics have documented that the majority of mathematical approaches use mathematical programming and simulation techniques that approximate the real value. To optimize the project's net present value, Cafaro and Grossmann (2014) presented a model for planning the shale gas supply chain. This model helped determine the quantity of freshwater needed for hydraulic fracturing

J. M. Ponce-Ortega et al., *Sustainable Water Systems*, Water Resources Development and Management, https://doi.org/10.1007/978-3-031-72498-5_16

and drilling, the size of processing plants, the number of wells, the sections and length of pipes for collecting crude gas, and the supply of processed gas. Taking into account the unpredictability of the shale gas supply rate, Wang and Xu (2014) suggested a new methodology for the simultaneous operation of shale gas, NGL recovery, and LNG regasification. A mathematical description of the well programming utilizing Lagrangian relaxation was published by Knudsen et al. (2014). Under uncertainty, Gao and You (2015) devised and planned the drilling, production, processing, and transportation of the shale gas supply.

Fracking, also referred to as hydraulic fracturing, is a process used to extract oil and shale gas from unconventional sources, usually shale formations (Fig. 16.1). For every shale gas well, hydraulic fracturing requires 1–3 million gallons of water; in certain situations, this amount might reach up to 5 million gallons (Anderson and Theodori 2009). Following each pumping stage, the pressure is released, allowing the generated water to return to the surface together with the natural gas flowback. The water in the subsurface is referred to as "produced water", whereas the injected water that is retrieved after the hydraulic fracturing process is known as "flowback water" (Veil 2010). Total dissolved solids (TDS) and other pollutants are typically present in high concentrations in both generated and flowback waters. Issues regarding the negative consequences of hydraulic fracturing have been raised by the scientific community in several nations, including the US, Canada, South Africa, Australia, France, Spain, and England (Zeng et al. 2013). These issues primarily relate to the consumption and contamination of water. Different issues arise as a result, and they require excessive attention. These include water-related issues associated with shale gas production, like meeting the water requirements for each extraction stage, collecting flowback water, designing and operating water treatment and storage equipment optimally, organizing wastewater streams optimally, and, lastly, optimizing transportation duties. Reusing treated and stored flowback water can lower the amount of freshwater and wastewater utilized, as well as the associated expenses.

Water management and energy production system design and planning are inherently fraught with uncertainty. As a result, this feature has received a lot of attention while creating water networks. A comprehensive model for the Marcellus shale play supply chain design was created in 2018 by Chen et al. (2018). It takes uncertainty in life cycle economic and environmental performance into account. A two-stage stochastic mixed-integer linear fractional programming model was introduced by Gao and You (2015) to maximize the levelized cost of energy produced from shale gas. This model takes into account decisions about digging shale wells, producing shale gas, processing, having numerous uses, and transportation, as well as design and planning considerations. A methodology for optimization was also published by Zhang et al. (2016) to assess various wastewater treatment and disposal choices for water management during hydraulic fracturing (HF) operations. This methodology accounts for system uncertainty as well as cost-effectiveness. A methodology for the sustainable use of water under parametric uncertainty in macroscopic water networks was proposed by Napoles-Rivera et al. (2015). A mathematical programming formulation for the best management of water in shale gas wells was published by Lira-Barragán et al. (2016). This formulation takes into account the possibilities

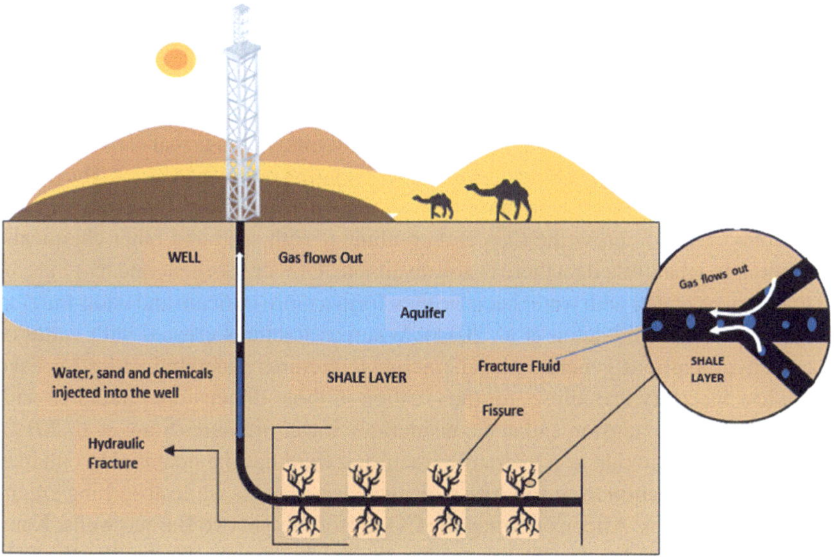

Fig. 16.1 Schematic representation of the hydraulic fracture process. Printed with permission of "*Optimization of the water-energy nexus in hydraulic fracturing processes using CO_2 and water as fracturing fluids*", *ACS Sustainable Chemistry and Engineering*, Copyright (2022), American Chemical Society

for treatment, storage, reuse, and disposal as well as the formation of flowback water. Uncertainty was included in a methodical approach for water distribution systems by Broad et al. (2015). On the other hand, Khor et al. (2014) integrated risk management and the synthesis of a water network under uncertainty. In addition, Ahn and Kang (2014) presented a long-term strategy for a sustainable water supply chain that takes future fluctuations in source availability and demand into account. Islam et al. (2014) looked at traditional water supply systems' uncertainties. Yang et al. (2014) created a method for water networks that accounts for the uncertainty in the enormous pollutant load as well as realistic and unit-specific shortcut models. A mixed-integer nonlinear programming model was suggested by Lira-Barragán et al. (2011a) to determine the best location for a new industrial plant that will have an impact on the water quality of the surrounding watershed. Furthermore, two mathematical programming programs with water networks as the primary criterion were reported by Lira-Barragán et al. (2011b, 2013). The first model presented a strategy for the best possible synthesis of recycling and reusing networks while simultaneously taking into account the integration of the surrounding watershed and the water network system to meet process and environmental constraints. The second model suggested a method for taking water usage and discharge into account when choosing installation sites for industrial facilities.

The scientific community has searched for alternate fracture fluids that can replace water entirely or in part because of the high water consumption and all the issues that come with it. This will improve the hydraulic fracturing process and lessen its negative effects on the environment. One option to partially or even entirely replace water in fracture fluid applications is to employ carbon dioxide. The San Antonio International firm cracked another unconventional well in Agueda del Cajon, Neuquén, according to the Petroquímica magazine. The main innovation is that carbon dioxide (CO_2) was utilized this time to fracture the gas formation rather than water. The procedures for CO_2 fracturing are the same as those for water fracturing, but different tools are employed to compress the CO_2 and combine it with sand and other chemicals. According to what little data there is now available, CO_2 cracks generate more gas or oil than those created with water because they form a more intricate and wide-ranging fracture network (Middleton et al. 2014). Whereas fractures created with water or viscous hydrocarbons generate planar fractures, fractures generated with CO_2 have a very low viscosity that allows for the creation of three-dimensional fractures with a greater vertical extension and at a considerably lower pressure (Song et al. 2019). Injecting carbon dioxide as a hydraulic fracturing fluid has the potential to enhance hydrocarbon extraction from soil-based oil and gas resources while absorbing greenhouse gas emissions. After examining the CO_2 adsorption in two Barnett wells, Kang et al. (2011) showed that the CO_2 adsorption was five to ten times higher than the CH_4 adsorption, and that both adsorptions were adjusted to the Langmuir equation. Shale gas reserves are therefore promising places to store CO_2. Based on the premise that the carbon dioxide matrix transport is a viscous fluid, Dahaghi (2010) and Schepers et al. (2009) employed COMET3 and Eclipse to simulate CO_2 injection reservoirs in the Devonian shale gas wells of Eastern Kentucky. Figure 16.2 illustrates CO_2 absorption and management in shale gas.

On the other hand, new studies to minimize CO_2 emissions in the atmosphere and its reuse in other processes have made carbon dioxide capture and storage crucial topics because of the significant influence that increased CO_2 emissions have on global warming. Fuss et al. (2008) provided an illustration of market uncertainty using a mathematical model, specifically focusing on changes in the trajectory of the CO_2 price and policy uncertainty, which have the potential to significantly alter the CO_2 price and impact investment choices in the power industry. At three scales: globally, national (Canada), and provincial (Alberta), Gunter et al. (1998) gave an initial look at big carbon-dioxide sinks about their capacities, retention/residence periods, rates of uptake, and relative costs of utilization. Numerical studies of CO_2 storage with enhanced CH_4 recovery in large stratified reservoirs were developed by Luo et al. (2013). The study neglected the absence of an aqueous phase in the reservoir and concentrated on the impact of injection/production well perforation placement and reservoir vertical permeability heterogeneity. Numerical studies of CO_2 storage with enhanced CH_4 recovery (CSEGR) in large stratified reservoirs were developed by Luo et al. (2013). The study neglected the absence of an aqueous phase in the reservoir and concentrated on the impact of injection/production well perforation placement and reservoir vertical permeability heterogeneity.

It is crucial to note that, as of right now, no prior research has used an optimization strategy to test the viability of using a water-carbon dioxide mixture as a fracturing fluid. The novel aspect of this chapter is, therefore, the mathematical programming

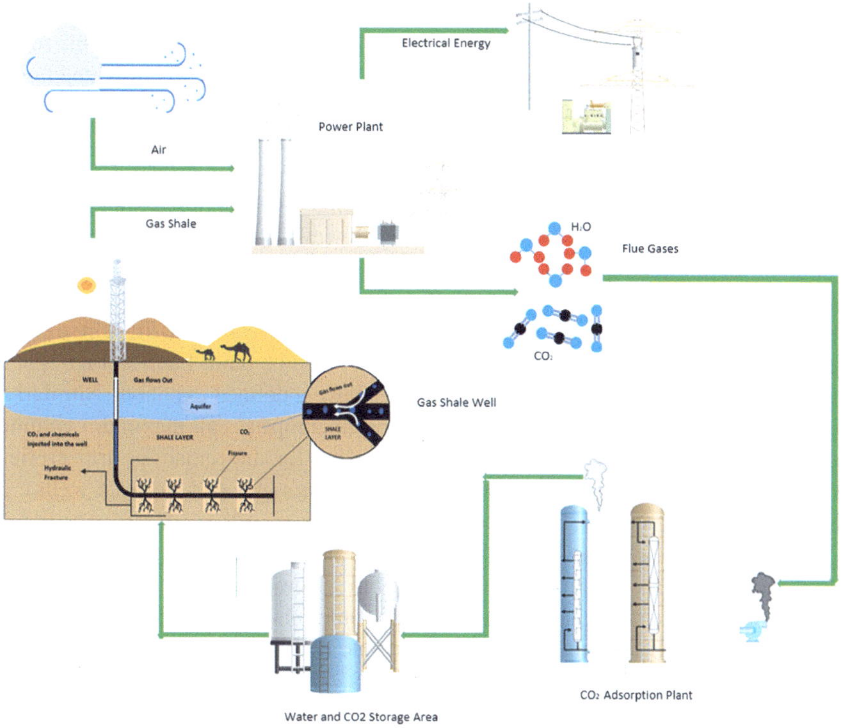

Fig. 16.2 CO_2 management in shale gas including CO_2 adsorption. Printed with permission of "*Optimization of the water-energy nexus in hydraulic fracturing processes using CO_2 and water as fracturing fluids*", *ACS Sustainable Chemistry and Engineering*, Copyright (2022), American Chemical Society

model that it suggests for the best possible control of the amount of water and CO_2 to be utilized as fracturing fluid in shale gas wells. Notably, a power station that employs shale gas as fuel to produce electricity, CO_2, and water also provides the generated water and CO_2. This makes the process cyclical. The availability of water and CO_2, flowback water, CO_2 adsorption methods, water treatment, disposal, and storage units are all taken into consideration in this model (Ferreyra-Quiroz et al. 2022).

16.2 Problem Statement

This chapter presents a mathematical approach for the completion phase of a collection of wells that will be completed simultaneously by one or more fracturing crews, depending on a particular scheduled time horizon. Three distinct situations for the well-completion phase of multiple sets are taken into account in this model. The known parameters are the fracture fluid, the flowback water, and the operation time. Freshwater expenses are contingent upon seasonal constraints and geographic location, including unit and transit costs. Additionally, flowback water is provided treatment technologies; nevertheless, these technologies must meet specific requirements to be reused or disposed of in the environment in the end. Every treatment unit has a unit operating cost that is linked to the maximum flowrate that may be treated in each unit, as well as fixed and variable costs for the capital cost function (which accounts for the exponent in the economies of scale). However, the technology needed to adsorb CO_2 from power plant exhaust gas comes at a cost when it comes to CO_2 capture. CO_2 takes into consideration the maximal adsorption capacity, fixed and variable costs, and unit operating expenses in addition to the freshwater. To store treated water and adsorbed CO_2, sets of storage/pits must also be built. Each storage unit's maximum volume and the capital cost function have a defined exponent. Similarly, a collection of disposals is provided; each disposal comprises all maximum volume restrictions as well as capital cost expenses. Lastly, for the remaining trajectories (wells-treatments, treatments-disposals, treatments-storage, and storages-wells), a transportation cost needs to be factored in.

The approach can obtain the optimal configuration for the water network associated with shale gas operations integrated with a power plant in terms of water, CO_2, and shale gas, taking into account all possible interconnections among them. The proposed MINLP formulation can determine the freshwater consumption, the amount of CO_2, and the total annual cost (TAC). The fracturing fluid is a mixture of water and CO_2, but to allow all of the wells to reach the completion phase while also lowering the overall annual cost, it is required to calculate the percentage of H_2O–CO_2 in the feed that will be pumped into the wells. However, it is commonly recognized that using CO_2 is more costly than using freshwater due to the equipment used to capture CO_2 and combine it with the compression process. Therefore, the TAC will significantly increase if a high percentage of CO_2 is introduced into the fracture fluid. Freshwater availability is crucial to the project in this context because it determines how much CO_2 is required for hydraulic fracturing in the event of limited freshwater availability. Lastly, three distinct scenarios are examined, each with varying water availability.

16.3 Mathematical Model Formulation

The superstructure of the process is directly tied to the model formulation. All of the processing units and the flow separation involved in each of the shale gas activities are shown in Fig. 16.3. The primary process of hydraulic fracturing, seen in Fig. 16.3, involves the injection of freshwater, produced water, and CO_2 into the well together with fracking fluid. The flowback water is transported to the treatment unit after this operation. The power plant's generated water and CO_2 are also routed to the treatment and adsorption units; the water, on the other hand, is transported to the storage system if it meets the requirements for reuse, and to its ultimate disposal otherwise.

Counting the number of wells for a given field and a time horizon, previous works have calculated the water used to complete each well. Then, to develop this methodology, this chapter considers that the estimated water needed for each well will be the same for the mixture of CO_2–water. Therefore, the flowrate to the well n in the time t ($FT_{n,t}^{well\_in}$) is equal to the sum of the water and CO_2 flowrates entering the well ($Fw_{n,t}^{well\_in}$ and $FCO_{2n,t}^{well\_in}$). On the other hand, at the end of the fracturing process, the flowback water and CO_2 will leave the well; thus, the rates of return from well n during period t are represented by $Fw_{n,t}^{well\_out}$ and $FCO_{2n,t}^{well\_out}$. It should be noted that the flowback water can be treated and stored to be reused or sent to final disposal whereas the CO_2 is released into the environment (Fig. 16.2). The presented optimization approach involves a set of mathematical relationships to simulate all the processes needed by the shale gas extraction using water and CO_2 as fracture fluid, and the

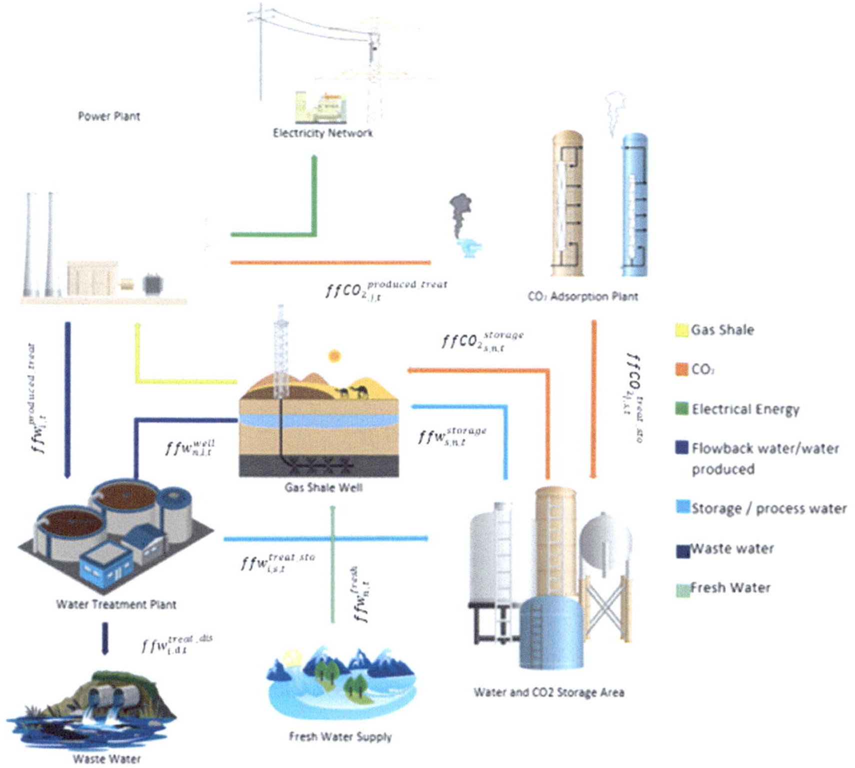

Fig. 16.3 Superstructure for hydraulic fracturing involving H_2O and CO_2 as fracture fluid. Printed with permission of "*Optimization of the water-energy nexus in hydraulic fracturing processes using CO_2 and water as fracturing fluids*", *ACS Sustainable Chemistry and Engineering*, Copyright (2022), American Chemical Society

model can consider freshwater constraints and CO_2 availabilities, relationships will be needed to meet these limitations. The mathematical mode is presented as follows.

16.3.1 Availability Constraints for Freshwater and CO_2

The main problem and limitation for the shale gas industry is the accessibility of the freshwater requiered to complete each well, principally in regions with major water shortage problems. This way, a significant contribution of this chapter is that it accounts for the water and the CO_2 available to be employed during the hydraulic fracturing phase of each well.

$$Fw_t^{fresh} \leq \left(\frac{AvailVw_t^{fresh\_max}}{H^{time}} \right), \quad \forall t \tag{16.1}$$

Fw_t^{fresh} is the freshwater consumption over a period t, $AvailVw_t^{fresh\_max}$ represents the freshwater availability over a period t and H^{time} corresponds to a time conversion factor. Likewise, CO_2 availability is given as the next relationship:

$$FCO_{2t}^{used} \leq \left(\frac{AvailVCO_{2t}^{used\_max}}{H^{time}} \right) \quad \forall t \tag{16.2}$$

FCO_{2t}^{used} is the used CO_2 by power plant over a period t, $AvailVCO_{2t}^{used\_max}$ represents CO_2 availability over a period t and H^{time} corresponds to a time conversion factor.

16.3.2 Availability of Generated Water

The water generated by the power plant is sent to the treatment plant:

$$Fw_t^{generated} \leq \left(AvailVw_t^{generated\_max} \right) \quad \forall t \tag{16.3}$$

$Fw_t^{generated}$ is the water generated by power plant over a period t, $AvailVw_t^{generated\_max}$ represents the generated water availability over a period t.

16.3.3 Split Freshwater Generated Water and CO_2

Freshwater is split and sent to each well n whereas the generated water and CO_2 are segregated and sent to treatment plants i, adsorption plants j, for the completion process.

$$Fw_t^{fresh} = \sum_n ffw_{n,t}^{fresh}, \quad \forall t \tag{16.4}$$

$$FCO_{2t}^{used} = \sum_j ffCO_{2j,t}^{used}, \quad \forall t \tag{16.5}$$

$$Fw_t^{generated} = \sum_i ffw_{i,t}^{generated}, \quad \forall t \tag{16.6}$$

16.3.4 Water and CO_2 Provided to Wells

The required water during the hydraulic fracturing stage in each well n ($Fw_{n,t}^{well\_in}$) are provided by freshwater ($ffw_{n,t}^{fresh}$) plus the water coming from the storage system ($ffw_{sw,n,t}^{storage}$).

$$Fw_{n,t}^{well\_in} = ffw_{n,t}^{fresh} + \sum_{sw} ffw_{sw,n,t}^{storage}, \quad \forall n, \ \forall t \tag{16.7}$$

Similarly, the CO_2 required by each well n ($FCO_{2n,t}^{well\_in}$) is expressed as the CO_2 source coming from the storage system ($ffCO_{2sg,n,t}^{storage}$).

$$FCO_{2n,t}^{well\_in} = \sum_{sg} ffCO_{2sg,n,t}^{storage}, \quad \forall n, \ \forall t \tag{16.8}$$

16.3.5 Flowback Water

After the completion stage is finished, flowback water is acquired from each well ($Fw_{n,t}^{well\_out}$), which is split and sent to a set of treatment units ($ff_{n,i,t}^{well}$).

$$Fw_{n,t}^{well\_out} = \sum_i ffw_{n,i,t}^{well}, \quad \forall n, \ \forall t \tag{16.9}$$

It should be noted that all flowback water is treated, avoiding the storage or disposal of the untreated hydraulic fracturing fluid.

16.3.6 Water and CO_2 Inlet to Treatment and Adsorption Units

The model involves an interception network made up of several treatment units typically used to treat hydraulic fracturing fluid. Similarly, scientific reports with relevant information on the treatment units used in existing shale gas fields can be found in the literature. Nonetheless, the requirement to consider in this model for any treatment technology is that the outflows present a proper quality and the appropriate conditions to be reused or disposed of, this will avoid the generation of environmental problems or technical problems in the process.

Also, this approach includes adsorption units to capture the CO_2, generally considered to treat flue gases from power plants and the constraints considered for this technology are the state of matter (liquid) and purity of CO_2 (90%). Accordingly, all efficiently proven technologies for treating flowback water and capturing CO_2 can be considered even if they involve different costs and capacities. Lastly, the model can select the required technologies (if it involves different units) and determine the number of treatment/adsorption technologies and size for each one.

After the flowback water flow is split, the balance for the mixer at the inlet of each treatment unit founds that the inlet flow for unit i during the period t ($Fw_{i,t}^{treat\_in}$) is provided by the sum of the flowback water aplit at the outlet of the wells at the same time and sent to the unit ($ffw_{n,i,t}^{well}$) plus the flow of generated water in the power plant ($ffw_{i,t}^{generated}$).

$$Fw_{i,t}^{treat\_in} = \sum_{n} ffw_{n,i,t}^{well} + \sum_{i} ffw_{i,t}^{generated}, \quad \forall i, \ \forall t \tag{16.10}$$

Likewise, the CO_2 balance for the mixer at the inlet of each adsorption unit states that the inlet flow for unit j during the period t ($FCO_{2j,t}^{treat\_in}$) is given as the sum of CO_2 in the flue gas stream produced by the power plant ($ffCO_{2j,t}^{used}$).

$$FCO_{2j,t}^{treat\_in} = \sum_{j} ffCO_{2j,t}^{used}, \quad \forall j, \ \forall t \tag{16.11}$$

16.3.7 Water and CO_2 Exit from Treatment/Adsorption Units

Based on the presentedsuperstructure, there is a splitter at the output of each treatment unit *i*, where the output stream ($Fw_{i,t}^{treat\_out}$) should be sent to the storage units ($ffw_{i,s,t}^{treat\_sto}$) or final disposal ($ffw_{i,d,t}^{treat\_dis}$) during period *t*:

$$Fw_{i,t}^{treat\_out} = \sum_{sw} ffw_{i,sw,t}^{treat\_sto} + \sum_{d} ffw_{i,d,t}^{treat\_dis}, \quad \forall i, \ \forall t \tag{16.12}$$

Likewise, there is a CO_2 output at the adsorption unit *j*, where the exit stream ($FCO_{2j,t}^{treat\_out}$) is directed to storage units ($ffCO_{2j,sg,t}^{treat\_sto}$).

$$FCO_{2j,t}^{treat\_out} = \sum_{sg} ffCO_{2j,sg,t}^{treat\_sto}, \quad \forall j, \ \forall t \tag{16.13}$$

16.3.8 Water Balance for Waste Disposal and Discharge

The wastewater streams generated by the project ($Fw_{d,t}^{waste}$) come from the interception network, where the water quality complies with environmental regulations for pollutants:

$$Fw_{d,t}^{waste} = \sum_{i} ffw_{i,d,t}^{treat\_dis}, \quad \forall d, \ \forall t \tag{16.14}$$

16.3.9 Water and CO_2 Inlet to Storage Units

The storage water flowrate ($Fw_{s,t}^{storage\_in}$) is fed into the storage system by the waste streams treated in the interception network, whereas the storage flowrate for CO_2 ($FCO_{2sg,t}^{storage\_in}$) is supplied by the CO_2 adsorbed in the adsorption unit, fulfilling the quality requirements for their reuse:

$$Fw_{sw,t}^{storage\_in} = \sum_{i} ffw_{i,sw,t}^{treat\_sto}, \quad \forall s, \ \forall t \tag{16.15}$$

$$FCO_{2sg,t}^{storage\_in} = \sum_{j} ffCO_{2j,sg,t}^{treat\_sto}, \quad \forall s, \ \forall t \tag{16.16}$$

16.3.10 Water and CO_2 Outlet from Storage Units

After the water and CO_2 are stored, the exit streams ($ffw_{sw,n,t}^{storage} and ffCO_{2sg,n,t}^{storage}$) from the storage units *sw* and *sg*, for water and CO_2 respectively, over a period *t* are given as follows:

$$Fw_{sw,t}^{storage\_out} = \sum_{n} ffw_{sw,n,t}^{storage}, \quad \forall s, \ \forall t \tag{16.17}$$

$$FCO_{2sg,t}^{storage\_out} = \sum_{n} ffCO_{2sg,n,t}^{storage}, \quad \forall s, \ \forall t \tag{16.18}$$

16.3.11 Balances in Treatment and Adsorption Units

Many treatment units involve water losses during their operation, whereas adsorption units accounts for the amount of CO_2 that can be adsorbed, these two aspects can be modeled as follows:

$$Fw_{i,t}^{treat\_out} = \alpha w_i^{treat} Fw_{i,t}^{treat\_in}, \quad \forall i, \ \forall t \tag{16.19}$$

$$FCO_{2j,t}^{treat\_out} = \alpha CO_{2j}^{treat} FCO_{2j,t}^{treat\_in}, \quad \forall j, \ \forall t \tag{16.20}$$

αw_i^{treat} and αCO_{2j}^{treat} are the efficiency factors in volumetric terms for units *i* and *j*, respectively.

16.3.12 Water and CO_2 Balances in Storage Units

The usual development of a well platform includes the completion stage where the hydraulic fracturing process takes place. This step requires the injection of large quantities of water and CO_2 into the wells as a hydraulic fracturing fluid, which takes between 1 and 3 months. Once this stage is finished, a significant part of the total injected water is obtained in the wells as return water, while the CO_2 is released into the environment. Flowback water can be reused in hydraulic fracturing processes to reduce freshwater consumption; thus, this stream must be treated and stored. Therefore, considering a schedule for the completion phases of a set of wells implemented by various hydraulic fracturing equipment, the importance of storage consists of saving the treated flowback water and CO_2 adsorbed to be used during the periods where the fracturing process is implemented; this water also represents a backup resource for periods with low accessibility of freshwater.

This way, the accumulation balances for water in the storage pits/impoundments stated that the volume in the pit sw at time t ($Vw_{sw,t}^{storage}$) is equal to the one at end of the previous period ($Vw_{sw,t-1}^{storage}$), plus the difference of the inlet and outlet flows multiplied by the time conversion factor:

$$Vw_{sw,t}^{storage} = Vw_{sw}^{storage\_initial} + H^{time}\left(Fw_{sw,t}^{storage\_in} - Fw_{sw,t}^{storage\_out}\right), \quad \forall sw, \quad \forall t = 1 \tag{16.21}$$

$$Vw_{sw,t}^{storage} = Vw_{sw,t-1}^{storage} + H^{time}\left(Fw_{sw,t}^{storage\_in} - Fw_{sw,t}^{storage\_out}\right), \quad \forall sw, \quad \forall t > 1 \tag{16.22}$$

For the first stage, a given value of the initial volume of the pit is considered. Furthermore, it is needed to assure the continuity of the cycles, this way, the initial volume is equal to the last volume in the well/storage sw:

$$Vw_{sw}^{storage\_initial} = Vw_{sw,t}^{storage}, \quad \forall t = t^{final} \tag{16.23}$$

Likewise, for CO_2, the accumulation balance of the storage tanks states that the volume in the tank sg at period t $\left(VCO_{2sg,t}^{storage}\right)$ is equal to the one at the end of the previous period ($VCO_{2sg,t-1}^{storage}$), plus the difference of the inputs and outputs multiplied by the time conversion factor:

$$VCO_{2sg,t}^{storage} = VCO_{2sg}^{storage\_initial} + H^{time}\left(FCO_{2sg,t}^{storage\_in} - FCO_{2sg,t}^{storage\_out}\right), \quad \forall sg, \quad \forall = 1 \tag{16.24}$$

$$VCO_{2sg,t}^{storage} = VCO_{2sg,t-1}^{storage} + H^{time}\left(FCO_{2sg,t}^{storage\_in} - FCO_{2sg,t}^{storage\_out}\right), \quad \forall sg, \quad \forall t > 1 \tag{16.25}$$

Guaranteeing the continuity of the cycles, the initial volume is equal to the last volume in the well sg:

$$VCO_{2sg}^{storage\_initial} = VCO_{2sg,t}^{storage}, \quad \forall t = t^{final} \tag{16.26}$$

16.3.13 Capacity and Existence of Treatment and Adsorption Units

This optimization approach determines the number of units, technologies and final disposals for water treatment and CO_2 adsorption, and the size of each one. Furthermore, for treatment and adsorption units the following relationships select whether units i and j are needed or not (the existence of the treatment and adsorption units is simulated by the binary variables: yw_i^{treat} and yCO_{2j}^{treat}, thus, the operational capacity for each treatment ($Fw_i^{treat\_cap}$) and adsorption ($FCO_{2j}^{treat\_cap}$) unit is given as follows:

$$Fw_i^{treat\_cap} \geq Fw_{i,t}^{treat\_in}, \quad \forall i, \ \forall t \tag{16.27}$$

$$Fw_i^{treat\_cap} \leq Fw_i^{treat\_max} yw_i^{treat}, \quad \forall i \tag{16.28}$$

$$FCO_{2j}^{treat\_cap} \geq FCO_{2j,t}^{treat\_in}, \quad \forall j, \ \forall t \tag{16.29}$$

$$FCO_{2j}^{treat\_cap} \leq FCO_{2j}^{treat\_max} yCO_{2j}^{treat}, \quad \forall j \tag{16.30}$$

$Fw_i^{treat\_max}$ and *abo* are upper limits for the operational capacity associated with units i and j.

16.3.14 Capacity and Existence of Storage Units

Likewise, for the storage units, the next relationships are required:

$$Vw_{sw}^{storage\_cap} \geq Vw_{sw,t}^{storage}, \quad \forall s, \ \forall t \tag{16.31}$$

$$Vw_{sw}^{storage\_cap} \leq Vw_{sw}^{storage\_max} yw_{sw}^{storage}, \quad \forall s \tag{16.32}$$

$$VCO_{2sg}^{storage\_cap} \geq VCO_{2sg,t}^{storage}, \quad \forall s, \ \forall t \tag{16.33}$$

$$VCO_{2sg}^{storage\_cap} \leq VCO_{2sg}^{storage\_max} yCO_{2sg}^{storage}, \quad \forall s \tag{16.34}$$

$Vw_{sw}^{storage\_cap}$ and $VCO_{2sg}^{storage\_cap}$ are the optimal capacity for the storage units *sw* and *sg*. Similarly, $Vw_{sw}^{storage\_max}$ and $VCO_{2sg}^{storage\_max}$ are the maximum storage capacities for units *sw* and *sg*. The binary variables $yCO_{2sg}^{storage}$ and $y_s^{storage}$ are used to simulate the existence of storage units.

16.3.15 Capacity and Existence of Disposals

Lastly, the final disposals accumulate all the flowrate discharged on them during all the given time horizon. Therefore, the operational capacity for the disposal ($Vw_d^{waste\_cap}$) is calculated as follows:

$$Vw_d^{waste\_cap} = H^{time} \sum_t Fw_{d,t}^{waste}, \quad \forall d \tag{16.35}$$

$$Vw_d^{waste\_cap} \leq Vw_d^{waste\_max} y_d^{waste}, \quad \forall d \tag{16.36}$$

$Vw_d^{waste\_max}$ represents the maximum capacity for each disposal and the existence of each simulated is modeled by the binary variable y_d^{waste}. Therefore, it can be calculated the number of treatment technologies, storages/pits and disposals in the optimal solution and their capacities.

16.3.16 Total Water Used

Even though water constraints can be overcome, it is needed to calculate the total freshwater required to complete all proposed wells. Thus, the following relationship calculates the total freshwater needs (TWR):

$$TWR = H^{time} \left(\sum_t Fw_t^{fresh} \right) \tag{16.37}$$

16.3.17 Total CO_2 Used

The total CO_2 needed for the completion stage during a period t is calculated by the next equation (TCR)

$$TCR = H^{time} \sum_t FCO_{2_t}^{used} \tag{16.38}$$

Since current water concerns, it is imperative to determine the total water needs of the shale gas industry to exploit a given number of wells since from an environmental point of view the minimization of freshwater has greater relevance than the economic aspects. Therefore, the best economic solution usually involves to the minimum consumption of CO_2; nevertheless, for cases where freshwater accessibility is low, environmental and economic objectives may contradict each other, and as a consequence, Pareto curves should be useful to show solutions that compensate the contradictory objectives.

16.3.18 Operating Costs

The operating costs considered in this model correspond to the cost of freshwater, transportation costs (including all routes) and operating costs of the treatment/ adsorption units, which are presented in the next section.

16.3.18.1 Freshwater and CO_2 Costs

The cost of freshwater ($Costw^{fresh}$) is calculated by multiplying the sum of freshwater required during all periods times the freshwater unit cost (UCw^{fresh}) and the time conversion factor:

$$Costw^{fresh} = H^{time} UCw^{fresh} \sum_{t} Fw_{t}^{fresh} \tag{16.39}$$

CO_2 cost is obtained similarly as freshwater using the unit cost of CO_2 ($UCCO_2^{produced}$):

$$CostCO_2^{used} = H^{time} UCCO_2^{used} \sum_{t} FCO_{2t}^{used} \tag{16.40}$$

16.3.18.2 Operating Costs of the Treatment and Adsorption Uns

Treatment and adsorption operating costs are calculated as follows:

$$Costw^{op\_treat} = H^{time} \sum_{i} \sum_{t} UOCw_{i}^{treat} Fw_{i,t}^{treat\_in} \tag{16.41}$$

$$CostCO_2^{op\_treat} = H^{time} \sum_{j} \sum_{t} UOCCO_{2j}^{treat} FCO_{2j,t}^{treat\_in} \tag{16.42}$$

$UOCw_{i}^{treat}$ is the unit operating cost for the treatment unit i and $UOCCO_{2j}^{treat}$ for adsorption unit j.

16.3.18.3 Freshwater and CO_2 Transportation Costs

The freshwater transportation cost ($Costw^{trans\_fresh}$) should be obtained by multiplying the flow rates of freshwater sent to well n in period t times the unit transportation cost ($UTCw_n^{fresh}$):

$$Costw^{trans\_fresh} = H^{time} \sum_n \sum_t UTCw_n^{fresh} ffw_{n,t}^{fresh} \tag{16.43}$$

Furthermore, CO_2 transportation cost is obtained by multiplying the flowrates of CO_2 sent to well n from the storage tank sg in period t times the unit transportation cost:

$$CostCO_2^{trans\_sto} = H^{time} \sum_{sg} \sum_n \sum_t UTCCO_{2n}^{sto} ffCO_{2sg,n,t}^{storage} \tag{16.44}$$

The unit costs are different for each well considering the geographical distance between freshwater, CO_2 and the well. For some shale gas fields, the freshwater resource is available at substantial distances from the wells; thus, in cases like the Marcellus shale gas region, the transport of freshwater is several times greater than the cost of freshwater (Yang et al. 2014).

16.3.18.4 Flowback Water Transportation Cost

The flowback water obtained from the wells is transported to the interception network and the cost involved with this task is calculated through the following equation:

$$Costw^{trans\_fb} = H^{time} \sum_n \sum_i \sum_t UTCw_{n,i}^{used} ffw_{n,i,t}^{well} \tag{16.45}$$

$UTCw_{n,i}^{used}$ represents the unit cost of transporting the flowback water in the route selected for well n and treatment i. Thus, the unit cost may vary for each route.

16.3.18.5 Transportation Cost of the Treated Water to Disposals

The flow that exists the treatment should be sent to the disposals and the transport costs that are involved are the following:

$$Costw^{trans\_treat\_dis} = H^{time} \sum_i \sum_d \sum_t UTCw_{i,d}^{treat\_dis} ffw_{i,d,t}^{treat\_dis} \tag{16.46}$$

$UTCw_{i,d}^{treat\_dis}$ represents the unit transportation cost for the treated water from treatment i to disposal d.

16.3.18.6 Transportation Cost for Treated Water/CO_2 Adsorbed to Storage

Treated water and CO_2 adsorbed should be sent to the storage units considering the following costs:

$$Costw^{trans\_treat\_sto} = H^{time} \sum_{i} \sum_{sw} \sum_{t} UTCw_{i,sw}^{treat\_sto} ffw_{i,sw,t}^{treat\_sto} \quad (16.47)$$

$$CostCO_2^{trans\_treat\_sto} = H^{time} \sum_{j} \sum_{sg} \sum_{t} UTCCO_{2j,sg}^{treat\_sto} ffCO_{2j,sg,t}^{treat\_sto} \quad (16.48)$$

$UTCw_{i,sw}^{treat\_sto}$ is the unit transportation cost for treated water from treatment i to storage pits sw, whereas $UTCCO_{2j,sg}^{treat\_sto}$ is the unit transportation cost for CO_2 adsorbed from treatment j to storage tanks sg.

16.3.18.7 Transportation Cost for Reused Water to Wells

The water stored in the tanks/storages is sent to the wells for reuse and the involved cost is calculated for in the following equation:

$$Costw^{trans\_sto\_well} = H^{time} \sum_{sw} \sum_{n} \sum_{t} UTCw_{sw,n}^{sto\_well} ffw_{sw,n,t}^{storage} \quad (16.49)$$

$UTCw_{sw,n}^{sto\_well}$ represents the unit transporion costs for the reused water from storage area sw to well n.

16.3.18.8 Transportation Cost for Generated Water to the Treatment Unit

The flow of water leaving the power plant should be sent to the treatment unit and the transport costs involved are calculated as follows:

$$Costw^{trans\_pp\_treat} = H^{time} \sum_{i} \sum_{t} UTCw_{i}^{generated\_treat} ffw_{i,t}^{generated} \quad (16.50)$$

$TCw_{i}^{generated\_treat}$ represents the unit transportation cost for generated water by a power plant that is sent to treatment unit i.

16.3.18.9 Transportation Cost for Used CO_2 to the Adsorption Unit

Likewise, the unit transportation cost for CO_2 is determined as follows:

$$CostCO_2^{trans\_pp\_treat} = H^{time} \sum_j \sum_t UTCCO_{2,j}^{used\_treat} ffCO_{2,j,t}^{used} \tag{16.51}$$

$UTCCO_{2j}^{used\_treat}$ represents the unit transportation cost for CO_2 produced by the power plant that is sent to treatment unit j.

16.3.18.10 Total Supply to Wells

The fracture fluid needed (mixture of CO_2 and water) should be determined by the sum of the water plus the CO_2 that will be injected into the wells n during a period t.

$$FT_{n,t}^{well\_in} = Fw_{n,t}^{well\_in} + FCO_{2n,t}^{well\_in}, \quad \forall n, \ \forall t \tag{16.52}$$

16.3.18.11 Total Operating Cost

Lastly, the sum of all the operating costs presented above yields the total operating cost (TOC) for the project:

$$\begin{aligned} TOC = {} & Costw^{fresh} + Costw^{op\_treat} + Costw^{trans\_fresh} + Costw^{trans\_fb} \\ & + Costw^{trans\_treat\_dis} + Costw^{trans\_treat\_sto} + Costw^{trans\_sto\_well} \\ & + Costw^{trans\_pp\_treat} + CostCO_2^{used} + CostCO_2^{op\_treat} + CostCO_2^{trans\_sto} \\ & + CostCO_2^{trans\_treat\_sto} + CostCO_2^{trans\_pp\_treat} \end{aligned} \tag{16.53}$$

16.3.19 Capital Costs

The acquisition of treatment units, adsorption units, storage units, and the construction of final disposal generates capital costs, which are included in the economic objective function.

16.3.19.1 Capital Cost for Water Treatment Units and CO_2 Adsorption Units

The capital cost for the water treatment units includes a fixed charge (FCw^{treat}) and a variable charge ($VCw^{trea}t$) that depends on the flow treated, the cost is calculated as follows:

$$CapCostw^{treatment} = k_F w \sum_i \left[FCw^{treat} y_i^{treat} + VCw^{treat} \left(Fw_i^{treat\_cap} \right)^{\beta w^{treat}} \right] \quad (16.54)$$

k_{Fw} is a factor used to annualize the investment, whereas βw^{treat} is the exponent linked with economies of scale for the treatment units.

Furthermore, the capital cost for CO_2 is determined similarly as water:

$$CapCostCO_2^{treatment} = k_F CO_2 \sum_j \left[FCCO_2^{treat} y_j^{treat} + VCCO_2^{treat} \left(FCO_{2j}^{treat\_cap} \right)^{\beta CO_2^{treat}} \right] \quad (16.55)$$

16.3.19.2 Capital Cost for Water Storage, CO_2 Storage and Disposal Units

Correspondingly for the storage units and disposals, the following equations are needed:

$$CapCostw^{storage} = k_F w \sum_s \left[FCw^{storage} y_{sw}^{storage} + VCw^{storage} \left(Vw_{sw}^{storage\_cap} \right)^{\beta w^{storage}} \right] \quad (16.56)$$

$$CapCostw^{waste} = k_F w \sum_d \left[FCw^{waste} y_d^{waste} + VCw^{waste} \left(Vw_d^{waste\_cap} \right)^{\beta w^{waste}} \right] \quad (16.57)$$

$$CapCostCO_2^{storage} = k_F CO_2 \sum_s \left[FCCO_2^{storage} y_{sg}^{storage} + VCCO_2^{storage} \left(VCO_{2sg}^{storage\_cap} \right)^{\beta CO_2^{storage}} \right] \quad (16.58)$$

Here, βw^{treat}, $\beta w^{storage}$, βw^{waste}, βCO_2^{treat} and $\beta CO_2^{storage}$. are the only nonlinear terms included in the entire presented mathematical model.

16.3.19.3 Total Capital Cost

Finally, the total capital cost (TCC) is obtained by the sum of all the capital costs mentioned above (treatment units, adsorption units, storage units, and disposals).

$$TCC = CapCostw^{treatment} + CapCostCO_2^{treatment} + CapCostw^{storage} + CapCostw^{waste} + CapCostCO_2^{storage} \quad (16.59)$$

16.3.20 Total Annual Cost

The total annual cost (*TAC*) is calculated by the sum of total operating cost and total capital cost as follows:

$$TAC = TOC + TCC \quad (16.60)$$

16.4 Results and Discussion

The presented mathematical model is a mixed-integer nonlinear programming program (MINLP), which was resolved with the software General Algebraic Modeling System (GAMS). The model includes 16 binary variables, 15,291 continuous variables, and 7684 constraints. DICOPT, CONOPT, and CPLEX solvers were implemented within GAMS (Brooke et al. 2015).

Currently, shale gas production has grown significantly around the world, for example, the United States is the number one natural gas producer in the world having Marcellus (located mainly in Pennsylvania) and Barnett (located in Texas) as the largest shale gas reserves. Therefore, most of the information used to implement the presented model in this chapter was taken from technical reports for the Marcellus and Barnett regions. Mexico's estimated resource of technically recoverable shale gas (545 Tcf) is the 6th largest shale gas resource in the World. Most of Mexico's shale gas resource is in the Eagle Ford Shale, which is a mature shale gas and oil play in the US. Nonetheless, the geographical location chosen for this work was the Northwest region in Mexico (Ruud et al. 2017). Concerning this chapter, one of the main objectives is water management, which is optimized in periods of a week with a time horizon of 52 weeks (a year), and the completion for each well takes 5 weeks. The first city proposed in this region was Reynosa, Tamaulipas, which is a border city in the northern part of the state of Tamaulipas and it is located on the southern bank of the Rio Grande having one of the biggest aquifers located in this region. The aquifer is located within Hydrological Region No. 24 "Río Bravo". It comprises the northern part of the State of Tamaulipas and a small part of the State of Nuevo León covering an approximate area of 17,500 km^2. Ojinaga is a rural border town in the northern Mexican state of Chihuahua. Ojinaga is situated where the Río Conchos drains into the Río Grande (known as the Rio Bravo in Mexico), an area called La Junta de los Rios. Besides, the Alamo Chapo aquifer is located in the northeast portion of the state of Chihuahua and covers an area of 4883.5 km^2. The climate is very dry and semi-warm, the dry climate continues in importance because of its influence and extension, with an average annual rainfall of 316 mm. Therefore, the National Water Commission (CONAGUA) indicates that for the aquifer of Alamo Chapo, there is an annual average groundwater availability of millions of m^3/year, nevertheless, the use of this water could interfere with essential activities for the society and also it could affect the ecosystem (CONAGUA 2020a, b). Nonetheless, this water is available only if it is extracted from the subsoil, the availability of fresh surface water may be limited because of the seasonal droughts of the designated geographic locations. The National Meteorological Service (SMN) is the official agency of the Mexican government in charge of providing meteorological and climatological information. One of the climatic phenomena that most affect the economic activities of the country is drought, the SMN is in charge of detecting the current state and evolution of this phenomenon, its methodology is based on obtaining and interpreting various indices or indicators of drought such as the Standardized Precipitation Index (SPI) that quantifies the conditions of deficit or excess of precipitation (30, 90, 180, 365 days), rain anomaly in percent of normal (30, 90, 180, 365 days), Vegetation Health Satellite Index (VHI) that measures the degree of vegetation stress through observed radiance, the CPC-NOAA leaky bucket soil moisture model that estimates soil moisture using a one-layer hydrological model, the Normalized Vegetation Difference Index (NDVI), the average temperature anomaly, the percentage of water availability in the country's dams and the contribution of local experts. These indices are displayed in layers through a Geographic Information System (GIS) and by consensus, the regions affected by drought are determined, according to the intensity scale that is common in the three countries, ranging from abnormally dry (D0), moderate drought (D1), severe drought (D2), extreme drought (D3) to exceptional drought (D4). Therefore, polygons are drawn for each drought intensity, generating shapefile files, and when the polygons correspond to the mid-month analysis (issued on the 15th of each month) they are used to quantify the drought over the national territory and when they correspond to the final evaluation of any month. Table 16.1 presents the seasonal droughts in the locations of Reynosa, Tamaulipas and Ojinaga, Chihuahua.

Hayes (2009) presented the amount of water needed to perform hydraulic fracture, which is in a range of 11,335–15,114 m^3. Slutz et al. (2012) reported a range of 12,700–19,000 m^3, whereas Shaffer et al. (2013) indicated that it is between 8000 and 27,000 m^3 for each well in the Marcellus shale region. This chapter accounts for the injection of CO_2 as an alternative to fracturing fluid. In 2014, Middleton et al. (2014) reported that using supercritical CO_2 as a working fluid, where supercritical CO_2 has many potential benefits and drawbacks compared with water as a fracturing fluid. This way, the average value used in this work for the fracturing fluid (mixture of CO_2–water) required in the completion of each well is 15,000 m^3. Hayes (2009)

Table 16.1 Results of the drought monitor at the municipal level

			2020																								2021	
			Jan		Feb		Mar		Apr		May		Jun		Jul		Aug		Sep		Oct		Nov		Dec		Jan	
City	State	Basin	15	31	15	28	15	31	15	30	15	31	15	30	15	31	15	31	15	30	15	31	15	30	15	31	15	31
Reynosa	TAM	Rio Bravo	D0	D1	D2	D2	D2	D3	D2	D2	D2											D0	D1	D1	D1	D2	D2	D0
Ojinaga	CHH	Rio Bravo		D0		D0				D0	D0	D1	D1	D1	D1	D2	D3	D3	D3	D3	D3	D3	D3	D4	D4	D4	D4	D4

D0: Abnormally Dry | D1: Moderate Drought | D2: Severe Drought | D3: Extreme Drought | D4: Exceptional Drought

Printed with permission of "*Optimization of the water-energy nexus in hydraulic fracturing processes using* CO_2 *and water as fracturing fluids*", *ACS Sustainable Chemistry and Engineering*, Copyright (2022), American Chemical Society

reported that approximately 25% of the water injected during the hydraulic fracturing is collected over several days following the hydraulic fracturing. Thus, it was established that the flow rate of the fracturing fluid in each well is 428.57 m^3/d, however, all the scenarios involve different percentages for the feeding. The first percentage of feeding was 100–0% water and CO_2 respectively, this increased a 10% for CO_2 while for the water decreased by 10% until reaching 0% water and 100% CO_2. The flowback water recovered after each phase had a value of 25% of the water injected into the well; 80%, 35%, and 10% only for the first three weeks, respectively.

This way, it is considered that the weeks that do not have a drought are the ones that have maximum freshwater accessibility (428.57 m^3/d for each well). Furthermore, the weeks that present an abnormal drought only have 80% of the maximum freshwater accessibility. Likewise, the weeks with a moderate drought, severe drought, extreme drought, and exceptional drought have 60%, 40%, 20% and 0% of maximum freshwater accessibility, respectively.

Additionally, according to the scheduling presented in Fig. 16.4, three sets of wells (20 wells in total) should be completed by three hydraulic fracturing crews. The scheduling performance is not dependent on freshwater availability because if freshwater is scarce, the CO_2 and water generated by the power plant will replace the freshwater.

Furthermore, this example uses storage units, disposal units, and treatment/adsorption units that are normally used in shale gas fields. Table 16.2 presents some data related to the treatment and storage units and disposals. Because of the treatment and adsorption units are designed to ensure a required quality for the outlet stream for its reuse or disposal, pits and tanks are used to store water and CO_2 respectively (Song et al. 2019).

Figure 16.5 shows that the first subset of wells $n1$ corresponds to wells 1–8 (operated by hydraulic fracturing crew 1); while the subsets $n2$ and $n3$ are the wells 9–15 (related to crew 2) and 16–20 (associated to crew 3), respectively. Therefore, the cost for the unit of freshwater is 3.23 \$/m^3 and 4.65 \$/m^3 for Reynosa and Ojinaga, respectively (CACMAPAMR 2020; JMASC 2020). Therefore, unit transportation costs are associated with the supply of freshwater into the wells. Likewise, there are other transportation costs involved in this model, for example, transportation costs

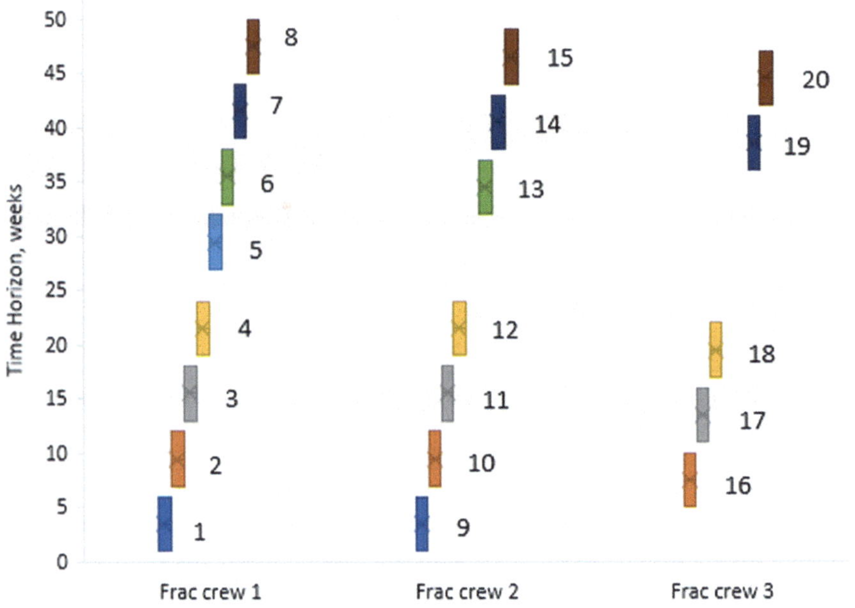

Fig. 16.4 Scheduling for the completion phase in each well. Printed with permission of "*Optimization of the water-energy nexus in hydraulic fracturing processes using CO_2 and water as fracturing fluids*", *ACS Sustainable Chemistry and Engineering*, Copyright (2022), American Chemical Society

Table 16.2 Data for storage, disposals and treatment/adsorption units

Concept/case	Water			CO_2	
	Treatment	Storage	Disposal	Treatment	Storage
Fixed cost, \$	95,000	10,000	52,000	0	10,000
Variable cost, \$/m^3	4.1	6.3	13.5	0	6.3
Maximum capacity, m^3/d or m^3	3200	12,000	20,000	3200	12,000
Exponent for the economies of scale	0.65	0.72	0.9	0.65	0.72

Printed with permission of "*Optimization of the water-energy nexus in hydraulic fracturing processes using CO_2 and water as fracturing fluids*", *ACS Sustainable Chemistry and Engineering*, Copyright (2022), American Chemical Society

from the wells ($n1$, $n2$, and $n3$) to the interception network, costs for the segments from treatment/adsorption units, and finally unit transportation costs from the storage to the wells. These costs are presented in Table 16.3. The cost associated with CO_2 regarding the transportation from storage to wells is the same as the water because supercritical CO_2 behaves as a low-viscosity fluid with a density similar to water (Kang et al. 2011).

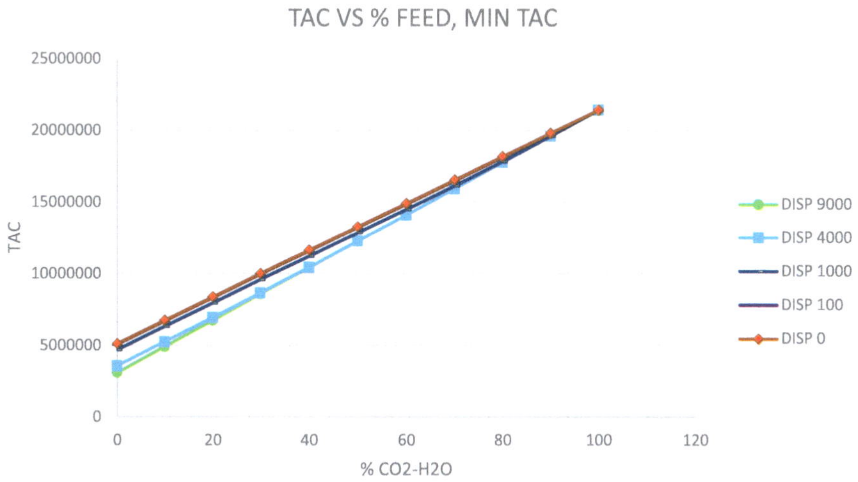

Fig. 16.5 TAC for different freshwater availability per week. Printed with permission of "*Optimization of the water-energy nexus in hydraulic fracturing processes using CO_2 and water as fracturing fluids*", *ACS Sustainable Chemistry and Engineering*, Copyright (2022), American Chemical Society

Table 16.3 Unit transportation costs associated with water and CO_2

	Water			CO_2		
Transportation cost	*n*1	*n*2	*n*3	*n*1	*n*2	*n*3
Freshwater supply to wells/ flue gases to adsorption unit, $/m^3	4	7	12	–	–	–
Storage to wells, $/m^3	0.98	1.23	1.65	0.98	1.23	1.65
Wells to interception network, $/m^3	0.39	0.72	0.94	–	–	–
Treatment/adsorption units to storage, $/m^3	1.41			1.41		
Treatment/adsorption units to disposal, $/m^3	8.3			0		

Printed with permission of "*Optimization of the water-energy nexus in hydraulic fracturing processes using CO_2 and water as fracturing fluids*", *ACS Sustainable Chemistry and Engineering*, Copyright (2022), American Chemical Society

The transportation cost for the water generated by the power plant is considered zero because the CO_2 adsorption and water treatment plants are planned to be near the power plant. Furthermore, there are costs associated with the water treatment and CO_2 capture (adsorption plant). The unit operating cost for the water treatment is 12.50 \$/m^3 while for CO_2 capture is 61.92 \$/m^3. The fixed and variable costs are equal to zero because both are included in the total operational cost (Ali et al. 2019; Ho et al. 2008).

16.4.1 Case Study

This example involves several cases to evaluate the best scheduling to perform hydraulic fracturing. Also, three cases were considered with different freshwater availabilities. The parameters previously indicated were used for **Cases A**, **B,** and **C**.

16.4.1.1 Case A

Case A considers the constant availability of freshwater during a time horizon for the completion phase of a set of wells. This is because the water is extracted from the aquifers. In this case, five amounts of freshwater availability were contemplated, TACs are presented in Fig. 16.5.

When water availability decreases, the *TAC* increases but reaches a point where it is similar to each other; t This because the use of CO_2 is higher than water and the cost of the water used is small compared to CO_2. To contrast the impact of freshwater availability, it is necessary to take the highest value (9000 m^3/week) against the lowest one (0 m^3/week); thus, the *TAC* is determined for every percentage accounted for in the model.

Figure 16.5 shows that for the highest availability (9000 m^3/week), the *TAC* increases almost 59%, each time the CO_2 percentage increases by 10%. Similarly, for the lowest availability, the *TAC* increases by almost 32%. Even though the *TAC* for the lowest availability increases in less proportion compared to the highest availability, the *TAC* for the highest availability will always be less than for the lowest availability, this is because more freshwater is available, and is cheaper the use of freshwater than CO_2 or treated water. Furthermore, it is important to mention that when there is no fresh water accessible, the water used to fracture the well comes from the water treated and stored. Finally, Table 16.4 shows the total freshwater needed for the availabilities accounted in the model.

16.4.1.2 Case B

Case B presents different freshwater availabilities through a year, which means that a certain amount of water will be delivered for each year. Figure 16.6 presents the *TAC* at separate freshwater accessibility per year considering all percentages mentioned in ***Case A***.

Figure 16.6 allows determining that at a higher percentage of CO_2, the *TAC* increases but the total water needed is fewer to perform hydraulic fracture. At higher percentages of CO_2, the total water needed will be a smaller amount than the water given each year, reaching a point where it keeps constant.

Table 16.4 Total freshwater required

% CO_2–H_2O	TWR, m^3				
0–100	237,090	171,570	45,569.998	4600	0
10–90	213,380	168,810	45,512.994	4600	0
20–80	189,670	165,170	45,456.004	4600	0
30–70	165,960	156,060	45,399	4600	0
40–60	142,250	140,700	45,341.996	4600	0
50–50	118,540	118,040	45,284.999	4600	0
60–40	94,836	94,836.79	45,228	4600	0
70–30	71,126.21	71,126.21	44,270.908	4600	0
80–20	47,415.629	47,415.629	41,292.635	4600	0
90–10	23,710	23,710.581	23,710.581	4557.004	0
100–0	0	0	0	0	0
Availability, m^3/week	9000	4000	1000	100	0

Printed with permission of "*Optimization of the water-energy nexus in hydraulic fracturing processes using* CO_2 *and water as fracturing fluids*", *ACS Sustainable Chemistry and Engineering*, Copyright (2022), American Chemical Society

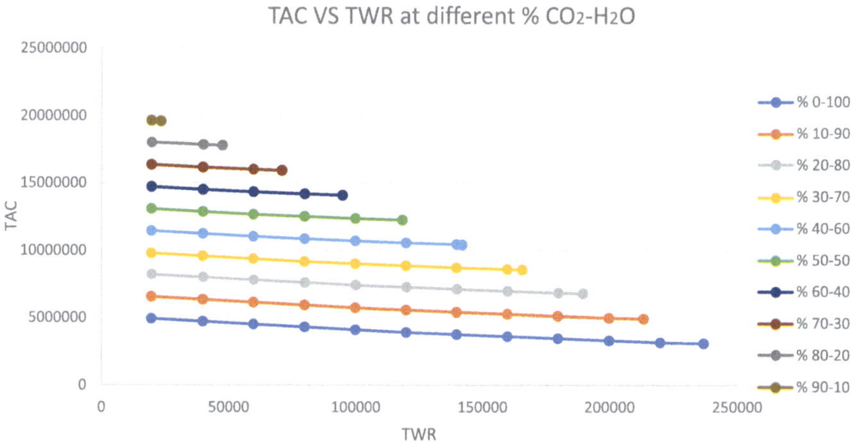

Fig. 16.6 *TAC* for different water availability per year. Printed with permission of "*Optimization of the water-energy nexus in hydraulic fracturing processes using* CO_2 *and water as fracturing fluids*", *ACS Sustainable Chemistry and Engineering*, Copyright (2022), American Chemical Society

16.4.1.3 Case C

Lastly, the third scenario presents constraints for the accessibility of freshwater based on geographic location and seasonal limitations. Table 16.1 shows the droughts considered in this model for both locations. This case was separated into two scenarios; the first one involves the maximum water availability (9000 m^3/week)

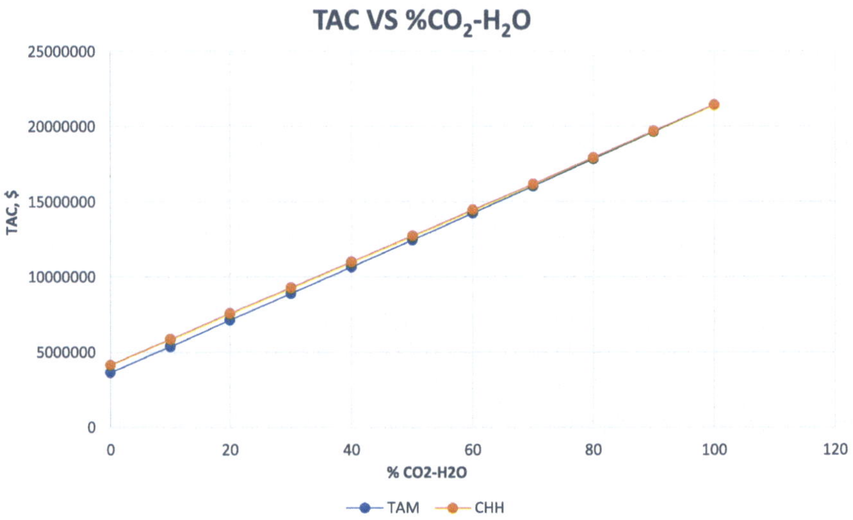

Fig. 16.7 TAC at different percentages of CO_2–H_2O with the same freshwater availability for the locations of Ojinaga and Reynosa. Printed with permission of "*Optimization of the water-energy nexus in hydraulic fracturing processes using CO_2 and water as fracturing fluids*", *ACS Sustainable Chemistry and Engineering*, Copyright (2022), American Chemical Society

is the same for both locations, whereas the second one considers that the location of Ojinaga Chihuahua has a maximum water availability of 4000 m^3/week whereas the maximum water availability for Reynosa Tamaulipas is 9000 m^3/week. The water availabilities considered before were based on the annual average groundwater availability for both locations and data reported by Mexico Drought Monitor in 2021 (CONAGUA 2021). Figure 16.7 presents the *TAC* for the first sub scenario, where the maximum water availability is the same for both locations. Table 16.5 presents the *TWR* for the geographical locations of Ojinaga and Reynosa.

Figure 16.8 shows the *TAC* when the maximum water availabilities are 9000 and 4000 m^3/week for Tamaulipas and Ojinaga, respectively; whereas Table 16.6 presents the *TWR* for the maximum availabilities.

16.5 Concluding Remarks

This chapter presented a novel mathematical programming model for the best possible management of CO_2 and water used as fracturing fluid in shale gas production, taking into consideration both environmental and economic factors. The overall goal of the model is to minimize the total annual cost (economic aspect), which is the total of the capital and operational costs. Comparably, the environmental considerations involved employing both as the fracturing fluid to execute hydraulic fracture rather than merely freshwater to minimize the total amount of freshwater

Table 16.5 TWR for the locations of Ojinaga and Reynosa with the same water availability

% CO_2–H_2O	TWR for Tamaulipas, m^3	TWR for Ojinaga, m^3
0–100	199,470	177,000
10–90	184,560	167,400
20–80	168,460	156,300
30–70	152,350	142,200
40–60	136,180	127,800
50–50	117,170	109,500
60–40	94,836	91,200
70–30	71,126	71,126
80–20	47,415	47,415
90–10	23,710	23,710
100–0	0	0

Printed with permission of "*Optimization of the water-energy nexus in hydraulic fracturing processes using CO_2 and water as fracturing fluids*", *ACS Sustainable Chemistry and Engineering*, Copyright (2022), American Chemical Society

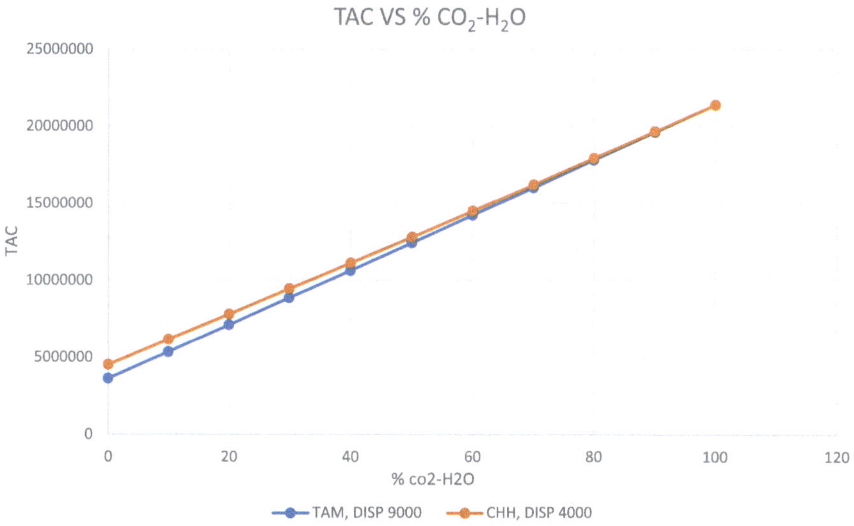

Fig. 16.8 *TAC* at different percentages of CO_2–H_2O with different freshwater availability for the locations of Ojinaga and Reynosa. Printed with permission of "*Optimization of the water-energy nexus in hydraulic fracturing processes using CO_2 and water as fracturing fluids*", *ACS Sustainable Chemistry and Engineering*, Copyright (2022), American Chemical Society

Table 16.6 TWR for the locations of Ojinaga and Reynosa with different water availability

% CO_2–H_2O	TWR for Tamaulipas, m^3	TWR for Ojinaga, m^3
0–100	199,470	89,799.97
10–90	184,560	88,899.91
20–80	168,460	88,000.06
30–70	152,350	87,100
40–60	136,180	84,999.82
50–50	117,170	81,499.915
60–40	94,836	74,400.32
70–30	71,126	61,599.52
80–20	47,415	44,598.17
90–10	23,710	23,710.581
100–0	0	0

required and to mitigate the release of CO_2 into the environment. Although using CO_2 indeed reduces the amount of freshwater used, this result required the employment of contemporary technologies (water treatment, CO_2 capture, and recycle/reuse water networks), in addition to appropriately disposing of waste streams following legal and regulatory requirements. Thus, the number of storage pits/tanks, water treatment, and adsorption technologies, as well as their capacities, are determined by the presented mathematical programming model. The model takes into consideration freshwater constraints such as the weekly maximum accessible water, the amount of freshwater provided annually, and the weekly water availability based on location (drought-affected areas).

Two significant Mexican shale gas reserves were the subject of a case study. The findings showed that a significant reduction in the amount of freshwater used in the hydraulic fracturing process occurs when CO_2 is added as a fracturing fluid. Nonetheless, because the overall cost of operating CO_2 is far higher than that of operating water, the total yearly cost rises significantly about the amount of CO_2 present in the fracture fluid. The data in Case A indicates that while the freshwater required (TWR) lowers by up to 10%, the TAC increases by 59–32% for every 10% increase in the CO_2 percentage. Similar to Case B, when TAC rises by 33% and TWR falls by 10–50%. In Case C, the TAC rises for Tamaulipas from 48 to 9% and for Ojinaga from 36 to 9%. It is interesting to notice that, despite Tamaulipas experiencing a higher proportion of TAC rises than Ojinaga, Ojinaga's TAC is consistently higher than Tamaulipas' due to Ojinaga's lower water availability. In the same vein, the TWR for Ojinaga drops from 1 to 47%, and for Tamaulipas drops from 7.5 to 50%.

Nomenclatures

Parameters

$AvailVCO_{2t}^{produced\_max}$	Maximum CO_2 availability, m^3/week
$AvailVw_t^{fresh\_max}$	Maximum water availability, m^3/week
$FCCO_2^{storage}$	Fixed charge for the CO_2 storage unit in the capital cost function, \$
$FCCO_2^{treat}$	Fixed charge for the CO_2 adsorption unit in the capital cost function, \$
$FCw^{storage}$	Fixed charge for the water storage unit in the capital cost function, \$
FCw^{treat}	Fixed charge for the water treatment unit in the capital cost function, \$
FCw^{waste}	Fixed charge for the disposal in the capital cost function, \$
$FCO_{2n,t}^{well\_in}$	Inlet CO_2 flowrate required in the wells, m^3/d
$FT_{n,t}^{well\_in}$	Total inlet flowrate required in the wells, m^3/d
$Fw_{n,t}^{well\_in}$	Inlet water flowrate required in the wells, m^3/d
$Fw_{n,t}^{well\_out}$	Flowback water leaving from the wells, m^3/d
H^{time}	Time conversion factor, d/ week
$UCCO_2^{used}$	Unit cost for the CO_2 produced, \$/m^3
$UOCCO_{2j}^{treat}$	Unit operating cost for the CO_2 adsorption units, \$/m^3
$UTCCO_{2,j}^{used\_treat}$	Unit transport cost for CO_2 produced from power plant to adsorption unit, \$/m^3
$UTCCO_{2n}^{sto}$	Unit transport cost for CO_2 from the storage tanks to the wells \$/m^3
$UTCCO_{2j,s}^{treat\_sto}$	Unit transport cost for the CO_2 adsorbed from the treatment to storage, \$/m^3
UCw^{fresh}	Unit cost for the freshwater, \$/m^3
$UOCw_i^{treat}$	Unit operating cost for the water treatment units, \$/m^3, \$/m^3
$UTCw_n^{fresh}$	Unit transport cost for the freshwater, \$/m^3
$UTCw_{s,n}^{sto\_well}$	Unit transport cost for the reused water from the storage to wells, \$/m^3
$UTCw_{i,d}^{treat\_dis}$	Unit transport cost for the treated water from the treatment to disposal, \$/m^3
$UTCw_{i,s}^{treat\_sto}$	Unit transport cost for the treated water from the treatment to storage, \$/m^3
$UTCw_{n,i}^{used}$	Unit transport cost for the flowback water from wells to treatment, \$/m^3
$VCCO_2^{storage}$	Variable charge for the CO_2 storage unit in the capital cost function, \$/m^3

$VCCO_2^{treat}$	Variable charge for the CO_2 adsorption unit in the capital cost function, \$/m^3
$VCw^{storage}$	Variable charge for the water storage unit in the capital cost function, \$/m
VCw^{treat}	Variable charge for the water treatment unit in the capital cost function, \$/m^3
VCw^{waste}	Variable charge for the disposal in the capital cost function, \$/m^3

Greek Symbols

αw_i^{treat}	Volumetric efficiency factor for the water treatment unit
αCO_{2j}^{treat}	Volumetric efficiency factor for CO_2 adsorption unit
$\beta CO_2^{storage}$	Exponent associated with the economies of scale for the CO_2 storage units
βCO_2^{treat}	Exponent associated with the economies of scale for the CO_2 adsorption units
βw^{disp}	Exponent associated with the economies of scale for the disposals
$\beta w^{storage}$	Exponent associated with the economies of scale for the water storage units
βw^{treat}	Exponent associated with the economies of scale for the water treatment units
$\gamma_j\vert$	CO_2 adsorption efficiency factor

Variables

$CapCostCO_2^{storage}$	Capital cost for CO_2 storage, \$
$CapCostCO_2^{treatment}$	Capital cost for CO_2 adsorption, \$
$CapCostw^{storage}$	Capital cost for water storage, \$
$CapCostw^{treatment}$	Capital cost for water treatment, \$
$CapCostw^{waste}$	Capital cost for disposal, \$
$CostCO_2^{op\_treat}$	Operating costs for CO_2 adsorption, \$
$CostCO_2^{produced}$	Operating costs for CO_2 production, \$
$CostCO_2^{trans\_pp\_treat}$	Transportation cost for the CO_2 from power to adsorption units, \$
$CostCO_2^{trans\_sto}$	Transportation cost for the CO_2 from storage units to wells, \$
$CostCO_2^{trans\_treat\_sto}$	Transportation cost for the CO_2 from adsorption units to storage units, \$

$Costw^{fresh}$	Freshwater cost, $
$Costw^{op\_treat}$	Operating costs for the water treatment, $
$Costw^{trans\_fb}$	Transportation cost for the flowback water from wells to treatment, $
$Costw^{trans\_fresh}$	Freshwater transportation cost, $
$Costw^{trans\_sto\_well}$	Transportation cost for reused water from storage to wells, $
$Costw^{trans\_treat\_dis}$	Transportation cost for treated water from treatment to disposals, $
$Costw^{trans\_treat\_sto}$	Transportation cost for treated water from treatment to storage, $
FCO_{2t}^{used}	CO_2 produced flowrate, m^3/d
$FCO_{2s,t}^{storage\_in}$	CO_2 flowrate entering the storage units, m^3/d
$FCO_{2s,t}^{storage\_out}$	CO_2 flowrate leaving the storage units, m^3/d
$FCO_{2j}^{treat\_cap}$	Capacity of CO_2 flowrate for the adsorption units, m^3/d
$FCO_{2j}^{treat\_max}$	Maximum CO_2 flowrate for the adsorption units, m^3/d
$FCO_{2j,t}^{treat\_in}$	CO_2 flowrate entering the adsorption units, m^3/d
$FCO_{2j,t}^{treat\_out}$	CO_2 flowrate leaving the adsorption units, m^3/d
Fw_t^{fresh}	Freshwater flowrate, m^3/d
$Fw_{s,t}^{storage\_in}$	Water flowrate entering to the storage units, m^3/d
$Fw_{s,t}^{storage\_out}$	Water flowrate leaving the storage units, m^3/d
$Fw_i^{treat\_cap}$	Capacity of water flowrate for the treatment units, m^3/d
$Fw_{i,t}^{treat\_in}$	Inlet water flowrate to the treatment units, m^3/d
$Fw_i^{treat\_max}$	Maximum water flowrate for the treatment units, m^3/d
$Fw_{i,t}^{treat\_out}$	Water flowrate leaving the treatment units, m^3/d
$Fw_{d,t}^{waste}$	Flowrate entering to the treatment units, m^3/d
$ffCO_{2,j,t}^{used\_treat}$	Segregated CO_2 flowrate from power plant to adsorption units, m^3/d
$ffCO_{2s,n,t}^{storage}$	Segregated CO_2 flowrate from storage system to wells, m^3/d
$ffCO_{2j,s,t}^{treat\_sto}$	Segregated CO_2 flowrate from treatment system to storage system, m^3/d
$ffw_{n,t}^{fresh}$	Segregated flowrate of freshwater to wells, m^3/d
$ffw_{s,n,t}^{storage}$	Segregated water flowrate from storage system to wells, m^3/d
$ffw_{i,d,t}^{treat\_dis}$	Segregated water flowrate from treatment system to disposals, m^3/d
$ffw_{i,s,t}^{treat\_sto}$	Segregated water flowrate from treatment system to storage system, m^3/d
$ffw_{n,i,t}^{well}$	Segregated flowrate from wells to treatment system, m^3/d
TAC	Total annual cost, $
TCC	Total capital cost, $

TCR	Total CO_2 required, m^3
TOC	Total operating cost, $
TWR	Total freshwater required, m^3
$VCO_{2s,t}^{storage}$	CO_2 storage/pits volume, m^3
$VCO_{2s,t-1}^{storage}$	CO_2 storage/pits volume for the previous period, m^3
$VCO_{2s}^{storage\_cap}$	Capacity for CO_2 storage/pits, m^3
$VCO_{2s}^{storage\_initial}$	Initial CO_2 storage/pits volume, m^3
$VCO_{2s}^{storage\_max}$	Maximum volume for CO_2 storage/pits, m^3
$Vw_{s,t}^{storage}$	Water storage/pits volume, m^3
$Vw_{s,t-1}^{storage}$	Water storage/pits volume for the previous period,
$Vw_{s}^{storage\_cap}$	Capacity for water storage/pits, m^3
$Vw_{s}^{storage\_initial}$	Initial water storage/pits volume, m^3
$Vw_{s}^{storage\_max}$	Maximum volume for water storage/pits, m^3
$Vw_{d}^{waste\_cap}$	Capacity for disposals, m^3
$Vw_{d}^{waste\_max}$	Maximum volume for disposals, m^3
$yCO_{2sg}^{storage}$	Binary variable used to model the existence of the CO_2 r storage units
$yw_{sw}^{storage}$	Binary variable used to model the existence of the water storage units
yw_{i}^{treat}	Binary variable used to model the existence of the water treatment units
yCO_{2j}^{treat}	Binary variable used to model the existence of the CO_2 adsorption units
y_{d}^{waste}	Binary variable used to model the existence of the disposals

Sets

I	$\{i \mid i$ is a water treatment unit$\}$
J	$\{j \mid j$is an adsorption unit$\}$
D	$\{d \mid d$ is a disposal unit$\}$
N	$\{n \mid n$ is well$\}$
S	$\{s \mid s$ is a storage tank/pit$\}$
T	$\{t \mid t$ is a period$\}$

Subscripts and Superscripts

i	Water treatment unit
j	CO_2 adsorption unit
d	Waste
n	Well

sw	Storage/pit for water
sg	Storage/tank for CO_2
t	Period

References

Ahn J, Kang D (2014) Optimal planning of water supply system for long-term sustainability. J Hydro-Environ Res 8(4):410–420. https://doi.org/10.1016/j.jher.2014.08.001

Ali H, Eldrup NH, Normann F, Skagestad R, Øi LE (2019) Cost estimation of CO_2 absorption plants for CO_2 Mitigation—method and assumptions. Int J Greenhouse Gas Control 88:10–23. https://doi.org/10.1016/j.ijggc.2019.05.028

Almutlaq A, Kazantzi V, El-Halwagi MM (2005) An algebraic approach to targeting waste discharge and impure-fresh usage via material recycle/reuse networks. Clean Technol Environ Policy 7(4):294–305. https://doi.org/10.1007/s10098-005-0005-8

Anderson BJ, Theodori GL (2009) Local leaders' perceptions of energy development in the Barnett Shale. J Rural Soc Sci 24:113–29. https://egrove.olemiss.edu/jrss/vol24/iss1/7. Accessed 15 Nov 2021

BP (2019) Centre for energy economics research and policy, UK. BP Stat rev world energy. https://www.bp.com/en/global/corporate/energy-economics/statistical-review-of-world-energy.html. Accessed 12 Nov 2021

Broad DR, Dandy GC, Maier HR (2015) A systematic approach to determining metamodel scope for risk-based optimization and its application to water distribution system design. Environ Model Softw 69:382–395. https://doi.org/10.1016/j.envsoft.2014.11.015

Brooke A, Kendrick D, Meeruas A, Raman R (2015) GAMS: a users guide. GAMS Development Corporation, Washington, D.C. https://www.gams.com/35/docs/UG_MAIN.html. Accessed Dec 2021

CACMAPAMR (2020) Consejo de Administración de la Comisión Municipal de Agua Potable y Alcantarillado del Municipio de Reynosa, Tamaulipas. Actualización de las tarifas del servicio de agua para el ejercicio fiscal 2020. Periódico Oficial del Estado de Tamaulipas, 2020. http://po.tamaulipas.gob.mx/wp-content/uploads/2020/02/cxlv-25-260220F.pdf. Accessed Dec 2021

Cafaro DC, Grossmann IE (2014) Strategic planning, design, and development of the shale gas supply chain network. AIChE J 60(6):2122–2142. https://doi.org/10.1002/aic.14405

Chen Y, He L, Li J, Zhang S (2018) Multi-criteria design of shale-gas-water supply chains and production systems towards optimal life cycle economics and greenhouse gas emissions under uncertainty. Comput Chem Eng 109:216–235. https://doi.org/10.1016/j.compchemeng.2017.11.014

CONAGUA (2020a) Subdirección general técnica gerencia de aguas subterráneas. Actualización de la disponibilidad media anual de agua en el acuífero bajo Río Bravo (2801). Estado de Tamaulipas, 2020. https://sigagis.conagua.gob.mx/gas1/Edos_Acuiferos_18/tamaulipas/DR_2801.pdf. Accessed Dec 2021

CONAGUA (2020b) Subdirección general técnica gerencia de aguas subterráneas. Actualización de la disponibilidad media anual de agua en el acuífero Alamo Chapo (0842). Estado de Chihuahua, 2020. https://sigagis.conagua.gob.mx/gas1/Edos_Acuiferos_18/chihuahua/DR_0842.pdf. Accessed Dec 2021

CONAGUA (2021) Monitor de Sequia de Mexico (MSM). Gobierno de Mexico, 2021. https://smn.conagua.gob.mx/es/climatologia/monitor-de-sequia/monitor-de-sequia-en-mexico. Accessed Dec 2021

Dahaghi AK (2010) Numerical simulation and modeling of enhanced gas recovery and CO_2 sequestration in shale gas reservoirs: a feasibility study. Paper presented at the SPE international conference on CO_2 capture, storage, and utilization, New Orleans, Louisiana, USA, Nov 2010. https://doi.org/10.2118/139701-MS

Durand B (2019) Petroleum, natural gas and coal: nature, formation mechanisms, future prospects in the energy transition. In: EDP sciences 2019. Les Ulis, pp 101–52

EIA (2012) U.S. Energy information administration (EIA). U.S. 2012 Annual Energy Outlook with Projects to 2035. US Department of Energy, Washington, DC. https://www.eia.gov/outlooks/aeo/er/pdf/0383er(2012).pdf. Accessed 19 Apr 2019

Ferreyra-Quiroz M, Lira-Barragán LF, El-Halwagi MM, Ponce-Ortega JM (2022) Optimization of the water-energy nexus in hydraulic fracturing processes using CO_2 andwater as fracturing fluids. ACS Sustain Chem Eng 10:17043–17058

Fuss S, Szolgayova J, Obersteiner M, Gusti M (2008) Investment under market and climate policy uncertainty. Appl Energy 85:708–721. https://doi.org/10.1016/j.apenergy.2008.01.005

Gao J, You F (2015) Deciphering and handling uncertainty in shale gas supply chain design and optimization: novel modeling framework and computationally efficient solution algorithm. AIChE J 61(11):3739–3755. https://doi.org/10.1002/aic.15032

Gunter WD, Wong S, Cheel DB, Sjostrom G (1998) Large CO_2 sinks: their role in the mitigation of greenhouse gases from an international, national (Canadian) and provincial (Alberta) perspective. Appl Energy 61:209–227. https://doi.org/10.1016/s0306-2619(98)00042-7

Hayes T (2009) Sampling and analysis of water streams associated with the development of Marcellus shale gas. In: Marcellus shale initiative publications database, vol 10. http://energyindepth.org/wp-content/uploads/marcellus/2012/11/MSCommission-Report.pdf . Accessed Dec 2021

Ho MT, Allinson GW, Wiley DE (2008) Reducing the cost of CO_2 capture from flue gases using pressure swing adsorption. Ind Eng Chem Res 47(14):4883–4890. https://doi.org/10.1021/ie070831e

Islam SM, Sadiq R, Rodriguez MJ, Najjaran H, Hoorfar M (2014) Reliability assessment for water supply systems under uncertainties. J Water Resour Plan Manag 140(4):468–479. https://doi.org/10.1061/(ASCE)WR.1943-5452.0000349

JMASC (2020) Junta Municipal de Agua y Saneamiento de Chihuahua. Tarifa para el cobro del servicio público para el año 2021 de la junta municipal de agua y saneamiento de Chihuhua. Periódico Oficial del Estado de Chihuahua, 2020. http://www.puentelibre.mx/noticia/jmas_tarifas_actualizacion_2021_metros_cubicos_usuarios_diciembre_2020/. Accessed Dec 2021

Kang SM, Fathi E, Ambrose RJ, Akkutlu IY, Sigal RF (2011) Carbon dioxide storage capacity of organic-rich shales. SPE J 16(04):842–855. https://doi.org/10.2118/134583-pa

Karavalakis G, Durbin TD, Villela M, Miller JW (2012) Air pollutant emissions of lightduty vehicles operating on various natural gas compositions. J Nat Gas Sci Eng 4:8–16. https://doi.org/10.1016/j.jngse.2011.08.005

Khor CS, Chachuat B, Shah N (2014) Fixed-flowrate total water network synthesis under uncertainty with risk management. J Clean Prod 77:79–93. https://doi.org/10.1016/j.jclepro.2014.01.023

Knudsen BR, Grossmann IE, Foss B, Conn AR (2014) Lagrangian relaxation based decomposition for well scheduling in shale-gas systems. Comput Chem Eng 63(5):234–249. https://doi.org/10.1016/j.compchemeng.2014.02.005

Lira-Barragan LF, Ponce-Ortega JM, Serna-González M, El-Halwagi MM (2011a) An MINLP model for the optimal location of the new industrial plant with simultaneous consideration of economic and environmental criteria. Ind Eng Chem Res 50(2):953–964. https://doi.org/10.1021/ie101897z

Lira-Barragan LF, Ponce-Ortega JM, Serna-González M, El-Halwagi MM (2011b) Synthesis of water networks considering the sustainability of the surrounding watershed. Comput Chem Eng 35:2837–2852. https://doi.org/10.1016/j.compchemeng.2011.03.021

Lira-Barragan LF, Ponce-Ortega JM, Nápoles-Rivera F, Serna-Gonzalez M, El-Halwagi MM (2013) Incorporating property-based water networks and surrounding watersheds in site selection of industrial facilities. Ind Eng Chem Res 52(1):91–107. https://doi.org/10.1021/ie3003792

Lira-Barragán LF, Ponce-Ortega JM, Serna-González M, El-Halwagi MM (2016) Optimal reuse of flowback wastewater in hydraulic fracturing including seasonal and environmental constraints. AIChE J 62(5):1634–1645. https://doi.org/10.1016/j.jclepro.2014.05.002

Luo F, Xu RN, Jiang PX (2013) Numerical investigation of the influence of vertical permeability heterogeneity in stratified formation of injection/production well perforation placement on CO_2 geological storage with enhanced CH_4 recovery. Appl Energy 102:1314–1323. https://doi.org/10.1002/aic.15032

Melikoglu M (2014) Shale gas: analysis of its role in the global energy market. Renew Sustain Energy Rev 37:460–468. https://doi.org/10.1016/j.rser.2014.05.002

Middleton R, Viswanathan H, Currier R, Gupta R (2014) CO_2 as a fracturing fluid: potential for commercial-scale shale gas production and CO_2 sequestration. Energy Procedia 63:7780–7784. https://doi.org/10.1016/j.egypro.2014.11.812

Napoles-Rivera F, Rojas-Torres MG, Ponce-Ortega JM, Serna-Gonzalez M, El-Halwagi MM (2015) Optimal design of macroscopic water networks under parametric uncertainty. J Clean Prod 88:172–184. https://doi.org/10.1016/j.jclepro.2014.05.002

Petrov AW, Ferri D, Krumeich F, Nachtegaal M, Van Bokhoven JA (2018) Stable complete methane oxidation over palladium based zeolite catalysts. Nat Commun 9(1):2545–2553. https://doi.org/10.1038/s41467-018-04748-x

Qin X, Gabriel F, Harell D, El-Halwagi MM (2004) Algebraic techniques for property integration via componentless design. Ind Eng Chem Res 43(14):3792–3798. https://doi.org/10.1021/ie034183k

Ruud W, Nadav S, Deepthi S, Walter AB (2017) Eagle ford shale play economics: U.S. versus Mexico. J Nat Gas Sci Eng 38:345–372. https://doi.org/10.1016/j.jngse.2016.12.009

Schepers KC, Nuttall BC, Oudinot AY, Gonzalez RJ (2009) Reservoir modeling and simulation of the Devonian gas shale of Eastern Kentucky for enhanced gas recovery and CO_2 storage. Paper presented at the SPE international conference on CO_2 capture, storage, and utilization, San Diego, California, USA, Nov 2009. https://doi.org/10.2118/126620-MS

Shaffer DL, Arias-Chavez LH, Ben-Sasson M, Romero-Vargas S, Yip NY, Elimelech M (2013) Desalination and reuse of high-salinity shale gas produced water: drivers, technologies, and future directions. Environ Sci Technol 47(17):9569–9583. https://doi.org/10.1021/es401966e

Slutz J, Anderson J, Broderick R, Horner P (2012) Key shale gas water management strategies: an economic assessment tool. SPE/APPEA international conference on health, safety, and environment in oil and gas exploration and production, Perth, Australia. https://doi.org/10.2118/157532-MS

Song X, Guo Y, Zhang J, Sun N, Shen G, Chang X, Yu W, Tang Z, Chen W, Wei W, Wang L, Zhou J, Li X, Li X, Zhou J, Xue Z (2019) Fracturing with carbon dioxide: from microscopic mechanism to reservoir application. Joule 3:1913–1926. https://doi.org/10.1016/j.joule.2019.05.004

Veil JA (2010) Water management technologies used by Marcellus shale gas producers, Argonne National Laboratory Argonne, IL. http://www.evs.anl.gov/pub/doc/Water%20Mgmt%20in%20Marcellus-final-jul10.pdf. Accessed 10 Sept 2021

Wang M, Xu Q (2014) Optimal design and operation for simultaneous shale gas NGL recovery and LNG re-gasification under uncertainties. Chem Eng Sci 112:130–142. https://doi.org/10.1016/j.ces.2014.03.021

Yang L, Salcedo-Diaz R, Grossmann IE (2014) Water network optimization with wastewater regeneration models. Ind Eng Chem Res 53(45):17680–17695. https://doi.org/10.1021/ie500978h

Zeng G, Chen M, Zeng Z (2013) Shale gas: Surface water also at risk. Nature 499(7457):154–154. https://doi.org/10.1038/499154c

Zhang X, Sun AY, Duncan IJ (2016) Shale gas wastewater management under uncertainty. J Environ Manage 165:188–198. https://doi.org/10.1016/j.jenvman.2015.09.038

Zou C, Zhao Q, Zhang G, Xiong B (2016) Energy revolution: from a fossil energy era to a new energy era. Nat Gas Ind B 3(1):1–11. https://doi.org/10.1016/j.ngib.2016.02.001

Chapter 17
Optimal Configuration of Water Networks in Shale Gas Hydraulic Fracturing Involving Economic and Environmental Considerations

17.1 Introduction

It is commonly known that a nation's level of development and its energy consumption are closely related; that is, nations with higher levels of progress require greater amounts of primary energy, and as a nation's development progresses, so does its energy consumption with this, and with the recent finding of fresh reserves, natural gas presents an appealing alternative to meet future energy demands (Melkoglu 2014). Some environmental benefits, such as the decrease in greenhouse gas emissions, and increased process efficiency when using this energy alternative can also be cited as additional benefits (Kuuskraa and Cleveland 2004). Natural gas's value as a fuel and its uses in the creation of electricity are widely acknowledged. According to the Energy Information Administration (EIA), the United States possesses natural and shale gas reserves that can sustain domestic electricity consumption for 575 years at current levels of electricity generation. Additionally, it is projected that by 2035, shale gas will account for nearly 50% of the nation's natural gas supply (EIA 2015a, 2017). The six major reserves countries—China, Argentina, Algeria, the United States, Canada, and Mexico—have more than two-thirds of the world's technically recoverable shale gas resources, which presents a significant opportunity for the remaining countries with the largest shale gas reserves (EIA 2015b). While the United States and Canada are the only countries producing large amounts of shale gas, it is expected that the other countries will begin industrial production in the coming years.

When compared to other fossil fuels, this resource has a few benefits, including a lower carbon impact, efficient energy production, a large supply, affordable pricing, and pipelines that make transportation and control simple. However, concurrently, recent research has concentrated on the environmental issues brought on by the extraction of shale gas while attempting to approximate sustainability methods (Fig. 17.1). De Melo-Martín et al. (2014) examined the health and environmental issues related to shale gas in this regard. Nicot and Scanlon (2012) investigated

J. M. Ponce-Ortega et al., *Sustainable Water Systems*, Water Resources Development and Management, https://doi.org/10.1007/978-3-031-72498-5_17

the effects that the methods of exploitation have on the environment because they utilize a lot of water and inject hazardous chemicals underground, which pollutes the soil and water more. Annevelink et al. (2016) investigated the contamination of the environment caused by the growth of shale gas production. Zhang and Yang (2015) established the harm associated with the hydraulic fracturing procedure used in US shale gas well operations. On the other hand, Chang et al. (2014) and Zou (2015) assessed water and energy usage about the pollution that results from hydraulic fracturing. Large amounts of water must be taken out of surface and underground freshwater reservoirs to meet requirements, which could have unintended consequences that need to be considered (Best and Lowry 2014). Due to concentrated brine and other hazardous substances, poorly treated wastewater streams and/or ultimate disposal procedures for hydraulic fracturing fluids and effluents can provide a risk for contamination (Vengosh et al. 2014).

Additionally, there have been some initiatives in the literature to address the environmental issues surrounding the extraction of shale gas to quantify and mitigate their consequences, which are connected to how the public views the exploitation processes (Yu et al. 2018). For instance, life cycle analyses of shale gas operations have been conducted by certain writers. In this context, Gao and You proposed several optimization approaches to designing shale gas supply chains involving economic and environmental criteria, where it was included the water management in shale gas production (Gao and You 2015a), and involving the water-energy nexus (Gao and

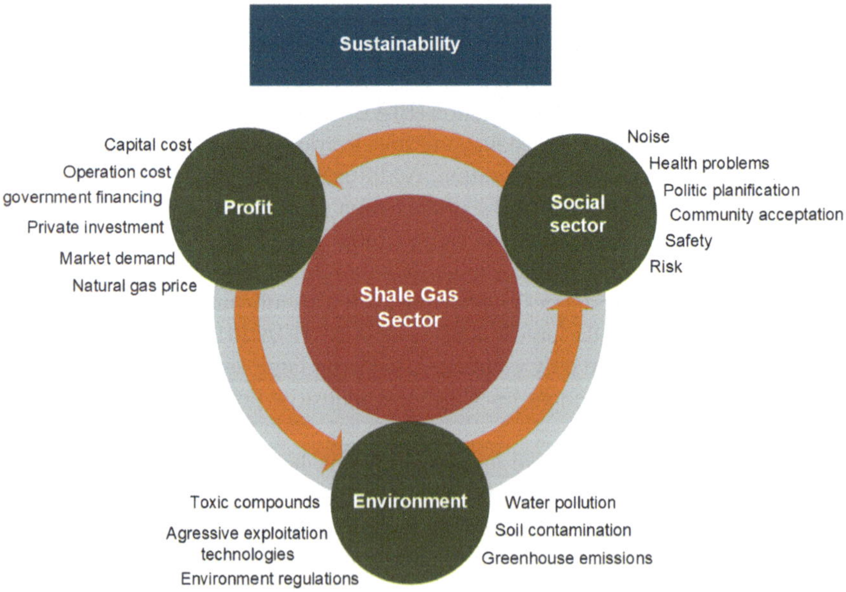

Fig. 17.1 Sustainability in the shale gas sector. Printed with permission of "*Optimal design of water networks for shale gas hydraulic fracturing including economic and environmental criteria*", *Clean Technologies and Environmental Policy*, Copyright (2018), Springer

You 2015b); Tagliaferri et al. (2015) implemented a life cycle analysis for shale gas processes to determine the real damage. Using a game theory approach, Gao and You (2017a) assessed the environmental impact through life cycle analysis, while Gao and You (2017b) investigated whether modular manufacturing could be applied to shale gas operations and design to create a sustainable process. Gao and You (2017c) also presented a study to implement the application of economic and environmental life cycle optimization using robust algorithms (Gao and You 2018). Lastly, Gao and You (2017d) looked at the prospects and difficulties facing this energy industry. Chang et al. (2015) included a comparison between water use and greenhouse gas emissions for a particular scenario in China in their special difficulties. In this regard, Jiang et al. (2013) suggested using ion exchange and ceramic membrane technologies to treat the produced water, while Nichols and Victor (2015) investigated the environmental advantages of implementing CO_2 emission capture and storage technologies. Chen (2015) evaluated the application of forward osmosis while taking the mitigation of membrane fouling into account. Advanced technologies have also demonstrated notable benefits, such as the utilization of novel microbial capacitive distillation cells (Stoll et al. 2015). Racharaks et al. (2015) presented a compelling plan that uses flowback water to cultivate microalgae, hence lowering the need for both water and nutrients. Despite these developments, no meticulous mathematical approach has been used in any of the earlier studies to optimize the water management chores associated with shale gas extraction.

Because of the high concentration of total dissolved solids that results in high salinity (Engelder et al. 2014; Rahm et al. 2013) presented a few strategies for the best selection of treatment technologies for hazardous compounds removal from the flowback water. This presents a significant challenge to producing wastewater streams with sufficient quality to be reused or disposed of. According to Yang et al. (2014), the use of optimization models to create effective management systems for the shale gas process has been shown to have both economic and environmental benefits. A mathematical approach for the best planning and investment in shale gas exploitation, including drilling and water management, was presented by Grossmann et al. (2014). To commercialize shale gas, Tan and Barton (2015) assessed a model that included the dynamic position of mobile plants. Arredondo-Ramírez et al. (2016) looked at a methodology for the best planning and the related infrastructure development in shale gas production, whereas Yuan et al. (2015) evaluated certain policies to support the development of shale gas based on technical and economic evaluations.

One of the first studies to incorporate the idea of a nexus in the context of shale gas production was Kharaka et al. (2013), who integrated the energy-water nexus when assessing the groundwater degradation in shale gas production. To approach the sustainability aspects, Savacool (2014) investigated the technical, economic, environmental, and social costs associated with hydraulic fracturing. Meanwhile, Gao and You (2017b) proposed a robust approximation to generate sustainable operations for shale gas systems, identifying the sector's challenges and future directions. Al-Douri et al. (2017) provided an overview of the monetization of the shale gas industry in this regard, outlining the main products and intermediates that can be obtained from the hydrocarbon components of shale gas in addition to the conversion technologies that offer opportunities, challenges, statistics, and insights. Recently, Oke et al. (2018) proposed a mathematical model to optimize the use of energy and water in hydraulic fracturing through the recycling and reuse of fracturing water implementing membrane distillation (MD). Al-Aboosi and El-Halwagi (2018) proposed a work whose objective is to study developing a design framework to integrate water and energy including multiple energy resources, cogeneration processes, and desalination, where solar energy is used to satisfy thermal power demands by the multi-effect distillation plant (MED).

However, water management in the shale gas extraction process is linked to most environmental concerns. Regarding this, Clark et al. (2013) calculated the amount of water used during the life cycle of shale gas and conventional gas production, taking into consideration flowback water reuse as well as the various stages of production. A mathematical programming approach was presented by Lira-Barragán et al. (2016) to determine the best flowback wastewater treatment system while minimizing the annual cost. Due to the high risk of spills and leaks to the surface and subsurface environments associated with large volumes of produced water (Clark et al. 2013), the appropriate treatment of the flowback fluid has raised serious concerns about the potential for contaminating reservoirs and water bodies, including surface water and groundwater, during the effluent disposal process. It is crucial to distinguish between flowback and produced water in this situation. After the completion phase is complete, the former is obtained quickly at a high rate and with an increasing concentration of pollutants, while the latter gradually rises to the surface at a low constant rate and with a high concentration of pollutants. However, prior studies have not included the environmental evaluation with financial goals and the handling of water during hydraulic fracturing operations.

This chapter presents an optimization approach for designing a water network for shale gas-related hydraulic fracturing operations (López-Díaz et al. 2018). It uses an effective treatment system with a recycle and reuse strategy for water streams, considering both economic and environmental goals at the same time. While the economic objective is to minimize the total annual cost (TAC), the environmental target is to maximize the removal of pollutants, which is like minimizing the environmental impact. In addition to determining the ideal size for each of the storage/pits, disposals, and treatment technologies needed for the solution, the model also establishes the internal layout, with particular attention to flowrates and concentrations.

17.2 Problem Statement

The problem addressed in this chapter can be stated as follows (see Fig. 17.2).

Given is a water reservoir (such as a lake or river) accessible to extract freshwater to be employed in the hydraulic fracturing operation and the unit and unit transportation costs (involving the distance between the locations for freshwater source and the wells). The approach considers a set of well pads ($N = \{n \mid n = 1, 2, \ldots, N_n\}$) prepared

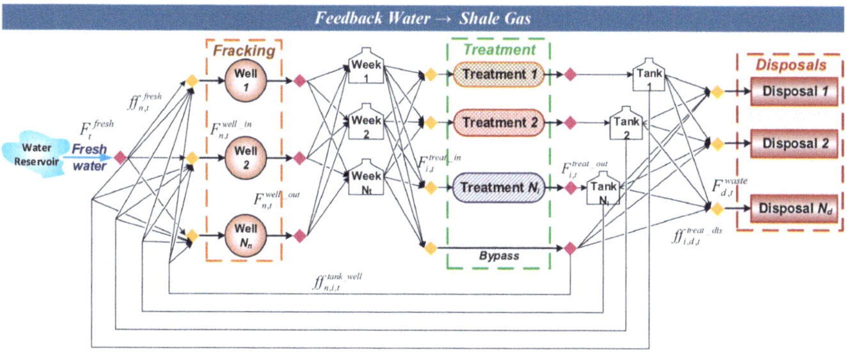

Fig. 17.2 Superstructure for the presented water network. Printed with permission of "*Optimal design of water networks for shale gas hydraulic fracturing including economic and environmental criteria*", *Clean Technologies and Environmental Policy*, Copyright (2018), Springer

to implement the completion process (hydraulic fracturing), which is implemented by a set of crews operating at the same time (the available equipment determines the maximum number of wells that can be completed concurrently). This way, the scheduling to implement the fracking process for each well pad, the water requirements for each step, the available freshwater in water reservoirs, the portion of the injected water that returns to the surface, the maximum number of wells that can be simultaneously completed, the environmental constraints for wastewater discharges, the maximum capacities for the storage and treatment units and the operation time for completion per well are given. After the flowback fluid returns to the surface, it is collected through a set of storage/pits ($J = \{j \mid j = 1, 2, \ldots, N_j\}$) installed on the site of exploitation to store the flowback water before being treated. This way, the produced water is stored according to the period (in weeks) where it is collected (because the water quality decreases as a function of time, due to the contact with toxic chemicals in the extracted wells); therefore, the scheme involves the possible existence of three pits (most of the flowback water is generated during the first weeks). Thus, it avoids the mixing of water streams with different qualities. The exponent, fixed and variable costs for the capital cost function, and the maximum volume for each storage unit are known. Furthermore, there is known a set of treatment technologies ($I = \{i \mid i = 1, 2, \ldots, N_i\}$) to treat the flowback water and meet the quality conditions for its reuse or to be disposed of. Each treatment unit has a linked storage tank (second storage system) to collect the treated hydraulic fracturing fluid before being reused or disposed of, and a bypass stream is also included. The presented model needs a unit operating cost, fixed and variable costs (accounting for the exponent in the economies of scale) for the capital cost function as well as the maximum capacity of the flowrate that can be treated for each treatment and storage unit. Lastly, it is given a set of disposals ($D = \{d \mid d = 1, 2, \ldots, N_d\}$), where the treated flowback water can be disposed of. Likewise, each disposal includes the parameters required in the capital cost function and a maximum volume. There is needed the unit transportation cost for all the trajectories considered by the superstructure.

Next, the problem consists of determining the optimal configuration for the water management to carry out the completion step for all the target wells minimizing the total cost and at the same time maximizing the pollutant removal, which is equivalent to minimizing the environmental impact. Consequently, the optimization process is intended to calculate the number of treatment technologies, storage/pits (for the two storage systems), and disposals required in the solution as well as their optimal sizes in addition to the internal configuration concerning flowrates and concentrations. The objective function consists of the minimization of the total annual cost (TAC), which is composed of the total operating cost (TOC) and the total capital cost (TCC). The TOC involves the freshwater costs, the operating costs for treatment units, and the transportation costs for all the considered routes; whereas the TCC considers the capital costs associated with the treatment units, storage/pits units, and disposals.

17.3 Mathematical Model Formulation

This chapter presents an optimization model for a recycle and reuse water network in shale gas hydraulic fracturing operations to generate significant reductions in freshwater consumption. Hence, the scheme involves a set of wells available to carry out the hydraulic fracturing, which demands important amounts of water. The water needs can be fulfilled through freshwater sources or even with treated water (recycled water). After the water is injected into the wells, the flowback water can be collected to be sent to a set of storage tanks; this way, the storage tanks can collect the flowback water according to different qualities for the flowback fluid, which depend strongly on period once the hydraulic fracturing is completed (the water properties are getting worse according to the week where is collected). This configuration is implemented to avoid the mixing of water with drastically different conditions. At the exit of the first storage, it is placed a set of treatment technologies, and a bypass is also considered for the case that the water streams can meet the needed quality to be disposed of or reused in the wells. In this case, the reused streams can be mixed with freshwater, and the corresponding properties are modified. Afterward, there is a set of tanks to present residence time to the water streams that can be reused (the hydraulic fracturing operation for each well is previously scheduled) and lastly, it is considered the final disposal (Fig. 17.2).

17.3.1 Optimization Model

The presented optimization model corresponds to the following mathematical formulation.

17.3.1.1 Availability Limitations for Freshwater

An important challenge for the shale gas industry is the limitations in water availability because this sector demands huge amounts of water for the hydraulic fracturing process (Myers 2012; Vidic et al. 2013). Therefore, this issue can be simulated by the following relationship:

$$F_t^{fresh} \leq \left(\frac{AvailV_t^{fresh\_max}}{H^{time}} \right), \quad \forall t \tag{17.1}$$

F_t^{fresh} represents the freshwater consumption during period t, $AvailV_t^{fresh\_max}$ corresponds to the water availability over period t and H^{time} signifies a time conversion factor.

17.3.1.2 Segregation of Freshwater

Freshwater is split and sent to each well n during their completion phase as follows:

$$F_t^{fresh} = \sum_n ff_{n,t}^{fresh}, \quad \forall t \tag{17.2}$$

17.3.1.3 Water Supply to Wells

The water needed for hydraulic fracturing in each well n ($F_{n,t}^{well\_in}$) is satisfied by freshwater ($ff_{n,t}^{fresh}$) plus the water coming from the tanks located at the exit of the treatment ($ff_{n,i,t}^{tank\_well}$) and the bypass stream ($ff_{n,t}^{bypass\_well}$):

$$F_{n,t}^{well\_in} = ff_{n,t}^{fresh} + \sum_i ff_{n,i,t}^{tank\_well} + ff_{n,t}^{bypass\_well}, \quad \forall n, t \tag{17.3}$$

To determine the concentration entering each well ($C_{c,n,t}^{well\_in}$), it is needed to multiply each term of the previous relationship times its concentration:

$$F_{n,t}^{well\_in} C_{c,n,t}^{well\_in} = ff_{n,t}^{fresh} C_c^{fresh} + \sum_i ff_{n,i,t}^{tank\_well} C_{c,i}^{treat\_out} + ff_{n,t}^{bypass\_well} C_{c,t}^{bypass}, \quad \forall c, n, t \tag{17.4}$$

C_c^{fresh} is the concentration of the freshwater, $C_{c,i}^{treat\_out}$ the concentration leaving the treatment units, and $C_{c,t}^{bypass}$ the concentration of the bypass stream. $C_{c,i}^{treat\_out}$ does not change concerning the time (this is contemplated that the treatment technology *i* always provides the same water quality).

17.3.1.4 Flowback Water

Numerous studies pointed out that flowback water should be obtained once the completion phase has finished; nonetheless, the majority of this fluid can be collected just during the first weeks. Consequently, flowback water ($F_{n,t}^{well\_out}$) is sent to a set of storage tanks before being treated ($ff_{n,j,t}^{well\_storage}$):

$$F_{n,t}^{well\_out} = \sum_{j} ff_{n,j,t}^{well\_storage}, \quad \forall n, t \tag{17.5}$$

17.3.1.5 Water Inlet to Storage Units Per week

The pollutants concentration for the flowback water raises during the time (Haluszczak et al. 2013; Marcon et al. 2017); therefore, the water captured during the first week is cleaner concerning the second week and afterward, the produced water in the second week is collected at better conditions than the third week (there is a difference between the flowback and produced water respect to the quality and the proposed methodology is intended to deal with flowback water due to the produced water encompasses gas and oil). The cleanest water is collected during the first week (after the completion) and the worst quality water is obtained in the third week. This way, the presented configuration involves the possible existence of a set of tanks (and each one of them is employed for each week); nevertheless, it is not allowed the mixing of water streams with different qualities. This way, the scheme can save the flowback water according to the period when it has been gathered. Also, in the first tank $j = 1$ (called "Week 1" in the proposed superstructure) only flowback water is fed during the first week $t = fbw1$ (after the hydraulic fracturing process has finished):

$$F_{j,t}^{storage\_in} = \sum_{n} ff_{n,j,t}^{well\_storage}, \quad \forall j = 1, \quad t = fbw1 \tag{17.6}$$

Likewise, for the second and third weeks,

$$F_{j,t}^{storage\_in} = \sum_{n} ff_{n,j,t}^{well\_storage}, \quad \forall j = 2, \quad t = fbw2 \tag{17.7}$$

$$F_{j,t}^{storage\_in} = \sum_{n} ff_{n,j,t}^{well\_storage}, \quad \forall j = 3, \quad t = fbw3 \tag{17.8}$$

17.3.1.6 Water Outlet from Storage Units

The exit water streams the impoundments or pits ($F_{j,t}^{storage\_out}$) are segregated to be sent to all the treatment technologies ($ff_{j,i,t}^{storage\_treat}$) or to the bypass stream ($ff_{j,t}^{storage\_bypass}$):

$$F_{j,t}^{storage\_out} = \sum_i ff_{j,i,t}^{storage\_treat} + ff_{j,t}^{storage\_bypass}, \quad \forall j, t \tag{17.9}$$

17.3.1.7 Water Inlet to Treatment Units

The used superstructure incorporates an interception network constituted by a set of proven treatment technologies to process the flowback water, which have been previously reported and recognized to efficiently process this type of stream. For each of the potential treatment units, the approach needs the unit capital and operating costs, volumetric efficiencies, and maximum inlet and outlet concentrations. Furthermore, the optimization process chooses the required technologies and determines their optimal size. The balance for the mixer at the inlet of each treatment unit states that the inlet flowrate for the technology i ($F_{i,t}^{treat\_in}$) is calculated by the water streams exit the storage tanks ($ff_{j,i,t}^{storage\_treat}$):

$$F_{i,t}^{treat\_in} = \sum_j ff_{j,i,t}^{storage\_treat}, \quad \forall i, t \tag{17.10}$$

The following component balance is used to determine the concentration entering each treatment unit ($C_{c,i,t}^{treat\_in}$):

$$F_{i,t}^{treat\_in} C_{c,i,t}^{treat\_in} = \sum_j ff_{j,i,t}^{storage\_treat} C_{c,j}^{storage}, \quad \forall c, i, t \tag{17.11}$$

$C_{c,j}^{storage}$ is the concentration at the storage tanks placed before the treatment.

Similarly, the flowrates (F_t^{bypass}) and concentrations ($C_{c,t}^{bypass}$) for the bypass stream are determined as follows:

$$F_t^{bypass} = \sum_j ff_{j,t}^{storage\_bypass}, \quad \forall t \tag{17.12}$$

$$F_t^{bypass} C_{c,t}^{bypass} = \sum_j ff_{j,t}^{storage\_bypass} C_{c,j}^{storage}, \quad \forall c, t \tag{17.13}$$

17.3.1.8 Water Outlet from Tanks Belonging to Treatment Units

The considered model involves a tank per treatment unit to store the fluid. Tank 1 stores all the water coming from Treatment 1, Tank 2 just receives fluid from Treatment 2 and the same sequence is employed for all the treatment technologies considered. The bypass stream does not require a tank to store the fluid. However the outlet streams for these tanks ($F_{i,t}^{tank\_out}$) are segregated and sent to wells to be reused ($ff_{n,i,t}^{tank\_well}$) and to the waste disposals ($ff_{i,d,t}^{tank\_dis}$):

$$F_{i,t}^{tank\_out} = \sum_{n} ff_{n,i,t}^{tank\_well} + \sum_{d} ff_{i,d,t}^{tank\_dis}, \quad \forall i, t \tag{17.14}$$

Likewise, the bypass stream is directed to the same destinations:

$$F_{t}^{bypass} = \sum_{n} ff_{n,t}^{bypass\_well} + \sum_{d} ff_{d,t}^{bypass\_dis}, \quad \forall t \tag{17.15}$$

17.3.1.9 Water Balance for Waste Disposal and Discharge

The wastewater streams ($F_{d,t}^{waste}$) are coming from the tanks located after the treatment ($ff_{i,d,t}^{tank\_dis}$). Likewise, the bypass stream can send wastewater to be disposed of ($ff_{d,t}^{bypass\_dis}$):

$$F_{d,t}^{waste} = \sum_{i} ff_{i,d,t}^{tank\_dis} + ff_{d,t}^{bypass\_dis}, \quad \forall d, t \tag{17.16}$$

To determine the concentration of pollutants at the final disposals ($C_{c,d,t}^{waste}$), it is included the component balance:

$$F_{d,t}^{waste} C_{c,d,t}^{waste} = \sum_{i} ff_{i,d,t}^{tank\_dis} C_{c,i}^{treat\_out} + ff_{d,t}^{bypass\_dis} C_{c,t}^{bypass}, \quad \forall c, d, t \tag{17.17}$$

Previous relationships is used to determine the operation for all the mixers and splitters considered by the superstructure; nonetheless, the mathematical model must include relationships to model the operation for treatment units, storage tanks (prior and after the treatment), equipment capacities, as well as the economic (i.e., the minimization of the total annual cost) and environmental (minimization of the environmental impact including the consumption of freshwater) objective functions. These mathematical relationships are incorporated as follows.

17.3.1.10 Balances in Treatment Units

Typically, the operation of treatment technologies includes flowrate losses (considering outlet streams concerning inlet streams), this aspect is simulated as follows:

$$F_{i,t}^{treat\_out} = \alpha_i^{treat} F_{i,t}^{treat\_in}, \quad \forall i, t \tag{17.18}$$

α_i^{treat} is the volumetric efficiency of the treatment technologies because sometimes the total wastewater that the treatment units receive just a fraction can be treated for different reasons, for example, losses for transportation or low efficiencies in the processes, etc. $F_{i,t}^{treat\_out}$ represents the treated water flowrate from the treatment units and $F_{i,t}^{treat\_in}$ the untreated water flowrate that requires treatment in the treatment units.

17.3.1.11 Balances in Storage Units

It is forbidden to mix streams with different values for pollutant concentrations for storage tanks (before and after treatment) to set their concentrations as given data. Nevertheless, this set of tanks consists of saving the fluid to be employed during the optimal periods where the water streams are required (the reused water represents a backup resource for periods with low freshwater availability). The mixing at the inlet for treatment, disposal, and reuse (directly in wells) is allowed. This last point is useful to satisfy the pollutant constraints included at the inlet of the key activities for the proposed water network (treatment, disposal, and reuse) due that the fluid streams with different qualities can be mixed among them to achieve a lower value than the maximum allowed concentration in each task (even in the reuse of freshwater can be employed).

The accumulation balances for the storage pits state that the volume in the pit j at the time t ($V_{j,t}^{storage}$) is equal to the one at the end of the previous period ($V_{j,t-1}^{storage}$), plus the difference between the inlet ($F_{j,t}^{storage\_in}$) and outlet ($F_{j,t}^{storage\_out}$) flowrates multiplied by a time conversion factor (H^{time}),

$$V_{j,t}^{storage} = V_j^{storage\_initial} + H^{time}\left(F_{j,t}^{storage\_in} - F_{j,t}^{storage\_out}\right), \quad \forall j, \quad t = 1 \tag{17.19}$$

$$V_{j,t}^{storage} = V_{j,t-1}^{storage} + H^{time}\left(F_{j,t}^{storage\_in} - F_{j,t}^{storage\_out}\right), \quad \forall j, \quad t > 1 \tag{17.20}$$

At the beginning (first period), there is considered a known initial pit volume ($V_j^{storage\_initial}$). Likewise, the following equation is needed to ensure the continuity of the cycles, thus the initial volume is equal to the last volume:

$$V_{j,t}^{storage} = V_j^{storage\_initial}, \quad \forall j, t = t^{final} \tag{17.21}$$

17.3.1.12 Balances in Tanks Belonging to Treatment Units

A similar balance is needed to model the operation of the tanks located at the exit of treatment:

$$V_{i,t}^{tank} = V_i^{tank\_initial} + H^{time}\left(F_{i,t}^{treat\_out} - F_{i,t}^{tank\_out}\right), \quad \forall i, \quad t = 1 \tag{17.22}$$

$$V_{i,t}^{tank} = V_{i,t-1}^{tank} + H^{time}\left(F_{i,t}^{treat\_out} - F_{i,t}^{tank\_out}\right), \quad \forall i, \quad t > 1 \tag{17.23}$$

$$V_{i,t}^{tank} = V_i^{tank\_initial}, \quad \forall i, \quad t = t^{final} \tag{17.24}$$

The normal storage units used in shale gas industry are storage pits, impoundments or even above ground storage tanks can be used.

17.3.1.13 Capacity and Existence of Treatment Units

The mathematical model is targeted to determine the capacity and potential existence of each unit (i.e., treatment technologies including tanks and final disposals) needed in the optimal solution. Then again, for the treatment units the next relationships determine if each technology is needed or not (the existence of the treatment units is simulated by the binary variable y_i^{treat}); also, the operational capacity for each treatment is represented by $F_i^{treat\_cap}$:

$$F_i^{treat\_cap} \geq F_{i,t}^{treat\_in}, \quad \forall i, t \tag{17.25}$$

$$F_i^{treat\_cap} \leq F_i^{treat\_\max} y_i^{treat}, \quad \forall i \tag{17.26}$$

$F_i^{treat\_\max}$ is an upper limit for the operational capacity associated with unit i (this information can be provided by sellers of treatment technologies). The optimal capacity ($F_i^{treat\_cap}$) must be used in the cost functions related to the capital costs for the treatment units.

17.3.1.14 Capacity and Existence of Storage Units

For the storage/pits units, the following relationships calculate the operational capacity and the existence of the storage units:

$$V_j^{storage\_cap} \geq V_{j,t}^{storage}, \quad \forall j, t \tag{17.27}$$

$$V_j^{storage\_cap} \leq V_j^{storage\_\max} y_j^{storage}, \quad \forall j \tag{17.28}$$

$V_j^{storage\_cap}$ is the optimal size for the storage unit, $V_j^{storage\_\max}$ the maximum allowed capacity for the pit, and $y_j^{storage}$ is a binary variable employed to simulate the existence of the storage/pits unit.

17.3.1.15 Capacity and Existence of Tanks Belonging to Treatment Units

The following relationships are added to simulate the tanks located at the exit of the treatment:

$$V_i^{tank\_cap} \geq V_{i,t}^{tank}, \quad \forall i, t \tag{17.29}$$

$$V_i^{tank\_cap} \leq V_i^{tank\_\max} y_i^{treat}, \quad \forall i \tag{17.30}$$

17.3.1.16 Capacity and Existence of Disposals

The operational capacity for final disposal d ($V_d^{waste\_cap}$) is calculated by the sum of all the flowrates disposed of by the project ($F_{d,t}^{waste}$) and involving the time conversion factor (H^{time}):

$$V_d^{waste\_cap} = H^{time} \sum_t F_{d,t}^{waste}, \quad \forall d \tag{17.31}$$

$$V_d^{waste\_cap} \leq V_d^{waste\_\max} y_d^{waste}, \quad \forall d \tag{17.32}$$

The last equation is included to avoid the optimal size (determined by the optimization process) exceeding the maximum capacity for each disposal ($V_d^{waste\_\max}$). The existence of each disposal is simulated by the binary variable y_d^{waste}. Therefore, the implementation of the proposed approach should calculate the number of units (treatment technologies, storages/pits, and disposals) needed in the optimal solution and their capacities.

17.3.1.17 Upper Limits for Concentrations

The next constraints are required to guarantee that water streams are in acceptable conditions to be processed for hydraulic fracturing, to be treated, as well as to be disposed of:

$$C_{c,n,t}^{well\_in} \leq C_c^{well\_\max}, \quad \forall c, n, t \tag{17.33}$$

$$C_{c,i,t}^{treat\_in} \le C_{c,i}^{treat\_\max}, \quad \forall c, i, t \tag{17.34}$$

$$C_{c,d,t}^{waste} \le C_{c}^{waste\_\max}, \quad \forall c, d, t \tag{17.35}$$

17.3.1.18 Total Water Used

It is needed to calculate the total freshwater needed to complete all the target wells (this issue can be considered as an environmental objective function):

$$TWR = H^{time} \sum_{t} F_{t}^{fresh} \tag{17.36}$$

17.3.1.19 Operating Costs

The involved operating costs correspond to the freshwater cost, operating costs for treatment units, and transportation costs (including all the trajectories), which are described in detail as follows.

17.3.1.20 Freshwater Cost

The freshwater cost ($Cost^{fresh}$) is calculated by multiplying the sum of the water required during all periods (F_t^{fresh}) times the unit cost for the freshwater (UC^{fresh}) including the time conversion factor (H^{time}):

$$Cost^{fresh} = H^{time} UC^{fresh} \sum_{t} F_{t}^{fresh} \tag{17.37}$$

17.3.1.21 Operating Costs for Treatment Units

The operation of treatment units considers the acquisition of chemicals, use of external energy utilities (i.e., electricity or steam), workers, etc. These costs need to consider the efficiency for the operation of the treatment technologies, which are involved in the unit cost UOC_i^{treat}; next, to obtain the operating costs for treatment units ($Cost^{op\_treat}$), the unit cost is multiplied by the flowrate entering each treatment unit in all the periods:

$$Cost^{op\_treat} = H^{time} \sum_i \sum_t UOC_i^{treat} F_{i,t}^{treat\_in} \tag{17.38}$$

17.3.1.22 Freshwater Transportation Cost

The freshwater transportation cost ($Cost^{trans\_fresh}$) is calculated as follows:

$$Cost^{trans\_fresh} = H^{time} \sum_n \sum_t UTC_n^{fresh} ff_{n,t}^{fresh} \tag{17.39}$$

UTC_n^{fresh} is the unit transportation cost for freshwater from its source (lake or river) to shale play (well pad). The unit costs to transport freshwater are different considering the geographical distance between the fresh sources concerning each pad. For some existing shale plays, the freshwater resource is available at significant distances from the well pad, and even in some relevant cases like the Marcellus shale region, the freshwater transportation is significantly (several times) higher than freshwater cost.

17.3.1.23 Flowback Water Transportation Cost

The flowback water produced at wells is gathered and transported to the first storage (before treatment); this way, the cost associated with this transportation path ($Cost^{trans\_fb}$) is calculated by the next equation:

$$Cost^{trans\_fb} = H^{time} \sum_n \sum_j \sum_t UTC_{n,j}^{well\_storage} ff_{n,j,t}^{well\_storage} \tag{17.40}$$

$UTC_{n,j}^{well\_storage}$ represents the unit transportation cost for the flowback water in the route between the well n and the storage unit j (which are called Week 1, Week 2, and Week 3). This cost may fluctuate considering each trajectory.

17.3.1.24 Transportation Cost for the Stored Water to Treatment

The produced water is directed to a set of treatment units; subsequently, a new transportation process is required, which yields the following cost ($Cost^{trans\_storage\_treat}$):

$$Cost^{trans\_storage\_treat} = H^{time} \sum_j \sum_i \sum_t UTC_{j,i}^{storage\_treat} ff_{j,i,t}^{storage\_treat} \tag{17.41}$$

$UTC_{j,i}^{storage\_treat}$ represents the unit transportation cost from the storage pit j to the treatment unit i.

17.3.1.25 Transportation Cost for the Treated Water to Disposals

The flowrate leaving the set of tanks belonging to the second storage may be sent to the final disposals and the transportation cost associated with this process ($Cost^{trans\_tank\_dis}$) is calculated by the following relationship:

$$Cost^{trans\_tank\_dis} = H^{time} \sum_i \sum_d \sum_t UTC_{i,d}^{tank\_dis} ff_{i,d,t}^{tank\_dis} \tag{17.42}$$

$UTC_{i,d}^{tank\_dis}$ represents the unit transportation cost from tank i to disposal d. In some existing shale regions, the distances between the well pads and the final disposals are considerable, and the cost to transport the wastewater streams represents an important concern in the shale gas industry.

17.3.1.26 Transportation Cost for Reused Water to Well

The water streams leaving the set of tanks may be sent to the wells (to be reused) and the involved transportation cost ($Cost^{trans\_tank\_well}$) is calculated as follows:

$$Cost^{trans\_tank\_well} = H^{time} \sum_n \sum_i \sum_t UTC_{n,i}^{tank\_well} ff_{n,i,t}^{tank\_well} \tag{17.43}$$

$UTC_{n,i}^{tank\_well}$ represents the unit transportation cost for the reused water from tank i to well n. The unit transportation costs depend on each trajectory according to geographical locations.

17.3.1.27 Total Operating Cost

The total operating cost (TOC) for the project is determined by the sum of all operating costs:

$$\begin{aligned} TOC = {} & Cost^{fresh} + Cost^{op\_treat} + Cost^{trans\_fresh} + Cost^{trans\_fb} \\ & + Cost^{trans\_storage\_treat} + Cost^{trans\_tank\_dis} + Cost^{trans\_tank\_well} \end{aligned} \tag{17.44}$$

17.3.1.28 Capital Costs

The acquisition of the treatment and storage units and the construction of final disposals generate capital costs, which are considered in the economic objective function.

17.3.1.29 Capital Cost for Treatment Units

The capital cost function associated with treatment technologies involves fixed (FC_i^{treat}) and variable (VC_i^{treat}) charges; and the last term depends on the flowrate to be processed:

$$CapCost^{treat} = k_F \sum_i \left[FC_i^{treat} y_i^{treat} + VC_i^{treat} \left(F_i^{treat\_cap} \right)^{\beta^{treat}} \right] \tag{17.45}$$

k_F is a factor used to annualize the investment, while β^{treat} is the exponent associated with the economies of scale for the treatment units. This way, k_F depends on the useful life of the plant and the interest rate.

17.3.1.30 Capital Cost for Storage Units and Disposals

Analogous capital cost functions are included for the storage units and disposals:

$$CapCost^{storage} = k_F \sum_j \left[FC^{storage} y_j^{storage} + VC^{storage} \left(V_j^{storage\_cap} \right)^{\beta^{storage}} \right] \tag{17.46}$$

$$CapCost^{tank} = k_F \sum_i \left[FC^{tank} y_i^{treat} + VC^{tank} \left(V_i^{tank\_cap} \right)^{\beta^{tank}} \right] \tag{17.47}$$

$$CapCost^{waste} = k_F \sum_d \left[FC^{waste} y_d^{waste} + VC^{waste} \left(V_d^{waste\_cap} \right)^{\beta^{waste}} \right] \tag{17.48}$$

The previous equations include the exponents (β), which is needed to consider the economies of scale in the capital costs; nevertheless, at the same time, this term incorporates nonlinearities in the model formulation.

17.3.1.31 Total Capital Cost

The total capital cost (TCC) is composed of the capital costs for treatment, storage, and tank units, and final disposals:

$$TCC = CapCost^{treat} + CapCost^{storage} + CapCost^{tank} + CapCost^{waste} \tag{17.49}$$

17.3.1.32 Total Annual Cost

The total annual cost (TAC) is constituted by the sum of TOC and TCC:

$$Min\ TAC = TOC + TCC \tag{17.50}$$

17.3.2 Environmental Analysis

The environmental objective is referred to as the environmental benefit obtained by the implementation of the proposed project; precisely by the operation of the interception network to remove the toxic compounds contained in the flowback water. At the beginning of the shale gas industry, the flowback water was discharged without adequate treatment (Wang et al. 2014; Theodori et al. 2014; Estrada and Bhamidimarri 2016). Nonetheless, the environmental constraints are becoming stricter, and this situation leads to the implementation of treatment systems in the shale gas industry. This way, the life cycle analysis (LCA) methodology is employed to determine the total environmental impact generated by the operation of this industry. Additionally, an important challenge in the shale gas sector is the generation of an adequate qualitative analysis for flowback water streams that represents an average study for this type of stream (it is known that the composition of flowback water is significantly different even among adjacent wells). Nonetheless, after this disadvantage is overcome, the environmental impact may be calculated considering that each compound yields a different value for the environmental impact.

17.3.2.1 Total Environmental Impact

The principal goal of the treatment system is the elimination of toxic compounds contained in the produced water streams. Next, the environmental objective consists of maximizing the total removal of pollutants in the treatment system (*IMPTOL*), which is calculated as the difference between the environmental impact produced for the flowback water with ($IMPIN_{p,t}$) and without ($IMPOUT_{p,t}$) treatment as follows:

$$Max\ IMPTOL = \sum_{p} \sum_{t} (IMPIN_{p,t} - IMPOUT_{p,t}) \tag{17.52}$$

Furthermore, the environmental impact generated by the flowback water stream without treatment is determined by the sum of individual unit impacts (IMP_p) times the pollutant concentration ($C^{wastep}_{c,d,t,p}$) and the disposed flowrate ($F^{waste}_{d,t}$):

$$IMPOUT_{p,t} = \sum_{c} \sum_{d} IMP_p C^{wastep}_{c,d,t,p} F^{waste}_{d,t}, \quad \forall p, t \tag{17.53}$$

Because of the flexible composition of the flowback fluid that depends on the geological conditions in the reserves, the use of a global composition can be employed. The use of the concentration of the total dissolved solids is the most popular in flowback fluid but other properties such as PH as well as turbidity may be used. The concentration of each compound in the total concentration ($C^{wastep}_{c,d,t,p}$) is calculated by multiplying the total concentration by the fraction of each chemical (Xp_p) in the fluid that enters the treatment units:

$$C^{well\_in}_{c,i,t,p} = Xp_p C^{treat\_in}_{c,i,t}, \quad \forall c, i, t, p \tag{17.55}$$

For the fluid that leaves the system:

$$C^{wastep}_{c,d,t,p} = Xp_p C^{waste}_{c,d,t}, \quad \forall c, d, t, p \tag{17.54}$$

The environmental impact for initial conditions (before treatment) is defined as the total eco-points that may be produced by the toxic chemicals in the discharge stream:

$$IMPIN_{p,t} = \sum_c \sum_i IMP_p C^{well\_in}_{c,i,t,p} F^{treat\_in}_{i,t}, \quad \forall p, t \tag{17.56}$$

The substantial amount of freshwater that is consumed for the shale gas exploitation processes has a specific impact on the extracted water from the reservoir, which is calculated as follows:

$$IFreshwater_t = IMP^{freshwater} F^{fresh}_t, \quad \forall t \tag{17.57}$$

Thus, the environmental objective function is greatly influenced by the efficiency of the treatment system and the flexibility between the operation of different technologies and the relationship with the selection of the final disposal.

17.4 Results and Discussion

The recent advances in drilling technologies and hydraulic fracturing processes have led to a growing interest in shale gas, which corresponds to an attractive energy resource. Nevertheless, this alternative comprises important concerns. For example, the enormous amounts of freshwater used by hydraulic fracturing, and the final disposal of wastewater streams (which are discharged with high levels of pollution) represent a relevant challenge to avoid the contamination of superficial and underground water reservoirs. Furthermore, the environmental regulations for these streams are becoming stricter to reduce hazardous effluents; this task can be carried out through treatment systems with proven efficiency in this type of residual stream. This way, this chapter aims to deal with these challenges in the shale gas industry. Afterward, the following case study is presented to evaluate the capabilities of the proposed model formulation.

17.4.1 Case Study

The case study is in northeastern Mexico; in this region, it is estimated an amount of technically recoverable shale reserves of 545 Tcf for natural gas. Nevertheless, in Mexico the shale plays have not been exploited yet and consequently, there is more uncertainty in comparison with the documented shale plays in the United States. The investment, development, and future production of this resource is led by PEMEX (Mexican governmental petroleum agency). Figure 17.3 presents the location for the potential shale reserves considered; nevertheless, this region presents serious conditions for water scarcity and the study for the future implementation of hydraulic fracturing operations, where the freshwater resources are restricted, like in the proposed methodology, which represents a valuable tool to be considered. In this example, it considers the availability of 3 fracking crews to complete 20 wells during a time horizon of 52 weeks (a year). Figure 17.4 shows the well exploitation schedule according to a previous geographic and operation analysis where the first well-pad has 8 wells, the second is formed by 7 wells, and finally the third is composed of 5 wells.

The hydraulic fracturing process needs a mixture of water, proppant materials, and thickening agents to be pumped into the cracks of the fractures in the deep reserves to release the trapped natural gas. Nonetheless, depending on the geographical and

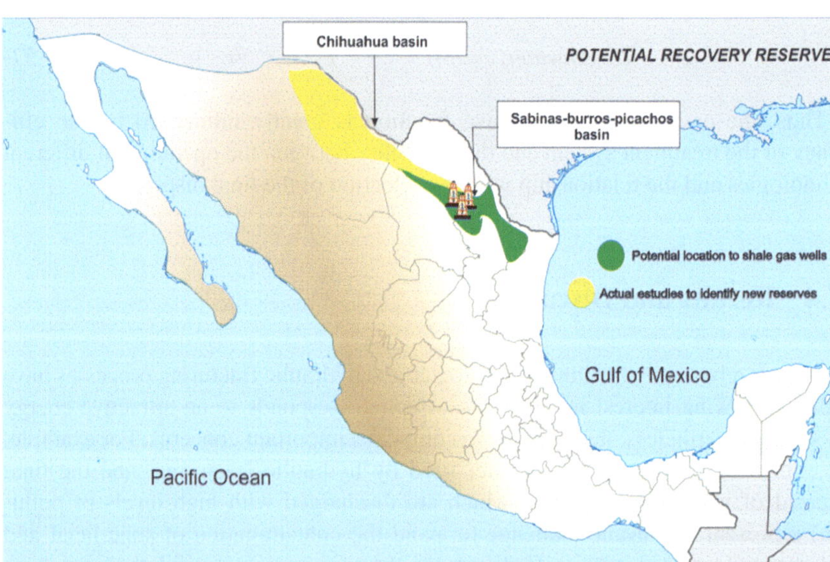

Fig. 17.3 Location of the potential shale gas reserves according to studies by PEMEX. Printed with permission of "*Optimal design of water networks for shale gas hydraulic fracturing including economic and environmental criteria*", *Clean Technologies and Environmental Policy*, Copyright (2018), Springer

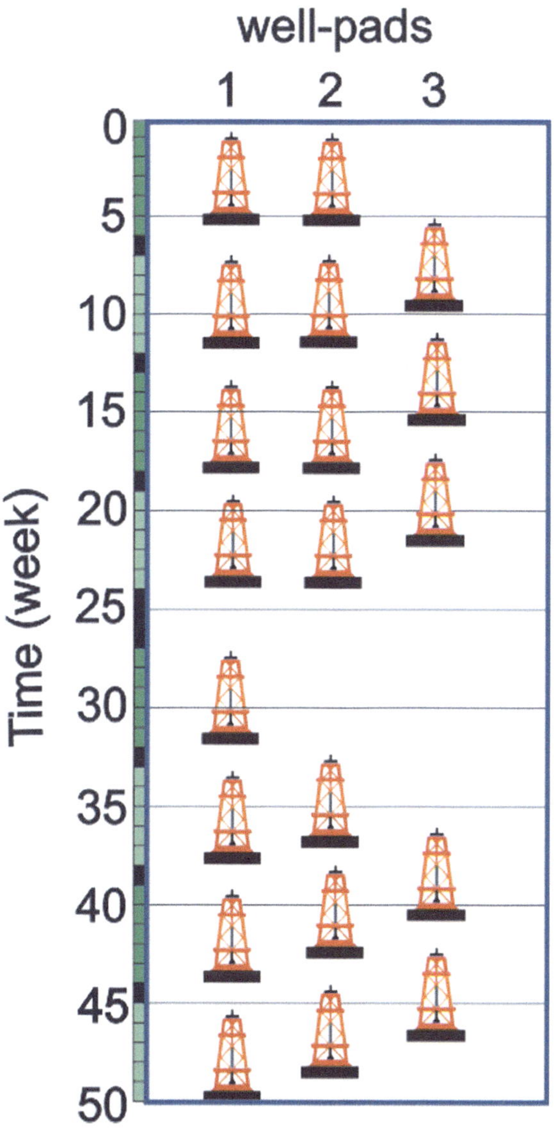

Fig. 17.4 Schedule of exploitation for the shale gas pits. Printed with permission of "*Optimal design of water networks for shale gas hydraulic fracturing including economic and environmental criteria*", *Clean Technologies and Environmental Policy*, Copyright (2018), Springer

geological conditions, the fracturing fluid should be prepared at different compositions. Usually, the consumption of water to be injected is calculated at 15,000 m^3 per well and it is assumed that 35% of this amount returns to the surface (produced water) during the first three weeks after the completion phase; next, the flowback water is collected and sent to a storage system. Furthermore, the freshwater presents a TDS concentration of 500 ppm; while the concentrations of TDS for the produced water in the first, second, and third week once the completion phase has finished are 75,000, 110,000, and 170,000 ppm, respectively. To obtain proper quantification

Table 17.1 Environmental impact in eco-points for each component in the flowback fluid

Compound	Environmental impact, E99/m^3
Sodium	3.23×10^{-2}
Chloride	1.14×10^{-1}
Calcium	1.02×10^{-2}
Strontium	1.73×10^{-4}
Barium	9.02×10^{-4}
Sulfate	2.33×10^{-4}
Bromide	6.79×10^{-4}

Printed with permission of "*Optimal design of water networks for shale gas hydraulic fracturing including economic and environmental criteria*", *Clean Technologies and Environmental policy*, Copyright (2018), Springer

of the environmental impact associated with wastewater streams, the composition of the flowback water streams based on TDS is not enough; subsequently, there is needed to know a wider chemical analysis. According to several studies, the average produced water is composed of 33% of sodium, 54% of chloride, 9% of calcium, 2.7% strontium, 0.6% of barium, 0.6% of bromide, and finally 0.1% of sulfate (EPA 2011). Table 17.1 shows the individual environmental impact of each compound before mentioned and defined by the life cycle analysis methodology.

In addition, the most conventional technologies used in the treatment of produced water (with proven efficiency in shale gas industry) are forward osmosis, reverse osmosis, multi-effect flash, multi-effect distillation, and mechanic vapor compression. Figure 17.5 presents a description of the performance employed by each technology; while Table 17.2 shows the maximum TDS concentration at the inlet streams, maximum capacities, and quality of outlet streams, and the capital costs and operating cost for each of the abovementioned treatment units. Forward osmosis has the most expensive costs (the capital cost is 5030 \$/m^3/day and the operation cost is 1.70 \$/m^3/day), and the capacity is low, but this unit involves the best recovery rates. Also, some treated water can be recycled to wells to satisfy the consumption of water, but when the treated flowrate does not satisfy the regulations some possibilities are considered to dispose of these streams to the environment as: additional treatment units, injection disposal wells, industrial use and discharging to a watershed; nonetheless, the limitation for effluents is fixed in 60,000 ppm of TDS when the final disposal corresponds to additional treatment units. Whereas injection disposal in wells requires a maximum TDS value of 40,000 ppm to mitigate the risks associated with the underground water reservoir contamination. The wastewater streams must not exceed a TDS concentration of 2000 ppm to avoid some operating problems such as brine incrustation in pipes and lastly, the environmental constraint to discharge the produced water over a watershed is 800 ppm (EPA 2015).

Thus, previous information must be specified to implement the proposed approach in the optimization software. This way, the optimization approach is a MINLP (Mixed Integer Nonlinear Programming) problem, and the presented model was coded in

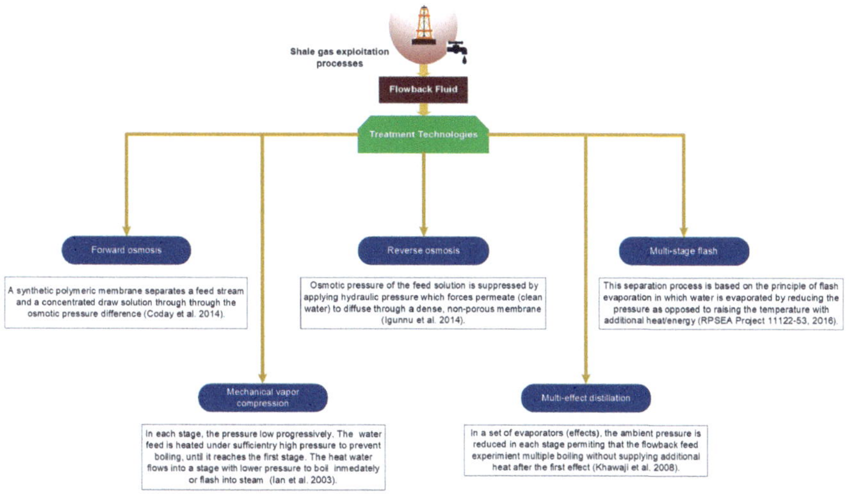

Fig. 17.5 Treatment technologies for produced water. Printed with permission of "*Optimal design of water networks for shale gas hydraulic fracturing including economic and environmental criteria*", *Clean Technologies and Environmental Policy*, Copyright (2018), Springer

Table 17.2 Water treatment technologies (Coday et al. 2014; Arthur et al. 2005; Darwish et al. 2003; Drewes et al. 2009; El-Dessouky 2004; Ettouney et al. 2002; Igunnu and Chen 2014; Khawaji et al. 2008; McGinnis et al. 2013; Matz and Fisher 1981; Ophir and Lokiec 2004; Veza 1995; Wade 1993, 2001; Xu and Drewes 2006)

	Technology	Maximum TDS concentration (mg/L)	Reference capacity (bpd)-m^3/day	Recovery rates	Capital cost U$/$m^3$/day	Operating cost \$/m^3
2	Forward osmosis	250,000	8	90% @ TDS 17,000	5030	1.70
4	Reverse osmosis	45,000	25,440	85% @ TDS 40,000	1905	1.38
3	Multi-stage flash	40,000	48,140	20% @ TDS 52,000	1830	1.45
5	Multi-effect distillation	100,000	48,140	35% @ TDS 58,000	2132	1.26
1	Mechanical Vapor compression	200,000	5090	80% @ TDS 60,000	1510	2.83

Printed with permission of "*Optimal design of water networks for shale gas hydraulic fracturing including economic and environmental criteria*", *Clean Technologies and Environmental Policy*, Copyright (2018), Springer

GAMS® (General Algebraic Modeling System) using SBB solver (Brooke et al. 2018). This solver is based on the combination of the standard Branch and Bound (B&B) method for the mixed integer linear programming part and some of the standard NLP solvers such as CONOPT, MINOS, or SNOPT. The optimization model involves 26,944 continuous variables, 14,568 constraints, and 17 binary variables. The CPU time depends on the analyzed scenario; nevertheless, an adequate interval is 2.5–15 s to obtain most solutions. Likewise, the optimization process was carried out with a processor Intel Core i7-3612QM at 2.10 GHz and 6 GB of RAM.

The multi-objective optimization problem involves two objectives that contradict each other (minimization of *TAC* vs maximization of *IMPTOL*). *IMPTOL* represents an indirect measure of the pollutants removed by treatment technologies. It indicates that when this objective is increased, then the environmental impact reduces, but at the same time the *TAC* is increased (owing to the selection of more effective treatment units, and typically these technologies are more expensive). In the other case, when the *TAC* is reduced, it generates a worse performance for *IMPTOL* (reducing its value) causing a higher environmental impact. This is represented in Fig. 17.6, which presents the Pareto curve generated in this example. **Point A** (green point) is the solution with the best performance for *TAC* (and at the same time the worst for *IMPTOL*). Though the opposite extreme solution is highlighted with **Point C** (red point), where *TAC* has the worst value, *IMPTOL* is notably improved. Lastly, the Pareto curve considers an intermediate solution with **Point B** (yellow point) to discuss a non-extreme solution. Also, the freshwater consumption (this is a key criterion that can play an important role in the selection of the final design) associated with each solution belonging to the Pareto curve is presented in Fig. 17.6. Table 17.3 presents the values of *TAC*, *IMPTOL*, *TWR*, and the number of treatment, storage, and disposal units required in **solutions A**, **B,** and **C**. Consequently, each point of the Pareto curve (even the non-highlighted solutions) has involved different designs and configurations.

The configuration for **solution A** is presented in Fig. 17.7, where this scheme requires two storage pits with capacities of 2100 m^3 and 4950 m^3 to save the produced water leaving the well pads. The treatment system is composed of two forward osmosis units and a plant of mechanical vapor compression; whereas the second storage system requires three storage units with capacities of 670, 4950 and 700 m^3, respectively, and the final disposal selected is additional treatment. Figure 17.8 presents the optimal distribution for the water streams to properly operate in **Point A**. The freshwater is the main source to carry out the completion; nonetheless, in some periods (particularly in the middle of the year) the freshwater consumption is considerably reduced, because in these weeks the reused streams are employed in hydraulic fracturing processes. Another relevant event is that the wastewater streams are only disposed of in the last weeks. Likewise, Fig. 17.9 presents the optimal design for **solution C**, where *IMPTOL* is maximized (with the best environmental performance). Consequently, more effective treatment technologies are chosen; precisely, the interception network is constituted by forward osmosis, mechanical vapor compression, reverse osmosis, and multi-effect distillation to improve, as much as possible, the quality of these water streams. Therefore, the wastewater streams can be disposed

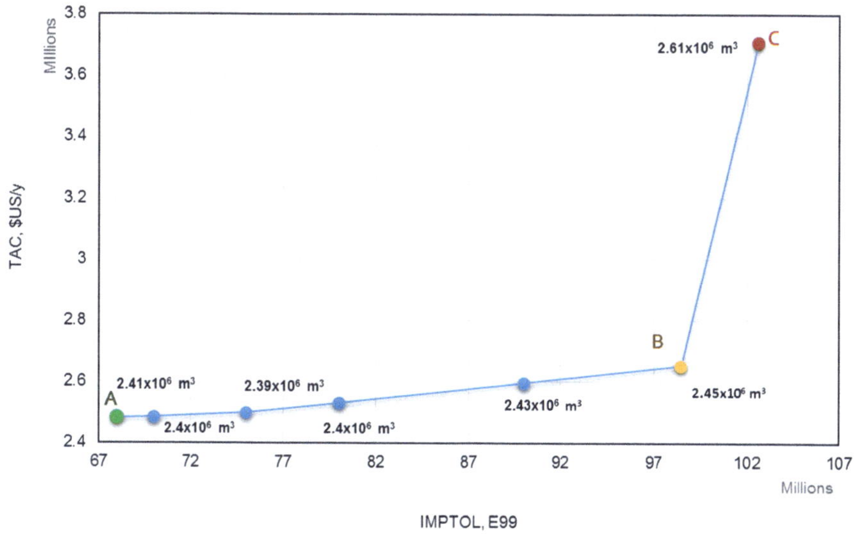

Fig. 17.6 Pareto graph for the proposed optimization problem. Printed with permission of "*Optimal design of water networks for shale gas hydraulic fracturing including economic and environmental criteria*", *Clean Technologies and Environmental Policy*, Copyright (2018), Springer

Table 17.3 Results of the Pareto solution for the proposed case study

Pareto solutions	A	C	B
TAC (US$/year)	2.49×10^6	3.75×10^6	2.68×10^6
IMPTOL (E99)	6.65×10^7	1.02×10^8	9.60×10^7
TWR (m^3/year)	2.41×10^5	2.63×10^5	2.47×10^5
Treatment units	3	8	3
Storage units	2	3	1
Final disposals	1	2	1

Printed with permission of "*Optimal design of water networks for shale gas hydraulic fracturing including economic and environmental criteria*", *Clean Technologies and Environmental Policy*, Copyright (2018), Springer

of in injection disposal wells (this option requires strict environmental regulation) as well as in industrial uses. Figure 17.10 presents the water distribution for **Point C**, where it does not use the bypass stream to prioritize pollutant removal and to increase the *IMPTOL* value. Lastly, the optimal configuration and water stream distribution for **Point B** are presented in Figs. 17.11 and 17.12, respectively. This way, this solution is analyzed to be considered an intermediate solution or breaking point because the *TAC* is reduced by 40%, the *IMPTOL* is diminished by 8.7% and *TWR* is 6.5% lower concerning **solution C**; else, when **Point B** is compared with **Point A**, the *TAC* augments 7.6%, while *IMPTOL* is increased 44.36%. In the optimal design for this case, only a storage pit is required (in the first storage system), two tanks (in the second storage system) two treatment technologies (forward osmosis and mechanical vapor compression) and the final disposal selected is additional treatment. The water distribution seems like the first solution with minor differences.

Finally, there are important differences in the optimal configurations, the water network, the objective functions, and the total water requirements. This way, the Pareto curve is selected to show all the optimal solutions to the proposed methodology to quickly visualize the main differences among them. The selection of the final configuration must be taken by decision-makers considering the most important

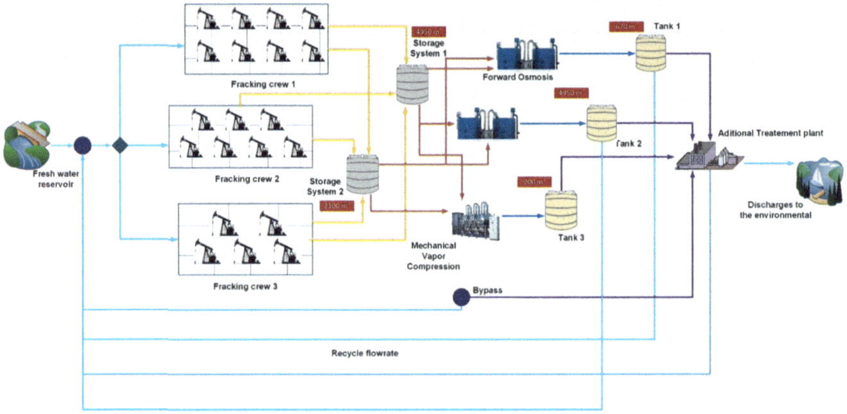

Fig. 17.7 Configuration for solution A. Printed with permission of "*Optimal design of water networks for shale gas hydraulic fracturing including economic and environmental criteria*", *Clean Technologies and Environmental Policy*, Copyright (2018), Springer

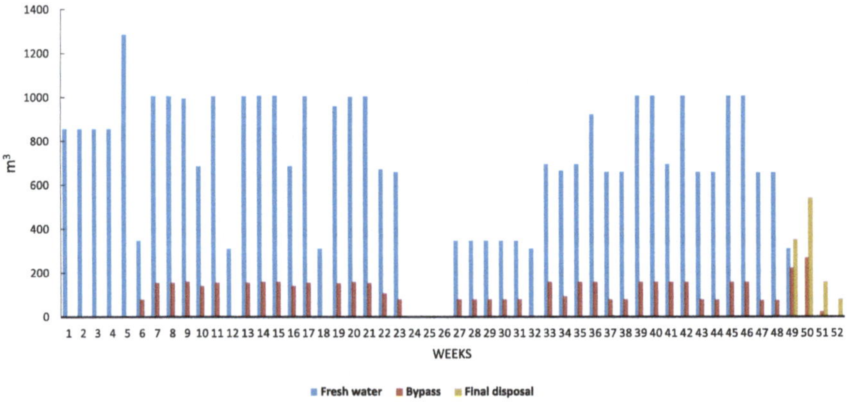

Fig. 17.8 Water distribution in the water network for solution A. Printed with permission of "*Optimal design of water networks for shale gas hydraulic fracturing including economic and environmental criteria*", *Clean Technologies and Environmental Policy*, Copyright (2018), Springer

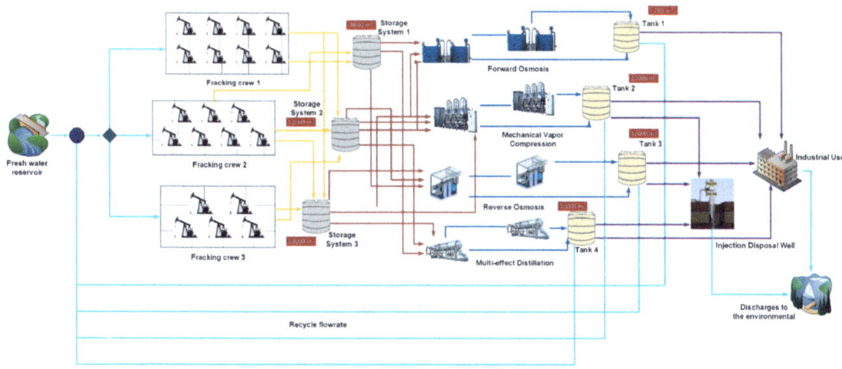

Fig. 17.9 Configuration for solution C. Printed with permission of "*Optimal design of water networks for shale gas hydraulic fracturing including economic and environmental criteria*", *Clean Technologies and Environmental Policy*, Copyright (2018), Springer

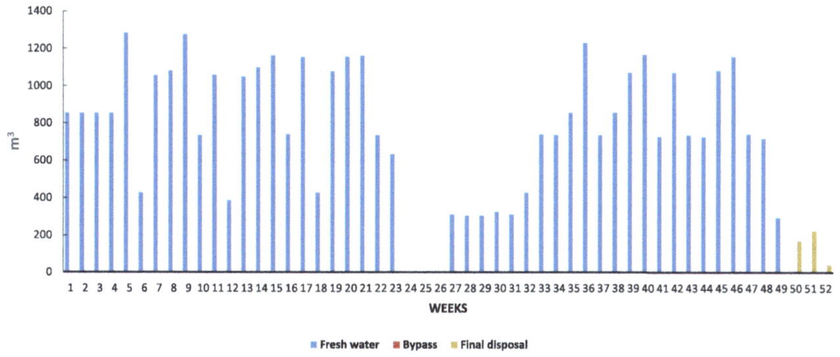

Fig. 17.10 Water distribution in the water network for solution C. Printed with permission of "*Optimal design of water networks for shale gas hydraulic fracturing including economic and environmental criteria*", *Clean Technologies and Environmental Policy*, Copyright (2018), Springer

aspects (such as total costs, environmental impact, and freshwater consumption), especially to the investors, government, and society.

17.5 Concluding Remarks

This chapter presented a multi-objective optimization approach for designing water networks for the key hydraulic fracturing activities. A Pareto set of optimal solutions is shown for the presented mixed-integer non-linear programming problem, illustrating the trade-offs between the environmental and economic goals of shale gas

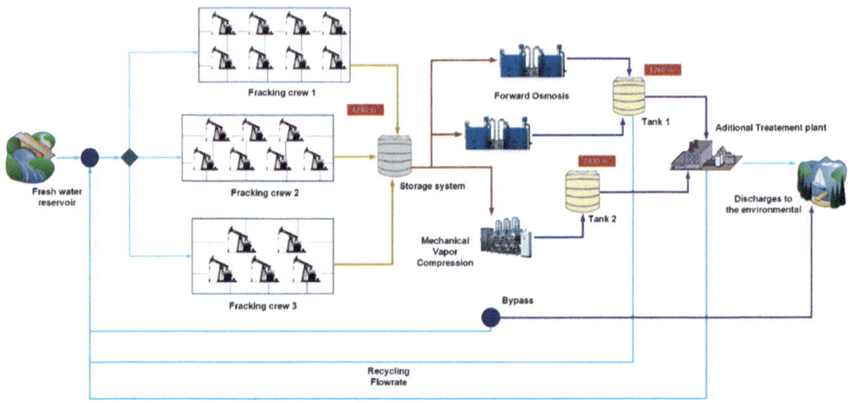

Fig. 17.11 Configuration for solution B. Printed with permission of "*Optimal design of water networks for shale gas hydraulic fracturing including economic and environmental criteria*", *Clean Technologies and Environmental Policy*, Copyright (2018), Springer

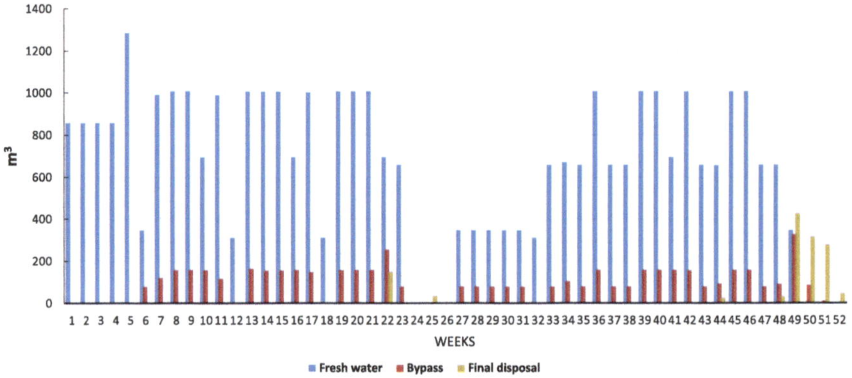

Fig. 17.12 Water distribution in the water network for solution B. Printed with permission of "*Optimal design of water networks for shale gas hydraulic fracturing including economic and environmental criteria*", *Clean Technologies and Environmental Policy*, Copyright (2018), Springer

exploitation. The model's formulation specifically aims to minimize overall expenditures while also maximizing pollution removal (this last effect decreases the environmental impact). This method is used to calculate the amount of freshwater needed, and the volume of flowback fluid that is collected, treated, and then disposed of. It also determines the best capacities for storage and treatment units, including the choice of treatment technologies, the flowrates for all the water streams that the superstructure takes into consideration, and the choice of final disposals. This chapter is also helpful in scheduling and planning hydraulic fracturing operations to address issues like the availability of freshwater due to the high demand for water and the management of wastewater disposal to prevent the release of contaminated effluents into the environment, which can contaminate nearby areas and subterranean and superficial water reservoirs.

A case study was also included in this chapter to demonstrate the efficacy of the presented methodology. When economic or environmental concerns are given priority, the optimal solutions in this scenario differ most significantly. To emphasize the configuration modifications involved in each scenario, an intermediate solution was thus explored. The Pareto curve is a useful way to display the findings in these kinds of approaches (when there are two competing objectives) because it makes it easier to see the key differences between the solutions. It is important to note that decision-makers should select the final design after considering several factors, including total costs, freshwater consumption, and environmental impact. They should also give special consideration to investors, the government, and society. Lastly, the suggested mathematical formulation is a general model that, with the right data, may be applied to any situation.

Nomenclatures

$AvailV_t^{fresh\_\max}$	Availability of freshwater in the reservoirs, m^3/s
$CapCost^{storage}$	Capital cost for the installation of tanks in the storage system, US$/year
$CapCost^{tank}$	Capital cost for the installation of tanks of the flowback recovery system, $US/year
$CapCost^{treat}$	Capital cost for the installation of the treatment system, $US/year
$CapCost^{waste}$	Capital cost for the piping and pumping system to send the final disposal, $US/year
$Cost^{fresh}$	Cost of the freshwater considering the price for industrial use, $US/year
$Cost^{op\_treat}$	Cost associated to the operation of the treatment units that involves the acquisition of chemical, use of external energy services, etc., $US/year
$Cost^{trans\_fb}$	Cost for transport the flowback fluid to the interception water network, mainly defined for pipping and pumping costs, $US/year
$Cost^{trans\_fresh}$	Transportation cost to send freshwater from the reservoirs to the water network feed, $US/year
$Cost^{trans\_storage\_treat}$	Transportation cost to send the flowback fluid from the recovery storage system to the treatments units, $US/year
$Cost^{trans\_tank\_dis}$	Transportation cost to send the treated flowrate from the treatment units to the final disposals, $US/year
$Cost^{trans\_tank\_well}$	Transportation cost to recycle the treated flowrate from the storage tanks through the feed of the water network $US/year

$C_{c,t}^{bypass}$	Concentration of total dissolved solids in the bypass flowrate, ppm
C_{c}^{fresh}	Concentration of total dissolved solids in the feeding of the freshwater in the water network, ppm
$C_{c,j}^{storage}$	Concentration of total dissolved solids at the storage tanks before the treatment system, ppm
$C_{c,i,t}^{treat\_in}$	Concentration of total dissolved solids that enters to the treatment system, ppm
$C_{c,i}^{treat\_\max}$	Maximum concentration of total dissolved solids to be processed by each treatment technology, ppm
$C_{c,i}^{treat\_out}$	Concentration of total dissolved solids from the treatment units to be disposed, ppm
$C_{c,d,t}^{waste}$	Concentration of total dissolved solids in the wastewater effluents to be disposed, ppm
$C_{c}^{waste\_\max}$	Maximum concentration limit of total dissolved solids in the effluents to be discharged to the disposals, ppm
$C_{c,n,t}^{well\_in}$	Concentration of total dissolved solids that enter to the wells as fracturing fluid, ppm
$C_{c}^{well\_\max}$	Maximum concentration limit of total dissolved solid that is permitted as feed to the shale gas wells, ppm
$IMPIN_{p,t}$	Environmental damage produced if the effluents are be discharged without treatment, eco-points
$IMPOUT_{p,t}$	Environmental damage that can be produced by the discharge of effluents with treatment
$IMPTOL$	Total environmental damage for the implementation of the treatment system for the effluents
$ff_{n,t}^{bypass\_well}$	Flowrate that is recycled from the derivation to the system feed, m^3/s
$ff_{d,t}^{bypass\_dis}$	Flowrate in the bypass that is sent to the disposals directly, m^3/s
$ff_{n,t}^{fresh}$	Flowrate of freshwater that enters the wells, m^3/s
$ff_{j,t}^{storage\_bypass}$	Flowrate from the recovery tanks to the bypass, m^3/s
$ff_{j,i,t}^{storage\_treat}$	Flowrate from the recovery tanks to the treatment units, m^3/s
$ff_{n,i,t}^{tank\_well}$	Flowrate from the storage system to the wells as recycled fluid, m^3/s
$ff_{i,d,t}^{tank\_dis}$	Flowrate from the storage system to the disposal alternatives, m^3/s
$ff_{n,j,t}^{well\_storage}$	Flowrate from the wells to the storage system, m^3/s
$FC^{storage}$	Fixed cost to install the storage system, \$US/year
FC_{i}^{treat}	Fixed cost to install the treatment system, \$US/year
$F_{d,t}^{waste}$	Total flowrate that is discharged to the final disposals, m^3/s
$F_{n,t}^{well\_in}$	Total flowrate that enters the wells as fracturing fluid, m^3/s
$F_{i,t}^{treat\_in}$	Total flowrate that enters the treatment system, m^3/s

$F_{i,t}^{tank\_out}$	Total flowrate that leaves each tank in the storage system, m^3/s
$F_{j,t}^{storage\_in}$	Total flowrate that enters each tank in the storage system, m^3/s
$F_{j,t}^{storage\_out}$	Total flowrate that leaves each tank in the storage system, m^3/s
FC^{tank}	Fixed cost to install tanks in the recovery system, m^3/s
FC^{waste}	Fixed cost for sending the effluents to the final disposals, m^3/s
$F_i^{treat\_cap}$	Total flowrate that the treatment technologies can be processed, m^3/s
$F_{i,t}^{treat\_in}$	Flowrate that is fed in the treatment units, m^3/s
$F_i^{treat\_\max}$	Maximum flowrate limit for the treatment units, m^3/s
$F_{i,t}^{treat\_out}$	Flowrate that leaves the treatment units, m^3/s
$F_{j,t}^{storage\_in}$	Flowrate that enters the recovery tanks during the shale gas exploitation, m^3/s
$F_{j,t}^{storage\_out}$	Flowrate that leaves the recovery storage system, m^3/s
$F_{n,t}^{well\_out}$	Total flowrate of flowback fluid that leaves each pit, m^3/s
F_t^{bypass}	Total flowrate in the bypass stream, m^3/s
F_t^{fresh}	Freshwater fed to the water network, m^3/s
H^{time}	Time conversion factor, time of unit operation, s
k_F	Annualization factor, year^{-1}
TAC	Total annual cost, \$US/year
TCC	Total capital cost, \$US/year
TOC	Total operational cost, \$US/year
TWR	Total water requirements, m^3
UC^{fresh}	Unit cost for freshwater, \$US/m^3/s
UOC_i^{treat}	Unit cost for treatment technology, \$US/m^3/s
UTC_n^{fresh}	Freshwater unit transportation cost, \$US/m^3/s
$UTC_{n,j}^{well\_storage}$	Unit transportation cost of flowrate from the wells to the recovery storage system, \$US/m^3/s
$UTC_{j,i}^{storage\_treat}$	Unit transportation cost of flowrate from the recovery system to treatment units, \$US/m^3/s
$UTC_{i,d}^{tank\_dis}$	Unit transportation cost of flowrate from the recovery storage system, \$US/m^3/s
$UTC_{n,i}^{tank\_well}$	Unit transportation cost of flowrate from the treated storage system recycled to the pit, \$US/m^3/s
$VC^{storage}$	Variable cost for storage units, \$US/m^3
VC^{tank}	Variable cost for tanks, \$US/m^3
VC^{waste}	Variable cost for discharging the effluents to the disposals, \$US/m^3
VC_i^{treat}	Variable cost for installing treatment units, \$US/m^3
$V_d^{waste\_cap}$	Volume of effluents that can be discharged to the final disposal alternatives, m^3

$V_d^{waste\_\max}$ Maximum volume that can be discharged to final disposal alternatives, m^3
$V_i^{tank\_cap}$ Capacity of tanks in the storage system, m^3
$V_i^{tank\_\max}$ Maximum capacity of tanks in the storage system, m^3
$V_{i,t}^{tank}$ Volume of the tanks after the treatment unit in certain period, m^3
$V_j^{storage\_cap}$ Capacity of storage units, m^3
$V_j^{storage\_initial}$ Capacity of storage units at initial conditions, m^3
$V_j^{storage\_\max}$ Maximum capacity of storage units, m^3
$V_{j,t}^{storage}$ Capacity of storage units in certain period, m^3
$V_{j,t-1}^{storage}$ Volume of the storage units in a previous period, m^3
y_d^{waste} Binary variable to decide the optimal final disposal
y_i^{treat} Binary variable to decide the installation of treatment units
$y_j^{storage}$ Binary variable for the installation of tanks in the storage system

Greek Symbols

α_i^{treat} Conversion factor for the flowrate in the treatment units
β^{tank} Factor that represents the economies of scale for tanks
β^{waste} Factor that represents the economies of scale in cost for discharge of effluents to a specific final disposal
$\beta^{storage}$ Factor that represents the economies of scale in cost for storage units
β^{treat} Factor that represents the economies of scale in cost for treatment units

References

Al-Aboosi FY, El-Halwagi MM (2018) An integrated approach to water-energy nexus in shale-gas production. Processes 6(5):52

Al-Douri A, Sengupta D, El-Halwagi MM (2017) Shale gas monetization—a review of downstream processing to chemicals and fuels. J Nat Gas Sci Eng 45:436–455

Annevelink MPJA, Meesters JAJ, Hendriks AJ (2016) Environmental contamination due to shale gas development. Sci Total Environ 550:431–438

Arredondo-Ramírez K, Ponce-Ortega JM, El-Halwagi MM (2016) Optimal planning and infrastructure development for shale gas production. Energy Convers Manage 119:91–100

Arthur JD, Langhus BG, Patel C (2005) Technical summary of oil & gas produced water treatment technologies. All Consulting, LLC

Best LC, Lowry CS (2014) Quantifying the potential effects of high-volume water extractions on water resources during natural gas development: Marcellus shale, NY. J Hydrol: Regnal Stud 1:1–16

Brooke A, Kendrick D, Meeruas A, Raman R (2018) GAMS-language guide. GAMS Development Corporation, Washington, D.C., USA

Chang Y, Huang R, Masanet E (2014) The energy, water, and air pollution implications of tapping China's shale gas reserves. Resour Conserv Recycl 91:100–108

Chang Y, Huang R, Ries RJ, Masanet E (2015) Life cycle comparison of greenhouse gas emissions and water consumption for coal and shale gas fired power generation in China. Energy 86:335–343

Chen G (2015) Treatment of shale gas drilling flowback fluids (SGDFs) by forward osmosis: membrane fouling and mitigation. Desalination 366:113–120

Clark CE, Horner RM, Harto CB (2013) Life cycle water consumption for shale gas and conventional natural gas. Environ Sci Technol 47(20):11829–11836

Coday BD, Xu P, Beaudry EG, Herron J, Lampi K, Hancock NT, Cath TY (2014) The sweet spot of forward osmosis: treatment of produced water, drilling wastewater, and other complex and difficult liquid streams. Desalination 333(1):23–35

Darwish MA, Al-Asfour F, Al-Najem N (2003) Energy Consumption in equivalent work by different desalting methods: case study for Kuwait. Desalination 152(1–3):83–92

De Melo-Martín I, Hays J, Finkel ML (2014) The role of ethics in shale gas policies. Sci Total Environ 470:1114–1119

Drewes JE, Cath TY, Xu P, Graydon J, Veil J, Snyder S (2009) An integrated framework for treatment and management of produced water. RPSEA Project, pp 07122–12

EIA (2015a) Annual energy outlook 2015 with projections to 2040 U.S. Energy Information Administration Web, Washington, D.C., 20585. http://www.eia.gov/forecasts/aeo/pdf/0383(2015).pdf. Accessed Aug 2017

EIA (2015b) Technically recoverable shale oil and shale gas resources: an assessment of 137 shale formations in 41 countries outside the United States. U.S. Energy Information Administration

EIA (2017) U.S. Energy Information and Administration. U.S. Department of Energy, Washington, D.C., USA

El-Dessouky HT (2004) Review of VC fundamentals and costing. In: Proceedings of the international conference on desalination costing, pp 79–94

Engelder T, Cathles LM, Bryndzia LT (2014) The fate of residual treatment water in gas shale. J Unconv Oil Gas Reserv 7:33–48

EPA (2015) Technical development document for proposed effluent limitations guidelines and standards for oil and gas extraction. U.S. Environmental Protection Agency

EPA (2011) Characterization of Marcellus shale and Barnett shale flowback waters and technology development for water reuse. Hydraulic fracturing technical workshop #4, USEPA meeting facilities

Estrada JM, Bhamidimarri R (2016) A review of the issues and treatment options for wastewater from shale gas extraction. Fuel 182:292–303

Ettouney HM, El-Dessouky HT, Faibish RS, Gowin PJ (2002) Evaluating the economics of desalination. Chem Eng Prog 98(12):32–40

Gao Y, You F (2015a) Shale gas supply chain design and operations toward better economic and life cycle environmental performance: MINLP model and global optimization algorithm. ACS Sustain Chem Eng 3:1282–1291

Gao J, You F (2015b) Optimal design and operations of supply chain networks for water management in shale gas production: MILFP model and algorithms for the water-energy nexus. AIChE J 61(4):1184–1208

Gao Y, You F (2017a) Game theory approach to optimal design of shale gas supply chains with consideration of economics and life cycle greenhouse gas emissions. AIChE J 63:2671–2693

Gao J, You F (2017b) Can modular manufacturing be the next game-changer in shale gas supply chain design and operations for economic and environmental sustainability? ACS Sustain Chem Eng 5(11):10046–10071

Gao J, You F (2017c) Economic and environmental life cycle optimization of noncooperative supply chains and product systems: modeling framework, mixed-integer bilevel fractional programming algorithm, and shale gas application. ACS Sustain Chem Eng 5(4):3362–3381

Gao Y, You F (2017d) Design and optimization of shale gas energy systems: overview, research challenges, and future directions. Comput Chem Eng 106:699–718

Gao J, You F (2018) Integrated hybrid life cycle assessment and optimization of shale gas. ACS Sustain Chem Eng 6(2):1803–1824

Grossmann IE, Cafaro DC, Yang L (2014) Optimization models for the optimal investment, drilling, and water management in shale gas supply chains. In: 8th International conference on foundations of computer-aided process design-FOCAPD

Haluszczak LO, Rose AW, Kump LR (2013) Geochemical evaluation of flowback brine from Marcellus gas wells in Pennsylvania, USA. Appl Geochem 28:55–61

Igunnu ET, Chen GZ (2014) Produced water treatment technologies. Int J Low Carbon Technol 9(3):157–177

Jiang Q, Rentschler J, Perrone R, Liu K (2013) Application of ceramic membrane and ion-exchange for the treatment of the flowback water from Marcellus shale gas production. J Membr Sci 431:55–61

Kharaka YK, Thordsen JJ, Conaway CH, Thomas RB (2013) The energy-water nexus: potential groundwater-quality degradation associated with production of shale gas. Procedia Earth Planet Sci 7:417–422

Khawaji AD, Kutubkhanah IK, Wie JM (2008) Advances in seawater desalination technologies. Desalination 221(1):47–69

Kuuskraa VA, Cleveland CJ (2004) Natural gas resources, unconventional. In: Encyclopedia of energy. Elsevier, pp 257–272

Lira-Barragán LF, Ponce-Ortega JM, Serna-González M, El-Halwagi MM (2016) Optimal reuse of flowback wastewater in hydraulic fracturing including seasonal and environmental aspects. AIChE J 62(5):1634–1645

López-Díaz DC, Lira-Barragán LF, Rubio-Castro E, You F, Ponce-Ortega JM (2018) Optimal design of water networks for shale gas hydraulic fracturing including economic and environmental criteria. Clean Technol Environ Policy 20:2311–2332

Marcon V, Joseph C, Carter KE, Hedges SW, Lopano CL, Guthrie GD, Hakala JA (2017) Experimental insights into geochemical changes in hydraulically fractured Marcellus shale. Appl Geochem 76:36–50

Matz R, Fisher UA (1981) Comparison of the relative economics of sea water desalination by vapour compression and reverse osmosis for small to medium capacity plants. Desalination 36(2):137–151

McGinnis RL, Hancock NT, Nowosielski-Slepowron MS, McGurgan GD (2013) Pilot demonstration of the NH_3/CO_2 forward osmosis desalination process on high salinity brines. Desalination 312:67–74

Melkoglu M (2014) Shale gas: analysis of its role in the global energy market. Renew Sustain Energy Rev 37:460–468

Myers T (2012) Potential contaminant pathways from hydraulically fractured shale to aquifers. Ground Water 50(6):872–882

Nichols C, Victor N (2015) Examining the relationship between shale gas production and carbon capture and storage under CO_2 taxes based on the social cost of carbon. Energ Strat Rev 7:39–54

Nicot JP, Scanlon BR (2012) Water use for shale gas production in Texas, US. Environ Sci Technol 46:3580–3586

Oke D, Majozi T, Mukherjee R, Sengupta D, El-Halwagi MM (2018) Simultaneous energy and water optimisation in shale exploration. Process 6(7):86. https://doi.org/10.3390/pr6070086

Ophir A, Lokiec F (2004) Review of MED fundamentals and costing. IDE Technologies Ltd, pp 69–78

Racharaks R, Ge X, Li Y (2015) Cultivation of marine microalgae using shale gas flowback water and anaerobic digestion effluent as the cultivation medium. Biores Technol 191:146–156

Rahm BG, Bates JT, Bertoia LR, Galford AE, Yoxtheimer DA, Riha SJ (2013) Wastewater management and Marcellus shale gas development: trends, drivers and planning implication. J Environ Manage 120:105–113

Savacool BK (2014) Cornucopia or curse? Reviewing the cost and benefits of shale gas hydraulic fracturing (fracking). Renew Sustain Energy Rev 37:249–264

Stoll ZA, Forrestal C, Ren ZJ, Xu P (2015) Shale gas produced water treatment using innovative microbial capacitive desalinations cell. J Hazard Mater 283:847–855

Tagliaferri C, Lettieri P, Chapman C (2015) Life cycle assessment of shale gas in the UK. Energy Procedia 75:2706–2712

Tan SH, Barton PI (2015) Optimal dynamic allocation of mobile plants to monetize associated or stranded natural gas, part I: Bakken shale play case study. Energy 93:1581–1594

Theodori GL, Luloff AE, Willits FK, Burnett DB (2014) Hydraulic fracturing and the management, disposal, and reuse of frac flowback water: views from the public in the Marcellus shale. Energy Res Soc Sci 2:66–74

Vengosh A, Jackson RB, Warner N, Darrah TH, Kondash AA (2014) Critical review of the risk to water resources from unconventional shale gas development and hydraulic fracturing in the United States. Environ Sci Technol 48:8334–8348

Veza J (1995) Mechanical vapour compression desalination plants—a case study. Desalination 101(1):1–10

Vidic RD, Brantley SL, Vandenbossche JM, Yoxtheimer D, Abad JD (2013) Impact of shale development on regional water quality. Sci 340(6134):1235009

Wade NM (1993) Technical and economic evaluation of distillation and reverse osmosis desalination processes. Desalination 93(1–3):343–363

Wade NM (2001) Distillation plant development and cost update. Desalination 136(1–3):3–12

Wang Q, Chen X, Jha AN, Rogers H (2014) Natural gas from shale formation—the evolution evidences and challenges of shale gas revolution in United States. Renew Sustain Energy Rev 30:1–28

Xu P, Drewes JE (2006) Viability of nanofiltration and ultra-low pressure reverse osmosis membranes for multi-beneficial use of methane produced water. Sep Purif Technol 52(1):67–76

Yang L, Grossmann IE, Manno J (2014) Optimization models for shale gas water management. AIChE J 10:3490–3501

Yu CH, Huang SK, Qin P, Chen XL (2018) Local residents' risk perceptions in response to shale gas exploitation: evidence from China. Energy Policy 113:123–134

Yuan J, Luo D, Xia L, Feng L (2015) Policy recommendations to promote shale gas development in China based on a technical and economic evaluation. Energy Policy 85:194–206

Zhang D, Yang T (2015) Environmental impacts of hydraulic fracturing in shale gas development in the United States. Pet Explor Dev 42(6):876–883

Zou C (2015) Shale gas in China: characteristics, challenges and prospects (I). Pet Explor Dev 42(6):753–767

SPRINGER NATURE

GPSR Compliance

The European Union's (EU) General Product Safety Regulation (GPSR) is a set of rules that requires consumer products to be safe and our obligations to ensure this.

If you have any concerns about our products, you can contact us on ProductSafety@springernature.com

In case Publisher is established outside the EU, the EU authorized representative is:

Springer Nature Customer Service Center GmbH
Europaplatz 3
69115 Heidelberg, Germany

The manufacturer's authorised representative in the EU is Springer Nature Customer Service Centre GmbH, Europaplatz 3, 69115 Heidelberg, Germany. If you have any concerns regarding our products, please contact ProductSafety@springernature.com

Printed and bound by CPI Group (UK) Ltd, Croydon, CR0 4YY
01/10/2026
02225915-0018